THE ELEMENTS

O

	IIIA	IVA	VA	VIA	VIIA	2 **He** 4.0026 ± 0.00005

8

IB	IIB	5 **B** 10.811 ± 0.003	6 **C** 12.01115 ± 0.00005	7 **N** 14.0067 ± 0.00005	8 **O** 15.9994 ± 0.0001	9 **F** 18.9984 ± 0.00005	10 **Ne** 20.183 ± 0.0005
		13 **Al** 26.9815 ± 0.00005	14 **Si** 28.086 ± 0.001	15 **P** 30.9738 ± 0.00005	16 **S** 32.064 ± 0.003	17 **Cl** 35.453 ± 0.001	18 **Ar** 39.948 ± 0.0005

28 **Ni** 58.71 0.005	29 **Cu** 63.54 ± 0.005	30 **Zn** 65.37 ± 0.005	31 **Ga** 69.72 ± 0.005	32 **Ge** 72.59 ± 0.005	33 **As** 74.9216 ± 0.00005	34 **Se** 78.96 ± 0.005	35 **Br** 79.909 ± 0.002	36 **Kr** 83.80 ± 0.005
46 **Pd** 106.4 ± 0.05	47 **Ag** 107.870 ± 0.003	48 **Cd** 112.40 ± 0.005	49 **In** 114.82 ± 0.005	50 **Sn** 118.69 ± 0.005	51 **Sb** 121.75 ± 0.005	52 **Te** 127.60 ± 0.005	53 **I** 126.9044 ± 0.00005	54 **Xe** 131.30 ± 0.005
78 **Pt** 195.09 ± 0.005	79 **Au** 196.967 ± 0.0005	80 **Hg** 200.59 ± 0.005	81 **Tl** 204.37 ± 0.005	82 **Pb** 207.19 ± 0.005	83 **Bi** 208.980 ± 0.0005	84 **Po** (210)	85 **At** (210)	86 **Rn** (222)

63 **Eu** 151.96 ± 0.005	64 **Gd** 157.25 ± 0.005	65 **Tb** 158.924 ± 0.0005	66 **Dy** 162.50 ± 0.005	67 **Ho** 164.930 ± 0.0005	68 **Er** 167.26 ± 0.005	69 **Tm** 168.934 ± 0.0005	70 **Yb** 173.04 ± 0.005	71 **Lu** 174.97 ± 0.005

95 **Am** (243)	96 **Cm** (247)	97 **Bk** (247)	98 **Cf** (249)	99 **Es** (254)	100 **Fm** (253)	101 **Md** (256)	102 **No** (253)	103 **Lr** (257)

Atomic Weights are based on C^{12} . . . . the 1961 Values

U.S.A.

D1472618

a brief introduction to
General,
Organic
and
Biochemistry

SECOND
EDITION

JOSEPH I. ROUTH
DARRELL P. EYMAN
DONALD J. BURTON

University of Iowa, Iowa City, Iowa

SAUNDERS GOLDEN SUNBURST SERIES

1976
W. B. SAUNDERS COMPANY · Philadelphia · London · Toronto

W. B. Saunders Company: West Washington Square
                           Philadelphia, PA 19105

                           12 Dyott Street
                           London, WC1A 1DB

                           833 Oxford Street
                           Toronto, Ontario M8Z 5T9, Canada

Library of Congress Cataloging in Publication Data

Routh, Joseph Isaac, 1910–
   A brief introduction to general, organic, and
biochemistry.

   (Saunders golden sunburst series)
   Includes bibliographies.
   1. Chemistry.   I. Eyman, Darrell P., 1937–
joint author.   II. Burton, Donald Joseph, 1934–
joint author.   III. Title.   [DNLM: 1. Biochemistry.
2. Chemistry.   3. Chemistry, Organic. QD33 R869b]
QD31.2.R68   1976   540   75-10389
ISBN 0-7216-7769-X

The cover illustration is a ball and stick molecular model of the heme portion of the hemoglobin molecule. The largest ball which represents an iron atom is attached to four nitrogen atoms located at the corners of a square. The complete structure of hemoglobin is shown on page 280.

A Brief Introduction to
General, Organic and Biochemistry   ISBN   0-7216-7769-X

Last digit is the print number: 9   8   7   6   5   4   3   2   1

PREFACE

The teaching of chemistry at the elementary level in high schools, community colleges, and universities has, in the past ten years, been greatly influenced by changing student interest, increased awareness of science by society, and current research discoveries. Interest in the environment, pollution, energy resources, and toxic agents has resulted in new textbooks of elementary chemistry that stress problems of environmental pollution and poisoning, sometimes to the exclusion of fundamental knowledge of the subject. The changes in teaching emphasis and the specialized textbooks were carefully assessed in the preparation of this revision of our text. It was obvious that courses based on the increased awareness of research developments in the fields of medicine, environment, and energy, integrated with the fundamentals of general, organic, and biochemistry, would be solid choices of students and educators alike. This necessitated a complete change and rewriting of the material in the first edition, with the result that the present edition is essentially a new textbook.

In the first section, the coverage of general chemistry has been designed to better prepare the students for the organic and biochemistry material that follows. A careful selection of topics of current interest emphasizes the interrelationship between the three sections of the text. Building on the fundamentals presented in general chemistry, the organic section focuses attention on organic compounds, with particular emphasis on functional groups and their reactions. The special topics covered in this part of the text are a mixture of organic and biochemistry and carry the student forward to the final section. The biochemistry material has been rearranged to present a more acceptable teaching order and has been revised to conform with the rapid changes that are occurring in the field. New chapters containing material on vitamins and coenzymes, an introduction to metabolism, and the biochemistry of genetics have been added.

To further assist the student, each chapter is preceded by a list of objectives and is followed by important terms and concepts plus carefully selected questions and suggested readings. Every effort was made to achieve the goals and objectives of the revision and still retain the limited size of the text.

Many suggestions for improvement of the text have been received from instructors and students. The authors gratefully acknowledge the assistance of all who have made suggestions, especially the students who have used the first edition of the book in beginning chemistry courses. Finally, the confidence of the publishers and their willingness to invest considerable time and effort in the improvement of this edition are gratefully acknowledged.

CONTENTS

CHAPTER 7

REACTION RATES AND CHEMICAL EQUILIBRIUM 102

CHAPTER 8

SOME CHEMISTRY OF NONMETALLIC ELEMENTS 122

CHAPTER 9

SOME CHEMISTRY OF METALLIC ELEMENTS 139

SOME FUNDAMENTAL CONCEPTS

CHAPTER 1

The *objectives* of this chapter are to enable the student to:

1. Differentiate facts, laws, hypotheses, and theories.
2. Recognize the role of energy in chemical reactions and in our daily lives.
3. Classify matter according to its composition as elements, compounds, homogeneous mixtures, or heterogeneous mixtures.
4. Utilize metric and SI units when discussing length, mass, and volume.
5. Interconvert temperatures on the Fahrenheit and Celsius scales.
6. Recognize the atomic symbols of the more common elements.
7. Discuss the mole concept as applied to atoms and molecules.
8. Determine the number of atoms in a given sample of an element or the number of molecules or formula units in a given weight of a compound.
9. Determine the per cent composition of a compound given its formula.
10. Determine the molecular formulas of a compound given its empirical formula and its molecular weight.

An introduction to the study of chemistry requires the definition of numerous scientific terms including many which are specific to chemistry. This vocabulary development is essential to the presentation of the simplest and most fundamental concepts. These concepts will serve as a foundation for subsequent development of more detailed and sophisticated ideas in the study of inorganic, organic and biological chemistry.

SCIENTIFIC INVESTIGATION

Science comprises the observations by man of the natural phenomena in his environment followed by attempts to present consistent explanations. The level of sophistication of the observations ranges from casual to those made under artificially generated or controlled conditions. The latter type of observation, called an *experiment,* can be used to test the validity of explanations referred to as **theories** or *models*. Occasionally, results of several experiments may lead to a conjectural statement, a guess called a **hypothesis,**

1

which attempts to explain causes. It is only after the establishment of repeatable results that the experimenter considers observations to be *facts*. Universally accepted facts called **laws** usually induce the scientist to construct a theory consistent with observations.

The methods which contemporary scientists use in studying phenomena are many and varied in their nature. Very few scientists would agree that there is a single scientific method which is of greater value than another. For example, a scientist may make many observations of a phenomenon before he realizes its uniqueness or potential significance and in doing so he may have established a law before he has designed controlled experiments. On the other hand, it is possible for an accepted theory to suggest the possibility of an entirely new observation. It is generally true, however, that scientists have some specific theory or phenomenon in mind when designing controlled experiments. Often the experimenter is enticed or forced to redefine the objectives of his experiment because the results are not those desired or expected. Many scientific advances can be attributed to accidental and unintentional discoveries. Although many such discoveries may be considered as lucky accidents, it is important to realize that scientists making such observations could have failed to recognize the significance of unusual results.

The fundamental laws of chemistry accepted today were obtained by experiments devised and carried out by many investigators in many different laboratories. A modern investigator is never completely satisfied with his explanation of a process or a reaction until his experimental results have been confirmed in other laboratories. Undoubtedly, all scientists would agree that there is the necessity to test and retest the consistency of accepted theories with new experimental results. Many theories for the explanation of chemical processes have been proposed as the result of a series of experiments only to be altered as new experimental evidence is obtained. Practicing scientists must be continually willing to subject theories to the tests of newly presented experimental data with the attitude that no theory is final.

MATTER AND ENERGY

Chemistry is the science that deals with the composition of substances and with the changes that they may undergo. It is also concerned with the properties of substances and with their energy relationships. All substances are forms of matter, and **matter** is anything that possesses **mass** and occupies **space.** Matter exists in three states, **solid, liquid,** or **gas,** depending on the temperature and pressure. Matter in one physical state may often be changed into another by suitable energy changes.

Energy is defined as the ability to do work. It exists in many different forms, each of which may be converted into any of the other forms. Heat, light, motion, sound, and electricity are all familiar forms of energy. To measure the amount of energy in any of its forms, it is ordinarily converted to heat energy and expressed in calories. A **calorie** is a unit of heat energy that is defined as the quantity of heat that will raise the temperature of one gram of water one degree centigrade (at 15° Celsius). The most commonly used unit is the large calorie, or **kilocalorie,** which equals 1000 calories.

Chemical energy is the energy that is stored up in chemical substances, and is released or consumed during chemical changes. Chemical processes in which chemical energy is released as heat are called **exothermic,** and those in which heat is absorbed are called **endothermic.** Sometimes these energy changes become the most important part of a chemical reaction—for example, the combustion of fuel in the furnace to produce heat, or the "burning" of foods in the body to produce heat and energy. It should be kept in mind that every chemical change is accompanied by a change in energy.

In 1906 Albert Einstein proposed that matter and energy are readily converted into each other and are related by the following expression:

$$E = \Delta mc^2$$

In this equation, E represents energy, Δm is the change in mass, and c is the velocity of light. This equation is often quoted as the basis for the harnessing of atomic energy and the production of the atomic bomb. That matter can be converted into energy was proved in nuclear reactions. This relationship also affects the **law of conservation of mass** and the **law of conservation of energy**, which state that matter and energy can neither be created nor destroyed. A more appropriate statement might be: *matter and energy can neither be created nor destroyed; they can only be interconverted.*

TOPIC OF CURRENT INTEREST

ENERGY IN PERSPECTIVE

Man's increasing use of energy resources parallels the development of civilization and social interaction. Efficient use of energy has determined his ability to supply food and physical comforts and in many cases has been the determining factor in establishing one social or political group as dominant in a geographic region. Energy resources have always been available to man in the form of fuels for burning, wind, flowing or falling water, and even domesticated animals. The differences among developing civilizations have often arisen in the area of technical skills for converting available resources to useful energy forms such as heat.

As late as 1850, more than 90 per cent of all energy in the United States was furnished by burning wood. The major energy sources in 1970 were 20.1 per cent coal, 75.8 per cent oil and gas (these three make up the **fossil fuels**), 3.8 per cent hydropower, and 0.3 per cent nuclear power. Wood burning is no longer a necessity but a luxury, practiced for the most part in esthetically pleasing fireplaces. The decline in the use of wood for fuel and the decline in the use of coal from 70 per cent in 1900 to 20 per cent in 1970 serve to illustrate some of the prevailing problems which face man in energy resource utilization. There is not enough wood available to furnish sufficient energy on a continuing basis for the growing population of the world. However, if the supply were sufficient, the air pollution problems would be acute.

Man's increasing difficulties in environmental control, as well as with economic and even social problems, arise in large part because of the continual increase in generation and consumption of energy. The lifestyle of North Americans as well as that of the inhabitants of other highly developed regions of the world is dependent upon immediate and continued availability of energy. It is no difficult task to compare available energy resources with the energy requirements of the future and conclude that current resources and technologies are inadequate. For example, currently known sources of natural gas are expected to give peak production by 1980 but will be essentially depleted within 65 years. Currently known sources of petroleum will reach peak production by 2000 but will be in very short supply by the year 2075. The fossil fuel form which is still relatively abundant is coal. The known sources of coal will allow for a manifold increase in production and supply, peaking after 2100. Although its use in combustion is complicated by air pollution problems, the natural supply of fossil fuels dictates that man must extract energy from coal in ever-increasing amounts in the near future. Increasingly larger portions of the energy which we use must come from nuclear power (see Chapter 2) or other as yet undeveloped sources (solar energy or nuclear fusion) as the fossil fuels are gradually depleted.

THE CHARACTERIZATION OF MATTER

Substances are usually recognized by their appearance, taste, odor, feel, and other similar characteristics. Such characteristics are called the **properties** of the substances and are divided into two classes, physical and chemical. The **physical properties** of a substance are the characteristics other than those involved in chemical processes. The chemical properties of a substance, however, are made manifest only when the substance undergoes a chemical change. Such characteristics as the physical state (solid, liquid, or gaseous), crystalline form, density, hardness, color, and luster are common physical properties. The **chemical properties** of a substance are its characteristic reactions with other substances, such as oxygen, water, acids, and bases, or its decomposition.

Properties are the signs by which we recognize substances. If all the physical and chemical properties of two substances are studied and it is found that they are identical, then the substances must be the same. If the properties are different, two substances have been characterized.

Substances are constantly undergoing physical and chemical changes. **Physical changes** are changes in the condition or state of a substance. They do not result in the formation of new substances nor do they involve a change in composition. An example of a physical change would be the breaking of a bottle. Although there has been a marked change, the substance is still glass. No new substance has resulted, nor has there been a change in the composition of the glass. The boiling of water involves a physical change from the liquid to the gaseous state, but the composition of the matter is unchanged.

If a piece of iron is filed into small pieces, a definite change is observed, yet the particles are readily identified as iron, for they have the same properties as the original piece. If the iron filings are exposed to moisture, however, the iron will soon be changed into rust. A magnet will no longer attract the particles; the metallic luster is gone; the properties are different from those of the original substance; and it is therefore concluded that a new substance has been formed. When a piece of wood is heated in a test tube, it is observed that dense fumes are formed, and a black, charred mass remains behind. The rusting of iron and the destruction of wood by heat are examples of chemical changes. **Chemical changes** are defined as those changes that result in the formation of new substances and involve alterations in the composition of the substance.

Up to this point, "types of matter" have been discussed without any specific statements about which types of matter display which properties. The degree of differentiation of different types of matter is determined by the level of observation. Solid particles of sugar and table salt cannot be differentiated by casual observation, but close examination with the naked eye, or even better, with a magnifying glass or microscope reveals that each particle of table salt is a nearly perfect cube, whereas the sugar particles have irregular shapes. If both samples are crushed repeatedly to decrease the particle size below that readily seen with a microscope, a different type of observation would be necessary to distinguish these two types of matter. This point is pursued to emphasize that the apparent similarity or dissimilarity of two samples of matter is determined by the sensitivity of the method of observation.

All matter which occurs in our environment can be classified into one of several categories shown in Figure 1–1. In this figure, the simplest forms of matter are found at the top and the most complex forms at the bottom. The two broadest categories, *mixtures* and *pure substances*, are related by the fact that the former are composed of two or more of the latter. All mixtures have variable composition and can be separated into their individual components by physical means, but pure substances have definite invariant compositions and cannot be separated into simpler components by physical means.

Most matter in our everyday lives is a mixture of substances. The food we eat, the water we drink, the air we breathe, and the "pure" soap we use are all mixtures. Visual inspection is often sufficient to recognize that the material is composed of two or more substances. For example, a close examination of a mixture of powdered iron and powdered sulfur allows identification of individual particles of iron and sulfur. In this mixture the individual ingredients still retain their characteristic properties as indicated by the fact that iron can be separated from the mixture using a magnet, whereas sulfur can be separated by dissolving it in the solvent carbon disulfide. The resulting solution of sulfur

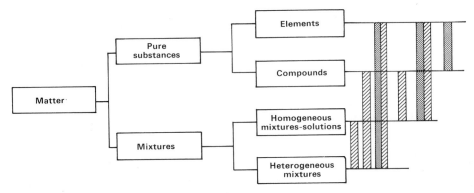

Figure 1-1   A schematic representation of the classification of matter in which vertical bars represent physical changes (cross hatched) and chemical changes (solid).

dissolved in carbon disulfide is a *homogeneous mixture* in which the particles of each pure substance are evenly distributed throughout the sample. In contrast, the mixture of powdered iron and powdered sulfur is a *heterogeneous mixture* with regions in which one or more pure substances are collected with the exclusion of other pure substances which are found in their own regions. Each region, called a *phase*, is homogeneous in that it has the same composition throughout. Thus, in this mixture, iron and sulfur are contained in different phases.

Pure substances can be classified as one of two types, *compounds* or *elements*. Elements are considered as basic units of matter that cannot be decomposed by ordinary chemical methods. There are 105 of these elements currently accepted, but a new manmade element has recently been claimed and is now being verified. The relative abundance by weight of elements in the earth's crust, in the **lithosphere,** in the oceans, lakes, rivers, and underground deposits collectively called the **hydrosphere,** and in the gases surrounding the earth, the **atmosphere,** is shown in Table 1-1. Eight elements—oxygen, silicon, aluminum, iron, calcium, sodium, potassium, and magnesium—make up 97 per cent of the composition of the earth. The remaining three per cent of the earth is composed of relatively small quantities of all the other naturally occurring elements.

By proper combination of elements, millions of more complex substances may be prepared. These substances, called compounds, are composed of two or more elements combined chemically and in definite proportions. They cannot be separated into simpler components by physical means but can be separated into component elements by a chemical decomposition. By identifying the decomposition products, the component elements, it is possible to determine the composition of the compound.

In Figure 1-1, the vertical cross-hatched and solid lines represent physical separations and chemical decompositions in going up and physical mixing and chemical combinations

TABLE 1-1   ELEMENTAL PERCENTAGE COMPOSITION OF THE EARTH'S LITHOSPHERE, HYDROSPHERE, AND ATMOSPHERE

Oxygen	49.5	Magnesium	1.9	Sulfur	0.05
Silicon	25.7	Titanium	0.6	Barium	0.05
Aluminum	7.5	Hydrogen	0.2	Chromium	0.03
Iron	4.7	Chlorine	0.2	Nitrogen	0.03
Calcium	3.4	Phosphorus	0.1	Fluorine	0.03
Sodium	2.6	Carbon	0.09	Nickel	0.02
Potassium	2.4	Manganese	0.08	Strontium	0.02
				All others	0.09

in going down. All transitions of matter except those involving nuclear processes are summarized in this figure.

MEASUREMENTS

In 1960 the international authority on units, the Conference Generale des Poids et Mesures, agreed to adopt the International System of Units, abbreviated **SI,** in all languages. In the past 15 years, many countries have made SI the only legally acceptable system of units. Many others are currently studying the details of a transition to the system or are actually preparing for the transition. It is likely that the United States and Canada will eventually use this system of units. The basic units of SI which will be of interest and used in this book are as follows:

Physical Quantity	Unit	Symbol
Length	Meter	m
Mass	Kilogram	kg
Time	Second	s
Electric current	Ampere	A
Temperature	Kelvin	K
Quantity	Mole	mol

From this list of SI units, many basic units can be derived, including units for energy, force, power, electric charge, and volume as well as others. The units of SI resemble in large part those of the metric system which are still used almost exclusively in chemical calculations and will be used in this book.

Length. The standard unit of length in the metric system is called the **meter,** which was originally based on one ten-millionth of the distance from the equator to the North

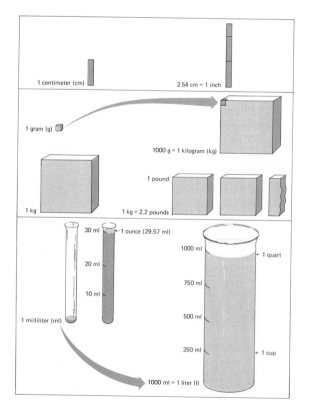

Figure 1–2   A comparison of the metric and English systems of measurement.

Pole. For many years the world relied on a material standard of length, the distance between two engraved lines on the International Meter Bar kept at Paris. The new definition established by the Eleventh General Conference states that the meter is 1,650,763.73 wavelengths of the orange-red line of krypton 86, which is equivalent to 39.37 inches. The **centimeter** is one-hundredth the length of a meter, and the **millimeter** is one-thousandth of a meter, or one-tenth of a centimeter. There are approximately 30 centimeters in a foot, or about 2.54 centimeters in an inch (Fig. 1–2).

These units of length are symbolized as m (meter), cm (centimeter), and mm (millimeter), and their relation to each other can be stated simply as follows: 1 m = 100 cm = 1000 mm.

Mass. The standard unit of mass is the **kilogram,** which is the mass of a block composed of platinum and iridium kept by the International Bureau of Weights and Measures. A kilogram weighs approximately 2.2 pounds. The **gram** is one-thousandth of a kilogram, and the **milligram** is one-thousandth of a gram (Fig. 1–2). These units are abbreviated as follows: kg (kilogram), g (gram), and mg (milligram). A **microgram** is one-thousandth of a milligram and is abbreviated μg. These relationships may be represented as follows: 1 kg = 1000 g; 1 g = 1000 mg; and 1 mg = 1000 μg.

Volume. The **liter** is a unit of volume exactly equal to a cubic decimeter. It is the volume occupied by a kilogram of pure water at 4° Celsius (the temperature at which a given volume of water weighs the most). A liter is slightly more than a quart (Fig. 1–2). A milliliter is one-thousandth of a liter and is the same as a **cubic centimeter.** A liter is commonly expressed as 1000 milliliters or 1000 cubic centimeters. A **microliter,** sometimes called a **lambda** (λ), is one-thousandth of a milliliter. Modern micro methods often require microliter quantities for analysis. A fluid ounce is approximately 30 cubic centimeters, and a teaspoon holds about 4 cubic centimeters. The units of volume are abbreviated as follows: l (liter), ml (milliliter), cm³ (cubic centimeter), and μl (microliter).

The units of the metric system are listed in Table 1–2, and some English metric equivalents are given in Table 1–3.

TABLE 1–2   UNITS OF THE METRIC SYSTEM

UNITS OF LENGTH		
Nanometer	=	0.001 micron
Micron	=	0.001 millimeter
Millimeter	=	0.001 meter
Centimeter	=	0.01 meter
Decimeter	=	0.1 meter
Meter	=	1.0 meter
Decameter	=	10.0 meters
Hectometer	=	100.0 meters
Kilometer	=	1000.0 meters
Megameter	=	1000000.0 meters
Gigameter	=	1000000000.0 meters
UNITS OF MASS		
Microgram	=	0.001 milligram
Milligram	=	0.001 gram
Gram	=	1.0 gram
Kilogram	=	1000.0 grams
UNITS OF VOLUME		
Microliter	=	0.001 milliliter
Milliliter	=	0.001 liter
Liter	=	1000 milliliters

TABLE 1–3   SOME ENGLISH-METRIC EQUIVALENTS

1 inch	=	2.54	centimeters
1 foot	=	30.5	centimeters
1.1 yard	=	1.0	meter
15 grains	=	1.0	gram
1 avoir. oz.	=	28.3	grams
1 pound	=	454.0	grams
2.2 pounds	=	1.0	kilogram
1 fluid oz.	=	29.6	cubic centimeters
1 quart	=	946.4	cubic centimeters
1 teaspoon	=	4.9	cubic centimeters
1 tablespoon	=	14.8	$cm^3 = \frac{1}{2}$ fluid ounce

Density.   The **density** of a material is the mass per unit volume. Two objects of equal volume but of different mass are said to have unequal densities. Air and other gases have very low densities, whereas metals have relatively high densities. Common units for expressing density include g/ml, g/l, and lbs/ft³. The densities of gases are usually expressed in units of g/l, whereas for liquids and solids g/ml is used.

The density of water is 1 g/ml. The weight, or mass, of a given volume of a material divided by the weight of an equal volume of water is the **specific gravity**. This ratio, which is really just the ratio of the density of a material to that of water, is dimensionless.

Temperature.   The degree of hotness, or the concentration of heat energy per unit mass, is the **temperature** of a body. In the United States temperature measurements are usually expressed on the familiar **Fahrenheit scale**. On this scale the temperature at which water freezes is 32 degrees (32°F), whereas water boils at 212°F. In scientific work this system has been replaced by the **centigrade**, or **Celsius, scale,** which is based on the freezing and boiling points of water. The freezing point is taken as 0°C and the boiling point as 100°C. A comparison of the thermometers associated with these two scales is shown in Figure 1–3.

A temperature reading on one scale may readily be converted to the corresponding temperature on the other scale. To convert degrees Fahrenheit to degrees Celsius add 40, multiply by $\frac{5}{9}$, and subtract 40 from the result.

As an example, the following is a conversion of 32°F to degrees Celsius:

$$32°F + 40 = 72$$
$$72 \times \tfrac{5}{9} = \tfrac{360}{9} = 40$$
$$40 - 40 = 0°C$$

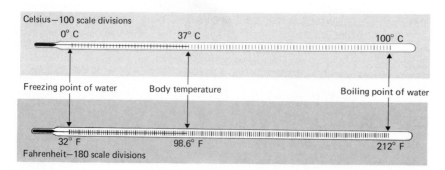

Celsius—100 scale divisions
0° C   37° C   100° C

Freezing point of water   Body temperature   Boiling point of water

32° F   98.6° F   212° F
Fahrenheit—180 scale divisions

Figure 1–3   Scale divisions of 100 on the Celsius thermometer = 180 on the Fahrenheit. Therefore, 1 scale division Celsius = $\frac{9}{5}$ scale division Fahrenheit, and 1 scale division Fahrenheit = $\frac{5}{9}$ scale division Celsius.

To convert degrees Celsius to degrees Fahrenheit, add 40, multiply by $\frac{9}{5}$, and subtract 40 from the result.

For example, the following is a conversion of $100°C$ to degrees Fahrenheit:

$$100°C + 40 = 140$$
$$140 \times \tfrac{9}{5} = \tfrac{1260}{5} = 252$$
$$252 - 40 = 212°F$$

In converting negative temperatures it is necessary to make sure that the algebraic sum is used.

For example, in the conversion of $-15°F$ to degrees Celsius:

$$-15°F + 40 = +25 \text{ (algebraic sum)}$$
$$+25 \times \tfrac{5}{9} = \tfrac{125}{9} = +14$$
$$+14 - 40 = -26°C \text{ (algebraic sum)}$$

Other methods of conversion from one temperature scale to another may be used, although the one given here seems easiest to remember. In both conversions 40 is added to the original temperature. This sum is multiplied by either $\frac{5}{9}$ or $\frac{9}{5}$, and 40 is subtracted from the result. Using a common point like the boiling point of water, $212°F$ and $100°C$, it can be seen that the Fahrenheit value is higher than the Celsius; therefore, it is necessary to use the larger factor $\frac{9}{5}$ to convert Celsius to Fahrenheit. Conversely, since the Celsius value is lower than the Fahrenheit, the factor $\frac{5}{9}$ can be used to convert Fahrenheit to Celsius.

Another method that is commonly used to convert readings on one scale to readings on the other is as follows:

$$C = \tfrac{5}{9}(F - 32°)$$
$$F = \tfrac{9}{5}C + 32°$$

THE ATOM: A NECESSARY CONCLUSION

Although the alchemists carried out many chemical reactions in their search for gold, they did not apply any systematic approach to the study of these reactions. In the seventeenth and eighteenth centuries a few chemists investigated the changes that occurred in chemical reactions. Several common elements were known early in the nineteenth century, and two laws of chemical change had been stated, in addition to the law of conservation of mass.

Before men became interested in the energy changes that accompany chemical reactions, several chemists had demonstrated the conservation of mass. By conducting reactions in closed containers, they showed that the products of a chemical change had the same mass as the starting materials. Careful experiments within the accuracy of the then available instruments involving combustion and precipitation, with the formation of new compounds, resulted in the formulation of the law of conservation of mass and subsequently the law of definite proportions and the law of multiple proportions.

The Law of Definite Proportions. *When two or more elements combine, they always combine in a fixed, or definite, proportion by weight.* For example, if water is formed from a mixture of hydrogen and oxygen gas, it will always contain 2 g of hydrogen for every 16 g of oxygen. When hydrogen is present in excess of the amount needed to combine with oxygen to form water, the resulting mixture will consist of water and the excess (uncombined) hydrogen gas.

It follows that each compound has a definite composition. The electrolysis of water always results in the formation of 2 parts of hydrogen gas and 16 parts of oxygen gas by weight.

The Law of Multiple Proportions.   *If an element unites with another element in more than one proportion by weight to form two or more compounds, these proportions by weight bear a ratio to one another that may be expressed in small whole numbers.* In the early studies of chemical change it was found that the same two elements would combine to form different compounds under different experimental conditions. For example, hydrogen and oxygen gas usually react to form water, but under conditions of a high-energy electric discharge they may form hydrogen peroxide. In a similar fashion it can be shown that carbon will combine with oxygen to form either carbon monoxide or carbon dioxide, depending on the experimental conditions. The proportions by weight of oxygen in the two examples bear a ratio to one another of 1:2.

In an attempt to explain the facts outlined in the laws of chemical change, John Dalton, an English schoolteacher, proposed his atomic theory in 1803. He began by assuming that all elements are composed of minute, invisible particles called **atoms.** The atom may then be considered the smallest unit of an element. Dalton stated further that all atoms of the same element had the same properties and the same weight, but that they differed from the atoms of all other elements in these respects. In chemical changes the atoms could combine to form small particles of compounds, or they could separate or change places in these compounds. When the atoms enter into chemical combination, the weight of each individual atom does not change.

Dalton's original theory has been modified, and many exceptions have been made to his statements. The importance of his theory should not be underestimated, however, since for the study of the structure of chemical compounds it marks the beginning of the modern era. It was over a century later that a clear conception of the internal structure of atoms was formulated.

ATOMIC SYMBOLS

In studying the elements and the chemical reactions that they undergo, it is often inconvenient to write out the complete name of each element every time it occurs. For this reason the chemist has assigned a symbol to each of the recognized 105 elements as a sort of chemical shorthand. The earlier symbols used by the alchemists were associated with heavenly bodies and were used to keep their discoveries secret. The modern system of atomic symbols has been kept as simple as possible to achieve common understanding of chemical reactions by all chemists. Some elements are represented by the first letter of their name; thus O stands for oxygen, N for nitrogen, C for carbon, and H for hydrogen. Since the names of several elements have the same first letter, in some instances another identifying letter has been added to distinguish these elements, for example, Ca for calcium, Ba for barium, Cl for chlorine, and Br for bromine. Some of the elements were known in ancient times and were given Latin names, since that language was then in more common usage. The symbols for these elements are derived from the Latin instead of the English name.

The symbols of the elements are given inside the back cover.

MOLECULES

In Dalton's atomic theory it is stated that the atom may be considered the smallest unit of an element that can take part in a chemical change. It is further stated that in chemical changes the atoms can combine to form small particles of compounds, or they can separate or change places in these compounds. Each small particle of a compound contains a definite number of atoms. This means that each unit particle of a compound must have the same number and kinds of atoms as all the other unit particles. These small unit particles of which every compound is composed are called **molecules.** The molecule can be considered the indivisible unit for compounds, much as the atom is the unit particle for elements.

The simplest compound would be one whose molecule contains 1 atom of each of the 2 elements that unite to form the compound. An example of such a simple compound is carbon monoxide, whose molecule is composed of 1 atom of carbon and 1 atom of oxygen (Fig. 1–4). Every molecule of water contains 2 atoms of hydrogen and 1 atom of oxygen.

The majority of molecules are composed of 2 or more different atoms, but some atoms of the same element are capable of uniting with each other to form a molecule of that element. This is particularly true of elements that are gases at room temperature (oxygen, hydrogen, nitrogen, and chlorine). Such gases always exist in molecular form when in the free state, each molecule containing 2 atoms of the element.

ATOMIC AND MOLECULAR WEIGHTS AND THE MOLE CONCEPT

Although it has never been possible to weigh a single atom of an element or a single molecule of a compound, it is possible to obtain the weights of equal numbers of different atoms or molecules and thus the relative weights of individual atoms or molecules. For example, by weighing the same number of hydrogen, oxygen, and sulfur atoms, it was found that the oxygen atoms weigh approximately sixteen times as much as the hydrogen atoms and one-half as much as the sulfur atoms. After several trials and errors, oxygen was assigned an atomic weight of 16 and was established as the reference element for the purpose of assigning atomic weights to all the elements. Atomic weights assigned in this manner have been used satisfactorily by chemists for many years. The physicists have used a slightly different scale based on a particular isotope of oxygen (isotopes will be discussed in Chapter 2). In 1961, however, the International Union of Pure and Applied Chemistry and the International Union of Pure and Applied Physics both agreed to use the same atomic weight values, based on the mass of a specific type of carbon atom, carbon-12.

The actual masses of atoms are very small compared to common mass units and thus is it convenient to introduce the **atomic mass unit, amu,** which is $\frac{1}{12}$ the mass of one carbon-12 atom. The mass of any atom can be expressed simply as the relative atomic weight in units of amu. For practical reasons, the most common mass unit used with the relative atomic weight scale is the gram. When the relative atomic weight of an element is given in units of grams, it is known as the **gram atomic weight.** A gram atomic weight of oxygen is 16 grams, whereas that of carbon is 12 grams. A gram atomic weight of each of these elements contains the same number of atoms. That number, traditionally referred to as **Avogadro's number,** has been experimentally determined in several ways

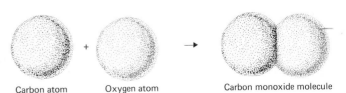

Figure 1–4   The union of an atom of carbon with an atom of oxygen.

Carbon atom   Oxygen atom   Carbon monoxide molecule

to be 6.023×10^{23}°. Avogadro's number of particles is referred to as a **mole.** Thus, a mole of atoms weighs one gram atomic weight. The terms **mole** and **gram atomic weight** are differentiated easily if the former is thought of as a specific number of things (e.g., atoms, molecules, particles), whereas the latter is thought of as a given quantity of matter (i.e., mass).

The mass of a carbon-12 atom in units of grams is calculated to be

$$\frac{12.0000 \text{ g C}}{6.023 \times 10^{23} \text{ atoms C}} = \frac{1.990 \times 10^{-23} \text{ g C}}{\text{atom C}}$$

The mass of one million oxygen atoms is

$$10^6 \text{ atoms O} \times \frac{16.0 \text{ g O}}{6.023 \times 10^{23} \text{ atoms O}} = 2.66 \times 10^{-17} \text{ g O}$$

In this calculation one million oxygen atoms is multiplied by a **conversion factor** which relates the mass of a mole of oxygen atoms to the number of atoms in a mole.

Using the same calculation technique, it is seen that the number of atoms in 96 grams of nitrogen is

$$96 \text{ g N} \times \frac{6.023 \times 10^{23} \text{ atoms N}}{14 \text{ g N}} = 4.13 \times 10^{24} \text{ atoms N}$$

The number of moles of helium which has a mass of 26 grams is

$$26 \text{ g He} \times \frac{1(\text{g atomic wt})\text{He}}{4 \text{ g He}} \times \frac{1 \text{ mol He}}{1(\text{g atomic wt})\text{He}} = 6.5 \text{ mol He}$$

In this calculation technique, called the **factor dimensional method,** conversion factors are applied to reach the desired units. The conversion factor applied must have a numerator and a denominator which are equivalent by definition. The conversion factor must be applied in such a way that after cancellation of units the desired units remain.

If the composition of a compound and the atomic weights of each atom in it are known, the molecular weight can be calculated readily. The **molecular weight** of a compound is the sum of the atomic weights of all the atoms present in 1 molecule of the substance. A molecule of carbon dioxide contains 1 atom of carbon and 2 atoms of oxygen. Since the atomic weight of carbon is 12 and the atomic weight of oxygen is 16, the molecular weight may be calculated as follows:

Element	Atomic Weight
Carbon	12
Oxygen	16
Oxygen	16
Carbon dioxide	44 amu = Molecular weight

°Extremely small values or large values are often represented as a number times ten to a power. For example, 10,000,000,000 is expressed as ten to the tenth power or 10^{10}. The fraction $1/10,000,000,000$ can be expressed as 0.0000000001 or $1/10^{10}$ or 1×10^{-10}.

If the molecular weight of a compound is given in units of grams, it is known as the **gram molecular weight.** Thus carbon dioxide has a gram molecular weight of 44 g; and water, 18 g. A gram molecular weight of a compound contains the same number of molecules as a gram molecular weight of any other compound, namely a mole, or 6.023×10^{23} molecules. The weight of a mole of molecules is one gram molecular weight.

FORMULAS

A **formula** expresses in symbols the composition of a substance. Since compounds are composed of atoms combined in definite proportions, they can be represented by a combination of the symbols of the atoms. A molecule of hydrogen chloride is composed of 1 atom of hydrogen and 1 atom of chlorine and is represented by the formula HCl. Where there is more than 1 atom of the same kind in the **formula unit** or in the molecule, the symbol is not repeated, but the number of atoms is indicated as a subscript to the symbol for the element. For example, the formula for water is written H_2O, meaning that 1 molecule of water contains 2 atoms of hydrogen and 1 atom of oxygen. The small subscript 2 which follows the H indicates that 2 atoms are present in the molecule. A molecule of sulfuric acid consists of 2 atoms of hydrogen, 1 atom of sulfur, and 4 atoms of oxygen. The formula H_2SO_4 represents 1 molecule, 1 molecular weight, 1 gram molecular weight, or 1 mole of sulfuric acid. The gram molecular weight equals the sum of the gram atomic weights of the atoms in the molecule, or 2 gram atomic weights of hydrogen plus 1 gram atomic weight of sulfur plus 4 gram atomic weights of oxygen. Again it should be observed that the subscript numbers are written after the atomic symbols to which they belong. Since the atoms of most gases do not exist by themselves, but unite to form molecules of the gas, the molecule of hydrogen is written H_2, of nitrogen N_2, and of oxygen O_2.

To designate more than 1 molecule of a substance, the appropriate number is placed in front of the formula. The term $3BCl_3$ represents 3 molecules of boron chloride, and $2O_2$ represents 2 molecules of oxygen gas. To avoid confusion of the subscript numbers and the numbers written in front of the formulas, it would be wise for the beginner to write the full meaning of the different terms. For example, $3H_2SO_4$ could read 3 molecules of sulfuric acid, each molecule consisting of 2 atoms of hydrogen, 1 atom of sulfur, and 4 atoms of oxygen.

The mass of 0.46 gram molecular weight of H_2SO_4 is calculated to be

$$0.46(\text{g molecular wt})H_2SO_4 \times \frac{98 \text{ g } H_2SO_4}{1(\text{g molecular wt})H_2SO_4} = 45 \text{ g } H_2SO_4$$

The number of formula units in 28 g of Na_2CO_3 is

$$28 \text{ g } Na_2CO_3 \times \frac{1(\text{g molecular wt})Na_2CO_3}{106 \text{ g } Na_2CO_3}$$

$$\times \frac{1 \text{ mol } Na_2CO_3}{1(\text{g molecular wt})Na_2CO_3} \times \frac{6.023 \times 10^{23} \text{ formula units } Na_2CO_3}{1 \text{ mol } Na_2CO_3}$$

$$= 1.59 \times 10^{23} \text{ formula units } Na_2CO_3$$

In this calculation the first and second conversion factors could have been combined since 1 mole of Na_2CO_3 has a mass of 106 g.

The formula of a chemical entity is correctly interpreted as a qualitative and quantitative statement of composition. For elements the symbol states qualitatively the identity of the atoms, whereas its quantitative interpretation involves numbers of atoms. The symbol of an element is commonly used to represent quantitatively either 1 atom or 1 mole of atoms as determined by the stated convention.

The formula of a chemical compound states qualitatively the identities of the atoms comprising the compound. In addition the formula states the relative number of each type of atom in the compound. The formula may state the actual number of each atom per molecule, or per formula unit, or the actual number of moles of each type of atom per mole of molecules or formula units. In this statement molecules and formula units are differentiated primarily as covalently bonded species and ionic species respectively. Covalently bonded species, which exist as discrete molecules under normal conditions, include CO_2, H_2 and H_2O, as examples. Ionic compounds and covalent network solid compounds do not exist as discrete "small" molecules under normal conditions, but have structures involving extensive three dimensional lattices as discussed in Chapter 4. For example, sodium chloride exists as a crystalline solid in which each sodium ion is surrounded by six chloride ions, and each chloride ion is surrounded by six sodium ions. It is impossible to identify a discrete sodium chloride "molecule" in this structure, and yet the formula NaCl qualitatively and quantitatively describes the composition. For this reason NaCl is referred to as a formula unit, and 6.023×10^{23} NaCl as a mole of formula units.

The **empirical formula** of a compound states the relative numbers of atoms of each element in the compound, whereas the **molecular formula** states the actual numbers of atoms of each element in a molecule of the compound. The empirical and molecular formulas are the same for some compounds, including HCl, CO_2, H_2O, and BF_3; but for many compounds the empirical formula and molecular formula are different as indicated in the following tabulation:

Empirical formula	Molecular formula
BCl_2	B_2Cl_4
P_2O_5	P_4O_{10}
NH_2	N_2H_4
CH_3O	$C_2H_6O_2$
CH	C_2H_2, C_6H_6

In order to establish the relationship between the empirical formula and the molecular formula, it is necessary to know the molecular weight. The molecular weight will always be a whole number multiple of the empirical formula weight. The whole number multiple which relates these formulas is the molecular weight divided by the empirical formula weight. The following calculation illustrates this:

A compound having the empirical formula CH has a molecular weight of 78 amu. What is the molecular formula?

The empirical formula weight is $12 + 1 = 13$ amu.

$$\frac{\text{molecular weight}}{\text{empirical formula weight}} = \frac{78 \text{ amu}}{13 \text{ amu}} = 6$$

The molecular formula is determined by multiplying each of the subscripts in the empirical formula by 6; thus, the molecular formula is C_6H_6.

IMPORTANT TERMS AND CONCEPTS

atomic mass unit
Avogadro's number
Celsius temperature scale
compound
element
empirical formula
energy

Fahrenheit temperature scale
gram atomic weight
gram molecular weight
law
law of conservation of energy
law of conservation of mass
law of definite proportions

law of multiple proportions
mass
mole
molecule
molecular formula
theory

QUESTIONS

1. Name several types of energy. Which form of energy is most easily measured?

2. Approximately how many calories of heat would be required to raise the temperature of a quart of water from $10°$ to $20°C$?

3. A popular professional baseball player can command an annual salary equivalent to his weight in gold. If gold sells at $1000 a kilogram and he weighs 176 pounds, what is his annual salary?

4. Spiders are said to have a top speed of 1.4 miles per hour. This speed is how many inches per second?

5. A child appeared to be running a high fever. The only thermometer available was calibrated in degrees Celsius and gave a reading of $40°C$. What was the child's temperature in degrees Fahrenheit?

6. How would you define a homogeneous mixture?

7. Why was it necessary to choose a reference element for atomic weights?

8. Why did chemists recently change the reference element for atomic weights from oxygen to carbon?

9. Classify each of the following as homogeneous or heterogeneous mixtures:

   (a) a pizza   (b) soda pop   (c) a milk shake   (d) water and sand

10. Calculate the weight of 2x atoms of element Y ($Z = 20$), if x atoms of element Z ($Z = 30$) weigh 2 grams. (Z = Atomic number.)

11. What is the molecular weight of PCl_5? of $Al(C_2H_5)_3$?

12. One formula-unit of $PbSO_4$ contains which of the following?

    (a) one atom of Pb
    (b) one gram atomic weight of Pb
    (c) one mole of sulfur
    (d) one mole of $PbSO_4$

13. A sample of 196 g of calcium contains how many moles?

14. Find the gram molecular weights of the following:

    H_2SO_4 NH_3 CCl_4 BBr_3

15. How many molecules are there in 105.74 grams of NH_3?

16. How many grams of nitrogen are there in 8.2 gram atomic weights of nitrogen?

17. Give the empirical formula for the following:

    (a) N_2O_4 (b) $C_{10}H_8$ (c) $Fe_3(CO)_{12}$ (d) $Al_2(SO_4)_3$

18. Given the empirical formula and molecular weight, find the molecular formulas of each of the following:

    (a) CH_2O, 90 amu
    (b) P_2O_5, 284 amu
    (c) CH_2, 84 amu
    (d) $AlCl_3$, 267 amu

SUGGESTED READING

Dinga: The Elements and the Derivation of their Names and Symbols. Chemistry, Vol. 41, No. 2, p. 20, 1968.

Dort: The Energy Cycle of the Earth. Scientific American, Vol. 223, No. 3, p. 54, 1970.

Hubbert: The Energy Resources of the Earth. Scientific American, Vol. 224, No. 3, p. 60, 1971.

Labbauf: The Carbon-12 Scale of Atomic Masses. Journal of Chemical Education, Vol. 39, p. 282, 1962.

Paul: International System of Units (SI). Chemistry, Vol. 45, No. 9, p. 14, 1972.

Perry: The Gasification of Coal. Scientific American, Vol. 230, No. 3, p. 19, 1974.

Sears: Unintentional Discoveries. Chemistry, Vol. 44, No. 16, p. 16, 1971.

Starr: Energy and Power. Scientific American, Vol. 224, No. 3, p. 36, 1971.

Zweifel and Guynn: Are There Viable Alternatives to Nuclear Power? Chemistry, Vol. 47, No. 8, p. 16, 1974.

ATOMS: STRUCTURES AND PROPERTIES

CHAPTER 2

The *objectives* of this chapter are to enable the student to:

1. List the subatomic particles and describe their properties.
2. Describe Rutherford's experiment and outline the conclusions reached.
3. Define isotopes and list examples.
4. Describe the Bohr atom.
5. Compare and contrast the Bohr atom and the wave-mechanics atom, including a differentiation of orbits and orbitals.
6. Recognize the significance of groups and periods in the periodic table as they relate to chemical and physical properties.
7. Indicate the positions in the long form of the periodic table of the nonmetals, the light metals, the transition metals, the rare earth elements, and the inert gases.
8. Define alpha, beta, and gamma rays and give their properties.
9. Explain radioactive decay and discuss why it occurs.
10. Calculate using Einstein's relationship the energy equivalent of a given mass of matter.
11. Discuss nuclear fission and fusion as sources of energy.

The acceptance and use of Dalton's atomic theory during the nineteenth century allowed chemists to use weight relationships in establishing the composition of many compounds and in characterizing many chemical reactions. The atom was thought of as a fundamental particle of matter and its actual nature was not of concern until the end of the century. By then the results of Michael Faraday's studies of electrical current passed through liquids and solutions were interpreted as evidence that the components of matter and thus of the atom were electrical in nature.

SUBATOMIC PARTICLES

As an aid in understanding the complexities of atomic structure, it is helpful to first consider the units, or subatomic particles, involved in the construction of an atom.

Electrons. The first of the subatomic particles to be discovered was the **electron.** The mass of an electron is 9.1×10^{-28} grams, or about $\frac{1}{1837}$ of the mass of the hydrogen atom, which is the lightest of all atoms. The charge on the electron is -4.8×10^{-10} electrostatic units, and since this is the smallest unit charge of electricity known, it is referred to as -1.

Protons. From common observation it has been assumed that all substances are electrically neutral. For example, a piece of metal, rock, or wood does not convey an electric shock when touched, nor does it indicate a flow of electric current when attached to delicate measuring instruments. For this reason, it probably can also be assumed that all substances or compounds are composed of an equal number of positive and negative electrical units.

A particle having nearly the same mass as a hydrogen atom, with a positive charge of 4.8×10^{-10} electrostatic units has been established as a fundamental unit of atomic structure and called a **proton.** An electrical charge of 4.8×10^{-10} electrostatic units has been adopted as the magnitude of a unit charge for the electron and the proton. Common usage refers to a charge of -1 for the electron and $+1$ for the proton.

Neutrons. For several years the electron and proton were thought to be the only subatomic particles. Until 1932, in order to account for the structure of atoms, a combination of protons and electrons was thought to exist in the nucleus. Such a combination would be electrically neutral and possess a mass approximately equal to that of a proton. In 1932 the discovery of a new structural unit, the **neutron,** increased our understanding of the nucleus. This third particle is not electrically charged.

The properties of the subatomic particles are summarized in Table 2–1.

TABLE 2–1   MASSES AND CHARGES
ON SUBATOMIC PARTICLES

PARTICLE	MASS (amu)	CHARGE
Electron	0.00055	−1
Proton	1.00759	+1
Neutron	1.00898	0

THE STRUCTURE OF ATOMS

Shortly after the discovery of protons in 1911 several investigators were attempting to describe the structure of atoms. One type of experiment involved the bombardment of very thin sheets of gold foil with **alpha particles,** which are helium ions with a mass of 4 amu and a charge of $+2$. It was known that alpha particles are given off by radium, and that they are positively charged and possess a small mass. When these particles struck the gold foil, several phenomena were observed. Some of the particles passed through the foil and continued in a straight line; others passed through but had their paths altered; in addition, others were deflected from the surface of the foil. The British physicist Rutherford explained this scattering of alpha particles by suggesting that the atoms of the metal foil consist mostly of space, with a small, positively charged, heavy nucleus surrounded by electrons at relatively large distances from the nucleus. Most of the alpha particles pass through the empty space of the atoms, but a few are deflected from their original paths. Deflection of the alpha particles occurs when they come too close to the positively charged nucleus and are repelled by it (Fig. 2–1).

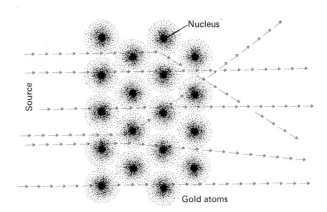

Figure 2-1 Alpha ray bombardment of atoms in a thin metal foil.

The nucleus contains the protons and neutrons, while the electrons are arranged at varying distances from the nucleus. Some conception of the relatively enormous space that exists between electrons and the nucleus of an atom may be gained from the following example. If the electrons and nucleus of each atom from every compound and substance on earth could be stripped of their space and combined in a dense, compact mass, the resultant body would be a sphere $\frac{1}{2}$ mile in diameter. This sphere would weigh as much as the earth and would possess such a high density that a cube of the material one centimeter on a side would weigh about 100 million tons. We may, then, summarize our concept of the atom as a particle of matter in which a cloud of electron density is located about a small, dense mass of protons and neutrons.

The Nucleus. The nucleus, which accounts for practically all the mass of the atom, is composed essentially of protons and neutrons. In size its diameter is approximately $\frac{1}{10,000}$ that of the atom. As stated earlier, the proton represents a unit charge of positive electricity, in contrast to the electron, which is the unit charge of negative electricity. The mass of the proton is similar to that of the hydrogen atom and is taken as one unit of atomic mass. The neutron has a charge of zero and a mass approximately the same as that of the proton. The number of positively charged protons in the nucleus is equal to the **atomic number, Z,** of the atom, which is also equal to the number of negatively charged electrons outside the nucleus. Since each proton or neutron contributes one unit to the weight of an atom, its **atomic weight** is equal to the total number of protons and neutrons in the nucleus.

ISOTOPES

For many years it was believed that all atoms of a given element had the same mass. In 1932 Harold Urey discovered some hydrogen with a mass approximately twice that of the mass of ordinary hydrogen. This heavy hydrogen was called **deuterium,** and water containing deuterium in place of hydrogen was called **heavy water.** Atoms of the same element but possessing different masses are called **isotopes.** The existence of many isotopes of different elements was proved by the use of the mass spectrograph. Nearly all elements possess at least two isotopes, while lead, for example, possesses 14 isotopes, whose atomic masses vary from 203 to 216. The atomic weight of an element reported in tables of atomic weights represents the naturally occurring mixture of isotopes.

Isotopes of an element can be more readily explained on the basis of the theory of atomic structure. For example, ordinary hydrogen has a nucleus containing 1 proton surrounded by a shell containing 1 electron, whereas deuterium has a nucleus that contains

1 proton and 1 neutron, accounting for its atomic weight of 2. Another example of an element with two naturally occurring isotopes is chlorine. One isotope contains 17 protons and 18 neutrons, and the second has 17 protons and 20 neutrons. The naturally occurring mixture of these two isotopes contains 23 per cent of the heavier isotope, resulting in an observed atomic weight of 35.46 amu.

On the basis of atomic structure, isotopes can be defined as atoms that have the same atomic number but different atomic weights. This means that isotopes have different numbers of neutrons in their nuclei. They have the same number of protons and electrons, however, and behave alike chemically. About 300 different isotopes similar to those of hydrogen and chlorine have been found occurring in natural mixtures.

THE ELECTRONIC STRUCTURE OF ATOMS

Having discussed the gross structure of the atom, it is now necessary to discuss the details of the positions, energies, and properties of the electrons in atoms. Since the electrons of an atom are determining factors of, and are involved in, all chemical reactions, it is necessary to consider the factors which influence their positions and energies. Just as in the case of Rutherford's studies of the structure of the atom, other experiments also can be used to test proposed models for the electronic structure of an atom.

THE BOHR ATOM

In 1913 Niels Bohr proposed a model for the electronic structure of the hydrogen atom, consistent with experimental facts. He assumed a Rutherford nuclear atom in which the electron moves about the nucleus in circular orbits of given radii. This means that the kinetic energy, or energy of motion, of the electron in the atom, which is directly related to the radius of the orbit, can assume only certain discrete values. That is, the possible energies of the electron are **quantized.** Bohr suggested that the single electron in hydrogen during thermal or electrical excitation can add only discrete quantities of energy and thus shift from one energy level to another. When the electron goes from a higher energy level to one of lower energy, it emits light of an energy corresponding to the difference between energy levels. That is, the energy emitted or absorbed, ΔE, is $E_{upper} - E_{lower}$. The energy levels of the atom are assigned **quantum numbers,** n, ranging from 1 to ∞, the relative energies of which are displayed in the energy level diagram in Figure 2–2. The **quantum levels** associated with the quantum numbers

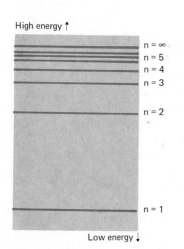

High energy ↑

n = ∞
n = 5
n = 4
n = 3

n = 2

n = 1

Low energy ↓

Figure 2–2   Energy levels for the hydrogen atom.

are often referred to as "**shells**," or "**orbits**," and are lettered K, L, M, N, and so forth, rather than $n = 1, 2, 3, 4, \ldots$. In the hydrogen atom, the radius of the "Bohr orbit" associated with a given quantum number, n, is $0.053n^2$ nm (nanometer). The size of the "shell," or orbit, increases with increasing n.

The Bohr theory of the hydrogen atom was truly a breakthrough in the attempt to find a suitable quantitative model for the atom. It successfully forecast many experimentally measured properties of hydrogen. However, the theory could not be successfully extended to other elements, and consequently it was abandoned for another model capable of incorporating elements of higher atomic number.

THE WAVE-MECHANICS ATOM

In 1924 Louis De Broglie proposed that electrons could have wave-like properties as well as particle properties. This proposal was experimentally verified by Davisson and Germer in 1927. In 1926 Erwin Schrödinger assumed that the motion, position, energy, and other properties of an electron in an atom could be related in a mathematical expression, the **wave equation,** which assumed wave-like properties for the electron. The solutions to the wave equation do not give the exact position of the electron when it has a given energy, but they do relate to the **probability** of finding the electron in a given position when it has a given energy. These solutions, called **wave functions,** or more commonly **orbitals,** contain a set of special numbers, the **quantum numbers.** These include the following:

The **principal quantum number,** n, which corresponds to the quantum number n in the Bohr atom and is generally related to the average distance of the electron from the nucleus. The electrons with a given n are said to occupy a given quantum level, the energy of which increases with increasing n. The values of n are $1, 2, 3, 4, \cdots \infty$.

The **azimuthal quantum number,** ℓ, which is related to the shapes of the orbitals. The values of ℓ are $0, 1, 2, 3, \cdots n - 1$; that is, the highest ℓ value for a given n is $n - 1$. For a given n, there are n values of ℓ. The orbitals with ℓ values of 0, 1, 2, and 3 are commonly referred to as s, p, d, and f orbitals, respectively. The orbital with $n = 1$ and $\ell = 0$ is denoted 1s, whereas the orbital with $n = 2$ and $\ell = 1$ is 2p. The relative energies of the orbitals in a given quantum level increase with increasing ℓ values, as is indicated in the energy level diagram in Figure 2–3.

One quantum number required to describe the electron in an atom does not arise in the solution of the wave equation. The **spin quantum number,** m_s, which is related to the spin of the electron, can have values of $+\frac{1}{2}$ and $-\frac{1}{2}$. Each orbital can contain two electrons with opposing spins. That is, the m_s values must have opposite signs. The number of electrons which can be contained in any quantum level, n, is $2n^2$, as the first level can contain 2 electrons, the second level 8 electrons, and the third level 18 electrons.

The orbital described by the quantum numbers is pictured as an electron cloud in Figure 2–4, which represents the 1s-orbital. In this cross-sectional representation of the electron cloud, the number of dots per unit area is related to the probability of finding the electron in that area, and can be thought of as an **electron density.** In the 1s-orbital the probability is greatest at the center of the atom, the nucleus, and decreases with increasing distance from the nucleus. The most common diagrammatic representation of the orbitals is that defined by a surface which encloses 95 per cent of the probabilities of the electron. This type of diagram is shown in Figure 2–5 for s- and p-orbitals. Notice that the p-orbitals consist of two **lobes** of electron density on either side of the nucleus, but the electron density at the nucleus is zero.

The schematic tabulation of the electrons in a given atom is the **electronic configuration.** The electronic configurations of the first ten elements, found in Table 2–2,

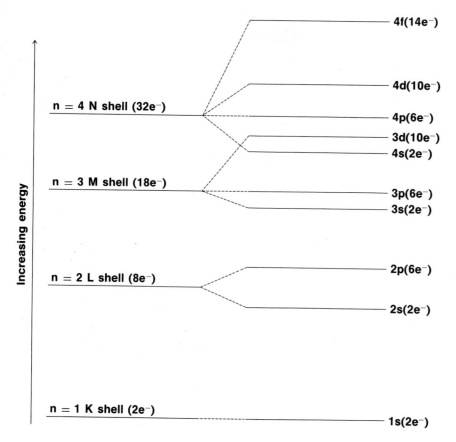

Figure 2-3   Energy levels and sublevels in the first four quantum levels or shells.

lead to several fundamental observations. The first of these observations, already mentioned, is that *only two electrons can occupy any orbital.* Also, it is observed that *electrons occupy an orbital only if all orbitals of lower energy are filled.* In addition, it can be seen that *an orbital is not occupied by a pair of electrons until other orbitals of equivalent energy are each occupied by one electron.*

A shorthand method of expressing electronic configuration is by indicating the principal quantum number and the number of electrons in each type of orbital as follows:

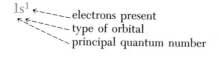

Figure 2-4   The electron cloud representation of an orbital.

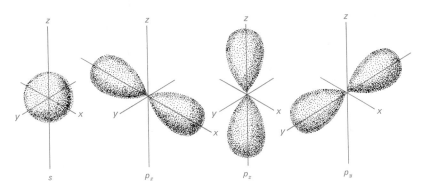

Figure 2–5   The shapes of the s and p atomic orbitals.

This notation is for the hydrogen atom which has one electron in an s-orbital in the first principal quantum level. The electronic configuration for helium with an atomic number of 2 is $1s^2$. For an atom the sum of the right superscripts always equals the atomic number. For ions the electronic configuration is indicated by simply increasing or decreasing the sum of the right superscripts to indicate the numbers of electrons added or lost. Some additional examples of the use of this scheme are:

$$
\begin{array}{ll}
\text{C } (Z = 6) & 1s^2 2s^2 2p^2 \\
\text{N } (Z = 7) & 1s^2 2s^2 2p^3 \\
\text{Mg}^{+1} (Z = 12) & 1s^2 2s^2 2p^6 3s^1 \\
\text{F}^{-1} (Z = 9) & 1s^2 2s^2 2p^6
\end{array}
$$

PERIODIC TABLES

Prior to 1850 about one-half of the elements now known had been discovered. Their chemical and physical properties and their combinations with other elements to form compounds were studied by many chemists. Considerable controversy existed as to the assignment of correct atomic weights for the elements. Most chemists used their own

TABLE 2–2   ELECTRONIC CONFIGURATIONS OF THE FIRST TEN ELEMENTS

	1s	2s	2p		
H	①				
He	⑪				
Li	⑪	①			
Be	⑪	⑪			
B	⑪	⑪	①		
C	⑪	⑪	①	①	
N	⑪	⑪	①	①	①
O	⑪	⑪	⑪	①	①
F	⑪	⑪	⑪	⑪	①
Ne	⑪	⑪	⑪	⑪	⑪

system of symbols for the elements, and a separate set of atomic weights was used for inorganic and organic chemistry.

The most extensive attempt to classify the elements was carried out by the Russian chemist Mendeleev (Fig. 2–6) in 1869. When he started to prepare a chemistry manual for his students at St. Petersburg University, his goal was to discover a logical interconnection between the properties of the chemical elements and their compounds. He arranged the elements in order of increasing atomic weights in such a way that elements with similar properties were placed in the same vertical columns. From his studies he concluded that *both the chemical and physical properties of the elements vary in a periodic fashion with their atomic weights*. His arrangement of elements in vertical columns was called a **periodic table.**

The vertical columns in the table contain elements with similar physical and chemical properties and are called **groups** or **families** of elements. Mendeleev named the horizontal rows in the table **periods.** A period contained the elements that occurred between two elements with similar properties or between two successive elements in the same group.

The periodic table as developed by Mendeleev enabled chemists to classify their knowledge and to concentrate their study on the physical and chemical properties of eight groups of elements rather than on each element individually. In the preparation of the table, if the element with the next higher atomic weight than the element just placed in the series did not fit in a group, Mendeleev left a blank space and moved the element up to the next higher group to test its similarity of properties to the elements of that group. This resulted in several blank spaces in the periodic table that he believed should contain elements that had not been discovered at that time. He predicted 3 of these elements and even listed the properties that they should possess. All 3 were discovered in his lifetime, which gave him considerable scientific satisfaction. On the one hundredth anniversary of his birth, the Russian government issued a postage stamp in his honor. Element No. 101 takes its name from him.

Some of the elements, however, seemed out of place in the table. When placed in the group with similar properties, the atomic weight of some of the elements was less than that of the element immediately preceding it. This resulted in redetermination of many of the atomic weights and correction of some experimental errors. In spite of the more accurate determination of atomic weights, there were still a few elements out of

Figure 2–6   Dmitri Mendeleev (1834–1907) made the most extensive early studies of chemical periodicity.

place in the table; for example, argon and potassium, and iodine and tellurium. After other difficulties had been encountered and repeated atomic weight determinations did not change the position of these elements that were out of place, it was finally admitted that the periodic grouping of elements according to their atomic weights might not be entirely correct.

In 1914 Moseley, a young English scientist, was studying the characteristics of x-rays given off by x-ray tubes. A simple x-ray tube is an evacuated electron tube in which cathode rays, or streams of electrons, strike the target, or anode, which sends out penetrating x-rays. Moseley constructed x-ray tube targets out of metallic elements. He then measured the frequencies of the x-rays emanating from these different metallic elements and recorded them as lines on a photographic plate. As he changed the target material from an element in the periodic table to the next highest member, the x-ray frequencies increased. When he arranged the elements in order of their x-ray frequencies, he established their position in the periodic table and assigned them atomic numbers that were dependent on their order in the table. His most important finding was that the square root of any given x-ray frequency for each element is very nearly proportional to the atomic number. By using this nearly constant increase in the square root of the x-ray frequency of elements, he was able not only to place elements in their proper position in the periodic table but also to locate the position of undiscovered elements. When a periodic table based on the atomic numbers was prepared, it was found that various properties agreed perfectly with the arrangement of the elements in the table. This new system places argon and potassium, and iodine and tellurium, in their proper positions in the table even though their atomic weights do not follow in proper progression. More accurately stated, then, **the physical and chemical properties of the elements are a periodic function of their atomic numbers.**

The most commonly used form of the periodic table, **the long form,** is shown as Table 2–3. In this table from left to right across a period the physical properties change from metal to nonmetal, while from top to bottom in a main group the elements become increasingly metallic. All of the subgroup elements are metals. Groups of similar elements are easier to locate; for example, the **nonmetals** are located in the upper right corner of the table to the right and above the colored line. The **light metals** are located in the upper left portion of the table, including the upper elements in Group IIIA. The best examples are elements with atomic numbers 3, 4, 11, 12, and 13. The **heavy metals** are located in the bottom half of the table. The most **active metals** are found at the bottom of the extreme left of the table in Groups IA, IIA, and IIIB. The most active **nonmetals** are found on the extreme right of the table in Groups VA, VIA, and VIIA. The **transition elements,** which are metals with more than one combining power, are located in the central portion of the bottom half of the table in Group IIIB to Group VIII inclusive.

The **rare earth elements** have very similar properties and are found in Group IIIB. For convenience, the complete list of rare earth elements is shown in a separate section at the bottom of the table. The heaviest elements, starting with actinium (with an atomic number of 89) and including uranium and the elements recently discovered, have been named **actinides** and are also located in Group IIIB, since they have very similar properties to the other elements in this group and to the rare earth elements. The actinides are also shown in a separate section at the bottom of the periodic table.

Although it is convenient to include the rare earth and actinide elements in separate sections in the table, the most accurate portrayal of the periodic table would represent the elements from 57 to 71 in a vertical column in Group IIIB. Elements from 72 to 89 would still proceed in a horizontal row to complete Period 6 and start Period 7. Elements from 89 to 103 would again be shown in a vertical column in Group IIIB. This form of the periodic table also possesses another advantage. One can more readily

TABLE 2-3   LONG FORM OF THE PERIODIC TABLE

IA	IIA	IIIB	IVB	VB	VIB	VIIB	VIII	VIII	VIII	IB	IIB	IIIA	IVA	VA	VIA	VIIA	O
1 H 1.00797 ±0.00001																	2 He 4.0028 ±0.00005
3 Li 6.939 ±0.0005	4 Be 9.0122 ±0.00005											5 B 10.811 ±0.003	6 C 12.01115 ±0.00005	7 N 14.0067 ±0.00005	8 O 15.9994 ±0.0001	9 F 18.9984 ±0.00005	10 Ne 20.183 ±0.0005
11 Na 22.9898 ±0.00005	12 Mg 24.312 ±0.0005											13 Al 26.9815 ±0.00005	14 Si 28.086 ±0.001	15 P 30.9738 ±0.00005	16 S 32.064 ±0.003	17 Cl 35.453 ±0.001	18 Ar 39.948 ±0.0005
19 K 39.102 ±0.0005	20 Ca 40.08 ±0.005	21 Sc 44.956 ±0.00005	22 Ti 47.90 ±0.005	23 V 50.942 ±0.0005	24 Cr 51.996 ±0.001	25 Mn 54.9380 ±0.00005	26 Fe 55.847 ±0.003	27 Co 58.9332 ±0.00005	28 Ni 58.71 ±0.005	29 Cu 63.54 ±0.005	30 Zn 65.37 ±0.005	31 Ga 69.72 ±0.005	32 Ge 72.59 ±0.005	33 As 74.9216 ±0.00005	34 Se 78.96 ±0.005	35 Br 79.909 ±0.002	36 Kr 83.80 ±0.005
37 Rb 85.47 ±0.005	38 Sr 87.62 ±0.005	39 Y 88.905 ±0.0005	40 Zr 91.22 ±0.005	41 Nb 92.906 ±0.0005	42 Mo 95.94 ±0.005	43 Tc (99)	44 Ru 101.07 ±0.005	45 Rh 102.905 ±0.0005	46 Pd 106.4 ±0.05	47 Ag 107.870 ±0.003	48 Cd 112.40 ±0.005	49 In 114.82 ±0.005	50 Sn 118.69 ±0.005	51 Sb 121.75 ±0.005	52 Te 127.60 ±0.005	53 I 126.9044 ±0.0005	54 Xe 131.30 ±0.005
55 Cs 132.905 ±0.0005	56 Ba 137.34 ±0.005	57 °La 138.91 ±0.005	72 Hf 178.49 ±0.005	73 Ta 180.948 ±0.0005	74 W 183.85 ±0.005	75 Re 186.2 ±0.05	76 Os 190.2 ±0.05	77 Ir 192.2 ±0.05	78 Pt 195.09 ±0.005	79 Au 196.967 ±0.0005	80 Hg 200.59 ±0.005	81 Tl 204.37 ±0.005	82 Pb 207.19 ±0.005	83 Bi 208.980 ±0.0005	84 Po (210)	85 At (210)	86 Rn (222)
87 Fr (223)	88 Ra (226)	89 †Ac (227)															

° Lanthanum Series

58 Ce 140.12 ±0.0005	59 Pr 140.907 ±0.0005	60 Nd 144.24 ±0.005	61 Pm (147)	62 Sm 150.35 ±0.005	63 Eu 151.96 ±0.005	64 Gd 157.25 ±0.005	65 Tb 158.924 ±0.0005	66 Dy 162.50 ±0.005	67 Ho 164.930 ±0.0005	68 Er 167.26 ±0.005	69 Tm 168.934 ±0.0005	70 Yb 173.04 ±0.005	71 Lu 174.97 ±0.005

† Actinium Series

90 Th 232.038 ±0.0005	91 Pa (231)	92 U 238.03 ±0.005	93 Np (237)	94 Pu (242)	95 Am (243)	96 Cm (247)	97 Bk (247)	98 Cf (249)	99 Es (254)	100 Fm (253)	101 Md (256)	102 No (253)	103 Lw (257)

Atomic Weights are based on C^{12}—12.0000 and conform to the 1961 Values

predict the properties of an element from its position in this type of table than in the older form of the table.

ATOMIC STRUCTURE AND PERIODIC PROPERTIES

The electronic configuration of an element determines its chemical properties as well as many of its physical properties. The electrons most important in determining chemical properties are those found in the outermost quantum level, or shell. These electrons are called the **valence electrons.** The periodically repeated occurrence of elements with similar properties is due to the periodically repeated occurrence of elements with the same number of valence electrons. For example, in all of the elements in Period 2, starting with lithium, the first shell is filled with 2 electrons. Lithium has 1 electron in the second shell; beryllium, 2; boron, 3; carbon, 4; and so on to neon, in which the second shell is completed with 8 electrons. The element with the next highest atomic number would naturally start a new period under lithium, since its first two shells are completely filled with electrons, and it has 1 electron in the third shell. Following sodium, magnesium has its first two shells completely filled with electrons and 2 extra electrons in the third shell; aluminum has 3 electrons in the third shell, and so on to argon, which has the third shell filled with 8 electrons. The element with the next highest atomic weight would then start a new row, or series, being placed under lithium and sodium. This is potassium, with the first two shells completely filled, with 8 electrons in the third shell and 1 electron in the fourth shell. Since the number of electrons necessary to fill a given shell (starting with shell number one) is 2, 8, 18, and 32, it is not surprising that 2, 8, 18, or 32 elements are needed to complete a horizontal row, or period, in the periodic table. The relationships just discussed may be seen in the following tabulation:

Electrons in	Li	Be	B	C	N	O	F	Ne
First shell	2	2	2	2	2	2	2	2
Second shell	1	2	3	4	5	6	7	8

	Na	Mg	Al	Si	P	S	Cl	Ar
First shell	2	2	2	2	2	2	2	2
Second shell	8	8	8	8	8	8	8	8
Third shell	1	2	3	4	5	6	7	8

	K	etc.
First shell	2	etc.
Second shell	8	etc.
Third shell	8	etc.
Fourth shell	1	etc.

Another symbolism for indicating the electronic configuration of an atom is very helpful in predicting chemical properties. The **Lewis symbol** is written as the letter(s) denoting the element surrounded by dots symbolizing the number of electrons in the outermost shell, the **valence electrons.** The letter(s) of the symbol represent the nucleus and all electrons in inner closed quantum levels, or shells. Examples of Lewis symbols are:

$$\text{Li}\cdot \quad \text{Be}: \quad \cdot\text{B}: \quad :\text{C}\cdot \quad :\overset{\cdot}{\text{N}}\cdot \quad :\overset{\cdot\cdot}{\text{O}}\cdot \quad :\overset{\cdot\cdot}{\text{F}}\cdot \quad :\overset{\cdot\cdot}{\underset{\cdot\cdot}{\text{Ne}}}:$$

The Lewis symbols for elements in one vertical column, or group, in the periodic table differ only in letters, since each has the same number of valence electrons. Thus, the Lewis symbols for nitrogen and phosphorus, both in Group VA, are $:\!\overset{\cdot}{\underset{\cdot}{N}}\!\cdot$ and $:\!\overset{\cdot}{P}\!\cdot$.

PERIODIC VARIATION OF ATOMIC PROPERTIES

Many properties of atoms are observed to vary periodically with their location in the periodic table. Three very important properties which show periodic variation are atomic size, ionization potential, and electron affinity.

The size of an atom is determined by the radius of a sphere whose surface defines the most probable location of the valence electrons. Several factors can influence atomic radii. However, for an atom outside of the influence of other atoms, electrical charges, and other forces, the major factors are three in number. Recalling that the quantum level, or shell, of the valence electrons increases in size with increasing principal quantum number, n, the sizes of atoms would be expected to increase as progressively higher quantum levels are occupied. This is found to be true. However, within a given principal quantum level, or period, the size of atoms decreases with increasing atomic number or nuclear charge. This occurs because the electron-nuclear force of attraction increases with increasing nuclear charge. The third factor which influences atomic sizes is the repulsion between valence electrons and electrons in inner filled quantum levels. This effect tends to "shield" the valence electrons from the nucleus and decreases the force of attraction toward the nucleus.

To summarize the periodic variation of atomic sizes, it can be stated that there is an increase in size within a group with increasing atomic number, and a decrease in size within a period with increasing atomic number. Thus, the smaller atoms are found in the upper right corner of the periodic table, and the larger atoms are in the lower left corner.

The actual force of interaction between the valence electrons and the nucleus can be measured by determining the energy involved in removing an electron from an atom. The process of removing an electron from an atom generates a positive ion and is called **ionization.** Using Lewis symbols, the removal of an electron from a neutral boron atom is shown as:

$$:\!\overset{\cdot}{B}\cdot \longrightarrow\ :\!B^{+1} + e^-$$

In this representation, the charge on the boron ion is indicated with a $+1$ right superscript, and the electron which the boron loses is represented by e^-. The *energy which must be added to a gaseous atom to remove a valence electron* is called the **ionization potential.** The ionization potential is largest for electrons which are held very tightly by the atom, and consequently increases with decreasing atomic radius. Ionization potentials are largest for elements in the upper right corner of the periodic table and smallest for those in the lower left corner. The periodic variation of the ionization potentials within the first 65 elements is shown in Figure 2–7.

In many chemical reactions an atom adds an electron to become a negative ion. This process, which might also be called ionization, is represented as follows for carbon adding an electron to become a negative ion.

$$:\!\overset{\cdot}{C}\cdot\ +\ e^- \rightarrow\ :\!\overset{\cdot}{\underset{\cdot}{C}}\cdot^{-1}$$

The tendency of an atom to gain an electron is called **electron affinity.** The electron affinity is the *energy released when an electron is added to a gaseous atom.* Since added electrons will be held most tightly if they enter an orbital near the nucleus, the highest

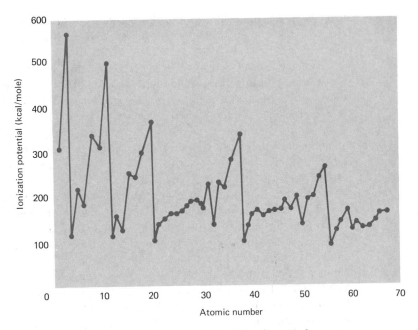

Figure 2-7   The ionization potentials of the first 65 elements.

electron affinities are associated with the smallest atoms. Electron affinities are largest for elements in the upper right corner of the periodic table and smallest for those in the lower left corner.

The ionization of an atom results in ions which have radii quite different from those of the parent atoms. Positive ions are always smaller than their parent atoms, whereas negative ions are always larger than their parent atoms.

The variation of ionization potentials and electron affinities among the elements serves to differentiate between two classes of elements, **metals** and **nonmetals.** In general, metals which are located in the left portion of the periodic table have low ionization potentials and electron affinities and, consequently, the highest tendencies to lose electrons and become positive ions. The number of electrons lost is most commonly that number which will remove all electrons from the highest occupied quantum level. The resulting ion will have an electronic configuration equivalent to that of an element in Group O. The gases in Group O of the periodic table, the **inert gases,** are very unreactive or chemically inert, but the heavier members are known to react with the very reactive element fluorine. The electronic configurations of the inert gases are unique in that each involves a set of p-orbitals which is completely filled with electrons. These electronic configurations are more stable than those of any other elements. The elements in Group IA of the periodic table form ions with a positive electrical charge of one, whereas the elements in Group IIA form ions with a positive electrical charge of two. Using atomic symbols, the common sodium and potassium ions in the first group would be represented by Na^+ and K^+. Magnesium and calcium ions would be written Mg^{+2} and Ca^{+2}. Occasionally, 3 electrons may be removed from the outer shell of an atom, but this is not common. The removal of 4 electrons from the outer shell, which is rare, would result in an ion with four positive charges.

In the nonmetallic elements on the right side of the periodic table there occurs an entirely different behavior. Nonmetals have high ionization potentials and high electron affinities. Instead of losing or giving up an electron, the **nonmetallic elements tend to gain electrons.** Chlorine, in Group VIIA is an example. Chlorine has 7 electrons in its outer shell. When the chlorine atom takes on an extra electron from a metallic atom, it is converted into a chloride ion. The **chloride ion,** written Cl⁻, has a negative electrical charge of one unit and a stable arrangement of 8 electrons in the third principal quantum level.

Going back to Group VIA in the table, an atom of sulfur can gain 2 electrons from a metallic atom to form a **sulfide ion.** Nonmetallic elements rarely gain 3 electrons, so there are few ions with 3 negative charges. The periodic variation of atomic properties is reflected in the chemical and physical properties of chemical compounds. For example, some elements form acids, whereas others have the property of forming bases. **Acids** are compounds that usually contain hydrogen, oxygen, and another element. The characteristics of an acid are due to the hydrogen ion, which is hydrogen with a unit positive electrical charge. **Bases** are compounds of hydrogen and oxygen with another element, in which the oxygen and hydrogen combine to form a hydroxide ion. The hydroxide ion (OH⁻) carries a unit negative charge and is responsible for the characteristics of a base. The property of forming an acid or base depends on the position of the element in the periodic table. For example, in the periodic table the elements of Group I form strong bases, those of Group II, moderately strong bases, and Group III and Group IV elements may form weakly basic or weakly acidic compounds. The elements in Groups V, VI, and VII form strong acids, such as phosphoric, sulfuric, and hydrochloric acids. In general, the elements on the left form strong bases, those on the right form strong acids, and those in the middle can form either weak acids or weak bases.

THE DISCOVERY OF RADIOACTIVITY

In 1895 Röntgen, a German physicist, discovered that the cathode-ray tube emitted invisible rays capable of penetrating opaque substances. These rays were called x-rays and their properties aroused considerable interest. In the following year the French physicist Becquerel investigated several fluorescent substances as possible sources of these penetrating rays. He thought that x-rays might have their origin in the greenish fluorescence produced by the cathode rays impinging on the glass of the tube. Of the substances tested, only a uranium compound affected a photographic plate that had been protected by a wrapping of black paper. Becquerel found that all uranium compounds gave off penetrating rays, which he called **Becquerel rays,** and the intensity of the rays was proportional to the amount of uranium contained in the compound. Becquerel called this production of radiation by uranium compounds **radioactivity.**

Madame Curie and her husband, Pierre, investigated this new property called radioactivity and tested all other known elements for radioactive properties. The only other radioactive element that they found was **thorium,** which, like uranium, possessed a very high atomic weight. In further investigations they discovered a sample of pitchblende, a uranium-containing ore, which exhibited four times the radioactivity of a similar quantity of a uranium or thorium salt. Starting with a ton of pitchblende ore residues, they separated the various elements in the ore and tested each for radioactivity. From the fraction containing the element bismuth, they isolated a new radioactive element that was named **polonium** in honor of Poland, Marie Curie's native country. The barium fraction was also radioactive, suggesting the presence of another new element. This new element, discovered in 1898, possessed radioactivity two to three million times that of uranium, and was named **radium.**

Rutherford studied the radiations given off by radium by placing some of the radioactive material in the bottom of a thick lead well. The lead shielded the apparatus from any other radiations and allowed him to focus the rays on a photographic plate. Under the influence of a magnetic field or an electric field, the rays were deflected in such a way that three types of radiations were observed (Fig. 2–8). Rutherford named

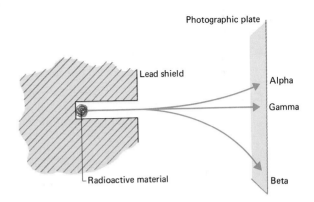

Figure 2-8   The effect of a magnetic field on the three types of radiation given off by radioactive material. In this illustration the magnetic field is perpendicular to the plane of the paper.

the rays alpha, beta, and gamma, and found that they were given off from all radioactive elements.

1. **Alpha rays,** which consist of **alpha particles,** are positively charged helium atoms. An alpha particle, therefore, is the nucleus of the helium atom, or the helium atom minus its pair of electrons, and consists of 2 protons and 2 neutrons with a positive charge of two units. When released from the nucleus of a radioactive element, alpha particles have a velocity of over 10,000 miles per second. They immediately collide with thousands of air molecules, which slow them down and finally stop them after they have gone about 8 cm. These collisions not only produce heat, but also produce ionized particles, since some of the air molecules have electrons knocked out of them by the alpha particles.

The symbol for the alpha ray is $_2^4\text{He}$. This symbol illustrates the general representation for various isotopes of all elements. It has a left subscript which is the atomic number of the isotope and a left superscript which is the **mass number,** the number of **nucleons** (protons and neutrons) in the nucleus. The alpha particle is also symbolized by α-particle.

2. **Beta rays** are made up of **beta particles,** which are streams of electrons. Their initial velocity approaches that of the speed of light, or over 150,000 miles per second. They are more penetrating than alpha rays and are of the opposite electrical charge. The beta particle is symbolized as $_{-1}^{0}e$ because of its very small mass and negative charge. The symbol β-particle is also used.

3. **Gamma rays,** γ, are similar to x-rays, but have an even greater power of penetration. They do not consist of individual particles, but are uncharged electromagnetic waves, and thus are not deflected in a magnetic field, as are alpha and beta rays. Although alpha and beta rays cause fogging of photographic plates, gamma rays are more penetrating and will fog films more rapidly.

It is of interest to compare the penetrating powers of these high-velocity particles or rays. From Rutherford's original experiments he learned that alpha particles could pass through very thin sheets of metal foil. They are larger particles than beta or gamma rays and can be stopped by a sheet of paper. Beta rays have about 100 times the penetrating power of alpha particles and will pass through thin sheets of metal. Gamma rays are difficult to stop, since they have 10,000 times the penetrating power of alpha particles, and about 2 inches of lead or 12 inches of iron are required to screen them out.

NATURAL RADIOACTIVE DECAY

As the number of protons in the nucleus increases, there is an increase in the number of neutrons required to counteract the proton–proton repulsions. Whenever the proton–proton repulsions exceed the stabilizing influence of the neutrons, the nucleus undergoes a change converting it to some stable nucleus. All elements with atomic numbers greater than 83, and some lower, undergo these changes called **radioactive decay.**

Natural radioactive decay usually proceeds by loss of an alpha or beta particle from the nucleus of an unstable isotope and is virtually always accompanied by γ-ray emission. **Alpha decay** always results in the formation of an isotope of a different element. The product has an atomic number two less and mass number four less than the original isotope. This is illustrated by the following reaction equation:

$$^{238}_{92}\text{U} \rightarrow\ ^{234}_{90}\text{Th} + {}^{4}_{2}\text{He}$$

The thorium isotope formed in this reaction is unstable and undergoes beta decay as follows:

$$^{234}_{90}\text{Th} \rightarrow\ ^{234}_{91}\text{Pa} + {}^{0}_{-1}\text{e}$$

The process of **beta decay** involves conversion of a neutron, $^{1}_{0}\text{n}$, into a proton by loss of an electron. The product has approximately the same mass as the starting isotope but the atomic number has increased by one.

In the nuclear reactions so far described, the electrons outside of the nucleus (extra-nuclear) have not been considered. Obviously the loss of an alpha particle must be accompanied by the loss of two extranuclear electrons to maintain electrical neutrality. Also, the loss of a beta particle involves the gain of an electron. Electrons apparently can be readily shifted back and forth in the surrounding environment to establish neutrality of newly formed elements or isotopes. For example, some of the alpha particles emitted from nuclei pick up 2 electrons and become helium atoms.

Figure 2–9 includes these examples of alpha and beta decay as the first two steps in a radioactive decay series. This series starts with $^{238}_{92}\text{U}$ and proceeds through 14 decay steps before the stable isotope of lead, $^{206}_{82}\text{Pb}$, is formed.

The rate at which various unstable nuclei undergo radioactive decay varies as a function of nuclear stability but is always a first order process. That is, it depends only on the quantity of radioactive matter. For such processes the time required for half of

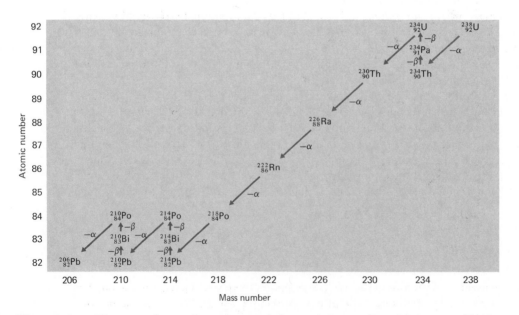

Figure 2–9 The series of natural transformations from uranium to the stable isotope of lead.

the nuclei of a given isotope to decay is a constant called the **half-life.** Half-lives of radioactive isotopes range from fractions of a second to thousands of years.

Radioactive decay occurs with a rate which is independent of temperature and the chemical nature of the radioactive isotope. Thus $^{238}_{92}U$ decays at the same rate in a sample of the element and in the compound UO_2. Consequently the amount of a radioactive isotope present in a given sample depends on the half-life, the elapsed time since the element was formed, and the rate at which the element has been formed by decay of other substances.

Radioisotope dating is a procedure for estimating the date of formation of objects by analyzing the isotope composition. The decay $^{238}_{92}U \rightarrow {}^{206}_{82}Pb$ which has a half-life of 4.5×10^9 years has been used to establish that uranium found on the earth is about 4.5×10^9 years old. However, this does not necessarily establish the time of formation of the earth's crust. Another important example of radioisotope dating involves the decay of $^{14}_{6}C$. This isotope is generated at a nearly constant rate in the upper atmosphere as follows:

$$^{1}_{0}n + {}^{14}_{7}N \rightarrow {}^{14}_{6}C + {}^{1}_{1}H$$

The radioactive $^{14}_{6}C$ thus produced is continually ingested into living plants and animals and consequently an equilibrium concentration of $^{14}_{6}C$ is established. At death the ingestion ceases and the $^{14}_{6}C$ content of the residual matter continually decreases due to $^{14}_{6}C$ radioactive decay with a half-life of 5760 years. Carbon isotope analysis allows estimation of the time of death of living organisms with a surprising degree of accuracy.

BOMBARDMENT REACTIONS

Rutherford, in 1919, conceived the idea of bombarding the nucleus of an element with alpha particles from radium, traveling at high velocities. He succeeded in knocking protons out of the nucleus of nitrogen atoms with the resultant formation of atoms of hydrogen. Later it was shown that his reaction also produced an isotope of oxygen:

$$^{14}_{7}N + {}^{4}_{2}He \longrightarrow {}^{1}_{1}H + {}^{17}_{8}O$$

Nitrogen atom α particle   Proton   Oxygen isotope

Since 1919, other transformations have been studied, and several radioactive isotopes have been produced by nuclear bombardment. The important subatomic particle, the neutron, was discovered in 1932 by the British physicist Chadwick in his study of the nuclear bombardment of beryllium with alpha particles:

$$^{9}_{4}Be + {}^{4}_{2}He \longrightarrow {}^{12}_{6}C + {}^{1}_{0}n$$

Beryllium atom α particle   Carbon isotope   Neutron

It was shown in 1934 by Frédéric and Irène Curie Joliot, son-in-law and daughter of Marie Curie, while studying the bombardment of aluminum, boron, and magnesium atoms with streams of alpha particles from polonium, that radioactive isotopes could be artificially produced:

$$^{10}_{5}B + {}^{4}_{2}He \longrightarrow {}^{13}_{7}N + {}^{1}_{0}n$$

Boron atom (stable) α particle   Nitrogen isotope (radioactive)   Neutron

The 88 elements known to man in 1930 did not include any with atomic numbers higher than 92. Since that time four elements with atomic numbers below 92, technetium (43), promethium (61), astatine (85), and francium (87), and twelve elements with atomic numbers above 92 have been made by bombardment reactions. The twelve, called **transuranium** elements, include the recently prepared element of atomic number 105 which as yet has not been well characterized and thus is not included in the periodic table or listings of elements in this text.

Transuranium elements were first prepared in 1940 by Seaborg and MacMillan by bombarding uranium-238, $^{238}_{92}U$, with neutrons of relatively low energy. Neutrons of low energy have no difficulty penetrating the nucleus since they experience no electrostatic repulsion. The product of this bombardment, another uranium isotope having a very short half-life, underwent beta decay to give an isotope of neptunium. This new element having a half-life of only 2.3 days also underwent beta decay to generate another new element, plutonium. This series of processes is represented as follows:

$$^{238}_{92}U + ^{1}_{0}n \rightarrow ^{239}_{92}U$$

$$^{239}_{92}U \rightarrow ^{239}_{93}Np + ^{0}_{-1}e$$

$$^{239}_{93}Np \rightarrow ^{239}_{94}Pu + ^{0}_{-1}e$$

The same products can be generated by bombarding uranium-238 with high energy deuterons giving the following reactions:

$$^{238}_{92}U + ^{2}_{1}H \rightarrow ^{239}_{92}U + ^{1}_{1}H$$

Uranium-238 has been used as the target in numerous other bombardment reactions to give synthetic elements.

NUCLEAR ENERGY

Radioactive decay is accompanied by the release of large amounts of energy. It is also true that the combination of protons and neutrons to generate new nuclei will be accompanied by the release of large amounts of energy. These two processes, the splitting of nuclei into smaller nuclei and other particles and the combination of protons and neutrons, are called **nuclear fission** and **nuclear fusion,** respectively.

As stated in Chapter 1, the energy released in a nuclear process can be calculated by considering Einstein's relationship,

in which energy, **E,** is related to the mass change, Δm, and the velocity of light, **c.** Using this equation it can be shown that a mass decrease of one gram in a nuclear process results in the liberation of 2.15×10^{10} kcal of energy. The energy released in chemical processes used as energy sources is about 8 kcal/g. The fission of 1 gram of uranium-235 releases about 20×10^{6} kcal. This comparison of chemical energies and nuclear energies serves to illustrate the tremendous potential which nuclear energy has to supply man's energy needs.

NUCLEAR FISSION

Many isotopes of heavy elements when bombarded by neutrons undergo fission. Uranium-235 has been used extensively as a fissionable isotope. This isotope, when bombarded by neutrons, undergoes several different fission reactions to give two different

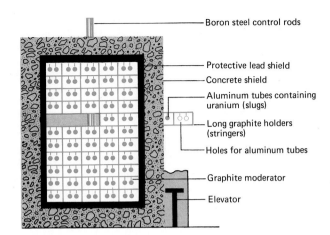

Boron steel control rods

Protective lead shield
Concrete shield
Aluminum tubes containing uranium (slugs)

Long graphite holders (stringers)

Holes for aluminum tubes

Graphite moderator

Elevator

Figure 2–10 A diagrammatic representation of a nuclear reactor, more commonly known as a pile. (Isotopes Division, U. S. Atomic Energy Commission.)

lower mass atoms, beta particles, and several neutrons. The released neutrons bombard other nuclei to induce the same types of decay processes leading to a chain reaction. The chain reaction continues only if the sample mass is large enough for most of the released neutrons to be captured within the sample. The sample mass required for propagation is called the **critical mass.** The chain reaction fission of uranium-235 was the basis for the first **atomic bomb.** In this bomb two samples of uranium-235, both of subcritical mass, were brought together to give a sample exceeding the critical mass. The ensuing fission generated so much heat that it exploded the mass.

Controlled fission reactions have found increased utility as sources of energy. Such controlled processes take place in a **nuclear reactor** called a **pile** because of the method of its construction (Fig. 2–10). It consists of a huge pile of graphite bricks piled in a honeycomb fashion interlaced with slugs of uranium sealed in aluminum cans. The uranium contains only about 0.7 per cent fissionable uranium-235 which acts as the **fuel.** The graphite acts as a **moderator** in that it causes the neutrons from uranium-235 fission to lose energy precluding the formation of $^{239}_{92}U$ but allows neutrons to be captured by other $^{235}_{92}U$. Additional moderation is imposed by insertion of boron steel or cadmium control rods which absorb neutrons very effectively. The extent of insertion or removal of these rods controls the net flow of neutrons in the pile, and thus the rate at which fission occurs.

During the fission of the nuclear fuel in a reactor, the large amount of heat released is transferred to a circulated coolant of water or liquid sodium. This heat can be transferred in another exchanger to generate steam for a steam turbine which drives an electric generator as shown in Figure 2–11.

NUCLEAR FUSION

The source of the energy of the sun and stars is nuclear fusion, which produces heavier nuclei from light nuclei. A three-step process is thought to convert protons into helium nuclei.

$$^1_1H + ^1_1H \rightarrow ^2_1H + ^0_{+1}e$$

$$^2_1H + ^1_1H \rightarrow ^3_2He$$

$$^3_2He + ^3_2He \rightarrow ^4_2He + 2\,^1_1H$$

The first step generates a deuteron and a **positron,** which is a particle like an electron except that it is positively charged. The amount of energy released in this fusion of hydrogen is greater than the energy released by the fission of an equal mass of a heavy element.

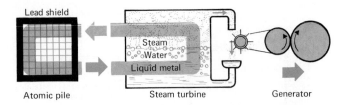

Lead shield

Steam
Water
Liquid metal

Atomic pile   Steam turbine   Generator

Figure 2–11   An artist's conception of an atomic power plant.

Fusion reactions require extremely high temperatures for initiation and have been started using the heat generated by a fission reaction. The hydrogen bomb involves a fusion reaction initiated by a nuclear fission-type bomb.

Because of the tremendous production of energy per unit of mass involved, controlled fusion reactions have an enormous potential in furnishing man with a virtually unlimited source of energy. To date, the technology involved in containing matter at the temperatures required for fusion (1,000,000°C) has not been developed but is under study. As man assesses energy needs of the future and compares the available energy sources, it becomes increasingly more obvious that even controlled nuclear fission must eventually be replaced. At present nuclear fusion appears as one of the best possible replacements.

USES OF RADIOISOTOPES

CHEMICAL, BIOLOGICAL, AND MEDICAL RESEARCH

It has been stated earlier that isotopes of an element, whether stable or radioactive, possess essentially the same chemical properties; consequently, radioisotopes placed in a reaction mixture can be traced by determining the location of the radioactivity in the products. This procedure, called a **tracer experiment,** has been used to assist in the establishment of reaction mechanisms.

Several isotopes used in chemical, biological, and medical research as well as in medicine are listed in Table 2–4. The types of experiments in which these isotopes have been used range from specific determination of the location of radioisotopes in a molecule to determination of mass transfer from one point to another by radioactivity detection.

When small amounts of radioisotopes are incorporated into a compound and the compound is fed to plants and animals, the biological systems involved do not differentiate the radioisotopes from stable isotopes. For example, biological systems do not distinguish

TABLE 2–4   COMMON ISOTOPES USED IN
CHEMICAL, BIOLOGICAL, AND MEDICAL
RESEARCH

ISOTOPE	NAME	HALF-LIFE	RADIATION
$^{3}_{1}H$	Tritium	12.5 years	Beta
$^{14}_{6}C$	Carbon-14	5760 years	Beta
$^{45}_{20}Ca$	Calcium-45	180 days	Beta, gamma
$^{24}_{11}Na$	Sodium-24	14.9 hours	Beta, gamma
$^{32}_{15}P$	Phosphorus-32	14.3 days	Beta
$^{35}_{16}S$	Sulfur-35	87.1 days	Beta
$^{59}_{26}Fe$	Iron-59	46.3 days	Beta, gamma
$^{60}_{27}Co$	Cobalt-60	5.2 years	Beta
$^{82}_{35}Br$	Bromine-82	36 hours	Beta, gamma
$^{131}_{53}I$	Iodine-131	8 days	Beta

radioactive $^{14}_{6}C$ and stable $^{12}_{6}C$ so that $^{14}_{6}C$ has been used extensively in studying mechanisms in biological processes.

In studies of photosynthesis, $^{14}_{6}CO_2$ used in place of $^{12}_{6}CO_2$ is incorporated into the sugars and starches produced. By tracing $^{14}_{6}C$ through the complex reactions of photosynthesis, considerable information was extracted about the overall process.

The pathways of metabolism in plants and animals are being studied by labeling simple sugars with $^{14}_{6}C$, amino acids with radioactive carbon, nitrogen, or sulfur, and lipid material with $^{14}_{6}C$, $^{32}_{15}P$, and $^{3}_{1}H$. The mechanism by which the animal body forms complex compounds such as cholesterol and the heme of hemoglobin has been established by the use of several radioisotopes.

An example of medical research that led to both diagnosis and treatment involves radioactive iodine, $^{131}_{53}I$. It has long been known that the thyroid gland takes up and utilizes most of the iodine in the body. When extremely small amounts of $^{131}_{53}I$ are administered by mouth to human subjects, the rate of iodine uptake by the thyroid gland may be determined. The results of such studies indicate normal thyroid function or conditions of hyper- or hypothyroidism. In patients with hyperthyroidism, the treatment with larger quantities of $^{131}_{53}I$ will effectively reduce the size of the thyroid gland by irradiation of the tissue involved in the disorder.

Radioactive phosphorus, $^{32}_{15}P$, has been used to study the formation and deposition of bone and teeth. Radioactive sulfur, $^{35}_{16}S$, has been incorporated into proteins and used to study various aspects of protein metabolism. It has also been employed in a determination of the turnover rate of plasma proteins in the body. Radioactive iron, $^{59}_{26}Fe$, has proved helpful in a study of the production of red blood cells and hemoglobin in anemia.

Living cells are destroyed by the radiation emitted by radioactive materials. For this reason the radiation emitted by radium has been used to treat malignant tissue, which seems more susceptible to radiation damage than normal tissue. Recently cobalt-60 which gives off higher energy radiation and is less expensive has been used in the treatment of cancer.

RADIOISOTOPES IN INDUSTRY

Since about 1947 the application of radioisotopes to industry has increased rapidly. Before radioisotopes were produced in nuclear reactors, both x-rays and the gamma rays of radium were used to detect flaws in metal castings. A flaw in a piece of metal absorbed less radiation than the sound metal. At present, the gamma rays from $^{60}_{27}Co$ are being used more successfully for this purpose.

The petroleum industry was an early user of radioisotopes, especially in pipelines used to transport various grades of oil and gasoline. By adding a small amount of radioactive antimony at the boundary between two batches, it is relatively easy to locate the boundaries and switch an individual batch of gasoline or oil into the proper section of the pipeline system. More recently, the radioactivity at the boundary has been used to initiate the action of automatic switches which divert each batch into the correct pipeline.

In the food industry, many applications are being explored. Bakery products are kept fresh longer by a sterilization process brought about by irradiation with a radioactive source. Cesium-137, produced in nuclear reactors, is used as a source of radioactivity to destroy trichinosis in fresh pork.

IMPORTANT TERMS AND CONCEPTS

alpha rays	beta rays	electron affinity
atomic number, Z	electron	gamma rays

ionization potential
isotopes
Lewis symbols
neutron
nuclear fission

nuclear fusion
nucleus
orbital
periodic table

proton
quantum number
radioactive decay
valence electron

QUESTIONS

1. Why do we commonly speak of a charge of -1 for the electron and $+1$ for the proton?

2. Compare and contrast electrons, protons, and neutrons.

3. Given the atomic structure of an atom, how could you determine the atomic weight? the atomic number?

4. Are isotopes expected to have similar chemical behaviors?

5. The only naturally occurring isotope of element X has 9 protons and 10 neutrons in its nucleus. What is its approximate atomic weight? How many electrons does it have? What is the Lewis symbol for this element?

6. Name and explain briefly the four quantum numbers.

7. Using Lewis symbols, illustrate the formation of a negative ion and a positive ion.

8. How do electron affinities vary with atomic size?

9. Use the shorthand method to express the electronic configurations for each of the following:

   (a) S   (b) Ca   (c) C   (d) O^{-2} (e) Be^{+2}

10. Briefly explain the contribution of Moseley to the construction of an accurate periodic table.

11. Should the properties of elements vary periodically as a function of atomic weight or atomic number? Explain.

12. What relationship exists between the electronic configurations of the elements in any one vertical column or group? in any one horizontal row or period?

13. In what way are the electron structures of ions and inert gases related?

14. The alkali metals never form negative ions in chemical reactions. Explain.

15. Indicate the position in the periodic table where the following can be found:

    (a) elements with largest electron affinities
    (b) elements with largest ionization potential
    (c) metals
    (d) nonmetals
    (e) inert gases

16. Name the two most important factors which influence the size of an atom and give examples.

17. Group VII atoms (halogens) tend to take on electrons to form negative ions; explain, giving examples.

18. What are the properties of alpha particles? How could you detect their presence?

19. What is meant by the half-life of a radioactive element?

20. In general, what happens to a radioactive element when it loses an alpha particle? a beta particle?

21. Explain the principle involved in the synthesis of new elements possessing atomic weights greater than that of uranium.

22. What is meant by nuclear fission?

23. Complete the following equations:

   (a) $_{11}^{23}Na + _{1}^{2}H \rightarrow _{11}^{24}Na + ?$
   (b) $_{20}^{44}Ca + ? \rightarrow _{21}^{44}Sc + _{0}^{1}n$
   (c) $_{82}^{214}Pb \rightarrow ? + _{-1}^{0}e$

24. Why do neutrons get involved in nuclear reactions even when they have relatively low energy?

25. When one mole of radium-226 decays by the reaction $_{88}^{226}Ra \rightarrow _{86}^{222}Rn + _{2}^{4}He$, the mass of the products is 0.0053 g less than that of the reactants. How much energy is liberated in this process?

SUGGESTED READING ▬▬▬▬▬▬▬▬▬▬▬▬▬▬▬▬▬▬▬▬▬▬

Berry: Atomic Orbitals. Journal of Chemical Education, Vol. 43, p. 283, 1966.
Hubbert: The Energy Resources of the Earth. Scientific American, Vol. 224, No. 3, p. 61, 1971.
Keller: Predicted Properties of Elements 113 and 114. Chemistry, Vol. 43, No. 10, p. 8, 1970.
Morrow: On the Discovery of the Electron. Journal of Chemical Education, Vol. 46, p. 584, 1969.
Seaborg: From Mendeleev to Mendelevium and Beyond. Chemistry, Vol. 43, No. 1, p. 6, 1970.
Seaborg: Some Recollections of Early Nuclear Age Chemistry. Journal of Chemical Education, Vol. 45, p. 278, 1968.
Thompson: Nuclear Power—Today and Tomorrow. CHEMTECH, p. 495, August 1971.

CHEMICAL BONDS

CHAPTER 3

The *objectives* of this chapter are to enable the student to:

1. Use the rule of octets to predict the combining capacities for each group of the representative elements.
2. Explain the formation of an ionic compound and an ionic crystalline solid.
3. Differentiate ionic bonding and covalent bonding.
4. Explain the difference between a sigma-bond and a pi-bond.
5. Determine the combining capacities of elements in various compounds.
6. Recognize the trends of electronegativity in the periodic table.
7. Illustrate and describe the most common molecular geometries.
8. Explain atomic orbital hybridization and its relationship to molecular geometry.
9. Draw contributing forms of the resonance hybrid of molecules not adequately described by a single Lewis formula.

Satisfied that the laws of chemical change could be explained using Dalton's atomic theory, chemists began to ask why and how atoms are held together in compounds. In the early 1900's chemists called attention to the stability of helium with its pair of valence electrons. It was also stated that many compounds contained an even number of valence electrons, with the inert gases (except helium) having a stable configuration of eight valence electrons. These observations led to the statement that when atoms combine to form molecules, the valence electrons tend to group in pairs or octets. A more useful statement, however, might be that *when elements combine to form compounds, the valence electrons of their atoms show a tendency to assume the stable configuration of an inert gas.* It is found that this **rule of octets** can be used to predict the combining capacities of most elements in the first three periods.

IONIC BONDS

The nature of ionic bonding may be illustrated by considering the formation of sodium chloride. In the overall reaction it appears that sodium atoms react with chlorine atoms to form the compound sodium chloride. As seen in Figure 3–1, however, an electron is transferred from the outer shell of the sodium atom to the outer shell of the chlorine atom. This loss of an electron from sodium occurs readily and requires a relatively small amount of energy, which is called the **ionization potential.** When this electron is presented to the chlorine atom, the chlorine atom readily accepts it, with the release of energy,

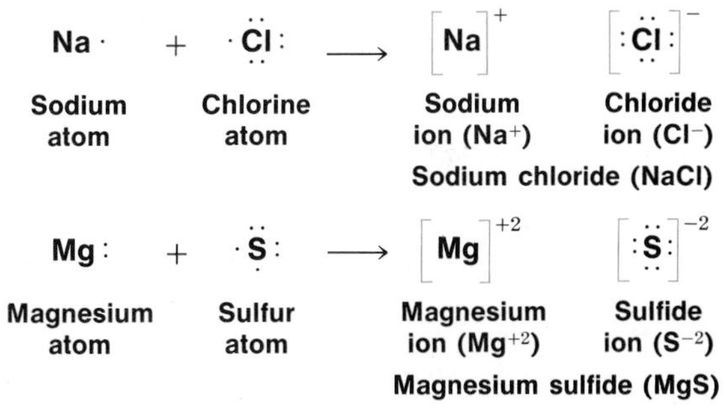

$$\text{Na} \cdot \;+\; \cdot \overset{\cdot\cdot}{\underset{\cdot\cdot}{\text{Cl}}} : \;\longrightarrow\; \left[\text{Na} \right]^{+} \quad \left[\overset{\cdot\cdot}{\underset{\cdot\cdot}{:\text{Cl}}} : \right]^{-}$$

| Sodium atom | Chlorine atom | Sodium ion (Na$^+$) | Chloride ion (Cl$^-$) |

Sodium chloride (NaCl)

Figure 3–1   The process of electron transfer commonly occurs in the formation of inorganic salts.

$$\text{Mg} : \;+\; \cdot \overset{\cdot\cdot}{\underset{\cdot}{\text{S}}} : \;\longrightarrow\; \left[\text{Mg} \right]^{+2} \quad \left[\overset{\cdot\cdot}{\underset{\cdot\cdot}{:\text{S}}} : \right]^{-2}$$

| Magnesium atom | Sulfur atom | Magnesium ion (Mg^{+2}) | Sulfide ion (S^{-2}) |

Magnesium sulfide (MgS)

which is called the **electron affinity.** The sodium ions and the chloride ions that are formed in the process are much more stable than the atoms, and resemble the inert gases neon and argon, respectively, in their valence electron configuration. The ions differ from the inert gas atoms in that they are no longer neutral but bear positive (Na) and negative (Cl) charges. These opposite charges attract each other and are responsible for the strong **ionic bond** between sodium ions and chloride ions in the compound sodium chloride. Compounds in which the atoms are held together by ionic bonds are called **ionic compounds.**

A compound formed by the transfer of 2 electrons to yield an outer shell of 8 valence electrons in each ion is magnesium sulfide. A relatively small amount of energy is required to remove the 2 outer electrons of magnesium. They are accepted by the outer shell of the sulfur atom with the release of energy. This electron transfer is illustrated in Figure 3–1. The ionic bond between magnesium and sulfide ions results from the strong electrostatic attraction between the oppositely charged ions.

In simple chemical reactions, atoms are commonly represented by the symbol alone. However, ions alone, or in compounds, are represented as the symbol bearing either positive or negative charges. For example, note ions such as Na$^+$ and Mg^{+2} and compounds such as Na$^+$Cl$^-$ and Mg^{+2}S^{-2}.

The number of electrons transferred in forming an ion corresponds to the **combining capacity** of an atom. The bonding resulting from electron transfer is called **electrovalence,** or **ionic bonding.** It can readily be seen that the combining capacity is equal to the number of electrons gained or lost by an atom when it is converted into an ion.

Ionic compounds are generally characterized as colorless, brittle, crystalline materials at room temperature. The melting points of these compounds are quite high because of their structures, which are discussed in Chapter 4. When dissolved in water, ionic compounds give solutions which conduct electricity.

COVALENT BONDS

About 1916 it was suggested that two atoms may combine by sharing valence electrons. The process of joining atoms to form molecules by the sharing of electrons is called **covalence.** As an example, hydrogen gas consists of molecules that contain 2 hydrogen atoms held together by a force resulting from the sharing of a pair of electrons, as shown in Figure 3–2. This figure illustrates the use of Lewis symbols for molecules with covalent bonds. In the symbol for the molecule, a pair of dots placed between the atoms represents a bond.

Hydrogen, chlorine, nitrogen, and other diatomic gaseous elements show similar behavior in sharing electrons to the extent that their valence shells are filled. In the chlorine molecule, each chlorine atom with 7 valence electrons is able to attain a complete

H· + ·H ⟶ H:H or H—H

Hydrogen   Hydrogen   Hydrogen
atom   atom   molecule

:Cl:Cl: or |Cl—Cl| :N:::N: or |N≡N|

Chlorine   Nitrogen
molecule   molecule

Figure 3–2 The sharing of electrons, or covalence, illustrated by hydrogen, chlorine and nitrogen molecules.

shell by sharing one pair of electrons with the other chlorine atom. In the nitrogen molecule, a filled shell is attained only if three pairs of electrons are shared. These examples are represented schematically in Figure 3–2 with Lewis formulas.

It is stated that the hydrogen and chlorine molecules are held together by a **single bond,** but that the nitrogen molecule has a **triple bond. Double bonds** are also known to exist in many molecules.

In some compounds, a single covalent bond exists in which one of the bonded atoms furnishes both of the electrons which are shared. Such a covalent bond is called a **coordinate covalent bond.** For example, the formation of the ammonium ion by the interaction of ammonia, NH_3, and a hydrogen ion, H^+, involves the formation of a coordinate covalent bond. This is represented in the following equation:

$$
\begin{array}{c}
\text{H} \\
\text{H}{-}\text{N}: + \text{H}^+ \rightarrow \\
\text{H}
\end{array}
\left[
\begin{array}{c}
\text{H} \\
| \\
\text{H}{-}\text{N}{-}\text{H} \\
| \\
\text{H}
\end{array}
\right]^+
$$

The rationalization of the existence of a force between atoms which share a pair of electrons is aided by the following hypothetical experiment. In such an experiment, diagrammatically represented in Figure 3–3 a, two hydrogen atoms are allowed to approach one another until their 1s atomic orbitals overlap and mutually occupy the space between the nuclei. The result of the overlap is an electron cloud associated with the molecule. An electron cloud associated with more than one nucleus is referred to as a **molecular orbital.** In the molecular orbital formed here by the combination of atomic orbitals, there is a high probability of finding the electrons in the region between the nuclei.

The results of a similar experiment involving the approach of fluorine atoms is shown in Figure 3–3 b, where it is seen that atomic 2p-orbitals on the approaching atoms overlap

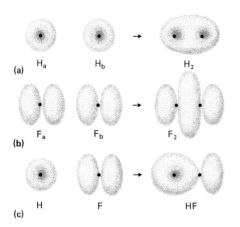

(a) H_a H_b H_2

(b) F_a F_b F_2

(c) H   F   HF

Figure 3–3 (a) Hypothetical approach of the electron clouds of two hydrogen atoms and the resulting electron cloud of the hydrogen molecule. (b) A similar approach of the p-orbitals of two fluorine atoms. (c) The approach of a 1s-orbital of a hydrogen and a 2p-orbital of a fluorine.

Figure 3-4   Side-by-side overlap of atomic p-orbitals to give a pi-bond.

to form a region between the nuclei where the probability of finding electrons is quite high. Figure 3–3 c shows a third type of experiment in which a 1s-orbital on hydrogen overlaps a 2p-orbital on fluorine to form a covalent bond.

In each of the three hypothetical experiments shown in Figure 3–3, the overlap of the electron clouds of approaching atoms results in an increased electron density along the internuclear axis. A covalent bond, resulting from increased electron density along the axis connecting the nuclei of adjacent atoms, is called a **sigma-bond** (σ-bond).

A covalent bond between adjacent atoms can also result from the overlap of p-orbitals in a side-by-side fashion, as shown in Figure 3–4. Overlap of electron clouds in this manner results in increased electron density above and below the internuclear axis. A bond of this type is called a **pi-bond** (π-bond).

The distance of internuclear separation in all molecules of which the atoms are covalently bonded is determined by the combination of attractive and repulsive forces. It is dependent on the sizes of the atomic orbitals which are being combined, the nuclear charges, the ionization potentials, the electron affinity, and other fundamental atomic properties. One half the internuclear distance of separation in homonuclear diatomic molecules (e.g., H_2, O_2, N_2, F_2, and so forth), as indicated in Figure 3–5, is defined as the **covalent radius.** Some covalent radii are given in Table 3–1.

COMBINING CAPACITY OR VALENCE

The concept of combining capacity as presented in the section on ionic bonding can be extended to include compounds involving covalent bonding. In general, the concept of combining capacity refers to the extent of involvement in bonding. In an ionic compound, the combining capacity of each ion corresponds to the number of electrons transferred to or from an atom in forming the ion. In covalently bonded compounds, the combining capacity is often called the valence and corresponds to the number of electron pairs shared by the atom of interest.

The element hydrogen only displays a combining capacity of 1 since its valence orbital, 1s, can contain only one pair of electrons, and since the atom can only lose or gain one electron. Any element which combines with hydrogen in a mole ratio of 1:1 also has a combining capacity of 1. For example, chlorine forms HCl, and thus displays a combining capacity of 1 as it does in NaCl, even though it can display other combining capacities. Carbon, in the compound CH_4, methane, has a combining capacity of 4, since each hydrogen has a combining capacity of 1.

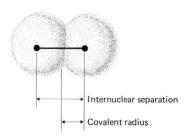

Figure 3-5   Covalent radius of a homonuclear diatomic molecule.

Internuclear separation

Covalent radius

TABLE 3–1   TRENDS IN THE COVALENT RADII OF
ELEMENTS IN NM(×10)

Period 2	Li	Be	B	C	N	O	F	Ne
	1.23	0.89	0.80	0.77	0.74	0.73	0.72	0.71
Group IA	H	Li	Na	K	Rb	Cs		
	0.37	1.23	1.57	2.03	2.16	2.35		

POLAR BONDS

"Pure" covalent bonds, as are found in homonuclear diatomic molecules, and "pure" ionic bonds, as are found in salts like NaCl, represent the extremes of bonding. In the former, the electron pair is shared equally between atoms, whereas in the latter there is no sharing of electrons, but rather a complete transfer of an electron from one atom to another. Bonding intermediate between "pure" covalent and "pure" ionic is very common. Since electrons may not always be shared equally between atoms, but yet are not necessarily completely transferred from one atom to another, there exists the possibility of unequal charge distribution in a covalent bond.

For example, in the compound HCl, hydrogen chloride, the shared pair of electrons is attracted more by the chlorine end of the molecule than by the hydrogen end. The result is an unequal charge distribution, with the chlorine end more negative and the hydrogen end of the molecule more positive. Hydrogen chloride may be represented as follows:

$$\text{HCl} \quad \text{or} \quad \text{H}\!:\!\overset{..}{\underset{..}{\text{Cl}}}\!: \quad \text{or} \quad \text{H}^+\!:\!\overset{..}{\underset{..}{\text{Cl}}}\!:^- \quad \text{or} \quad \boxed{+ \quad -}$$

Even though the molecule is electrically neutral, the center of the positive charge does not coincide with the center of the negative charge. The molecule is called a **dipole** and is said to possess a **dipole moment**. Dipole moments are often designated by the symbol ↦. In this symbol, the pointed end corresponds to the negative end of the dipole; the length is related to the magnitude; and the orientation of the dipole is indicated by the orientation of the symbol. When placed in an electrical field, such molecules will line up with their negative ends facing the positive plate (electrode) and their positive ends facing the negative plate (electrode). The dipole character of these compounds gives rise to the name **polar bonds** and **polar covalent compounds**.

In molecules with more than two atoms, there exists the possibility of having more than one polar bond. In such molecules, the dipole moment will be related to the summation of all polar bond contributions. In the water molecule, which is non-linear, each of the two bonds has an associated dipole. The sum of the bond dipoles is a dipole

Bond dipoles

Molecular dipole

Figure 3–6   Bond dipoles and the molecular dipole of water.

which is not aligned with either bond, but which bisects the HOH angle, as shown in Figure 3–6. The water molecule has a rather large dipole moment.

In the carbon dioxide molecule, CO_2, which is linear, each bond has a dipole; but the summation of these dipoles is zero, since they point in opposite directions. The dipoles are oriented as follows:

$$\overset{\longleftarrow\ +\ \longrightarrow}{\text{:O=C=O:}}$$

The carbon dioxide molecule has no net dipole moment.

ELECTRONEGATIVITY

It has already been stated that in the compound HCl the chlorine atom has a greater attraction for the electron pair than does the hydrogen atom. The attraction of an atom for shared electrons depends on the amount of energy required in the transfer of electrons from or to its outer electron shell. The attraction for valence electrons varies from element to element and is called its **electronegativity.** The relative electronegativity values that have been determined for some of the common elements follow the sequence $F > O > N \sim Cl > C > H$. These values are related to the ability of the atoms to attract shared electrons and thus increase the negative charge of their end of a molecule.

The electronegativity of an element is related to its ionization potential and its electron affinity. Elements which have high ionization potentials and high electron affinities exhibit high electronegativities. Elements with the highest electronegativities are found in the upper right corner of the periodic table. There is also a tendency toward increasing values across a period to the right and toward decreasing values moving down the elements of a group. If two elements with greatly different electronegativities combine, the bond will be highly polar or ionic in nature. Metals in Group IA and IIA combining with the nonmetals in group VIA and VIIA almost always form ionic compounds with ionic bonds. When the combination involves two elements with similar values, the bonds are usually covalent. If the compound is covalent, the atoms with the greatest difference in electronegativity will form the more polar bonds. Two like atoms with no difference in their electronegativity values will obviously form nonpolar covalent bonds.

MOLECULAR GEOMETRY

The shape of a molecule is determined by the distances between bonded atoms in the molecule, called **bond lengths,** and the angles between bonds, called **bond angles.** In simple diatomic molecules, such as H_2 and CO, the bond length completely describes the geometry of the molecule. In triatomic and higher polyatomic molecules, more than bond lengths must be stated to completely describe the geometry of the molecule. In such molecules the geometry is determined by the nature of the central atom, such as oxygen in H_2O and carbon in CCl_4.

In triatomic molecules, two molecular geometries are found, **linear** and **angular.** The use of Lewis formulas allows one to predict which of these two geometries a given molecule will exhibit. The molecule BeH_2, which has the Lewis formula H:Be:H, is linear, just as is CO_2, which has the Lewis formula $\ddot{\text{O}}{=}\text{C}{=}\ddot{\text{O}}$. In these Lewis formulas, the symbols (—) and (:) have been used interchangeably to represent pairs of electrons. In the BeH_2 molecule, all electrons are involved in bonding, whereas in CO_2 there are two pairs of electrons on each oxygen which are not involved in bonding. These electrons are called **non-bonding electrons,** as opposed to those involved in bonding, which are

calling **bonding electrons.** The molecule H_2O with the Lewis formula $H \overset{\displaystyle \diagup \overset{\textstyle \text{O}}{} \diagdown}{} H$, and the

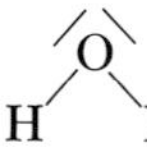

molecule NO_2 with the Lewis formula $|\underline{\text{O}} \overset{\dot{\text{N}}}{\diagup \diagdown} \underline{\text{O}}\rangle$ or $\langle \text{O} \overset{\dot{\text{N}}}{\diagup \diagdown} \underline{\text{O}}|$, are both angular.

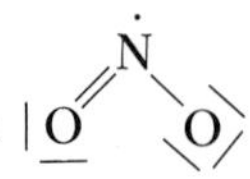

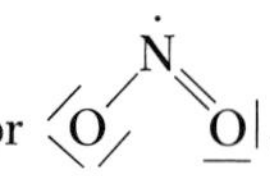

Notice that each of these molecules, in contrast to BeH_2 and CO_2, has non-bonding electrons on the central element. Since electrons repel one another, it is expected in these molecules that the bonding electrons and the non-bonding electrons associated with the central atom will orient themselves so that the repulsion interactions will be minimized. That is, they will seek the maximum distance of separation.

Typical examples of tetratomic molecules are BF_3 and NH_3 which have the Lewis formulas:

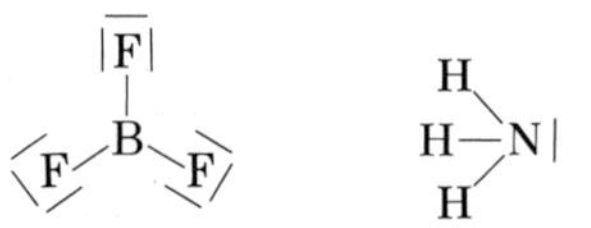

These formulas indicate that the electronic environment of the central atoms in these molecules is quite different. The BF_3 molecule has a central atom with three bonding pairs of electrons, whereas the NH_3 molecule has a central element with three bonding pairs and a non-bonding pair. The geometries of the two molecules are predicted by the application of the electron repulsion principle to be **planar triangular (trigonal)** for BF_3 and **triangular based pyramid (pyramidal)** for NH_3 (Fig. 3–7). In BF_3 electron repulsions are minimized by placing the bonding pairs of electrons at the corners of an equilateral triangle. The interaction of a fourth pair of electrons, the non-bonding pair, in NH_3 results in the non-planar geometry.

The predominant geometry found for molecules with four atoms bonded to a central atom is **tetrahedral,** as shown in Figure 3–7. In this type of molecule the central atom is located at the center of the tetrahedron and the bonded atoms are found at the apices. This geometry occurs when four bonding pairs of electrons seek the maximum distance of separation in order to minimize the repulsions. Another geometry observed for pentatomic molecules is **square planar,** in which the bonded atoms are located at the corners of a square with the central element located in the center of the square. This geometry is limited to transition metal compounds.

The most common geometry for molecules with five atoms around the central atom is **trigonal bipyramidal,** and that for six atoms around a central atom is **octahedral** (Fig. 3–7).

For simple molecules containing second period elements (i.e., B, C, and so forth) as the central atom, the approximate bond angles can be predicted by considering the sum of the non-bonding electron pairs and the bonding electron pairs associated with the central atom. For example, it was indicated that pentatomic molecules, such as CH_4, exhibit tetrahedral geometry. In this molecule the H—C—H bond angle is 109.5°, as the result of repulsions of the bonding electron pairs. However the ammonia molecule,

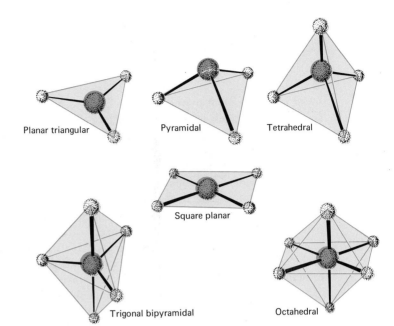

Figure 3-7 The most common geometries of simple molecules.

which has one non-bonding pair of electrons and three atoms bonded to nitrogen, and which has pyramidal geometry, has an H—N—H bond angle of 107°, nearly the same as that in CH_4. The water molecule, with two pairs of non-bonding electrons on the central atom and two bonded hydrogen atoms, displays an H—O—H bond angle of 105°. In general for second period elements, a total of four non-bonding electron pairs and atoms bonded to the central element will result in bond angles which equal or approximate those found in CH_4.

ATOMIC ORBITAL HYBRIDIZATION

The shape of a molecule can be enlightening as to the bonding interactions among the constituent atoms. In the cases of molecules having all single bonds, the relative orientations of the atoms establish the relative orientations of the sigma-bonds. For example in methane, CH_4, previously stated to have tetrahedral geometry, four sigma-bonds, bonding a carbon atom to each of four hydrogen atoms, are pointed toward the apices of a tetrahedron. The assumption of inter-bond angles of 109.5° is consistent with the experimental observation that each of the four hydrogen atoms bonded to carbon has the same chemical properties, and thus must be involved in identical bonding. These experimental observations are inconsistent with a bonding model which involves the use of pure atomic orbitals by carbon.

If the carbon atom in methane uses four pure atomic orbitals to bond to each of four hydrogen atoms, the molecule would not be tetrahedral. This is because the four orbitals of the valence shell of carbon, one s-orbital and three p-orbitals, are not equivalent. The use of pure atomic orbitals would result in three sigma-bonds which are mutually perpendicular, due to the overlap of the three 2p-orbitals of carbon with three 1s-orbitals of hydrogen. The fourth sigma-bond would arise by overlap of the carbon 2s-orbital with a hydrogen 1s-orbital. Clearly this bonding model would not produce a tetrahedral arrangement of four equivalent bonds, as is the case in methane.

A set of four equivalent sigma-bonds in methane is explained by considering the combination of the carbon 2s-orbital and the three carbon 2p-orbitals as generating a

Figure 3–8   An sp³ hybrid orbital.

set of four equivalent **hybrid orbitals.** These four hybrid orbitals, designated **sp³-orbitals,** are not like s-orbitals or p-orbitals. Each appears similar to a p-orbital with the two lobes being of different sizes (Fig. 3–8). The relative orientation of the four hybrid sp³-orbitals are such that the larger lobe of each points toward the apex of a tetrahedron (Fig. 3–9).

The H—N—H bond angle in ammonia, 107°, is best explained by considering the hybridization of a 2s-orbital and three 2p-orbitals on nitrogen to give a set of four sp³-hybrid orbitals. This model also explains the H—O—H bond angle of 105° in water.

For simple molecules having a total of three bonded atoms and non-bonding electron pairs around the central atom, the concept of atomic orbital hybridization can be used to explain the observed geometry. The bonded atoms or non-bonded electron pairs found at the corners of an equilateral triangle are bonded to a set of three equivalent **sp²-hybrid orbitals** on the central atom. This set of hybrid orbitals, arising from the combination of an s-orbital and two p-orbitals, results in bond angles of 120° (Fig. 3–9). In this bonding scheme, one p-orbital on the central element which is not included in the hybridization assumes an orientation perpendicular to the plane of the three hybrid orbitals.

Molecules having a total of two bonded atoms and non-bonding electron pairs on the central atom are linear. This geometry is explained by considering a set of two equivalent **sp-hybrid orbitals** on the central element and two unhybridized p-orbitals oriented perpendicular to the sp-hybrid orbitals and perpendicular to one another. The use of sp-hybrid orbitals by a central element results in a bond angle of 180° (Fig. 3–9).

In many simple molecules containing central elements from the third and higher periods, hybrid orbitals are not needed, and the bonding is more easily explained by the use of pure atomic orbitals. For example, the H—P—H bond angle in phosphine, PH_3, is 92° which is nearly that expected if each hydrogen is bonded to one of the three mutually perpendicular 3p-orbitals on phosphorus. The same is true for hydrogen sulfide, H_2S, in which the H—S—H bond angle is found to be 92°.

MULTIPLE BONDS

As indicated earlier, in many molecules more than one pair of electrons is shared between two atoms. In such multi-bonded molecules, there must be pi-bonds involved, since only one sigma-bond can form between any two atoms. In the double-bonded molecule carbon dioxide, CO_2, each carbon-oxygen interaction involves one sigma-bond and one pi-bond, as shown in Figure 3–10. The sigma-bond arises by overlap of the electron clouds of an sp²-hybrid orbital on oxygen with an sp-hybrid orbital on carbon, whereas the pi-bond involves overlap of p-orbitals from each atom.

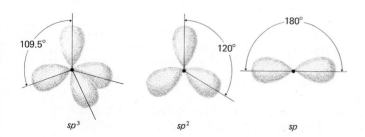

Figure 3–9   The relative orientations of equivalent hybrid orbitals.

sp³   sp²   sp

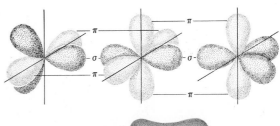

Figure 3–10 The approach of hybridized (shaded) and atomic (open) orbitals on carbon and oxygen to form sigma- and pi-bonds. The nonbonding electrons shown in the lower Lewis formula occupy hybrid sp^2-orbitals on oxygen.

In the triple-bonded nitrogen molecule, N_2, the sigma-bond can be ascribed to the overlap of the electron clouds of two sp-hybrid orbitals. The two pi-bonds arise by the overlap of two p-orbitals from each atom. This triple-bonded system is represented in Figure 3–11.

For many molecular species, it is not possible to define a single Lewis formula which is consistent with experimental facts. For example, the molecule SO_3 is planar triangular and has all sulfur-oxygen bonds of equal length, but the Lewis formula requires one sulfur-oxygen double bond. The following three Lewis formulas are equivalent in that each has one double bond.

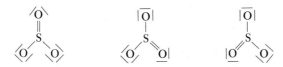

The real bonding in SO_3 can be described as a **resonance hybrid** of these three **contributing forms** and others. Any one contributing form inadequately describes the bonding, but SO_3 has characteristics of each of the contributors. Each bond is stronger than a single bond but weaker than a double bond.

When considering the bonding in molecules which are inadequately described by a single Lewis formula, the following principles are used:

1. Resonance contributors are limited to those Lewis formulas which differ only in the arrangement of electrons without shifting any atoms.
2. The hybrid of the resonance contributors represents the actual bonding of the molecule.
3. The resonance hybrid is more stable (lower in energy) than any of its contributing forms.
4. An increased stabilization is observed for molecules described as resonance hy-

Figure 3–11 The approach of hybridized (shaded) and atomic (open) orbitals on each of two nitrogen atoms to form a sigma-bond and two pi-bonds.

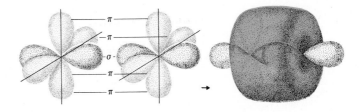

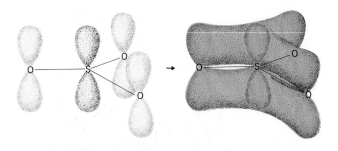

Figure 3–12   A delocalized π-bond in SO_3 arising from the overlap of four p-orbitals.

brids. This stabilization, called the **resonance energy,** is greatest when the contributing forms have equal energy.

5. Contributing forms which have the lowest energies (are most stable) make the greatest influence on the nature of the hybrid and on the resonance energy.

6. Separation of unlike charges to two identical atoms gives a high energy (less stable) resonance form with minimal contribution to resonance energy.

It is also possible to describe the bonding in SO_3 and other molecules, inadequately described by a single Lewis formula, by using σ-bonds and **delocalized π-bonds.** Delocalized π-bonds involve molecular orbitals which involve overlap of p-orbitals from more than two nuclei in a π-fashion. This is schematically represented for SO_3 in Figure 3–12. In this figure it is seen that p-orbitals on sulfur and each of the oxygen atoms, all perpendicular to the plane of the molecule, overlap to give a π-orbital spread over the four atoms. This orbital can contain only two electrons and results in one bond spread over three sulfur-oxygen interactions.

IMPORTANT TERMS AND CONCEPTS

bonding electrons	hybrid orbitals	polar bond
combining capacity	ionic bonds	pyramidal
coordinate covalent bond	multiple bonds	resonance hybrid
covalent bond	non-bonding electrons	sigma-bond
dipole	octahedral	tetrahedral
electronegativity	pi-bond	trigonal

QUESTIONS

1. Which of the electrons in an atom's structure are termed valence electrons? What relation do valence electrons have to atomic orbitals?

2. What is a covalent bond? Does the formation of this bond involve the transfer of valence electrons? Explain.

3. Describe a polar bond. What is meant by a dipole?

4. What is meant by the electronegativity of an atom?

5. Using Lewis formulas, illustrate the bonding in each of the following: CO_2, NH_4^+, H_2O, CH_4, OH^-, BCl_3, HCl, O_2, SO_2.

6. Predict which of the following molecules have dipole moments:

(a)

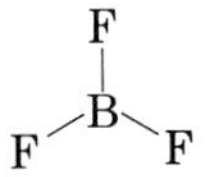

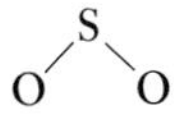

(c) $SiCl_4$ (tetrahedral)

(b)

(d) N—O

7. Assuming that all bonds are single bonds, predict the shapes of each of the following molecules or ions:

(a) AlH_3
(b) PCl_4^+
(c) BrF_5
(d) SbF_5

8. For molecules (a) and (b) in question 9, identify the type of hybrid orbitals used by the central atom.

9. Which of the following molecules must have at least one multiple (double or triple) bond? In each case determine whether the bond is double or triple.

(a) HCCH   (b) HNNH   (c) HOOH

10. What are the major differences between a sigma-bond and a pi-bond?

11. Using the rule of octets, predict the combining capacities of each of the following elements:

(a) Be   (b) Si   (c) P   (d) Al   (e) Na   (f) Cl   (g) O   (h) Ne

12. In the planar CO_3^{-2} ion the carbon-oxygen distances are all the same even though the Lewis formula predicts one double bond. Explain.

13. Draw contributing forms for the resonance hybrid of each of the following.

(a) 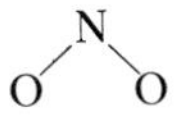   (b) 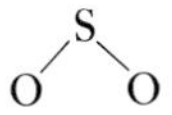   (c)

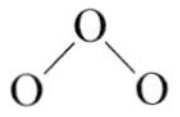

SUGGESTED READING

Ferreira: Molecular Orbital Theory, An Introduction. Chemistry, Vol. 41, No. 6, p. 8, 1968.

Gillespie: The Electron Pair Repulsion Model for Molecular Geometry. Journal of Chemical Education, Vol. 47, p. 18, 1970.

House: Ionic Bonding in Solids. Chemistry, Vol. 43, No. 2, p. 18, 1970.

Howald: Bond Energies in the Interpretation of Descriptive Chemistry. Journal of Chemical Education, Vol. 45, p. 163, 1968.

Sanderson: What Is Bond Polarity and What Difference Does It Make? Chemistry, Vol. 46, No. 8, p. 12, 1973.

THE STATES
OF MATTER

CHAPTER 4 ————————————————

The *objectives* of this chapter are to enable the student to:

1. Define pressure and describe how it is measured and expressed.
2. Describe the behavior of ideal gases by stating the gas laws.
3. Explain the origin of the Kelvin temperature scale and show its relationship to other temperature scales.
4. Calculate the effect upon a gas when pressure, volume, or temperature is varied.
5. Explain the origin of the gas constant, R.
6. Use the kinetic molecular theory to rationalize the gas laws, as well as the behavior of real gases.
7. Identify the regions of the atmosphere and explain the role of natural processes in altering the composition of the troposphere.
8. Apply the kinetic molecular theory to a differentiation of the properties of gases, liquids, and solids.
9. Describe the nature and origin of intermolecular forces.
10. Identify the types of solids and explain the distinguishing features of each.
11. Explain the nature of a dynamic equilibrium.
12. Discuss solid-liquid and liquid-gas changes of state and the influence of pressure and temperature.

   Because of the sizes and masses of atoms and molecules, man's study of matter always involves samples which contain large numbers of these units. The **macroscopic** properties of matter are determined by the properties of collections of large numbers of atoms and molecules. However, the properties of the individual atoms and molecules of a substance are significant in determining the properties of a collection. One of the most important considerations in predicting macroscopic properties is the forces between molecules, between ions, or between atoms which are not bonded. The nature and magnitudes of these interparticle interactions determine the **states of matter.**

STATES OF MATTER

   The states of matter are three in number: **solid, liquid,** and **gas.** These states are sufficiently different that only a few gross distinguishing features need be indicated to establish a clear differentiation.

Solids have a fixed shape and a fixed volume. They are incompressible; that is, they do not decrease their volumes when large forces are applied. The density of a solid generally is higher than that of a liquid or a gas.

Liquids have no fixed shape but assume the shape of their container. They do have a definite volume. Liquids are nearly incompressible and have densities which are usually lower than those of solids.

Gases have no fixed shape or volume but rather occupy the entire space afforded within the walls of their container. The densities of gases are much lower than those of liquids and solids. The ratio of the density of a gas to the density of a liquid or solid is usually less than $\frac{1}{1000}$. Gases expand when heated and contract when cooled or when external pressure is applied. Gases diffuse rapidly, filling any vessel and mixing with any other gas that is present in the vessel. Also, gases exert a pressure on the walls of any vessel in which they are stored. All the properties of gases can be explained on the basis of the kinetic molecular theory of matter to be presented later in this chapter.

Some gases possess characteristic colors or odors that aid in their detection. The majority of gases are colorless, however, and must be detected by one or more of their other characteristic properties.

PRESSURE AND ITS MEASUREMENT

Pressure is the property of matter which determines the direction of bulk flow, in that matter flows from regions of high pressure to regions of lower pressure. Pressure is expressed as force per unit area, often in units of pounds per square inches, lb/in^2. As stated previously, gases exert a pressure on the walls of the vessel in which they are stored. This pressure is most easily measured with a **barometer.**

A **Torricelli barometer** consists of a glass tube more than 760 mm in length, sealed at one end, completely filled with mercury, and inverted in a beaker also filled with mercury (Fig. 4–1). When inverted, the mercury drops somewhat in the tube, creating a vacuum in the upper portion of the tube. The extent to which the mercury drops is determined by the force per unit area which the earth's atmosphere exerts on the mercury in the beaker. The mercury drops until the pressure exerted by the atmosphere equals the pressure exerted by the column of mercury at the level of the mercury surface in the beaker. Under these conditions the pressure of the atmosphere is said to be "supporting" a column of mercury.

The standard unit of pressure used in studies of gases is **one atmosphere.** This is

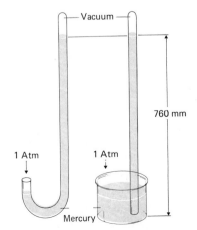

Figure 4–1   Simple barometers at sea level, or 1 atmosphere of pressure.

the average pressure which the earth's atmosphere exerts at sea level, and is equivalent to 14.7 lb/in^2, or a pressure which will support a column of mercury 760 mm in height. This pressure is commonly referred to as 760 torr. The pressure unit **torr,** which is $\frac{1}{760}$ atmosphere, is named in honor of the inventor of the barometer, Torricelli.

Almost any liquid can be used in a tube to make a barometer. One atmosphere of pressure will support a column of water 34 feet high, but this length is impractical to measure. Consequently the denser liquid mercury is used in most barometers.

If a barometer is carried to the top of a mountain, where the distance to the outer limit of the atmosphere is less than at sea level, it is observed that the atmosphere does not support a column of mercury 760 mm in length. At higher elevations the atmosphere supports even shorter columns of mercury.

BOYLE'S LAW

As early as 1660, Robert Boyle observed that *if the temperature of a sample of gas is kept constant, the volume varies inversely with the pressure applied to the gas.* In proportionality form, **Boyle's law** is given as $V_T \propto 1/P$. In this expression, V_T is the volume of the gas at a given temperature and P is the pressure.

Since gases are not always kept at atmospheric pressure, Boyle's law can be applied to calculate the volume of a gas at any known pressure.

As a problem, assume that a sample of gas has a volume of 240 ml at a pressure of 500 torr. If the pressure is reduced to 300 torr, keeping the temperature constant, what will be the new volume of the gas? First it is seen by simple reasoning that the new volume will be greater than the original, since the pressure is reduced, thereby increasing the volume of the gas. The new volume V is equal to the old volume, which is 240 ml, multiplied by the ratio of the two pressures, or stated numerically,

$$V = 240 \text{ ml} \times \frac{500 \text{ torr}}{300 \text{ torr}} = 400 \text{ ml}$$

As a check on the calculation, it is seen that the pressure has been approximately halved, so that the volume should be approximately doubled.

It can be seen that the most important point to be considered in problems of this type is whether the resulting volume will be larger or smaller than the original volume. If the final pressure is greater than the original pressure, then the resulting volume will be less than the original; however, if the resulting pressure is less than the original pressure, the resulting volume will be larger than the original.

CHARLES'S LAW AND THE ABSOLUTE TEMPERATURE SCALE

In 1787 the French physicist Charles observed that cooling a gas from 0°C to −1°C caused it to contract by $\frac{1}{273}$ of its volume. This immediately suggested that if it were possible to cool a gas to −273°C, the gas would have zero volume. In practice, however, gases become liquefied before they reach a temperature of −273°C, and therefore no longer follow the gas laws. Nevertheless, the value of −273°C was established as **absolute zero,** or the zero point of the **absolute,** or **Kelvin,** scale. One degree on the absolute or Kelvin scale is equal to one degree on the Celsius scale. To convert Celsius to absolute

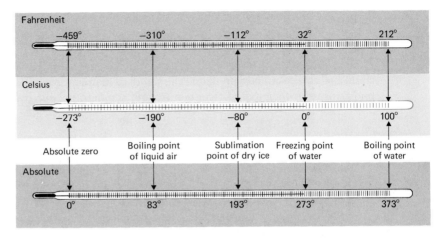

Figure 4–2 A comparison of the Fahrenheit, Celsius, and absolute temperature scales.

temperatures, merely add 273 to the value on the Celsius scale. Absolute zero is equivalent to a temperature of −273°C or −459°F as shown in Figure 4–2.

An apparatus similar to that shown in Figure 4–3A, fitted with a temperature controlling device, can be used to study the effect of the temperature on the volume of gas at constant pressure. As the temperature increases, the piston moves upward, increasing the volume of the gas (Fig. 4–3B). Cooling the gas causes the piston to move downward as the volume decreases (Fig. 4–3C).

Charles's Law states that *the volume of a gas varies directly as the absolute temperature, if the pressure is kept constant.* In proportionality form, Charles's law is given as $V_p \, \alpha \, T$.

By the application of Charles's law, it is possible to calculate the volume a sample of gas would occupy at a temperature different from the original, if the pressure is kept constant.

> As a problem, assume that a sample of gas has a volume of 100 ml at 30°C, and calculate the volume at 100°C. First, add 273° to each of the Celsius temperatures to convert them to absolute temperatures. This gives temperatures of 303°K and 373°K, respectively. Since the volume changes directly as the temperature, the new volume is equal to the old volume multiplied by the ratio 373/303, or the calculations may be expressed as:
>
> $$V = 100 \text{ ml} \times \frac{373°\text{K}}{303°\text{K}} = 123 \text{ ml}$$

In general, if the temperature of a gas increases, the volume increases, and in calculations,

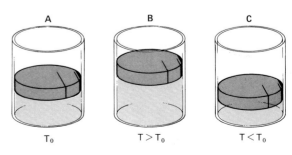

Figure 4–3 The effect of temperature changes on the volume of a gas at constant pressure.

the ratio of the two temperatures is written with the higher one over the lower. In problems in which the temperature is decreased, the volume decreases, and the ratio is written with the lower temperature over the higher.

Recall that the pressure of 760 torr, which is the average pressure at sea level, is established as the **standard pressure.** The **standard temperature** is $0°C$, or $273°K$. **Standard conditions** refer to standard temperature and standard pressure, **STP.**

GAY-LUSSAC'S LAW OF COMBINING VOLUMES

In 1808 the French chemist Gay-Lussac investigated the chemical reactions that occur between gaseous substances. He observed that *under similar conditions of tempera-ture and pressure the volumes of gases that react with each other are always in a ratio of small whole numbers.* Even though he did not recognize that gaseous elements formed molecules consisting of two atoms, his observation was confirmed by many investigators. The observation has now become **Gay-Lussac's law of combining volumes.** As an example of combining volumes, it has been shown experimentally that 1 liter of nitrogen gas will react with exactly 3 liters of hydrogen gas to form 2 liters of ammonia gas.

AVOGADRO'S HYPOTHESIS: THE MOLAR VOLUME

In 1811, Avogadro, an Italian physicist, proposed a fundamental explanation to account for Gay-Lussac's law of combining volumes. He stated that *equal volumes of all gases under the same conditions of temperature and pressure contain the same number of molecules.* In proportionality form this expression is $V_{T,P} \, \alpha \, n$ where n is the number of moles of molecules.

Knowing the gram molecular weight of oxygen, 32g/mole, and its density at $0°C$ and 760 torr, 1.429g/l, it is possible to calculate the volume occupied by one mole of oxygen at standard conditions. Thus,

$$\frac{32g \, O_2}{mol} \times \frac{1 \, l \, O_2}{1.429g \, O_2} = \frac{22.4 \, l}{mol}$$

One mole of oxygen gas occupies a volume of 22.4 liters at standard conditions. In fact a volume of 22.4 liters of any gas at standard conditions contains a gram molecular weight of the gas. For example, this volume of hydrogen weighs 2.016g. Therefore, the gram molecular weight of hydrogen is 2.0g, whereas 22.4 liters of nitrogen weighs 28g, giving nitrogen gas a gram molecular weight of 28g. The volume occupied by one mole of a gas, or 22.4 liters, is called the **molar volume** of a gas. A mole, or molar volume, of any gas under standard conditions contains the same number of molecules. This number of molecules equals 6.02×10^{23}, and is called **Avogadro's number.** In other words, 1 mole, or molar volume, or 22.4 liters, of a gas contains 6.02×10^{23} molecules of that gas.

IDEAL GASES AND THE EQUATION OF STATE

Restating Boyle's law, Charles's law, and Avogadro's hypothesis as

$$V_T \, \alpha \, 1/P, \quad V_P \, \alpha \, T, \quad \text{and} \quad V_{T,P} \, \alpha \, n$$

it is possible to write the following combined relationship:

$$V \propto nT/P$$

If this proportionality is changed to an equality using the proportionality constant R, and both sides of the equation are multiplied by P, the following relationship is obtained:

$$PV = RnT$$

This equation shows the relationship of all of the variables required to completely describe the behavior of a gas. A gas that rigorously obeys this equation is called an **ideal gas,** and the equation is called the **equation of state for an ideal gas.** The value of the **gas constant,** R, can be determined by using known values of P, V, n, and T.

> For example, it is known that one mole of a gas (n = 1) occupies the molar volume (22.4 liters) at standard temperature (273°K) and pressure (1 atmosphere or 760 torr). The constant R is thus found to be:
>
> $$R = \frac{PV}{nT} = \frac{(1 \text{ atm})(22.4 \text{ l})}{(1 \text{ mol})(273°K)} = \frac{0.082 \text{ l atm}}{\text{mol}°K} \quad \text{or} \quad \frac{62.3 \text{ l torr}}{\text{mol}°K}$$

The ideal gas equation can be used for many calculations already discussed.

> For example, the volume occupied by three moles of a gas at 4 atmospheres and 500°K is
>
> $$V = \frac{RnT}{P} = \frac{0.082 \text{ l atm} \times 3 \text{ mol} \times 500°K}{\text{mol}°K \times 4 \text{ atm}} = 30.7 \text{ l}$$

In this calculation, the procedure is to solve for the variable which is not given. As a second illustration, calculate the number of moles of gas which will exert a pressure of 16.6 atmospheres when contained in a 100 liter tank at 25°C.

$$n = \frac{PV}{RT} = \frac{16.6 \text{ atm} \times 100 \text{ l}}{0.082 \dfrac{\text{l atm}}{\text{mol}°K} \times 298°K} = 68.0 \text{ mol}$$

THE KINETIC MOLECULAR THEORY

The observation of many facets of the behavior of gases under many conditions has led to a theory which is consistent with most experimental facts. This theory, called the **kinetic molecular theory,** is based on a set of fundamental assumptions which serve as a model for predicting properties. These assumptions are:

1. Gases are composed of molecules or atoms which have negligible volume, and which are at relatively large distances of separation.
2. The molecules move at high velocities in straight paths, but in random directions.
3. The molecules collide with each other and with the container walls, but the collisions are totally elastic so that energy is conserved. As the result of these collisions, some molecules move faster than others, and a distribution of molecular velocities results.

4. The average kinetic energy of the molecules of all gases is the same at a given temperature, and is directly proportional to the absolute temperature.
5. There are no attractive or repulsive forces between molecules.

In addition to serving as a consistent explanation for the behavior of gases, this theory can be used to qualitatively differentiate the properties of materials in the three states of matter. Not all matter can exist in each of the three states, but most materials (e.g., elements or compounds) can. The physical conditions under which each state exists define a set of unique characteristics of the material.

The relative densities of the three states suggest a rather significant variation in the number of particles occupying a given volume, or in the average interparticle distance. The particles in a gas apparently have greater distances of separation than those of a liquid, and the particle separation in solids must be the smallest. The distance of separation of the particles in a material is a function of the magnitudes of the intermolecular forces of attraction. For compounds with very strong, attractive, intermolecular interactions, the distance of separation of the particles tends to be small, and this tendency is diminished as the attractive intermolecular interactions become smaller. In solids and liquids, for which the intermolecular forces are much larger than in gases, the average distance of separation of the molecules is quite small. The nature of the various types of intermolecular interactions in liquids and solids will be discussed later in this chapter.

The interactions between molecules in real gases are so weak that they do approximate the nonexistent forces assumed in the kinetic molecular theory. However, in reality, at least small interactions do exist.

Since in the gaseous state particles of matter are not closely packed, there must be some forces which tend to keep the particles separated. The forces opposing those of attraction are dependent on the kinetic energy of the particles. The **kinetic energy** of a particle is equal to one-half the product of the mass of the particle multiplied by the square of its velocity.

$$\text{K. E.} = \tfrac{1}{2}\,mv^2$$

If the effect of the kinetic energy of particles of matter is always greater than the forces of attraction, the particles will remain in continuous motion, as in a gas. The relationship between the forces of attraction and the kinetic energy of particles of matter will determine whether a substance exists as a solid, liquid, or gas.

The kinetic molecular theory supplies an explanation for Boyle's law. Figure 4–4A represents a cylinder fitted with a movable piston, and containing gas at a constant temperature under unit pressure. All the forces that tend to drive the piston down into the cylinder, including atmospheric pressure and the force of gravity on the weight and piston, are combined in the one unit of external pressure. This total external force is

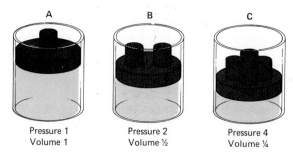

A

B

C

Pressure 1
Volume 1

Pressure 2
Volume ½

Pressure 4
Volume ¼

Figure 4–4   The effect of external pressure on the volume of a gas at constant temperature.

exactly balanced by the force of the impacts of the gas molecules on the lower surface of the piston. If a weight representing two units of external pressure is placed upon the piston, it sinks to a new level (Fig. 4–4B). When additional weight is added to the piston to make the total external force equal to four units of pressure, the piston moves downward to a new position (Fig. 4–4C).

Experimentally, it has been found that the volume of the gas under two units of external pressure is equal to one-half the volume under one unit of pressure. Since, at constant temperature, the velocities of the gas molecules do not change, and the masses of the particles remain the same, there is no change in the kinetic energy of the gas. The only way the force on the under surface of the piston can be increased to balance the two units of external pressure is for the molecules to strike the piston twice as often. This can be accomplished by moving the piston downward into the cylinder until the volume of the gas is half of the original volume.

The observations leading to Charles's law are also consistent with the kinetic molecular theory. An increase in temperature will increase the velocity and kinetic energy of the gas molecules. In effect this causes an increase in the internal forces, since the gas molecules strike the under surface of the piston more often and with greater force. The piston moves upward to increase the space between the molecules, thus reducing the number of impacts and the force exerted by the gas. The upward movement also increases the volume of the gas. A decrease in temperature causes a decrease in the volume of the gas, since the kinetic energy of the molecules is less and the internal forces are decreased.

The kinetic molecular theory is seen to be consistent with Avogadro's hypothesis if it is realized that increasing the number of molecules in a sample increases the frequency of collisions per unit area of the walls of the container. The frequency of collisions per unit area of the walls, and thus the pressure, can remain constant at a given temperature only if the surface area is increased along with increasing the sample size.

REAL GASES

There are no known gases which obey the ideal equation of state at all temperatures and pressures. As pointed out earlier, the forces of interaction between real gas molecules, although small, do exist. These forces become very significant under conditions when the average kinetic energy is small. Thus, at low temperatures or at high pressures, when these conditions exist, it is found that the ideal gas equation of state is not obeyed. That is, the ideal gas equation does not accurately describe the relationships among pressure, temperature, volume, and the number of moles for real gases.

Another property which distinguishes real and ideal gases is the real volumes of molecules. The kinetic molecular theory assumes that gaseous molecules occupy no space, but it is known that atoms and molecules do have finite volumes and do occupy space. Consequently, it is impossible for a real gas to adhere to Charles's law at very low temperatures, or to adhere to Boyle's law at very high pressures.

At high pressures or at low temperatures, attractive forces between gas molecules tend to pull the molecules together. If this effect is sufficiently large, the gas molecules coalesce to form a liquid. This process is called **liquefaction.** Liquefaction can be induced more readily at high pressures, since the average distance of molecular separation is smaller. At very high temperatures, gases often cannot be liquefied by increasing the pressure. The **critical temperature,** T_c, is the temperature above which a gas cannot be liquefied by increasing the pressure. The **critical pressure,** P_c, is the pressure required to liquefy a gas at the critical temperature.

TABLE 4–1   REGIONS OF THE ATMOSPHERE

REGION	AVERAGE MAX. ALTITUDE (km)	PRESSURE RANGE (torr)	AVERAGE TEMPERATURE RANGE (°K)
Troposphere	11	760–200	288–217
Stratosphere	48	200–1	217–270
Mesosphere	85	$1–5 \times 10^{-3}$	270–175

THE ATMOSPHERE

The air of the earth's atmosphere is a mixture of gases which has shown and still shows a fluctuating composition. This fluctuation, the result of chemical reactions both naturally occurring and induced by man, has become an aspect of concern because of man's dependence on oxygen and because of the role the atmosphere plays in controlling multitudes of systems which influence man's environment.

The regions of the atmosphere with their locations and properties are presented in Table 4–1. The **troposphere,** nearest the earth, extends to an average altitude of about 11 km (about 7 miles). It is characterized by thorough mixing due to shifting masses of air and fluctuating temperatures. Since it is the immediate environment at the surface of the earth, this region is of greatest importance to man. This is especially obvious if one realizes that the troposphere contains nearly 80 per cent of the mass of the entire atmosphere and virtually all its water vapor. It contains in addition to the gases listed in Table 4–2 many liquid and solid particles.

TOPIC OF CURRENT INTEREST

ATMOSPHERIC CHEMISTRY

Without the atmosphere, life would not be possible on earth. It supplies oxygen to man and carbon dioxide to plants. It shields all life on earth from the sun's lethal ultraviolet radiation and protects us from most of the cosmic rays from outer space. It is thought that there was no free oxygen in the earth's atmosphere until it was put there by plants through the decomposition of water molecules by way of light energy in

TABLE 4–2   COMPOSITION OF
CLEAN DRY AIR NEAR SEA LEVEL

COMPONENT	PER CENT BY VOLUME
Nitrogen	78.084
Oxygen	20.948
Argon	0.934
Carbon dioxide	0.0314
Neon, helium, krypton	0.0025
Methane (CH_4)	0.0002
Hydrogen	0.00005
Nitrous oxide (N_2O)	0.00005
Ozone (summer)	0 to 0.0001
(winter)	0 to 0.000007
Sulfur dioxide	0 to 0.000002
Nitrogen dioxide	0 to 0.000002

photosynthesis. Ultraviolet radiation from the sun dissociates molecular oxygen into atoms which combine with other O_2 molecules to give ozone, O_3.

$$O_2 \xrightarrow{\text{uv}} 2\,O$$

$$O + O_2 \longrightarrow O_3$$

The ozone is found in the atmosphere, in a layer about ten miles thick beginning at an elevation of about fifteen miles. This layer of ozone absorbs a very high percentage of the ultraviolet rays of the sun but transmits visible light. Once lethal ultraviolet radiation no longer reached the earth's surface, life was possible in locations other than deep in the ocean. Ozone shielding facilitated a rapid increase in the number of life forms on earth, resulting in an oxygen-carbon dioxide cycle. Plant life, by photosynthesis, converted water to oxygen, and animal life, in respiration, reconverted oxygen to carbon dioxide. Of course, other naturally occurring processes such as fire and volcano emission also converted oxygen to carbon dioxide.

The significance of the ozone layer is thus obvious. Man must continually guard against any process or material which would alter the ozone layer. Currently two types of materials are thought to have the potential for significantly altering the ozone layer by reacting with O_3. The first of these is the exhaust gases of supersonic transports which would fly much nearer the ozone layer than do slower jet aircraft. The second is the gaseous propellants of aerosol cans—compounds of carbon and halogens. These are known to react rapidly with ozone to form compounds which do not absorb ultraviolet radiation as efficiently as does ozone.

Man's burning of fossil fuels as a source of energy (p. 3) has resulted in a marked increase in the amount of atmospheric carbon dioxide since 1850. It is estimated that since 1890 the average concentration of carbon dioxide has risen from 0.0290 to 0.0314 per cent, and it is expected to be nearly 0.040 per cent by the year 2000. This is an alarming prediction if the "greenhouse effect" is real. That is, the increased concentration of carbon dioxide will result in an increased temperature at the earth's surface. The earth, as it is continually irradiated by the visible light of the sun, emits infrared radiation. Carbon dioxide absorbs the emitted infrared radiation and consequently destroys the earth's major mechanism of heat loss, resulting in a gradual rise in temperature. Imagine the eventual influence of increasing the average temperature of the troposphere. Sea levels would be influenced by melting glaciers and ice caps. Changes in climate could alter the ecology of many systems influencing man's existence. There is another condition on earth which apparently mediates the greenhouse effect to some extent. This effect, due to the reflection of the sun's rays by an increasing concentration of suspended particles in the atmosphere, would result in lower surface temperatures and eventually in the onset of an ice age.

INTERMOLECULAR FORCES

Although the assumptions are obviously invalid for solids and liquids, the kinetic molecular theory is quite useful in explaining many of the properties of these states. Considerations of the relative densities of gases, liquids, and solids lead to the conclusion that intermolecular distances are far greater in gases than in liquids and solids. Consequently, intermolecular interactions become an important factor in determining the properties of liquids and gases.

Several types of interactions give rise to the forces which cause gases to deviate from ideal behavior. These same interactions cause liquids and solids, which are composed of atoms or covalent molecules, to remain in the condensed state or **phase.** The most important of these interactions are **dipole-dipole** and those which give rise to **London forces.** These belong to a general class of forces called **van der Waal's forces.**

Dipole-dipole interactions develop from the attraction of oppositely charged portions of the dipoles of molecules. Figure 4–5 shows a representation of two orientations of

Figure 4-5   Two possible orientations of dipoles that lead to net attraction.

dipoles which lead to a net attraction. The magnitude of the attraction increases with increasing size of the dipoles of the molecules. In addition, it increases with decreasing distance of separation of the molecules. However, at very small distances of separation, the electron clouds on adjacent molecules repel one another.

London forces explain the fact that even atoms of the inert gases have intermolecular attractions and can be liquefied. If the electron cloud of an atom at some instant does not have spherical symmetry about the nucleus, but is distorted or **polarized,** the atom will have a net dipole moment. Such a dipole moment could influence a nearby atom by inducing a distortion in its electron cloud. Figure 4-6 represents the instantaneous shapes of the electron clouds of two adjacent atoms at three different times. In the center the atoms have spherical electron clouds, no dipole moments, and therefore no net force of attraction. In both the other configurations, the polarized atoms have aligned dipole moments, which is to be expected since these dipole moments are mutually influential. For any configurations in which the polarized atoms have aligned dipoles, there will be a net force of attraction. It can be concluded that all configurations, other than that of spherical electron clouds, will result in forces of attraction.

The magnitudes of London forces between atoms increase with increasing ease of electron cloud distortion or **polarizability.** Because of the distance of the valence electrons from the nucleus, and because of the screening effect of the inner filled shells of electrons, large atoms of the heavy elements have the largest polarizabilities, and hence the largest London forces. This accounts for the increase in boiling points of the inert gases in going from helium to xenon.

THE NATURE OF SOLIDS

Solids differ drastically from gases in that the rate of motion of molecules or ions is very slow. This is because the distance between molecules is virtually zero due to close packing. This arrangement, which results in large intermolecular and interionic interactions, does not allow rapid motion of particles through the solid object. Instead, the molecules or ions oscillate about an average rather stationary position. An increase in the temperature of the solid results in increased violence of the oscillation. Through interactions with neighboring molecules or ions in the solid, a distribution of oscillation violence arises. This distribution is analogous to the distribution of molecular velocities in a gas at a given temperature.

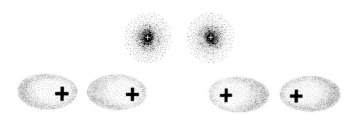

Figure 4-6   Instantaneous electron cloud shapes on neighboring atoms at three different times.

Figure 4-7   Three of the seven general classes of crystal systems.

Cubic
a = b = c

Orthorhombic
a ≠ b ≠ c

Hexagonal
a = b ≠ c

The Crystalline State

In many solids the atoms, molecules, or ions are arranged in a systematic, highly ordered pattern which repeats itself in three dimensions. Such solids are called **crystalline**. The three dimensional arrangement of the units is called the **space lattice** or **crystal lattice**. The smallest portion of a crystal which can be used to describe the crystal lattice is called the **unit cell.** The three dimensional extension of the unit cell results in highly regular macroscopic properties of crystals, such as angles between faces.

There are seven general classes of space lattices into which all crystal systems may be classified. Simple space lattices may be visualized as geometric entities in which the corners are occupied by the units of the solid. These space lattices vary in angles between faces and in the relative lengths of the edges. The simplest space lattice is the **cubic** lattice, in which all edges are the same length and all angles between faces are 90°. The **orthorhombic** space lattice has all edges unequal, but the interfacial angles are all 90°. The **hexagonal** space lattice has two edges equal and two angles equal. These three space lattices are represented in Figure 4-7.

Within each of the seven classes of space lattices there are several subclasses which are differentiated by units placed at the center or on the faces. Figure 4-8 shows the three subclasses within the cubic system: **simple cubic, body-centered cubic,** and **face-centered cubic.**

Types of Solids

Crystalline solids may be divided into four classes on the basis of the predominant forces holding the lattice together. These classes are **molecular, covalent network, ionic,** and **metallic.**

Molecular Solids.   Molecular solids are composed of discrete covalent molecules held together in the lattice by either dipole-dipole interactions or London forces. These

Figure 4-8   The subclasses of the cubic crystal system.

Simple cubic

Body-centered cubic

Face-centered cubic

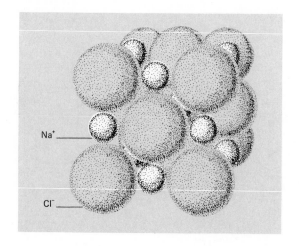

Na⁺ _____

Cl⁻ _____

Figure 4–9   A representative segment of crystalline NaCl.

are the weakest interparticle forces found in solids, and consequently, solids in this class have the lowest melting points and the highest vapor pressures. They tend to be soft, are not good thermal conductors, and act as electrical insulators. Examples include ice, dry ice, naphthalene (moth balls), sulfur, and white phosphorus. The exceptionally strong dipole-dipole interactions found in water give ice a set of physical properties which deviate from those of most molecular solids.

Covalent Network Solids.   Covalent network solids are composed of atoms which are covalently bonded with their neighbors to give a three dimensional network. A crystal of such a solid may be visualized as a gigantic molecule. The forces between atoms in these solids are due to very strong covalent bonds, and result in very high melting points, low vapor pressures, and hardness. Covalent network solids are poor conductors of electricity and heat. Examples include diamond and quartz.

Ionic Solids.   Ionic solids are composed of oppositely charged ions occupying sites in the crystal lattice. Each positive ion and each negative ion surrounds itself with the largest possible number of oppositely charged ions at the shortest possible interionic distance. The number of nearest neighbors is called the **coordination number.** Figure 4–9 shows a representative segment of crystalline NaCl in which both Na^+ and Cl^- have coordination numbers of six. The interionic forces in ionic crystals are quite strong and result in high melting points, low vapor pressures, and hardness. These properties are typical of salts which are ionic solids.

Metallic Solids.   Metallic solids are composed of metal atoms occupying the lattice sites. Over 60 per cent of all metallic solids have the **closest packed structures,** in which each atom touches its twelve nearest neighbors. Unlike the majority of other solids, metals are very good electrical and thermal conductors. The electrons of the metal atoms are thought to be responsible for carrying an electrical current. For this reason metallic solids are often thought of as a lattice of metal ions which is occupied by electrons which are free to move about in the lattice. This freedom of electron motion would account for the observed electrical conductivity. The properties of metallic solids show a wide variation. For example, the hardness ranges from very hard in the case of tungsten to very soft for the alkali metals. The melting points also show a wide variation, with mercury melting at $-38.9°C$ and tungsten melting at $3370°C$.

The Liquid State

The sequence in which we are discussing the states of matter—gases, solids, and finally liquids—is necessary because the liquid state has not been characterized nearly as well as the gaseous and solid states. Consequently, it is best to compare liquids to solids and gases so that they are characterized at least by contrast.

In liquids, the intermolecular distances of separation are much less than in gases and are probably only slightly greater than in solids. Consequently, as in solids, intermolecular forces hinder molecular motion so that rates of diffusion are considerably slower than in gases but faster than in solids. Random motion of molecules in liquids and the resulting collisions lead to distribution of molecular velocities and kinetic energies. The average kinetic energy of the molecules of a liquid is dependent on the temperature, but the nature of this dependence is not the same for all liquids. This variation is the result of differences in intermolecular forces characteristic of each type of molecule.

In different liquids, the variation of intermolecular forces leads to distinguishing characteristics. For example, the **viscosity,** or resistance to flow, of a liquid is largely due to attractions between adjacent molecules. The movement of molecules will be fastest when they have low masses, symmetrical shapes, and weak intermolecular interactions. As the temperature is increased, the increased kinetic energies of the molecules overcome the intermolecular forces and the viscosity decreases.

Another distinguishing characteristic of liquids which is determined by molecular forces is **surface tension.** A molecule in the center of a liquid sample experiences forces of attraction in all directions, whereas a molecule at or near the surface of the liquid experiences forces which are greatest in the direction toward the center of the sample (Fig. 4–10). This inward pull on the surface molecules results in surface tension, which is the resistance to expansion of the surface area. A small liquid sample in the absence of any external influence will assume the shape of a sphere, since this shape has the smallest surface area for a given volume. As expected, increasing the temperature of a liquid decreases the surface tension because of decreased intermolecular interaction.

CHANGES OF STATE

VAPORIZATION AND SUBLIMATION

In order to bring about liquefaction of a gas, it is necessary to decrease the average kinetic energy of the molecules sufficiently to allow intermolecular forces of attraction to pull the molecules together. In order to bring about the reverse process, **evaporation** or **vaporization,** it is necessary to increase the average kinetic energy of the molecules of a liquid until it is sufficiently large to overcome the intermolecular forces of attraction. Since there is a distribution of kinetic energies, some molecules will have kinetic energies higher than the average and some lower. The fraction of molecules with the highest kinetic energies will have the greatest tendency to overcome intermolecular forces and leave the liquid. If the highest energy molecules leave the liquid, the average kinetic energy of the molecules remaining behind will be decreased, and the temperature of the liquid will drop. Consequently, evaporation of a liquid always results in cooling the liquid, unless heat is supplied by the surroundings to maintain constant temperature.

Figure 4–10   The forces acting on molecules in a liquid.

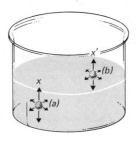

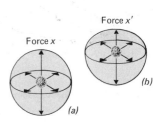

The amount of heat, or **thermal energy,** which must be supplied to a given quantity of a liquid to bring about vaporization is a characteristic property of the liquid. This property is called the **enthalpy of vaporization** (ΔH_{vap}). Refrigerators and air conditioners use a cyclic process of **liquefaction,** or **condensation,** and evaporation of a suitable refrigerant to bring about cooling. It is the evaporation part of the process which is used to cool. The compressor increases the pressure on the gaseous refrigerant in order to bring about liquefaction.

The process of molecules leaving the solid state and entering the gaseous state without going through the liquid state is called **sublimation.** In solids the intermolecular attractive forces are usually considerably larger than those found in liquids. Consequently, the tendencies of liquids to evaporate at a given temperature are usually greater than the tendencies of solids to sublime. The amount of thermal energy, or heat, required to sublime a given quantity of solid is called the **enthalpy of sublimation** (ΔH_{sub}). The enthalpy of sublimation of a material is always greater than the enthalpy of vaporization.

Vapor Pressure. Liquids and solids placed in closed containers do not completely evaporate or sublime. This is because the gaseous molecules are confined to the container and cannot diffuse away, and consequently eventually reenter the condensed state. The gaseous molecules moving about at random will reenter the condensed state at a rate which is dependent on the number of molecules in the gaseous state. Eventually the rate at which molecules are reentering the condensed state will equal the rate at which they are entering the gaseous state. Under these conditions, the number of gaseous molecules occupying the volume above the condensed state will be invariant with time, even though a dynamic process prevails. This situation is referred to as a **dynamic equilibrium** and the pressure exerted by the gas over the condensed state is called the **equilibrium vapor pressure.** The vapor pressure of a material does not depend upon the relative amounts of condensed state or gaseous state present, but it does depend upon the nature of the material.

BOILING

The vapor pressures of liquids and solids increase with increasing temperature, as is shown in Figure 4–11 for water, carbon tetrachloride, bromine, and iodine. The temper-

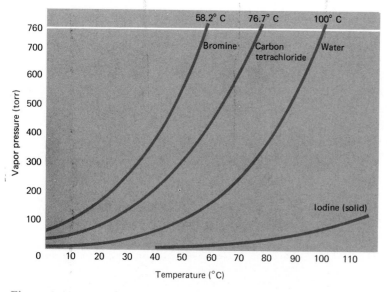

Figure 4–11   The variation of vapor pressure with temperature.

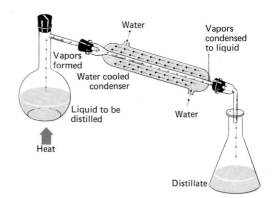

Figure 4–12   A simple distillation apparatus.

ature at which the vapor pressure of a material equals the pressure exerted by the surrounding atmosphere is called the **boiling point.** The temperature at which the vapor pressure of a substance is one atmosphere is called the **normal boiling point.** A liquid boils at lower temperatures if the pressure of its surrounding atmosphere is decreased. At higher altitudes, where the atmospheric pressure is relatively low, water boils at lower temperatures. Cooking food by boiling at high altitudes may be difficult, because water boils at a temperature which is not high enough for efficient cooking. This difficulty is overcome by the use of a pressure cooker, in which the steam that is formed in boiling is confined to increase the pressure. As the result of the increased pressure, the boiling point of the water is increased. For example, a steam pressure of 760 torr added to the atmospheric pressure will result in a boiling point of about 120°C.

Chemists often purify liquids by heating to convert the liquid to a gas and then cooling the gas to condense it back to a liquid. This process, called **distillation,** is performed using a distillation apparatus such as the one shown in Figure 4–12. Distillation results in separation of impurities from a sample because of differences in the boiling points. Occasionally it is necessary to distill a material at a relatively low temperature in order to avoid thermally induced decomposition. This can be accomplished through decreasing the pressure of the atmosphere surrounding the liquid by creating a partial vacuum in the distillation apparatus. The pressure in the system can be lowered until the desired boiling temperature is attained.

The change of a liquid to a gas at the boiling point requires the addition of the **enthalpy of vaporization,** ΔH_{vap}. At 760 torr and 100°C all the heat added (540 cal) to boil one gram of water is consumed in overcoming intermolecular attractions and increasing the average distance of separation between the molecules. The temperature and the average kinetic energy of the molecules do not change during the process. This is schematically represented in the heating curve in Figure 4–13. It is seen that as a liquid is boiled, its temperature remains constant. Before boiling starts and after boiling is completed, all heat added serves to increase the kinetic energy, which increases the temperature of the system. When a gas condenses to a liquid, it releases the enthalpy of vaporization to its surroundings. This is why steam is used so efficiently in many processes to transfer thermal energy.

The boiling points of liquids increase with increasing magnitudes of intermolecular or interatomic forces. The lowest boiling points are observed for non-polar species such as H_2, O_2, He, and the like. Somewhat higher boiling points are observed for liquids of non-polar molecules involving larger atoms whose electron clouds are more polarizable (e.g., Br_2, I_2, Xe, and so forth). Liquids of polar molecules show higher boiling points than liquids of non-polar molecules containing similar atoms. The highest boiling points are observed for the metallic elements and ionic compounds.

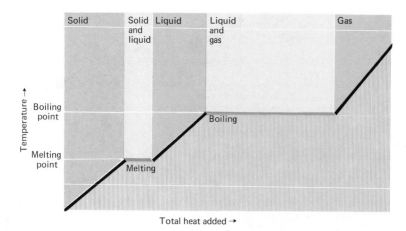

Figure 4–13   A heating curve showing the temperature of a system as it is influenced by the addition of heat at a constant rate.

MELTING AND FREEZING

Most solids, when heated sufficiently, undergo a solid to liquid transition. This transition is called **melting** or **fusion,** whereas the reverse process is called **freezing** or **crystallization.** The temperature at which this transition takes place is called the **melting point** or the **freezing point.** At this temperature, the vapor pressures of the solid and liquid are equal and the system is at equilibrium. Pure materials are thus predicted to remain at one temperature throughout a melting or freezing process. Impure materials have a gradual change in temperature during the solid-liquid transition. During the melting process, the temperature of an impure sample rises. This difference in the characteristics of pure and impure solids can be used as a simple test for purity. The melting points of solids are rather insensitive to pressure changes, although they do have a small dependence on them.

The heat added to bring about melting is called the **enthalpy of fusion,** ΔH_{fus}. The addition of 80 calories to 1 gram of ice at $0°C$ completely converts it to water at $0°C$. All the heat added during the melting process is consumed in liberating molecules from their rather static positions in the solid. Consequently during the melting process the average kinetic energy of the molecules, and therefore the temperature of the system, remains constant. This is represented in the heating curve in Figure 4–13, where it is seen that heat added is consumed in increasing the temperature of the system only before melting starts and after it is completed. The increase in average distance of separation of the molecules in going from solid to liquid is very small compared to that involved in vaporizing a liquid. Therefore, it is expected, as is indicated for water, that the enthalpies of fusion of solids are usually smaller than the enthalpies of vaporization of corresponding liquids.

The melting points of solids increase with increasing magnitudes of intermolecular and interatomic forces. In general they vary in the same way as the boiling points.

IMPORTANT TERMS AND CONCEPTS

absolute temperature scale	Avogadro's hypothesis	Boyle's law
atmosphere	barometer	Charles's law

crystal lattice	ideal gases	torr
distillation	kinetic energy	troposphere
dynamic equilibrium	kinetic molecular theory	van der Waals' forces
equation of state	pressure	vapor pressure
Gay-Lussac's law	standard conditions	vaporization

QUESTIONS

1. Atmospheric pressure is greater at sea level than in the mountains. Explain.

2. How does the kinetic molecular theory account for the pressure that a gas exerts on the walls of a container?

3. Explain why the volume of a balloon increases at higher elevations.

4. What is meant by "atmospheric pressure"? Name and describe the instrument that is used to measure atmospheric pressure.

5. If a sample of gas has a volume of 700 ml at 600 torr and a temperature of 60°C calculate the volume it would occupy at 400 torr and a temperature of 30°C.

6. Using the data in Table 4–2, calculate the partial pressure of oxygen at sea level at standard conditions.

7. Nitrogen gas has a molecular weight of 28. Calculate the weight of 1 liter of the gas under standard conditions of temperature and pressure. Express the answer as grams per liter.

8. Ammonia gas weighs 0.771 g per liter. On the basis of Avogadro's law, how would you calculate the molecular weight of the gas?

9. Using the ideal gas equation of state, calculate:

   (a) the pressure exerted by 2.6 moles of gas occupying 8.3 l at 15°C.
   (b) the moles of gas occupying 0.56 l at 4 atm and 0°C.
   (c) the volume occupied by 1.9 moles of gas at 750 torr and 96°K.

10. In the crystal structure of AlF_3, each Al^{+3} ion is surrounded by six F^-. What is the coordination number of F^-?

11. The enthalpy of vaporization of water is 540 cal/g. How much heat is required to vaporize one mole of water at its boiling point?

12. Why is the equilibrium between a liquid and its vapor called a dynamic equilibrium?

13. How is the enthalpy of fusion related to intermolecular forces?

14. Describe in terms of molecular motion the process of freezing.

15. From among the following, choose one from each pair which is appropriate for molecules having as their only intermolecular forces large London forces:

   (a) large polarizability; small polarizability
   (b) large atoms; small atoms
   (c) high boiling points; low boiling points
   (d) very viscous; free-flowing

16. Explain the relationship between intermolecular forces and surface tension.

17. When an impure liquid boils, does the temperature tend to increase, decrease, or remain constant?

18. The boiling point of water is always 100°C. True or false? Explain.

19. What role do intermolecular forces play in determining enthalpies of fusion and enthalpies of vaporization?

SUGGESTED READING

Cloud and Gibor: The Oxygen Cycle. Scientific American, Vol. 223, No. 3, p. 110, 1970.

Hall: Sulfur Compounds in the Atmosphere. Chemistry, Vol. 45, No. 3, p. 16, 1972.

Materials, A Scientific American Book. San Francisco, W. H. Freeman and Co., 1967.

McConnell: Crystals, Minerals, and Chemistry. Journal of Chemical Education, Vol. 40, p. 512, 1963.

Medeiros: Smog Formation Simplified. Chemistry, Vol. 45, No. 1, p. 16, 1972.

Neville: The Discovery of Boyle's Law, 1661–62. Journal of Chemical Education, Vol. 39, p. 356, 1962.

O'Sullivan: Air Pollution. Chemistry and Engineering News, p. 37, June 8, 1970.

Porter: The Laws of Disorder. Chemistry, Part 1, Vol. 41, No. 5, p. 23, 1968; Part 2, Vol. 41, No. 6, p. 21, 1968.

Research Reporter: Endangered Blanket of Ozone. Chemistry, Vol. 48, No. 1, p. 25, 1975.

Sanders: Chemistry and the Atmosphere. Chemistry and Engineering News, March 28, 1966.

Sanderson: The Nature of Ionic Solids. Journal of Chemical Education, Vol. 44, p. 516, 1967.

Strong and Stratton: Chemical Energy. New York, Reinhold Publishing Corp., 1965.

CHEMICAL REACTIONS AND THE NATURE OF COMPOUNDS

CHAPTER 5

The *objectives* of this chapter are to enable the student to:

1. Assign oxidation numbers to all elements in any compound.
2. Determine by inspection of the reaction equation whether a given reaction involves oxidation-reduction.
3. Determine by inspection whether a given reaction equation is balanced.
4. Identify the oxidizing agent and reducing agent in an oxidation-reduction reaction.
5. Determine the gram equivalent weight of an oxidizing agent or a reducing agent in a given chemical reaction.
6. Define acids and bases as presented in the concepts of Arrhenius, Brönsted-Lowry, and Lewis.
7. List the properties of acids and bases and define a neutralization reaction.
8. Determine the gram equivalent weight of an acid or a base in a given chemical reaction.
9. Name binary acids and oxyacids and the ions produced by their dissociation in water.
10. List the properties of normal and acid salts and relate formulas and names of both.

The multitudes of known chemical entities, including elements and compounds, are best differentiated and classified by characterizing their physical and chemical properties. There are far too many of these entities to discuss in a one semester chemistry course, and consequently, it must suffice in later chapters to present specific properties for a limited number. It is possible, however, to discuss general classes of compounds with similar compositional and structural features, and general classes of reactions which elements and compounds undergo.

CHEMICAL REACTIONS

A chemical reaction is said to have taken place when two or more substances combine to give new substances, or when one substance decomposes to give different substances.

In the broadest sense, these two types of reactions might be called **combination** and **decomposition** reactions, respectively.

Another classification of reactions, more useful than combination and decomposition, is based on the loss or gain of electrons by atoms. **Oxidation-reduction** reactions are those reactions in which an element gains or loses electrons. The element or elements which undergo a reduction or oxidation may be in a compound. All other chemical reactions are called **metathesis** reactions. These reactions always involve exchange of groups between two interacting substances, combination of two substances, or decomposition of some substance, all without a loss or gain of electrons.

The simplest example of an oxidation-reduction reaction is the combination of any two elements. In forming the product, one element must transfer one or more electrons to another element. The reaction of hydrogen and oxygen to produce water is an oxidation-reduction reaction.

The reaction of HCl and NaOH to produce NaCl and H_2O is an example of a metathesis reaction. In this reaction there is an exchange of groups, or partners, in that Cl^-, originally combined with H^+, is combined with Na^+ in the product, and H^+, originally combined with Cl^-, combines with OH^- to produce water.

The chemist makes use of symbols and formulas to state the facts of chemical reactions. To represent the formation of water, a **reaction equation** is written as follows:

$$2H_2 + O_2 \rightarrow 2H_2O$$

This equation states that two molecules (or two moles) of hydrogen and one molecule (or one mole) of oxygen combine, or react, to give two molecules (or two moles) of water. Hydrogen and oxygen are called **reactants** and water is called the **product.** Although, as indicated, the equation can be interpreted either on a molecular basis or on a mole basis, the latter of these interpretations has the advantage that the coefficients can immediately be interpreted in useable mass or volume units. One mole of oxygen, O_2, is 32 grams or 22.4 liters at standard temperature and pressure.

A chemical equation is a quantitative and qualitative statement of what happens in a chemical reaction. Any chemical reaction can be written in the form of a chemical equation if the correct formulas for all the reactants and products are known, and if the mass relationships of reactants and products are known. The chemical equation must obey the law of conservation of mass in that the reactants and products must contain the same total mass of each type of atom in the system. When this requirement is met, the equation is said to be **balanced.**

The simplest method of balancing chemical reactions is by inspection. This method always works for metathesis reactions, but it is often very difficult for oxidation-reduction reactions. The following equation, representing the interaction of phosphoric acid and calcium hydroxide, is readily balanced by inspection:

$$H_3PO_4 + Ca(OH)_2 \rightarrow Ca_3(PO_4)_2 + H_2O$$

Inspection of the products indicates that there are $3Ca^{+2}$ present, and therefore there must be $3Ca(OH)_2$ on the reactant side of the equation. This inspection also indicates that there must be $2H_3PO_4$ on the reactant side. These two adjustments lead to the following partially balanced equation:

$$2H_3PO_4 + 3Ca(OH)_2 \rightarrow Ca_3(PO_4)_2 + H_2O$$

A simple count of atoms indicates that in this equation there are far too many hydrogens and oxygens on the reactant side, $6H^+$ and $6OH^-$. To give complete balancing, water should have the coefficient 6:

$$2\,H_3PO_4 + 3\,Ca(OH)_2 \rightarrow Ca_3(PO_4)_2 + 6\,H_2O$$

The coefficients assigned to reactants and products in a balanced chemical equation are referred to as the **reaction stoichiometry.** These coefficients indicate the relative numbers of atoms, molecules, formula units, and ions involved in the reaction, and therefore can be interpreted as an indication of the mass relationships.

OXIDATION NUMBERS

In addition to being a statement of composition, a formula yields information which can be used to predict properties. The systematic assignment of **oxidation numbers** to the elements within a formula facilitates this prediction of chemical properties. Oxidation numbers are arbitrarily assigned to elements within a formula to indicate whether that element in the formation of the compound has lost or gained electron density. Elements which gain electron density are assigned negative oxidation numbers, whereas those which lose electron density are assigned positive values. The following set of rules is arbitrarily followed in assigning oxidation numbers:

1. Uncombined elements have an oxidation number of 0.
2. In a molecule the sum of the oxidation numbers of all the atoms must equal zero.
3. Simple ions, which are charged monatomic species, have oxidation numbers equal to the charge.
4. For complex ions, which are charged polyatomic species, the sum of the oxidation numbers of all the atoms must equal the charge on the complex ion.
5. Chemically combined hydrogen has an oxidation number of $+1$, except when combined with active metals such as Li, Na, Be, Mg, and so forth.
6. Chemically combined oxygen has an oxidation number of -2, except in peroxides where it is -1.

The commonly observed oxidation numbers of the elements show periodic variation as determined by that number of electrons which must be lost or gained to attain the electronic configuration of an inert gas. Consequently, the elements in Group I of the periodic table show only the $+1$ oxidation number when chemically combined. The elements of Group VII, the halogens, show only -1 oxidation numbers when combined with elements of lower electronegativity. Many elements show a variation of oxidation numbers in different compounds.

The following examples illustrate the assignment of oxidation numbers in some representative formulas:

1. K_2O Oxygen is -2 and potassium $+1$. The sum of the oxidation numbers is 0. The assignment is represented as follows:

$$\overset{2(+1)(-2)}{K_2O}$$

2. HNO_3 3 (oxygen at -2) $= -6$
1 (hydrogen at $+1$) $= +1$
total oxidation number $= -5$

So the oxidation number of nitrogen must be $+5$.

$(+1)(+5)(-2)3$
HNO_3

METATHESIS REACTIONS

Metathesis reactions, which involve no change in oxidation numbers, most commonly occur between ions in solutions. These reactions are most favorable and have the highest tendency to proceed when one of the products does not exist as discrete ions in solution. This happens when the product is a **gas**, a **precipitate** (insoluble solid species), or a **covalent undissociated molecule.** The following three metathesis reactions are examples of these:

$$2HCl + Na_2CO_3 \rightarrow 2NaCl + H_2O + CO_2\uparrow \text{ (gas)}$$
$$HCl + AgNO_3 \rightarrow HNO_3 + AgCl\downarrow \text{ (solid)}$$
$$HCl + NaOH \rightarrow NaCl + H_2O \text{ (covalent molecule)}$$

For reactions which occur in solution, reaction equations are often abbreviated to include those reactants which are directly involved in chemical change. This requires a knowledge of the nature of the reactants in solution, that is, whether or not they are dissociated into ions. The following equations are the **net equations** corresponding to those given above:

$$2H^+ + CO_3^{-2} \rightarrow H_2O + CO_2\uparrow \text{ (gas)}$$
$$Ag^+ + Cl^- \rightarrow AgCl\downarrow \text{ (solid)}$$
$$H^+ + OH^- \rightarrow H_2O \text{ (covalent molecule)}$$

Reactions which do not involve exchange of groups but rather involve combination of two substances or decomposition of some substance are also classified as metathesis reactions. The following reaction illustrates a non-oxidation-reduction combination reaction:

$$H^+ + :NH_3 \rightarrow NH_4^+$$

The reverse of this reaction, a decomposition, is also classified as a metathesis reaction.

OXIDATION-REDUCTION REACTIONS

In an oxidation-reduction reaction some species lose electrons and some species gain electrons; that is, some species have an increase in oxidation number and some have a decrease. The process of removing electrons from an atom, thereby giving a more positive oxidation number, is called **oxidation.** After oxidation has occurred, the atom is said to have been **oxidized.** The process of adding electrons to an atom, and so giving a more negative oxidation number, is called **reduction.** After reduction has occurred, the atom is said to have been **reduced.**

The element which gains electrons in an oxidation-reduction reaction is called the **oxidizing agent,** whereas the element which loses electrons is the **reducing agent.** In addition to the conservation of mass in a chemical reaction, there must also be conservation of electrons, electrical charges, and oxidation numbers. If electrons are lost by an element being oxidized, they must be gained by the oxidizing agent. For this reason reference is made less often to separate oxidation or reduction reactions than to combined oxidation-reduction reactions. In any reaction in which an oxidation occurs, it is always accompanied by a reduction, and the oxidation and reduction always take place to an equal degree.

The assignment of oxidation numbers to reactants and products in an equation enables immediate detection of the oxidizing agent and the reducing agent. The oxidation number of the oxidizing agent becomes more negative while that of the reducing agent becomes more positive. This is illustrated with the following equation in which oxidation numbers have been assigned:

$$\overset{(+4)(-2)2}{MnO_2} + \overset{(+4)(-2)2}{PbO_2} \rightarrow \overset{+2}{Pb^{+2}} + \overset{(+7)(-2)4}{(MnO_4)^{-1}}$$

> The oxidation number of manganese increases from $+4$ to $+7$ during the reaction, indicating that it is the reducing agent. Lead has its oxidation number decreased from $+4$ to $+2$, and it is the oxidizing agent. Manganese is oxidized while lead is reduced.

The mass of a reducing agent or oxidizing agent which is involved in the transfer of one mole of electrons is a **gram equivalent weight.** Thus a gram equivalent weight is the weight of one mole of the material divided by the total oxidation number change. Since one gram equivalent weight is involved in the transfer of one mole of electrons, one gram equivalent weight of reducing agent will react completely with one gram equivalent weight of oxidizing agent.

CLASSES OF CHEMICAL COMPOUNDS

Chemists classify compounds on the basis of composition, physical properties, and chemical properties. The broadest general classification, based only on composition, differentiates **inorganic compounds** and **organic compounds.** Inorganic compounds do not contain carbon whereas organic compounds always do. Although there are many exceptions, naturally occurring organic compounds can be thought of as originating from living matter, both plant and animal. Many additional organic compounds have been synthesized by chemists. The nature of organic compounds will be given special attention in a later section of this book. Most of the surface of the earth is composed of inorganic compounds. Some compounds containing carbon are classified as inorganic because of their chemical behavior. These include carbides such as CaC_2, carbonates such as $MgCO_3$, and cyanides such as KCN.

The nature of the chemical bonding in a compound is commonly used to differentiate between **covalent compounds** and **ionic compounds.** This differentiation is also established by considering chemical and physical properties of the compounds. Covalent compounds are more volatile than ionic compounds, although an extensive overlap of properties of the two classes is observed.

Another classification of compounds, which is based on a combination of chemical and physical properties, involves the ability of a water (**aqueous**) solution of the compound to conduct an electric current. If this ability exists, the compound is an **electrolyte,** whereas if the aqueous solution will not conduct, the compound is a **nonelectrolyte.** The nature of this differentiation is discussed more extensively in Chapter 6.

ACIDS AND BASES

The concept of acids and bases was originally severely limited to the special conditions prevailing in aqueous solutions. It has, however, evolved over many years to a concept which not only explains many reactions but facilitates the prediction of many others. In the evolution to more sophisticated theory, each new idea incorporated earlier ideas and extended the acid-base concept to new cases.

In 1887 Arrhenius, a Swedish chemist, proposed a theory of ionization (which will be discussed in Chapter 9) that provided answers to many chemical questions, and a more suitable definition of acids and of bases. An **acid** was defined as a substance containing hydrogen that dissociates to yield hydrogen ions in a water solution. A **base** was defined as a substance that dissociates to give hydroxide ions in a water solution.

The **Brönsted-Lowry concept** of an acid as a **proton donor** and a base as a **proton acceptor** is useful for work with either aqueous or nonaqueous solutions. It incorporates the Arrhenius concept as a special case where the base is always the hydroxide ion.

The most sophisticated concept proposed to date, which attempts to explain metathesis reactions, is the **Lewis concept.** According to Lewis, a base is any species capable of donating a pair of electrons for the formation of a coordinate covalent bond, and an acid is any species which is capable of accepting a pair of electrons to form a coordinate covalent bond. This concept incorporates the proton of the Arrhenius and Brönsted-Lowry concepts as an acid. It also considers metal ions and many molecules such as BF_3 as acids.

Acids are most commonly used in aqueous solution, although when pure they may exist as solids, liquids, or gases. For example, hydrogen chloride, HCl, is a gas; boric acid, H_3BO_3, is a solid; and sulfuric acid, H_2SO_4, is a liquid. When hydrogen chloride is present in a water solution it dissociates as follows:

$$HCl \rightarrow \underset{\text{hydrogen ion}}{H^+} + \underset{\text{chloride ion}}{Cl^-}$$

This equation illustrates the common definition of an acid as a substance that yields hydrogen ions in aqueous solution. According to the Brönsted and Lowry definition, the hydrogen chloride gas would be called an acid since it is capable of donating a proton to a base. The general concept of their definition would be represented by the equation:

$$\underset{\text{acid}}{HA} + \underset{\text{base}}{B} \rightleftharpoons \underset{\text{acid}}{HB^+} + \underset{\text{base}}{A^-}$$

Therefore, in H_2O the dissociation of hydrochloric acid would be represented as:

$$\underset{\text{acid}}{HCl} + \underset{\text{base}}{H_2O} \rightleftharpoons \underset{\text{acid}}{H_3O^+} + \underset{\text{base}}{Cl^-}$$

The species produced when an acid loses its proton is called the **conjugate base** of the acid. When a base gains a proton its **conjugate acid** is formed. In the above reaction, Cl^- is the conjugate base of HCl and H_3O^+ is the conjugate acid of H_2O. Two species such as HCl and Cl^-, related by a loss or a gain of a proton, are called a **conjugate pair.** The double arrows in these equations indicate that the forward and reverse reactions are both possible. In such a **reversible** reaction, if the forward reaction is proceeding at the same rate as the reverse reaction, the system is said to be in a state of **dynamic chemical equilibrium.**

The properties of Arrhenius acids in aqueous solutions include the following:

1. All acids in solution have a sour taste (use caution and taste only dilute solutions). Citrus fruits, for example, taste sour because of the presence of citric acid, an organic acid. The sour taste of vinegar is due to another organic acid, acetic acid.

2. Acids change the blue color of litmus dye to red. This is one of the simplest tests for the presence of an acid. A substance such as litmus that has one color in an acid solution and another color in a basic solution is called an **indicator.**

3. Acids react with many metals to form hydrogen gas. The metal reduces the hydrogen of the acid, liberating hydrogen gas. For example, in the following reaction, zinc reduces the hydrogen of sulfuric acid to form zinc sulfate and gaseous hydrogen:

$$Zn + H_2SO_4 \rightarrow ZnSO_4 + H_2\uparrow$$

In this reaction Zn is oxidized by H^+.

Not all metals possess enough chemical activity to replace the hydrogen of an acid. The hydrogen of an acid can be replaced by most metals indicating that it is **replaceable hydrogen.** The quantity of an acid which will yield one mole of protons, or which has one gram of replaceable hydrogen, is one **gram equivalent weight of the acid.**

4. Acids react with oxides and hydroxides to form water and a **salt.** A salt may be considered as a compound containing neither the initial hydrogen of an acid nor the hydroxide group of a base. The action of an acid on an oxide or hydroxide of a metal is illustrated by the following:

$$MgO + 2HNO_3 \rightarrow Mg(NO_3)_2 + H_2O$$
$$NaOH + HCl \rightarrow NaCl + H_2O$$

In the reaction between an acid and a metallic hydroxide, or base, both the acid and the base are neutralized. A reaction of an acid and a base to form water and a salt is therefore called a **neutralization reaction.**

5. Acids react with carbonates and bicarbonates to form carbon dioxide gas, as illustrated by the following equations:

$$Na_2CO_3 + H_2SO_4 \rightarrow CO_2\uparrow + H_2O + Na_2SO_4$$
$$NaHCO_3 + HCl \rightarrow CO_2\uparrow + H_2O + NaCl$$

Baking soda, $NaHCO_3$, is widely used for the neutralization of acid and for the production of carbon dioxide gas.

Baking powders contain a bicarbonate and some acid-forming substance, which, when moisture is added, release gaseous carbon dioxide throughout the cake batter, making it light. The lactic acid of sour milk produces the same effect when mixed with baking soda. The action of a common type of fire extinguisher depends on the reaction between sulfuric acid and sodium bicarbonate.

Acids are usually classified according to the number of hydrogen ions they yield per molecule, and by the number of elements they contain. Those that yield one hydrogen ion per molecule, such as HCl, are called **monoprotic;** those yielding two, such as H_2SO_4, are called **diprotic;** and those yielding three, such as H_3PO_4, are called **triprotic. Binary,** or hydro- acids, are composed of hydrogen and one other element, and **ternary,** or oxy-, acids, contain oxygen and another element in addition to hydrogen. The binary acids are named from the element that is combined with hydrogen. They begin with the prefix **hydro-** and end with the suffix **-ic.** Examples:

HCl HBr HF
Hydrochloric Hydrobromic Hydrofluoric

The most common ternary acids are named after the element other than hydrogen or oxygen, and they end with the suffix **-ic.** Examples:

$$H_2SO_4 \qquad HNO_3 \qquad H_3PO_4 \qquad HClO_3$$
Sulfuric Nitric Phosphoric Chloric

If the same elements unite to form more than one ternary acid, the acid with one less oxygen than the most common form ends in **-ous.** Examples:

$$H_2SO_3 \qquad HNO_2 \qquad H_3PO_3 \qquad HClO_2$$
Sulfurous Nitrous Phosphorous Chlorous

Acids of the halogens (chlorine, bromine, iodine) containing oxygen can exist as -ic or -ous acids; chloric and chlorous acids are examples. When these acids contain one more oxygen atom than the -ic acid they are named **per- -ic** acids, and if they contain one less oxygen atom than the -ous acid they are named **hypo- -ous** acids. Examples:

$$HClO_3 \qquad HClO_4 \qquad HClO_2 \qquad HClO$$
Chloric Perchloric Chlorous Hypochlorous

Bases are defined as compounds that contain the hydroxyl group, —OH, and ionize to form hydroxide ions in a water solution. This definition is by far the most practical and useful when dealing with aqueous solutions. In the Brönsted-Lowry concept, all negative ions, being proton acceptors to some degree, are considered to be bases, with the hydroxide ion being the strongest base that can exist in a water solution.

In the discussion of the properties and reactions of common bases in aqueous solutions, use of the term *base* will be confined to the metallic hydroxides and ammonia. Examples of common bases are: sodium hydroxide, $NaOH$; potassium hydroxide, KOH; calcium hydroxide, $Ca(OH)_2$; and ammonia (NH_3).

All of the metals in the bases listed above are located in Groups I and II on the left of the periodic table. The hydroxide ion is formed from hydrogen and oxygen by a covalent bond, whereas the metal is combined with the hydroxide by an ionic bond.

Ammonia is very soluble in water and gives a basic solution because of the reaction:

$$NH_3 + H_2O \rightleftharpoons NH_4^+ + OH^-$$

The properties of bases in aqueous solutions are as follows:

1. When a base is dissolved in water, the solution has a slippery feeling.

2. Solutions of bases have a bitter, metallic taste. (CAUTION: they should be tasted only in dilute solutions.)

3. Bases change the red color of litmus to blue. In general, bases reverse the color change that was produced by an acid in an indicator.

4. Bases react with acids to form water and a salt:

$$2KOH + H_2SO_4 \rightarrow 2H_2O + K_2SO_4$$

One **gram equivalent weight of a base** is that quantity which will react with one mole of H^+, or will neutralize one gram equivalent weight of acid.

Bases which consist of a metal combined with the hydroxide ion are named by starting with the name of the metal and ending with the word **hydroxide.** For example, $NaOH$ is sodium hydroxide, KOH is potassium hydroxide, and $Ca(OH)_2$ is calcium hydroxide.

SALTS

As stated before, when an acid and a base react with each other, acidic and basic properties of the solution disappear with the formation of water and a salt. Salts may also be formed by the displacement of hydrogen from an acid with a metal.

$$Zn + 2HNO_3 \rightarrow Zn(NO_3)_2 + H_2\uparrow$$

From this reaction a salt may be defined as a compound formed by replacing the hydrogen of an acid with a metal. A more general definition of a salt would be the combination of any negative ion, except hydroxide, with any positive ion other than hydrogen.

Inorganic salts consist of ions combined by ionic bonds in the solid or crystalline state. The atoms of a complex or polyatomic ion, such as NO_3^-, are usually held together by covalent bonds.

Salts react with themselves, with acids, with bases, and with water in ways which are usually predictable by considering the nature of the possible products. As indicated earlier, these metathesis reactions proceed if they result in the formation of a gas, a solid, or covalent undissociated molecules.

The properties of salts may be summarized as follows:

1. Salts react with each other to form new salts.

$$NaCl + AgNO_3 \rightarrow NaNO_3 + AgCl\downarrow$$

This is a common reaction in chemistry and is used as a test for the presence of a chloride. When silver nitrate is added to a solution that contains chlorides, a positive reaction is indicated by the formation of a white precipitate (insoluble AgCl).

2. Salts react with acids to form other salts and other acids.

$$BaCl_2 + H_2SO_4 \rightarrow BaSO_4\downarrow + 2HCl$$

3. Salts react with bases to form other salts and other bases.

$$MgSO_4 + 2KOH \rightarrow K_2SO_4 + Mg(OH)_2\downarrow$$

Magnesium hydroxide, the base formed in this reaction, is only slightly soluble in water.

4. Salts react with water by hydrolysis to give acidic and basic solutions. These reactions are discussed in Chapter 7.

A **normal salt** is one in which all the hydrogen of an acid has been replaced by another positive ion. For example, Na_2CO_3, $BaSO_4$, and KNO_3 are normal salts. Salts formed from acids that contain more than one replaceable hydrogen atom may retain one or more replaceable hydrogen atoms in their molecule, and are called **acid salts.** For example, sulfuric acid may react with sodium hydroxide to form sodium acid sulfate, which is commonly called sodium bisulfate or sodium hydrogen sulfate.

$$H_2SO_4 + NaOH \rightarrow NaHSO_4 + H_2O$$

If both hydrogen atoms are replaced by sodium, the normal salt sodium sulfate is formed.

$$H_2SO_4 + 2NaOH \rightarrow Na_2SO_4 + 2H_2O$$

Other examples of acid salts are sodium bicarbonate ($NaHCO_3$), sodium dihydrogen phosphate (NaH_2PO_4), and disodium hydrogen phosphate (Na_2HPO_4). The latter two compounds are made from phosphoric acid, which has three replaceable hydrogen atoms; therefore, two acid salts are possible.

Salts are named in two parts, the name of the positive ion and the name of the negative ion. Metallic positive ions are referred to by the name of the metal (i.e., sodium, potassium) unless the metal displays more than one oxidation number. For metals of variable oxidation number, the lower oxidation number is referred to by adding the suffix **-ous,** whereas the upper uses **-ic.** Often the Latin names of the elements are used as stems in this system (i.e., ferr- for iron, plumb- for lead). It is also common to name the element and indicate the oxidation number in parentheses with Roman numerals. The NH_4^+ ion is called the ammonium ion.

The negative ions of salts are named with reference to the acid from which they come. Simple negative ions are named by adding the suffix **-ide** to the stem of the name of the element. Negative ions derived from ternary acids ending in -ic are given the suffix **-ate,** and those derived from ternary acids ending in -ous are given the suffix **-ite.** The negative ions of acid salts are named by adding as a prefix **bi-,** the word hydrogen, or the word acid.

The following list includes the formulas and names of several common salts:

NaCl	sodium chloride	KNO_2	potassium **nitrite**
$FeBr_2$	ferrous bromide	$(NH_4)_2SO_3$	ammonium **sulfite**
	[iron(II) bromide]	$Ca(ClO)_2$	calcium **hypochlorite**
$Fe(NO_3)_3$	ferric nitrate	Na_2CO_3	sodium **carbonate**
	[iron(III) nitrate]	$NaHCO_3$	sodium **bicarbonate**
$Ba_3(PO_4)_2$	barium phosphate	$KHSO_3$	potassium **bisulfite**
$Al_2(SO_4)_3$	aluminum sulfate		

IMPORTANT TERMS AND CONCEPTS

acid	Lewis acid-base concept	reactants
Arrhenius acid-base concept	metathesis reaction	reaction products
base	neutralization reaction	reducing agent
Brönsted-Lowry acid-base concept	oxidation	reduction
electrolyte	oxidation number	reversible reaction
gram equivalent weight	oxidizing agent	salt

QUESTIONS

1. Give two definitions of acids and bases used by chemists.

2. Which of the definitions of an acid and a base are most frequently applied to aqueous solutions?

3. Name the following acids and bases:

   (a) HCl   (f) $HClO_4$
   (b) H_3PO_4 (g) H_2SO_3
   (c) HNO_2 (h) HNO_3
   (d) H_2SO_4 (i) $Ca(OH)_2$
   (e) NH_4OH (j) H_2CO_3

4. Write the formula for a normal salt and for an acid salt. Explain the difference between the two compounds.

5. Indicate whether each of the following reactions is oxidation-reduction or metathesis:

   (a) $2HNO_3 + 3H_2S \rightarrow 2NO + 4H_2O + 3S$
   (b) $HCl + H_2O \rightarrow H_3O^+ + Cl^-$
   (c) $4FeS_2 + 11O_2 \rightarrow 2Fe_2O_3 + 8SO_2$
   (d) $PCl_3 + 3H_2O \rightarrow P(OH)_3 + 3HCl$

6. Noting the properties, indicate which of the following are acids and which are bases:

   (a) a compound which reacts with zinc to form $H_2(g)$
   (b) a compound which changes red color of litmus to blue
   (c) a solution of this has a bitter taste
   (d) a compound which reacts with an oxide or a hydroxide to form water and a salt
   (e) a solution which reacts with a carbonate to form carbon dioxide gas

7. Give the products in each of the following reactions:

   (a) $HCl + NaOH \longrightarrow$
   (b) $HCl + KOH \longrightarrow$
   (c) $HNO_3 + NaOH \longrightarrow$
   (d) $CH_3CO_2H + NaOH \longrightarrow$
   (e) $HCl + NH_3 \longrightarrow$

8. Indicate the reducing agent and oxidizing agent in each of the following reactions:

   (a) $Zn + H_2SO_4 \rightarrow ZnSO_4 + H_2$
   (b) $2Fe^{+3} + 2I^- \rightarrow 2Fe^{+2} + I_2$
   (c) $2HNO_3 + 3H_2S \rightarrow 2NO + 4H_2O + 3S$
   (d) $Ca + 2H_2O \rightarrow Ca(OH)_2 + H_2$

9. Indicate the oxidation numbers of the following:

   (a) N in NO_2
   (b) Fe in Fe_2O_3
   (c) N in HNO_3
   (d) Mn in $(MnO_4)^-$
   (e) S in $(SO_4)^{-2}$
   (f) C in $(C_2O_4)^{-2}$

SUGGESTED READING

Drago and Matwyioff: Acids and Bases. Lexington, Mass., D. C. Heath and Co., 1968.
Robbins: Ionic Reactions and Equilibria. New York, The Macmillan Co., 1967.

WATER AND SOLUTIONS

CHAPTER 6

The *objectives* of this chapter are to enable the student to:

1. Define hydrogen bonding and explain the significance of the molecular structure of water in hydrogen bonding.
2. Explain how hydrogen bonding imparts some abnormal properties to water.
3. Define hydrates and discuss their formation and decomposition.
4. List several methods for purifying water.
5. Discuss the undesirables in hard water and several methods for their removal.
6. Define solute, solvent, solution, saturated solution, solubility, and supersaturated solution.
7. Discuss factors affecting solubility.
8. Express the concentration of a solution of known composition as percentage by weight or percentage by volume.
9. Describe a titration and calculate the quantity of a given reagent required to reach the equivalence point in a given titration.
10. Discuss Raoult's law and colligative properties.
11. Present the currently accepted theory of electrolytes.
12. Explain the difference between solutions and colloids.

WATER

The most abundant compound in man's environment is water, H_2O. The oceans, covering about 71 per cent of the earth's surface, comprise 97 per cent of all the earth's water. Another 2 per cent is contained in the ice of the polar caps. Of the remaining 1 per cent, called fresh water because it is sufficiently pure for human consumption, only about 60 per cent is involved in the evaporization-rainfall cycle which moves water to all parts of the earth. The remaining 40 per cent is ground water found at depths in excess of 1000 ft and thus practically unavailable. Although the amount of water in the atmosphere is only a small fraction of the available fresh water it is a major factor in weather systems.

Water is a necessity of life processes primarily because it serves as a transport medium. It has the unique property of forming solutions with many substances which can then be transported as water moves. Because of its importance in life processes and other processes at the surface of the earth, water and its solutions deserve the special attention found in this chapter.

OCCURRENCE

In addition to obvious bodies of water on the earth's surface, water is found mixed with almost all other matter in our environment. The soil contains large quantities of water, which are essential for the growth of plants. Its presence in the atmosphere is readily recognized, because it often condenses into dew, fog, rain, or snow. As a substance essential to our existence, water ranks next to oxygen in importance. The body can survive several weeks without food, but only a few days without water. The digestion of food, the circulation, the elimination of waste materials, and the regulation of acid-base balance and body temperature, as well as other vital functions, depend on an adequate supply of water. Approximately two-thirds of the body weight is water, and most of the foods we eat have a water content of from 10 to 90 per cent. Bread, for example, is about 35 per cent water, meat is about 70 per cent, and most vegetables are over 75 per cent.

PHYSICAL PROPERTIES

Pure water has no odor, taste, or color. It freezes at 0°C (32°F) and boils at 100°C (212°F). These values are abnormally high for a compound with such a low molecular weight. Other physical properties of water are also exceptional. Its maximum density occurs at 4°C, so that when water freezes, it becomes less dense, allowing ice to float. The volume increases by nearly one-tenth in changing from water to ice, which explains the cracked automobile blocks and broken water pipes that occur in freezing weather.

The universal distribution of water and its widespread use in the laboratory have caused many measurements to be based on its physical properties. For instance, the Celsius thermometer is based on the freezing and boiling points of water (Fig. 4–2). Densities of matter are often referred to that of water. The **specific gravity** of a substance is the ratio of the density of that substance to that of water at 4°C, and consequently is a dimensionless number. The specific gravity of water is 1.000, since 1 ml of water at 4°C weighs 1.000 g. If a solution weighed 1.030 g per ml (grams per milliliter), it would have a specific gravity of 1.030, and 1 ml of the solution would be 1.030 times as heavy as 1 ml of water.

In Chapter 3 it was indicated that water is a covalent compound whose molecules have highly polar bonds giving rise to a dipole moment. Careful studies of water have established that the bond angle is 104.5 degrees as seen in Figure 6–1. The non-bonding electrons on oxygen occupy orbitals which are nearly sp^3 as are the bonding orbitals. Consequently the non-bonding electrons project out of the molecular plane.

The abnormal physical properties of water including high boiling point, high melting point, high heats of vaporization and fusion, high surface tension, and the density maximum at 4°C are all attributable to abnormally high intermolecular attractive interactions. These interactions, called **hydrogen bonds,** arise because of the very strong attractions between the non-bonding electrons of the oxygen and partially positive-charged hydrogen atoms on adjacent molecules. The resulting molecular association which occurs in

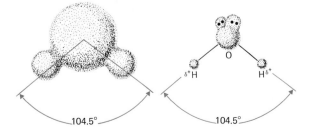

Figure 6–1   The structure of water.

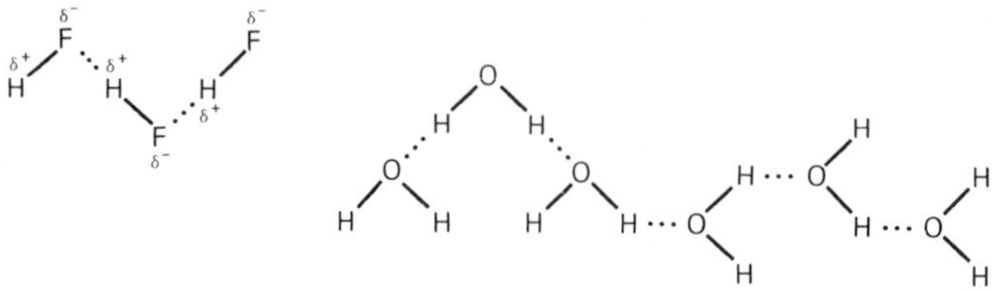

Figure 6-2   Hydrogen bonding in hydrogen fluoride and in water.

other compounds when hydrogen is directly bonded to a small electronegative atom is shown in Figure 6-2 for hydrogen fluoride and water.

The efficiency of hydrogen bonding decreases markedly when hydrogen is bonded to larger, less electronegative atoms. This is shown in Figure 6-3, in which it is seen that NH_3, H_2O, and HF, the lightest hydrogen compounds of groups V, VI, and VII, show much higher boiling points than the next heavier hydrogen compounds of their respective families. The compounds of group IV have no non-bonding electron pairs on the central atoms and thus cannot be involved in hydrogen bonding with one another. Thus they display the normal increase in boiling point with increasing mass.

CHEMICAL PROPERTIES

Since water is a polar compound, it has the property of attracting ions of ionic compounds and disrupting their crystal structure. When solid crystals of sodium chloride are placed in water, the forces holding the sodium ions and chloride ions together are overcome by the attraction of the water molecules for the ions. This process results in the solution of the sodium chloride as represented in Figure 6-4. In a similar fashion, water will also attract other polar molecules. Since so many substances are either ionic or polar in nature, water has the property of dissolving a large majority of common inorganic compounds. In the laboratory many chemical reactions are studied in aqueous solution. In the living organism the constituents of the cells are kept in solution by water, and most of the reactions that take place in the tissues will not take place in its absence.

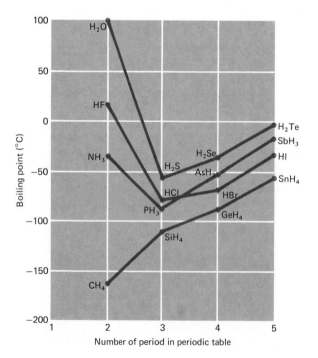

Figure 6-3   Illustration of the abnormally high boiling points of H_2O, HF, and NH_3.

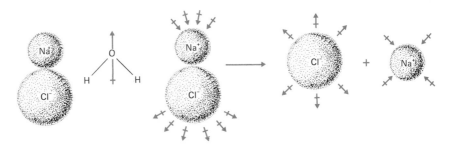

Figure 6-4   Sodium chloride going into solution illustrates the effect of water on ionic compounds.

Water is one of the most stable compounds known, and for many years it was thought to be an element. It may be heated to very high temperatures (2000°C) without appreciable decomposition. However, if an electric current is passed through it, decomposition occurs; two volumes of hydrogen are produced for each volume of oxygen.

Another interesting chemical property of water is its action with certain metals. If a small piece of metallic potassium is placed in water, a violent reaction takes place with the formation of a hydroxide and hydrogen gas:

$$2K + 2H_2O \rightarrow 2KOH + H_2\uparrow$$

The compound that is formed is called potassium hydroxide.

Water will combine with the oxides of some metals to form a metallic hydroxide, which is also called a base.

$$\underset{\substack{\text{Magnesium} \\ \text{oxide}}}{MgO} + H_2O \rightarrow \underset{\substack{\text{Magnesium} \\ \text{hydroxide}}}{Mg(OH)_2}$$

Certain oxides of nonmetals react with water to form acids.

$$\underset{\substack{\text{Sulfur} \\ \text{trioxide}}}{SO_3} + H_2O \rightarrow \underset{\substack{\text{Sulfuric} \\ \text{acid}}}{H_2SO_4}$$

One of the most important chemical properties of water is concerned with the process of hydrolysis. In **hydrolysis,** or breaking apart with water, a compound is split into two parts, the hydrogen of the water uniting with one part to make an acid, leaving the hydroxide ion, or the hydroxide uniting with one part to form a base leaving the hydrogen ion. An example of hydrolysis is the reaction of water and a compound such as ammonium sulfate.

$$\underset{\substack{\text{Ammonium} \\ \text{sulfate}}}{(NH_4)_2SO_4} + 2HOH \rightleftharpoons \underset{\substack{\text{Ammonium} \\ \text{hydroxide}}}{2NH_4OH} + \underset{\substack{\text{Sulfuric} \\ \text{acid}}}{H_2SO_4}$$

The double arrow indicates that the reaction can go in both directions. Actually, only a small amount of the ammonium sulfate reacts with water in this way. The quantitative aspects of these reactions will be discussed in Chapter 7.

HYDRATES

Water molecules combine with the molecules of certain substances, forming loose chemical combinations called **hydrates.** These hydrates form well defined crystals when their solutions are allowed to evaporate slowly. For example, copper sulfate forms blue crystals when an aqueous solution of this substance evaporates slowly. The formula for crystalline copper sulfate is $CuSO_4 \cdot 5H_2O$. The water held in combination is called **water of crystallization** and is written separately to indicate its loose chemical attachment. When this hydrate is heated, it loses its water of crystallization and changes into a white powder whose formula is $CuSO_4$. Examples of other common hydrates are washing soda, $Na_2CO_3 \cdot 10H_2O$; alum, $K_2Al_2(SO_4)_4 \cdot 24H_2O$; gypsum, $CaSO_4 \cdot 2H_2O$; and crystalline sodium sulfate, $Na_2SO_4 \cdot 10H_2O$.

When the water of crystallization has been removed from a hydrate, the resulting compound is said to be **anhydrous.** Substances which give up water of crystallization on exposure to air at ordinary temperatures are called **efflorescent.** Other substances take up water on exposure to atmospheric conditions and are said to be **hygroscopic.** If they take up so much water from the air that they are finally dissolved in it, they are called **deliquescent** substances. Compounds such as sodium hydroxide and calcium chloride are so hygroscopic that they take up water from other materials. Calcium chloride is commonly used as a drying agent or desiccating agent by the chemist.

A hydrate of special interest is gypsum, or calcium sulfate ($CaSO_4 \cdot 2H_2O$), which on heating gives up part of its water, undergoes **dehydration,** to form plaster of Paris, $(CaSO_4)_2 \cdot H_2O$. When the plaster of Paris is mixed with water, it "sets" in a few minutes to re-form hard crystalline gypsum. In setting, it expands slightly to form a tight cast or mold. Plaster of Paris is used extensively in making surgical casts.

PURIFICATION OF WATER

Naturally occurring water contains impurities dissolved from the rocks and soil. Even rain water contains particles of dust and dissolved gases from the air. The impurities present in water may be classified as either mineral or organic matter. The mineral matter found in natural water usually consists of common salt and various compounds of Ca^{+2}, Mg^{+2}, and Fe^{+3}. These ions form insoluble precipitates or "curd" with the negative ions in soap. Water that contains these dissolved salts does not readily form a lather with soap and is called **hard water,** whereas water with little or no mineral matter lathers easily and is called **soft water.**

The organic matter in water is derived from decaying animal and vegetable material. Bacteria utilize this type of material for food, and may cause diseases unless they are removed before the water is used for drinking purposes.

A source of pure drinking water is extremely important to the health of a community. For this reason it would be appropriate to consider several of the methods for the purification of water.

Distillation. In the process of distillation, water is boiled and the resulting steam is cooled and condensed in a different container. The condensed steam is called the distillate, or distilled water. The chemist uses distillation to produce water free from bacteria and dissolved mineral matter. Distilled water is used widely in the preparation of solutions in the laboratory and in the hospital. Distillation is the most effective method for the purification of water, but is too expensive to be employed by large towns or cities.

Boiling. Water from natural sources may be made safe for drinking by boiling for 10 to 15 minutes. This process does not remove the impurities but does kill any pathogenic bacteria that might be present. The flat taste of boiled water is due to the loss of dissolved gases; it may be improved by aeration, such as by pouring water from one vessel to another. This method of water purification is reliable in emergencies, but is not generally employed for civilian water supplies.

Aeration. Water may be purified by exposure to air for long periods. The oxygen of the air dissolves in the water and oxidizes organic material, thus depriving bacteria of their source of food. It also kills bacteria by direct chemical reaction. Most cities do not depend on this process alone for water purification, but use it to remove objectionable tastes and odors from the water. Aeration of water supplies is usually accomplished by spraying the water into the air from fountains, or by allowing it to flow in thin sheets over tiles.

HARD AND SOFT WATER

Since the common methods of water purification do not remove the dissolved inorganic matter, many cities have hard water. There are many disadvantages to the use of hard water, in addition to the previously mentioned difficulties with soap. Cooking with hard water has a toughening effect on foods. Also, the iron salts in hard water often discolor white fabrics, pottery, and enamelware. When hard water is boiled, the mineral salts deposit a scale on the sides of boilers, pipes, and utensils in which it is heated. This causes not only a waste of fuel but also a corrosion of the metal.

Methods for Softening Water. The inorganic matter present in hard water usually consists of bicarbonates, sulfates, or chlorides of calcium, magnesium, and iron. Water that contains only calcium or magnesium bicarbonate is called **temporary hard water,** because these salts can be removed by heating. When heated, they are converted into the insoluble carbonates that form most of the scale on boilers and teakettles. Temporary hard water can therefore be softened by boiling.

$$Ca(HCO_3)_2 \xrightarrow{\Delta} CaCO_3\downarrow + H_2O + CO_2\uparrow$$

Calcium   Calcium
bicarbonate   carbonate
(soluble)   (insoluble)

The small triangle is used by chemists as a symbol for heat.

Water that contains sulfates or chlorides of calcium, magnesium, or iron is called **permanent hard water,** because it does not lose these salts on heating. Permanent hard water can be softened by adding a chemical compound that will convert the soluble calcium, magnesium, or iron salts into insoluble precipitates, which may be removed by filtration. The following compounds are commonly used for water softening:

Sodium carbonate, Na_2CO_3 (washing soda)
Sodium tetraborate, $Na_2B_4O_7$ (borax)
Ammonium hydroxide, NH_4OH (household ammonia)
Trisodium phosphate, Na_3PO_4
Sodium hydroxide, $NaOH$ (caustic soda or lye)
Mixture of calcium hydroxide, $Ca(OH)_2$, and sodium carbonate, Na_2CO_3

The reaction between a soluble calcium salt and sodium carbonate may be used as a typical water softening reaction:

$$CaSO_4 + Na_2CO_3 \rightarrow CaCO_3\downarrow + Na_2SO_4$$

Removal of the insoluble calcium carbonate yields soft water.

Water softeners used in homes, hospitals, laundries, and small industries often employ synthetic **ion exchange resins** or a material called **Zeolite,** which is a natural sodium aluminum silicate such as $NaAlSi_{12}O_6$. As the water passes through a bed or column of the resin or Zeolite, the Ca^{+2}, Mg^{+2}, or Fe^{+3} ions are retained and exchanged for less objectionable ions, usually Na^+, that pass through the softener with the water. For example, the sodium ions in the Zeolite are exchanged for calcium ions, as represented in the following equation:

$$Ca^{+2} + Na_2Zeolite \rightarrow 2Na^+ + CaZeolite$$

Continual use of the Zeolite eventually places Ca^{+2} in all sites previously occupied by two Na^+ ions, resulting in a resin which can no longer be effective. The reverse reaction, liberating Ca^{+2}, can be effected by washing with a solution having high concentrations of Na^+.

Since about 1940 many types of ion exchange resins have been produced and studied. The two major types are the cation and anion exchangers, which are made with a great variety of resins and active groups. The most common cation exchangers contain sulfonic acid groups RSO_3H, whereas an important group of anion exchangers contains quaternary ammonium groups $R_4N^+Cl^-$. The resins are commonly polymers, such as polystyrene-divinylbenzene resins, which will be discussed in a later chapter. Water that is used in the laboratory and in many industries not only must be softened but must be free from inorganic ions. To produce this de-ionized or **demineralized water,** a bed of mixed resins is used. For example, a combination of a strongly acid cation exchanger and a strongly basic anion exchanger would remove sodium and chloride ions, as represented in the following reactions:

$$Na^+ + Cl^- + HSO_3Resin \rightarrow Na^+SO_3Resin^- + H^+ + Cl^-$$

$$H^+ + Cl^- + NH_2Resin \rightarrow Cl^-NH_3Resin^+$$

Since ion exchange resins will not remove nonelectrolytes or organic matter from water, for strict analytical purposes the water is first distilled and then passed through a mixed-bed, ion-exchange resin.

SOLUTIONS

The recognition of solutions and the fact that many substances dissolve in water and other liquids can be traced back to the beginning of chemistry. The chemical industries that existed as early as 3000 to 4000 B.C. utilized solutions in the preparation of glass, pottery, perfume, and dyes. The alchemists mixed so many curious substances in solution, often with heating, that their exact chemical analysis would pose a problem even for the modern chemist. In the seventeenth century chemists such as Robert Boyle were well aware of the value of solutions in their chemical investigations. At present one does not have to be a chemist to recognize solutions in the home, the supermarket, and the drug store. Water softeners, detergents and cleaning compounds, perfumes, lotions, and many medications are readily available in solution.

The common examples of sugar or salt dissolved in water illustrate a mixture that is called a solution. The sugar molecules or the sodium ions and chloride ions of salt become uniformly distributed among the water molecules during and after the dissolving process (Fig. 6–5). A **solution,** then, may be defined as a homogeneous mixture of two or more substances in which the particles are of atomic or molecular size. In solution, the substance that is dissolved is called the **solute,** whereas the substance in which the solute is dissolved is called the **solvent.** If there is any doubt as to which substance dissolves in the other, it is common practice to call the solvent the substance present in the greatest amount.

The composition of solutions can vary only within the limits of the ability of one

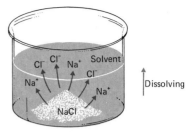

Figure 6-5 A representation of sodium chloride (the solute) dissolving in water (the solvent).

substance to dissolve in the other. If a solution is prepared by stirring in an excess of solute until no more will dissolve, it is said to be a **saturated solution.** At any given temperature a saturated solution will contain a definite quantity of solute in a given volume of solvent. This quantity is referred to as the **solubility.** For example, the addition of 40 g of sodium chloride to 100 g of water at 20°C would result in a solution that contains 36 g of sodium chloride and 4 g of undissolved crystals because the solubility of sodium chloride in water is 36 g NaCl per 100 g H_2O. A state of equilibrium would exist between the dissolved and undissolved solute. Particles of the crystalline solute would be continually going into solution at a rate that is exactly equal to the rate at which solute particles are crystallizing out of the solution (Fig. 6-6). Thus a saturated solution may be defined as a solution in which the dissolved solute exists in a state of equilibrium with the undissolved solute.

If a saturated solution is prepared at a higher temperature and is then allowed to cool, the extra solute that was dissolved at the higher temperature usually crystallizes out of the solution. If the hot solution is cooled slowly and is not disturbed, the excess solute may not crystallize out. In this case the solution will contain more solute than it can ordinarily dissolve at room temperature. Such a solution is called a **supersaturated** solution. If this solution is disturbed by the addition of a crystal of the solute, the material in excess of that required to saturate the solution at that temperature will immediately crystallize out.

Since matter exists as a solid, a liquid, or a gas, there are nine types of two-component solutions theoretically possible. These are listed in Table 6-1. The most common types of solutions are those in which a liquid is the solvent. In liquid solutions and gaseous solutions, the solute molecules diffuse, or move about freely, in the solvent. For example, if one drops a crystal of a colored solute, such as potassium permanganate ($KMnO_4$), into a vessel of water, a purple color is soon observed in the water immediately surrounding the crystal. In a few hours the purple color is evenly scattered throughout the entire solution.

By far the most common and most important solvent is water. Its solutions are called **aqueous solutions.** The oceans are gigantic aqueous solutions containing many dissolved solutes. Aqueous solutions play an important role in many naturally occurring processes by serving as a transport medium for solutes. When food is digested in the body, it is

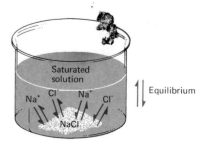

Figure 6-6 A state of equilibrium existing in a saturated solution of sodium chloride.

TABLE 6-1 TYPES OF SOLUTIONS

TYPES OF SOLUTIONS	EXAMPLES
Solid in liquid	Salt in water
Liquid in liquid	Gasoline
Gas in liquid	Ammonia in water
Solid in solid	Carbon in iron (steel)
Liquid in solid	Mercury in silver (amalgam)
Gas in solid	Hydrogen in palladium
Solid in gas	Iodine vapor in air
Liquid in gas	Water vapor in air
Gas in gas	Carbon dioxide in oxygen

dissolved and carried into the circulatory system in aqueous solution. The growth of plants depends on the transport of food and wastes in aqueous solutions.

FACTORS AFFECTING SOLUBILITY

Except for gases as solutes or solvents, the dissolving process, **dissolution,** involves increasing the average distance of separation between solute molecules and between solvent molecules. Each of these processes necessitates overcoming intermolecular forces of attraction and consequently consumes energy. If the dissolving process is to be at all favored, this consumption of energy must be compensated for in some way. If the separated solute molecules can have significant attractive intermolecular interactions with the solvent molecules, then the dissolving process may be energetically favorable. This solute-solvent interaction is called **solvation.** Solutes consisting of polar molecules are not highly soluble in solvents consisting of non-polar molecules, because the solute-solvent interaction is too weak to overcome the energy required to separate the solute molecules. Non-polar solute molecules, although easily separated, are not highly soluble in polar solvents, because the energy required to separate solvent molecules is not compensated for by the weak solute-solvent interaction. In general, it can be stated that ionic and polar solutes are most soluble in polar solvents, and non-polar solutes are most soluble in non-polar solvents.

Water is a common solvent for polar solutes, such as sugar and salts, but it is unsatisfactory for non-polar solutes, such as fat or oil. Iodine is only slightly soluble in water, but will dissolve readily in alcohol. Ether, carbon tetrachloride, and gasoline are good solvents for fatty material, whereas turpentine is used to dissolve paint.

The summation of the energy terms discussed in the previous paragraphs determines

TABLE 6-2 SOLUBILITY AT DIFFERENT TEMPERATURES

SUBSTANCE	GRAMS DISSOLVED BY 100 ml OF WATER AT		
	0°C	20°C	100°C
Potassium nitrate	13.3	31.6	246.0
Copper sulfate	14.3	20.7	75.4
Sodium chloride	35.7	36.0	39.8
Calcium hydroxide	0.185	0.165	0.077

whether the **enthalpy of solution,** $\Delta H_{sol'n}$, of a solute is positive or negative. For those solutes which undergo an endothermic dissolution, ΔH greater than zero, the solubility increases with increasing temperature. Solutes with exothermic dissolution have decreasing solubility with increasing temperature, but the solubility of most solutes increases with a rise in temperature. Table 6–2 illustrates the change in solubility of certain solutes at different temperatures.

The first two substances listed in the table show a definite increase in solubility as the temperature is raised, but the solubility of sodium chloride is only slightly affected by the change in temperature. The solubility of calcium hydroxide, on the other hand, decreases with a rise in temperature.

The solubility of gases in liquids and solids is decreased by a rise in temperature and is increased by an increase in pressure. In the preparation of carbonated drinks, large amounts of carbon dioxide are forced into solution by pressure at a low temperature. If a cold bottle of soda water is opened, the pressure is released and the gas escapes slowly from the solution, forming bubbles in the water. If a warm bottle is opened, the carbon dioxide escapes rapidly, causing foam to spurt out of the bottle.

CONCENTRATIONS OF SOLUTIONS

The **concentration** of a solution states the quantity of solute contained in a given amount of solution or solvent. When a solution contains a small amount of the solute, it is said to be **dilute;** when it contains a large amount, it is said to be **concentrated.**

It is often necessary to have a more quantitative expression of the concentration of a solution than that represented by the terms dilute, concentrated, saturated, or supersaturated. A common method of designating the concentration of an aqueous solution is based on **percentage by weight or by volume.** Unfortunately, there are so many ways to express percentage concentration that the term is often confusing. The three most commonly used are: (a) as grams of solute in 100 ml of solution, (b) as grams of solute in 100 g of solution, and (c) as ml of solute in 100 ml of solution. The weight-per-volume units, (a), are preferred by laboratories and industries where the solute is most important and the solvent is merely used as a vehicle for the solute. In analytical procedures the accurate measurement of a volume of the solution provides a definite weight of solute for a chemical reaction. The weight-per-weight units, (b), are preferred in many studies, because the concentration does not change with changes in temperature.

The volume-per-volume units, (c), are the least accurate and are often used when the solute is a liquid. For example, a 5 per cent solution of alcohol is prepared by diluting 5 ml of alcohol to a volume of 100 ml with water. This type of solution is often expressed as per cent by volume to distinguish it from (a) or (b).

A concentration expression which states the mass ratios of solutes to the total mass of the solution, but in units of moles, is called the **mole fraction (X).** The mole fraction of a component of a solution is the number of moles of that component divided by the total number of moles of all components. The following calculation illustrates the use of this concentration expression:

What is the mole fraction of H_2SO_4 in an aqueous solution which is 30% H_2SO_4 by weight? In 100 g of this solution there would be 30 g H_2SO_4 and 70 g H_2O. The number of moles of H_2SO_4 and H_2O respectively would be

$$30 \text{ g } H_2SO_4 \times \frac{1 \text{ mol } H_2SO_4}{98 \text{ g } H_2SO_4} = 0.306 \text{ mol } H_2SO_4$$

$$70 \text{ g } H_2O \times \frac{1 \text{ mol } H_2O}{18 \text{ g } H_2O} = 3.89 \text{ mol } H_2O$$

The mole fraction equals

$$\frac{\text{mol } H_2SO_4}{\text{mol } H_2SO_4 + \text{mol } H_2O} = \frac{0.306}{0.306 + 3.89} = 0.073$$

A commonly used mass per volume concentration expression is **molarity (M)**, which is the number of moles of solute per liter of solution. This method of expressing the concentration is convenient because the concentration bears a definite relationship to the molecular weight of the solute. A solution that contains 1 mole of the solute in 1 liter is called a **molar** solution. In calculating the amount of a compound that is used to prepare solutions of a given molarity, the following scheme may be used:

How much sulfuric acid, H_2SO_4, must be used to prepare 2 liters of 1M solution?

In this calculation, the quantity of solution is 2 liters and 1M is a conversion factor which is used as follows:

$$2\cancel{l} \times \frac{1 \cancel{\text{mol } H_2SO_4}}{\cancel{l}} \times \frac{98 \text{ g } H_2SO_4}{\cancel{\text{mol } H_2SO_4}} = 196 \text{ g } H_2SO_4$$

The last conversion factor necessitates determining the gram molecular weight of H_2SO_4.

By the use of molarity, it is possible to obtain any desired fraction of a mole of a substance by merely measuring a volume of the solution.

Another mass per mass concentration expression is **molality (m)**, which is the number of moles of solute per kilogram of solvent. For example a 1 **molal** solution contains 1 mole of the solute in 1 kg of solvent. For aqueous solutions, a comparison of a molal with a molar solution reveals two important differences. A molar solution, containing 1 mole of solute, has a final volume of 1000 ml, whereas the volume of the molal solution containing 1 mole of solute exceeds 1000 ml because the solute has been added to 1000g of solvent. Temperature changes affect the volume of a solution and will therefore slightly alter the molarity of a solution, but will not change the molality of a solution.

TITRATION

Titration is the process of adding a solution of one reactant to another reacting sample until the number of gram equivalent weights of the reactants are the same. If the concentration of the solution added is known, it is called a **standard solution,** and the quantity added can be used to determine the amount of the second reactant. The **equivalence point** of the titration is commonly detected using indicators which change color upon addition of excess standard solution. The most common indicator used in titrations of acids and bases is **red phenolphthalein,** which is red in basic solution and colorless in acidic solution. The following calculations illustrate the utility of data obtained in a titration.

What volume of 0.020M $Ca(OH)_2$ will react with 50.00 ml of 0.100M H_2SO_4?

The conversion sequence used in this calculation is volume of acid to moles of acid, moles of acid to moles of base, and moles of base to volume of base.

$$V_{(acid)} \quad \times \quad M_{(acid)} \quad \times \text{ Stoichiometry} \times \quad 1/M(\text{base})$$

$$50.00 \text{ ml } H_2SO_4 \times \frac{0.100 \text{ mol } H_2SO_4}{1000 \text{ ml } H_2SO_4} \times \frac{1 \text{ mol } Ca(OH)_2}{1 \text{ mol } H_2SO_4} \times \frac{1000 \text{ ml } Ca(OH)_2}{0.020 \text{ mol } Ca(OH)_2}$$

$$= 250 \text{ ml } Ca(OH)_2$$

What is the molarity of a $KMnO_4$ solution if 50.00 ml is required for the titration of 50.00 ml of 0.050M $K_2C_2O_4$ solution? The balanced reaction equation is:

$$2MnO_4^- + 5C_2O_4^{-2} + 16H^+ \rightarrow 2Mn^{+2} + 10CO_2 + 8H_2O$$

The starting point is the quantity of $K_2C_2O_4$ solution which is converted to moles. The ratio of the moles of $KMnO_4$ and $K_2C_2O_4$ at the equivalence point is obtained from the reaction stoichiometry. To obtain units of molarity, the number of moles of $KMnO_4$ must be divided by the volume involved.

$$V_{(red.)}/V_{(ox.)} \quad \times \quad M_{(red.)} \quad \times \text{ Stoichiometry} \times \text{ Volume conversion}$$

$$\frac{50.00 \text{ ml } K_2C_2O_4}{50.00 \text{ ml } KMnO_4} \times \frac{0.050 \text{ mol } K_2C_2O_4}{1000 \text{ ml } K_2C_2O_4} \times \frac{2 \text{ mol } KMnO_4}{5 \text{ mol } K_2C_2O_4} \times \frac{1000 \text{ ml } KMnO_4}{1 \text{ l } KMnO_4}$$

$$= \frac{0.02 \text{ mol } KMnO_4}{1 \text{ sol'n}} = 0.02M$$

PHYSICAL PROPERTIES OF SOLUTIONS

Vapor Pressure

The vapor pressure of a solution containing a nonvolatile solute is always less than that of the solvent. For nonelectrolyte solutes, the effect can be quantitatively predicted using **Raoult's law,** which states that the vapor pressure of a solution component will equal the mole fraction of that component multiplied by its vapor pressure in the pure state. Therefore, for the solvent, $P_{solv} = X_{solv} \times P_{solv}°$, where P_{solv} is the vapor pressure of the solvent in the solution, X_{solv} is the mole fraction of the solvent, and $P_{solv}°$ is the vapor pressure of the pure solvent. The vapor pressure of a solution is the sum of the vapor pressures of all of the components.

Many properties of solutions are quantitatively related to the mole fraction of the solute or the number of solute particles. These properties, called **colligative properties,** include, in addition to vapor pressure, boiling point, freezing point, and osmotic pressure.

Freezing and Boiling Points of a Solution

It can readily be shown that solutions of nonvolatile solutes freeze at lower temperatures and boil at higher temperatures than does the pure solvent. The **freezing point** is the temperature at which the vapor pressure of the solid and the liquid are the same, whereas the **boiling point** is the temperature at which the vapor pressure of the liquid reaches atmospheric pressure. The relation between the vapor pressure and the freezing and boiling points of the pure solvent water, and those of a 1 molal solution, is shown in Figure 6-7. For dilute solutions of nonvolatile solutes, the decrease in the freezing point and the increase in the boiling point are proportional to the molality. The dissocia-

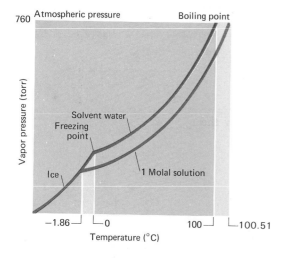

Figure 6–7 Relation of the vapor pressure, freezing points, and boiling points of water and a 1 molal solution.

tion of electrolytes in aqueous solution leads to more than one mole of solute particles per mole of solute so that colligative properties are larger than expected.

Practical use is made of the lowering of the freezing point of a solution. The antifreeze preparations used in automobile radiators are substances, such as methyl alcohol and the glycols, that lower the freezing point of the circulating water and prevent its freezing in the winter. When rock salt is sprinkled on ice, it forms a solution which freezes at a temperature below that of the ice, and therefore the ice melts.

OSMOTIC PRESSURE

Many plant or animal membranes are semipermeable in that they allow one component of a solution to pass through, while they hold back another component. The roots of a plant are covered with a semipermeable membrane that allows the passage of water into the plant but will not allow the substances in the sap to pass out into the ground. If the solutions on either side of a semipermeable membrane are unequal in concentration, there is a tendency to equalize the concentration.

The selective flow of a diffusible component through a membrane is called **osmosis.** The diffusible component, usually water, will tend to flow from the more dilute solution into the concentrated solution resulting in an increased pressure in the more concentrated solution. This pressure tends to force the solvent back out of the region of higher concentration, so that after a while an equilibrium is established by the balancing of the rate of outward and inward flow of solvent. When the solutions on each side of a semipermeable membrane have established equilibrium and have an equal concentration of components, they are said to be **isotonic.** This series of events is diagrammed in Figure 6–8. The pressure at which this equilibrium is reached is called the **osmotic pressure.** This is illustrated in Figure 6–9 where it is seen that a pressure (force per unit area) must

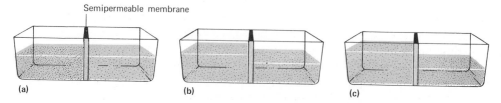

Figure 6–8 Osmosis. In *A* the solution on the left has a concentration twice that of the solution on the right. In *B* the solvent flows from right to left and in *C* the concentrations are equal.

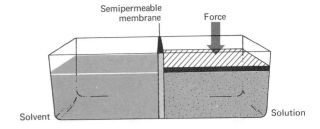

Figure 6-9   Pressure (force per unit area) applied to the right side stops flow through the semipermeable membrane.

be applied to the solution in order to counteract the flow of solvent across the semipermeable membrane. The osmotic pressure of a dilute solution is proportional to the molal concentration of the solution, and increases with the temperature. For example, the molal osmotic pressure for water at 0°C is about 22.4 atmospheres, compared to 24.4 atmospheres at 25°C.

SOLUTIONS OF ELECTROLYTES

As mentioned previously, substances whose solutions will conduct an electric current are called **electrolytes.** Those whose solutions will not conduct the current are **nonelectrolytes.** Solutions of electrolytes conduct an electric current because of the presence of ions in solution. In general, compounds other than acids, bases, and salts are classified as nonelectrolytes.

The following points summarize the currently accepted theory of electrolytes:

1. When an electrolyte is dissolved in water, it either dissociates into ions or reacts with water (hydrolyzes) to generate ions.

2. The sum of the positive charges that result from the dissociation of the electrolyte is equal to the sum of the negative charges.

3. Nonelectrolytes that fail to conduct an electric current when in solution do not dissociate to form ions.

4. Ions possess properties different from the corresponding uncharged atoms or molecules, and are responsible not only for the electrical properties but also for the chemical properties of a solution.

In general the dissociation of ionic compounds and dilute solutions of covalent strong electrolytes, such as hydrochloric acid, is nearly complete, whereas weak electrolytes and more concentrated electrolytes are only partially dissociated.

Many compounds, especially salts, are formed by electron transfer, and consist of ions even when they exist in the crystalline state. When they are dissolved in water, the forces holding them in the tightly packed crystalline state are overcome, and they gain freedom of movement. They then behave like ions in solution; for example, they will migrate under the influence of an electrical current. Covalent polar compounds, though not consisting of ions in the liquid or solid state, will often form ions by hydrolysis when dissolved in water, and are also classed as electrolytes.

Strong electrolytes are ionic compounds and hydrolyzable covalent polar compounds that dissociate completely into ions when in dilute solutions. **Weak electrolytes** are substances that dissociate only slightly into ions when in solution and exist essentially as undissociated molecules. Most salts, bases, and acids such as HCl, HNO_3, and H_2SO_4 are classed as strong electrolytes. Examples of weak electrolytes in water are the base ammonia and acids such as H_2CO_3, H_3BO_3, and acetic acid.

Hydrogen chloride exists as a gas and is a covalent polar compound. When added to water it undergoes hydrolysis to give the **hydronium ion,** H_3O^+, and the chloride ion, Cl^-, as indicated in the following equation:

$$HCl + H_2O \rightarrow H_3O^+ + Cl^-$$

In aqueous solutions discrete H^+ ions never exist, but are always hydrated.

The chemist is aware of the existence of hydrated hydrogen ions and also recognizes that many other ions exist in solution in a hydrated state. Since it is often difficult to determine the exact number of water molecules attached to an ion, and since even H^+ may exist with from one to three molecules of water of hydration, hydrogen and other ions are commonly written without their attached water molecules. The dissociation in aqueous solution of two strong acids and a weak acid may thus be represented as follows:

$$HCl \rightarrow H^+ + Cl^-$$
$$H_2SO_4 \rightarrow H^+ + HSO_4^-$$
$$CH_3CO_2H \rightleftharpoons H^+ + CH_3CO_2^-$$

The dissociation of the weak acid, acetic acid, is represented as a reversible reaction to indicate the existence of undissociated acetic acid molecules.

The common bases are metallic hydroxides, which are classified as ionic compounds since they exist as crystalline ionic solids with discrete metal ions and hydroxide ions. The dissociation in aqueous solutions of two strong bases is shown in the following equations:

$$NaOH \rightarrow Na^+ + OH^-$$
$$KOH \rightarrow K^+ + OH^-$$

The weakly basic properties of ammonia are explained by the following hydrolysis reaction:

$$NH_3 + H_2O \rightleftharpoons NH_4^+ + OH^-$$

As discussed previously, covalent strong electrolytes such as hydrochloric acid are completely dissociated in very dilute solution, but in more concentrated solutions, as a result of interionic attraction, they may not appear to be 100 per cent dissociated. In general, acids that dissociate almost completely into ions at ordinary concentrations are called **strong acids,** and those that dissociate to a small extent into ions are called **weak acids. Strong bases** are those that are highly dissociated in solution, and **weak bases** are only slightly dissociated and furnish relatively few hydroxide ions. Nearly all salts are almost completely dissociated into ions in solution and are usually classed as strong electrolytes. The degree of dissociation into ions or the extent of hydrolysis to give ions in dilute solutions is shown in Table 6–3 for some typical electrolytes.

TABLE 6–3   DISSOCIATION
OF TYPICAL ELECTROLYTES

	DISSOCIATION INTO IONS (PER CENT)
Hydrochloric acid	95.0
Nitric acid	92.0
Sulfuric acid	61.0
Acetic acid	1.3
Carbonic acid	0.17
Boric acid	0.01
Sodium hydroxide	91.0
Potassium hydroxide	91.0
Ammonium hydroxide	1.3
Most salts	70–100

TABLE 6-4   TYPES OF SOLUTIONS AND THEIR
PROPERTIES

TRUE SOLUTIONS	COLLOIDAL SOLUTIONS	SUSPENSIONS
1. Particle size less than 1 nm	Particle size 1 nm to 100 nm	Particle size 100 nm or more
2. Invisible	Visible only in ultra- or electron microscope	Visible to naked eye
3. Will pass through filters and membranes	Will pass through filters but not membranes	Will not pass through filters
4. Possesses molecular movement	Exhibits Brownian movement	Moves only by force of gravity

COLLOIDS

All the solutions described in previous chapters have consisted of soluble solutes dissolved in a solvent. Aqueous solutions have been stressed, in which the solutes have broken up into either molecules or ions that formed homogeneous mixtures with the solvent molecules. These solutions, whose particles are of molecular or ionic dimensions and will not settle out on standing, are called **true solutions.** If finely divided clay is shaken with water and allowed to stand, the particles will slowly settle to the bottom. The clay particles are insoluble in the solvent and can be seen by the naked eye. Such a mixture is called a **suspension.**

Some mixtures of insoluble solutes have particles so small that they will not settle out of the solvent on standing, and cannot be seen by the naked eye. These substances, whose particles are intermediate in size between those in true solutions and those in suspensions, are called **colloids.**

The particles in the three types of solutions just discussed vary from atomic or molecular size in the true solution to large, visible particles in the suspension. The sizes of the particles in a colloidal solution are usually considered to vary from 1 to 100 nanometers (nm) or 10^{-9} meters in diameter. The relations of these three types of solutions and some of their properties are shown in Table 6-4.

In colloid chemistry the particles are called the dispersed phase, and the fluid in which they are dispersed is called the dispersion medium. The correct term for a colloidal solution is therefore a colloidal dispersion, although the term "solution" is commonly used.

TYPES OF COLLOIDS

Colloidal solutions may be classified by the same system employed for true solutions. Instead of solute and solvent, the particles are called the dispersed or discontinuous phase, and the solvent in which they are dispersed is called the dispersion medium or continuous phase. Theoretically, there are nine possible types of colloidal solutions, but since a gas in a gas system would not produce a colloid, we shall consider only eight, as shown in Table 6-5. A colloidal solution is called a **suspensoid** if the dispersed phase is a solid, an **emulsoid** if the dispersed phase is a liquid.

A general classification of colloidal systems depends on the physical state of the dispersed phase relative to the dispersion medium. The most common systems are known as sols, gels, aerosols, and emulsions. Both sols and gels are formed by the dispersion of solids, such as proteins, starch, and soaps, in an aqueous medium. A **sol** is a colloidal solution that is liquid at room temperature, with water as the continuous phase and the solid as the discontinuous phase. A **gel** is a similar colloidal dispersion that is a solid at

TABLE 6–5   TYPES OF COLLOIDAL SYSTEMS

DISPERSED, OR DISCONTINUOUS, PHASE	DISPERSION MEDIUM, OR CONTINUOUS, PHASE	EXAMPLES
Solid	Solid	Gems, ruby glass
Solid	Liquid	Plasma, inks, gold sols, jello
Solid	Gas	Smoke, dust clouds
Liquid	Solid	Pearls, opals
Liquid	Liquid	Homogenized milk, mayonnaise
Liquid	Gas	Fog, sprays, mists
Gas	Solid	Pumice, meerschaum
Gas	Liquid	Foams, meringue, whipped cream

ordinary temperatures. Both the liquid and solid phases are continuous, with the solid forming a network of filaments, or a "brush heap" type of structure. An **aerosol** is a colloidal system in which either a solid or a liquid is dispersed in a gas. Smoke is an example of an aerosol and is a colloidal dispersion of solid ash in air. An **emulsion** consists of a liquid dispersed in a liquid. A common example of an emulsion is milk, in which butter fat is dispersed in an aqueous solution.

Another method of classification of colloidal systems is based on the attraction or affinity between the dispersed particles and the dispersion medium. If the mutual affinity is great, the system is called **lyophilic;** but if the mutual affinity is small, the system is **lyophobic.** In common colloidal systems in which water is the dispersion medium these terms become **hydrophilic** and **hydrophobic,** or literally, "water loving" and "water hating."

PROPERTIES OF COLLOIDS

Movement.   The molecules in a true solution are in a state of constant rapid motion. Both solute and solvent molecules exhibit this **molecular motion.** Since colloidal particles are composed of an aggregate of many molecules, the movement of the particles is very slow compared with that of an individual molecule. Apparently the major motion in a colloidal dispersion is caused by the bombardment of the particles by the molecules of the dispersion medium. This erratic movement of colloidal particles was first observed under an ultramicroscope by Robert Brown and is known as **Brownian movement** (Fig. 6–10).

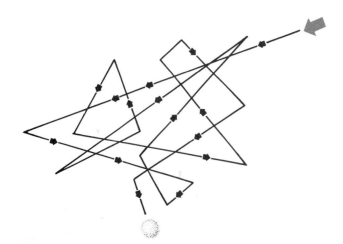

Figure 6–10   Erratic random movement of colloidal particles known as Brownian movement.

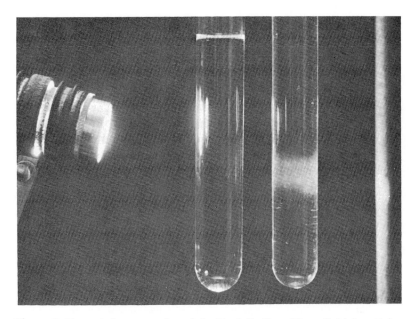

Figure 6–11   A demonstration of the Tyndall effect. The colloidal particles in the tube on the right reflect the beam of light and make its path visible. The tube at left contains a true solution.

If a strong beam of light is passed through a colloidal solution, the path of the beam is clearly outlined because of the reflection of the light from the surfaces of the moving colloidal particles (Fig. 6–11). This phenomenon is called the **Tyndall effect;** it may be used as a simple test to distinguish between true solutions and colloidal solutions. A similar effect is observed when a bright ray of sunlight enters a darkened dusty room, which is really an aerosol. The path of light is clearly outlined by the reflection from the surfaces of the dust particles. The scattering of light by dust and water particles in the earth's atmosphere accounts for the red color of the sun and sky at sunrise and sunset. If a vessel containing a colloidal solution is placed on the stage of a microscope and a strong beam of light passes through the vessel at right angles to the optical path, the reflection of light from the moving colloidal particles may be observed through the eyepiece. With an **ultramicroscope** arranged in this fashion, it is possible to observe the light reflected from the individual particles even though the particles themselves cannot be seen.

IMPORTANT TERMS AND CONCEPTS

colligative properties	hydrogen bonds	Raoult's law	solvent
colloids	molality	saturated solution	specific gravity
deuterium	molarity	solubility	theory of electrolytes
equivalence point	osmotic pressure	solute	titration
hydrates			

QUESTIONS

1. What is the basis of the Celsius or centigrade thermometer?

2. What is meant by the statement that "water is a polar molecule"?

3. Define and give an example of hydrolysis.

4. Write the formulas for and explain the difference between anhydrous and crystalline copper sulfate.

5. (a) How can water be purified in an emergency?
   (b) How is water treated to remove bacteria and dissolved mineral matter?

6. How is water ordinarily purified for use by the residents of a city?

7. Describe a method for softening (a) temporary hardness in water and (b) permanent hardness in water.

8. Place after the solution on the left the letter of each statement that correctly applies to it.

   Colloidal Solution

   True Solution

   Suspension

   a. Will pass through a membrane.
   b. Particles are invisible.
   c. Particles are less than 100 nm in diameter.
   d. Will not pass through filter paper.
   e. Exhibits Tyndall effect in strong beam of light.
   f. Particles will settle out on standing.

9. What is an electrolyte? a nonelectrolyte? Give examples of each.

10. How do the colligative properties of electrolytes compare with those of nonelectrolytes? Explain.

11. How many grams of glucose are there in 700 ml of a 10 weight per volume per cent solution?

12. How many grams of K_3PO_4 would be used to prepare 700 ml of a 2M solution?

13. Explain the process of osmosis. What is meant by the term "isotonic"?

14. Write the equation for the dissociation of (a) a strong acid and (b) a weak acid. Explain the difference between them.

15. Give an example of a solution of: (a) a gas dissolved in a liquid, (b) a solid dissolved in a liquid, and (c) a liquid dissolved in a liquid.

16. What are the main factors that affect the solubility of a solid solute?

17. For solutes which undergo an endothermic dissolution, does the solubility increase or decrease with increasing temperature?

18. What is the mole fraction of $KClO_3$ in an aqueous solution which is 25% $KClO_3$ by weight?

19. What is the molarity of a solution prepared by adding enough water to 194 g of H_3PO_4 to prepare 600 ml of solution?

20. A volume of 46.00 ml of 0.200M H_2SO_4 solution is required for the titration of 32.00 ml of a $Ca(OH)_2$ solution. What is the molarity of the $Ca(OH)_2$ solution?

21. What is the molarity of a $Na_2C_2O_4$ solution if 11.3 ml of this solution reacts with 47.4 ml of 0.05M $KMnO_4$ solution? (See reaction equation in text.)

22. What is the most important difference between solutions and colloidal dispersions? Explain.

23. What is the "Tyndall effect"? Why does a true solution fail to exhibit this phenomenon?

SUGGESTED READING

Boyd: The Wonder of Water. Chemistry, Vol. 47, No. 6, p. 6, 1974.

Hall: Water. Chemistry, Vol. 44, No. 8, p. 6, 1971.

Jensen: Lewis Acid-Base Theory: Part I. Development. Chemistry, Vol. 47, No. 3, p. 11, 1974.

Jensen: Lewis Acid-Base Theory: Part II. Development. Chemistry, Vol. 47, No. 4, p. 13, 1974.

Keller: What Is Happening To Our Drinking Water? Chemistry, Vol. 48, No. 2, p. 16, 1975.

Kettani: The Control of the Water Cycle. Scientific American, Vol. 228, No. 4, p. 46, 1973.

Morris: Stress, Collisions and Constants. Part I. Solutions. Chemistry, Vol. 44, No. 4, p. 10, 1971.

Othmer: Water and Life. Chemistry, Vol. 43, No. 10, p. 12, 1970.

Penman: The Water Cycle. Scientific American, Vol. 223, No. 3, p. 54, 1970.

Robbins: Ionic Reactions and Equilibria. New York, The Macmillan Co., 1967.

Webb: Hydrogen Bond, "Special Agent." Chemistry, Vol. 41, No. 6, p. 16, 1968.

REACTION RATES AND CHEMICAL EQUILIBRIUM

CHAPTER 7 ──────────────────────

The *objectives* of this chapter are to enable the student to:

1. State the factors which influence reaction rates and explain why for each.
2. Explain the significance of a rate law.
3. Define activation energy.
4. Describe the conditions which prevail when a system is in chemical equilibrium.
5. Apply Le Chatelier's principle to predict the influence of changing conditions for a system at chemical equilibrium.
6. Use the ion product constant for water to determine $[H^+]$ and $[OH^-]$ given the concentration of only one.
7. Calculate pH and pOH of an aqueous solution given $[H^+]$ or $[OH^-]$.
8. Predict whether a given salt hydrolyzes to give a pH greater or less than 7.
9. Explain the common ion effect.
10. Differentiate between an electrolytic cell and a voltaic cell.
11. Apply Faraday's law to predicting the amount of chemical change accompanying the passage of electrical current.
12. Describe the anode and cathode reactions in the lead storage cell when it is charging or discharging.

The nature of dynamic equilibrium in systems undergoing physical changes was discussed in Chapter 4. In such equilibria, it was pointed out that two opposing processes were proceeding at the same rate, so that there was no net change in the system. Many chemical reactions are also reversible processes, in that the products formed can react to regenerate the reactants. When the rates of the opposing reactions are the same, no net changes in chemical composition take place, and the chemical system is in a state of **dynamic chemical equilibrium.**

The composition of a chemical system at equilibrium is determined by the energetics of the opposing reactions. If a reaction has a high tendency to proceed, it will be reflected by the composition of the system when equilibrium prevails. In such a system, the concentration of the products will be greater than that of the reactants.

Although consideration of the energetics of a chemical reaction allows prediction of whether or not the reaction will proceed, it does not allow any prediction of how fast the reaction will proceed, or in what way it will proceed. The study of rates of chemical reactions and the actual steps involved in converting reactants to products is called **chemical kinetics.**

RATES OF REACTIONS

The rate of a reaction refers to the amount of reactant consumed or product generated in a given amount of time. The easiest way to express the rate is to state the change in concentration per unit time (i.e., concentration change/time). The rate of a reaction is determined experimentally by periodic measurements of the concentration of the reacting species or the products.

It has also been established experimentally that the primary factors influencing rates include **concentration of reactants, temperature, nature of the reactants,** and **catalysts.** In order to understand these factors, the actual step by step procedure of a reaction will be considered. The step by step description of a reaction is called the **reaction mechanism.**

The Rate Law

Reaction rates are influenced by how often reactant atoms or molecules collide. As the concentration of reactant increases, the number of collisions per unit time increases, and, as a result, the reaction rate increases. This simple model is consistent with experimental observations which indicate that reaction rates increase with increasing concentration of the reactants.

For *any reaction involving a single step,* as opposed to an overall reaction involving a number of steps, the rate is directly proportional to the concentrations of the reactants. Consider the general reaction

$$A + B \rightleftharpoons C + D$$

The rate of forward reaction, $\underset{\rightarrow}{R}$, is related to the concentration of A and B as follows:

$$\underset{\rightarrow}{R} \alpha [A][B]$$

where [A] and [B] are the concentrations of A and B. The proportionality constant which must be applied to convert this proportion to an equality is called the **rate constant.** The statement of the relationship between rate and concentration is called the **rate law.**

$$\underset{\rightarrow}{R} = k[A][B]$$

For many reactions which proceed by a single step, the form of the rate law can be predicted by considering the coefficients on the reactants in the balanced equation. This is seen in the following reaction and its associated rate law:

$$2NO_2 \rightarrow 2NO + O_2$$
$$\underset{\rightarrow}{R} = k[NO_2][NO_2] = k[NO_2]^2$$

Many chemical reactions do not occur in a single step, but may proceed through two or more steps, as illustrated by the following general reaction:

$$A + B \rightarrow C \text{ step 1}$$
$$\underline{C + A \rightarrow D \text{ step 2}}$$
$$A_2 + B \rightarrow D \text{ overall reaction}$$

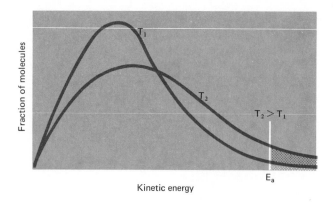

Figure 7–1   The distribution of molecular kinetic energies at two temperatures.

Rates of chemical reactions increase with increasing temperature, suggesting that molecules must have a certain amount of energy before chemical reactions will take place. Apparently the energy is furnished in the form of thermal energy as the temperature of the reacting system is increased.

Increasing the temperature increases both the number of collisions per unit time and the fraction of collisions which results in chemical reaction. At a given temperature there is a distribution of the kinetic energies of A and B, so that there will be some A and some B which possess the sufficient energy, E_a, the **activation energy,** to give chemical reaction. As shown in Figure 7–1 an increase in temperature increases the fraction of A and B having energy in excess of the activation energy. With increasing temperature, the number of effective A and B collisions increases, and consequently, the rate of reaction increases. It is also true that reactions having lower activation energies proceed faster at a given temperature than those having higher activation energies.

Occasionally an extra component added to a reaction system will take part in the reaction and result in an increased rate, but it will not be incorporated in the products. This component increases the rate of the reaction by furnishing an alternate reaction mechanism with lower activation energy. The phenomenon is called **catalysis** and the components which accomplish this are called **catalysts.**

The phenomenon of catalysis is basic to life. In animals, catalysts called **enzymes** play special roles in metabolism. These compounds and their catalytic role will be discussed in the biochemistry section of this book.

CHEMICAL EQUILIBRIUM

A reversible chemical reaction has a characteristic rate for the forward reaction and another for the reverse. For the reaction

$$A + B \rightleftharpoons C + D$$

the rate of the forward reaction will depend in some way on the concentrations of A and B, whereas the rate of the reverse reaction will depend in some way on the concentrations of C and D. If one performs an experiment in which A and B are mixed, and the rate of reaction is observed, it is found that the fastest rate prevails at the instant of mixing. The rate of disappearance of A and B decreases with time until eventually the concentrations of these reactants are constant. If, on mixing A and B, the concentrations of C and D are observed, it is found that they increase with the greatest rate at the moment of mixing. The rate of concentration increase slows with time, and eventually the concentration remains constant. The concentrations of A, B, C, and D become invariant with time simultaneously. When the concentrations of all species are invariant with time, chemical equilibrium prevails. The results of this hypothetical experiment are plotted in Figure 7–2.

If a reversible chemical reaction is at chemical equilibrium, the rates of the opposing reactions are equal. For the simple case where both the forward and reverse reactions have a one-step mechanism, the rate laws are

$$\underset{\rightarrow}{R} = k[A][B] \qquad \underset{\leftarrow}{R} = k'[C][D]$$

At chemical equilibrium these rates are equal, so that

$$k[A][B] = k'[C][D]$$
$$\frac{k}{k'} = \frac{[C][D]}{[A][B]}$$

At a given temperature k and k' do not change, and k/k' does not change. The ratio of the rate constants is called the **equilibrium constant,** K.

$$K = \frac{[C][D]}{[A][B]}$$

This treatment claims that, at a given temperature for a given reaction, the product of the concentrations of C and D divided by the product of the concentrations of A and B is a constant.

It can be shown that the results of the preceding treatment are valid regardless of the mechanism of the reaction. For the general reversible reaction

$$aA + bB + cC + \cdots \rightleftharpoons zZ + yY + xX + \cdots$$

the composition of the system at equilibrium must obey the following:

$$K = \frac{[Z]^z[Y]^y[X]^x \cdots}{[A]^a[B]^b[C]^c \cdots}$$

This expression is the **law of chemical equilibrium** in equation form. For a given reversible reaction system, the composition of the system at equilibrium must satisfy the conditions imposed by the equation. By convention, the equilibrium constant is written with the

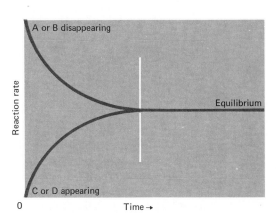

Figure 7–2 A chemical reaction attaining chemical equilibrium.

components on the right hand side of the reaction equation in the numerator. The equilibrium constant for the reaction written in the reverse direction is 1/K.

Equilibrium Constants

Equilibrium constants are fundamental characteristics of chemical reactions, in that they are dependent on the nature of the products and the reactants. They are also dependent on the temperature, but are independent of the concentration and the size of the system. The magnitude of the equilibrium constant indicates the extent to which a reaction proceeds. Larger equilibrium constants are associated with those reactions which proceed more nearly to completion. The following sample calculation illustrates the variation of K with extent of reaction.

At 50°C a mole of N_2O_4 gas in a 5 liter container is 15% dissociated, according to the equation

$$N_2O_4(g) \rightleftharpoons 2NO_2(g)$$

What is the equilibrium constant?

The equilibrium constant expression is $K = \dfrac{[NO_2]^2}{[N_2O_4]}$.

The first step in this problem is to determine the concentration of reactant and product at equilibrium. Before any N_2O_4 dissociates, its concentration is 1 mole/5 liters. If 15% is dissociated at equilibrium, then the concentration is

$$1 \text{ mol/5 liters} - 0.15 \text{ mol/5 liters} = \frac{0.85 \text{ mol}}{5 \text{ liters}}$$

The coefficients in the reaction equation indicate that 2 moles of NO_2 are formed from each mole of dissociated N_2O_4. If 0.15 mole of N_2O_4 dissociates, 0.30 mole of NO_2 is formed, and the concentration of NO_2 is 0.30 mole/5 liters. Substituting into the equilibrium expression

$$K = \frac{(0.30 \text{ mol/5 liters})^2}{(0.85 \text{ mol/5 liters})} = 0.021 \text{ mol/liter}$$

For this reaction which proceeds only 15%, the equilibrium constant is much less than one. It is seen in solving this problem that equilibrium constants occasionally have units. In fact, K will have units whenever the sum of the concentration exponents in the numerator does not equal that in the denominator. This situation prevails whenever there is a change in the number of moles going from reactants to products.

Le Chatelier's Principle

If a system is at chemical equilibrium, the composition of the system can be altered by certain perturbing influences. These include varying the temperature, changing the concentration of some component in the system, or, for systems of gaseous reactants, changing the total pressure of the system. In the case of gaseous systems, the total pressure may be changed by changing the volume. The general effects of perturbing influences are summarized in **Le Chatelier's principle,** which states that *a system at equilibrium will react to an applied stress so as to remove that stress.*

As an illustration of Le Chatelier's principle, consider the reaction used in the Haber process to make ammonia,

$$N_2(g) + 3H_2(g) \rightleftharpoons 2NH_3(g)$$

which proceeds with a decrease in the total number of moles and is exothermic. If N_2 is added to the system at equilibrium, $[N_2]$ is instantaneously increased, so that the reaction proceeding to the right progresses at a greater rate. At the moment N_2 is added, the rate of the reaction proceeding to the left is unaltered, since $[NH_3]$ has not been changed. If the reaction consuming N_2 proceeds for a short while at a rate greater than the reaction generating N_2, then the equilibrium "shifts." The concentration of NH_3 increases, and the concentration of H_2 decreases. A new equilibrium composition arises within the limits imposed by K, the equilibrium constant. In general if a component is added to a system at equilibrium, a reaction will take place to consume that component.

Increasing the temperature of the above system imposes a stress which can also be removed by shifting the equilibrium. The reaction producing NH_3 proceeds with the evolution of heat; it is exothermic. The reverse reaction is endothermic and consumes heat. Increasing the temperature of the system at equilibrium favors the endothermic reaction, which consumes the added heat, and the equilibrium is shifted so that the concentration of NH_3 decreases. In general for exothermic reactions, temperature increases decrease K, whereas for endothermic reactions, K increases with increasing temperature.

An increase in the total pressure of the above system imposes a stress which can be relieved by a reaction which decreases the total number of moles of gaseous material present, and thereby decreases the pressure. The reaction proceeding to the right to form NH_3 would relieve this stress. In general, for gaseous systems at equilibrium, the application of pressure will induce a net reaction which will decrease the pressure. Pressure has no effect on gaseous systems such as the following, in which the reaction involves no change in the number of moles of gaseous material.

$$H_2(g) + Cl_2(g) \rightleftharpoons 2HCl(g)$$

IONIZATION EQUILIBRIA

The Arrhenius theory of electrolytes, discussed in Chapter 5, qualitatively described the properties of certain solutes in aqueous solutions. The behavior of strong and weak electrolytes was explained by considering the extent of the dissociation of solute molecules into ions. For weak electrolytes, the partial dissociation can be treated as a chemical equilibrium of the following type:

$$MX \rightleftharpoons M^+ + X^-$$

For this equilibrium between the molecule MX and the ions M^+ and X^-, the **ionization or dissociation equilibrium constant, K_i,** is

$$K_i = \frac{[M^+][X^-]}{[MX]}$$

Almost all solutes in aqueous solution are either directly involved in, or at least influence, ionic equilibria.

Even in the absence of solutes, an ionic equilibrium prevails in water. This equilibrium involves the slight dissociation of water, which is represented in the following net equation:

$$H_2O \rightleftharpoons H^+ + OH^-$$

It must be emphasized that this reaction equation is a net ionic equation, and the H^+ and the OH^- are each associated with a number of water molecules by virtue of extensive hydrogen bonding. The H^+ is known to have at least one water molecule associated, so that it is usually represented as H_3O^+, **the hydronium ion.** The ionization equilibrium constant for this dissociation reaction is

$$K_i = \frac{[H^+][OH^-]}{[H_2O]}$$

Since the concentration of water molecules is constant at a given temperature (i.e., 55.55M at 4°C), the denominator of this expression is incorporated along with K_i into a new constant, **K_w, the ion product constant for water,** as follows:

$$K_i \cdot [H_2O] = K_w = [H^+][OH^-]$$

For water at 25°C, it is found experimentally that $[H^+] = [OH^-] = 10^{-7}$ moles/liter, so that

$$K_w = [H^+][OH^-] = [10^{-7}][10^{-7}] = 10^{-14}$$

The magnitude of K_w indicates the very small extent of the dissociation of water.

Even though the value of K_w was determined by studying pure water at 25°C, this product of $[H^+]$ and $[OH^-]$ is constant in the presence of solutes. For example, if the $[H^+]$ is increased by the addition of a solute, the $[OH^-]$ must decrease until $[H^+][OH^-] = 10^{-14}$. The decrease in $[OH^-]$ on the addition of H^+ is predicted by considering Le Chatelier's principle. The addition of H^+ places a stress upon the equilibrium system, and the stress is removed by a shift which consumes OH^-. The following calculation illustrates the quantitative treatment of this phenomenon:

It is observed that on the addition of 1×10^{-5} moles of gaseous HCl to 1 liter of pure water, the $[H^+]$ increases from 10^{-7}M to 1×10^{-5}M. What is the $[OH^-]$?

Recalling that $K_w = 10^{-14} = [H^+][OH^-]$, and solving for $[OH^-]$ gives

$$[1 \times 10^{-5}][OH^-] = 10^{-14}$$

$$[OH^-] = 10^{-14}/10^{-5} = 10^{-9}M$$

The answer indicates that while the $[H^+]$ increased, $[OH^-]$ decreased.

pH, pOH, pK

Chemists commonly use an alternate way of expressing $[H^+]$, $[OH^-]$, and K for various equilibria. This alternate expression involves the use of the negative logarithm of the quantity in question. The **pH** of an aqueous solution is defined as **the negative logarithm of $[H^+]$.** Thus,

$$pH = -\log[H^+] = \log 1/[H^+]$$

and $\quad [H^+] = 10^{-pH} = 1/10^{pH} = antilog(-pH)$

The **pOH** of an aqueous solution is defined as **the negative logarithm of $[OH^-]$**. Thus,

$$pOH = -\log[OH^-] = \log 1/[OH^-]$$

and $\quad [OH^-] = 10^{-pOH} = 1/10^{pOH} = antilog(-pOH)$

A similarly defined quantity **pK** is related to equilibrium constants. Thus, the ion product constant for water, K_w, has an associated **pK_w** defined as

$$pK_w = -\log K_w = \log 1/K_w$$

and $\quad K_w = 10^{-pK_w} = 1/10^{pK_w} = antilog(-pK_w)$

For the expression $K_w = [H^+][OH^-] = 10^{-14}$, taking the logarithm of each term and multiplying through by -1 gives

$$-\log K_w = -\log[H^+] - \log[OH^-] = -\log 10^{-14}$$

Substituting the definitions of pK, pH, and pOH into this expression gives

$$pK_w = pH + pOH = -\log 10^{-14}$$

The logarithm of 10^{-14} is -14, and thus,

$$pK_w = pH + pOH = 14$$

This expression leads to the conclusion that for an aqueous solution at 25°C the sum of pH and pOH must always be 14. The following sample calculations illustrate the utility of pH and pOH.

> What are the pH and the pOH of an aqueous solution containing 10^{-4}M HCl?
>
> $$pH = -\log[H^+] = -\log 10^{-4}$$
> $$= -(-4) = 4$$
>
> Since pH + pOH = 14, pOH = 14 − 4 = 10.

In this solution the $[H^+]$ is greater than that in pure water, 10^{-4}M as compared to 10^{-7}M, but the pH of the solution is smaller than that in pure water, 4 as compared to 7. Aqueous solutions in which the $[H^+]$ is greater than 10^{-7}M, and in which the pH is less than 7, are called **acidic solutions.** Aqueous solutions in which the $[H^+]$ is less than 10^{-7}M, and thus the $[OH^-]$ is greater than 10^{-7}M, have a pH which is greater than 7 and a pOH less than 7. Such solutions contain a preponderance of OH^- and are called **basic solutions. Neutral solutions** have a $[H^+] = [OH^-] = 10^{-7}$M, and a pH = 7.

> What are the $[H^+]$ and the $[OH^-]$ for a solution with a pH of 9? From the definition previously stated
>
> $$[H^+] = 10^{-pH} = antilog(-pH)$$

TABLE 7-1   pH AND pOH OF
AQUEOUS SOLUTIONS AT 25°C

$[H^+]$ mol/l	pH		$[OH^-]$ mol/l	pOH
1	0		10^{-14}	14
10^{-1}	1		10^{-13}	13
10^{-2}	2		10^{-12}	12
10^{-3}	3	acidic	10^{-11}	11
10^{-4}	4		10^{-10}	10
10^{-5}	5		10^{-9}	9
10^{-6}	6		10^{-8}	8
10^{-7}	7	neutral	10^{-7}	7
10^{-8}	8		10^{-6}	6
10^{-9}	9		10^{-5}	5
10^{-10}	10		10^{-4}	4
10^{-11}	11	basic	10^{-3}	3
10^{-12}	12		10^{-2}	2
10^{-13}	13		10^{-1}	1
10^{-14}	14		1	0

Thus, $[H^+] = 10^{-9}$M. Since pH + pOH = 14, pOH is 5, and $[OH^-] = 10^{-pOH} = 10^{-5}$M. The values of pH and pOH at various H^+ and OH^- are shown in Table 7-1. The usual pH ranges of several common liquids are listed in Table 7-2.

DISSOCIATION OF ACIDS AND BASES

In Chapter 6 it was indicated that acids and bases are electrolytes, but that many of these solutes behave as weak electrolytes and are therefore only partially dissociated into ions. The extent of the dissociation, which is dependent on the nature of the acid or base, can be treated as an ionic equilibrium.

Acetic acid, CH_3CO_2H, the acidic component of vinegar, is only partially dissociated in aqueous solution and is referred to as a weak acid. The following equation represents the dissociation of acetic acid in water to give hydronium ions and acetate ions.

$$CH_3CO_2H + H_2O \rightleftharpoons H_3O^+ + CH_3CO_2^-$$

TABLE 7-2   pH RANGES
OF COMMON LIQUIDS

LIQUID	pH RANGE
Human gastric juices	1.0–3.0
Lime juice	1.8–2.0
Lemon juice	2.2–2.4
Vinegar	2.4–3.4
Carbonated drinks	2.0–4.0
Orange juice	3.0–4.0
Tomato juice	4.0–4.4
Beer	4.0–5.0
Cow's milk	6.3–6.6
Human blood	7.3–7.5
Sea water	7.8–8.3
Household ammonia (1–5%)	10.5–11.5

The equilibrium constant for this reaction is

$$K = \frac{[H_3O^+][CH_3CO_2^-]}{[H_2O][CH_3CO_2H]}$$

For dilute solutions where the concentration of water is a constant, $[H_2O]$ is incorporated along with K into a new constant, K_a, the **acid dissociation constant**, as follows:

$$K \cdot [H_2O] = K_a = \frac{[H^+][CH_3CO_2^-]}{[CH_3CO_2H]}$$

In this expression $[H^+]$ has been substituted for $[H_3O^+]$, since the net reaction is

$$CH_3CO_2H \rightleftharpoons H^+ + CH_3CO_2^-$$

At 25°C, K_a for acetic acid is 1.8×10^{-5} and pK_a is 4.75.

The acid dissociation constants of several common acids are given in Table 7–3. **Polyprotic acids,** which yield more than one proton per acid molecule, have an acid dissociation constant for each dissociation step.

In theory the dissociation of weak hydroxy bases can be treated in the same way as weak acids. For the general weak base MOH, the net equation for the dissociation is

$$MOH \rightleftharpoons M^+ + OH^-$$

In reality there are few weak bases of the general formula MOH which have been well characterized in aqueous solutions. There are some with more than one hydroxide per mole of base, but the dissociation equilibria for these are complicated. There are, however, a large number of well characterized molecules which, when placed in water, undergo a hydrolysis reaction to generate OH^-. Among these compounds are ammonia, NH_3, and organic amines.

TABLE 7–3 ACID AND BASE DISSOCIATION CONSTANTS AT 25°C

ACID	K_a		pK_a
HCl		∞	—
HF		6.7×10^{-4}	3.2
HCN		4×10^{-10}	9.4
CH_3CO_2H		1.8×10^{-5}	4.7
HNO_2		4.5×10^{-4}	3.3
H_2SO_4	(K_{a1})	∞	—
	(K_{a2})	1.3×10^{-2}	1.9
H_2CO_3	(K_{a1})	4.2×10^{-7}	6.4
	(K_{a2})	4.7×10^{-11}	10.3
H_2S	(K_{a1})	1×10^{-7}	7.0
	(K_{a2})	1×10^{-15}	15.0
H_3PO_4	(K_{a1})	7.1×10^{-3}	2.2
	(K_{a2})	6.3×10^{-8}	7.2
	(K_{a3})	4×10^{-13}	12.4

BASE	K_b	pK_b
NaOH	∞	—
NH_3	1.8×10^{-5}	4.7

Gaseous ammonia, when dissolved in water, undergoes a hydrolysis reaction to give a basic solution. The hydrolysis reaction may be represented as follows:

$$NH_3 + H_2O \rightleftharpoons NH_4^+ + OH^-$$

The equilibrium constant for this reaction is

$$K = \frac{[NH_4^+][OH^-]}{[NH_3][H_2O]}$$

If the term $[H_2O]$, a constant in dilute solutions, is incorporated along with the equilibrium constant into a new constant K_b, the following expression is obtained:

$$K \cdot [H_2O] = K_b = \frac{[NH_4^+][OH^-]}{[NH_3]}$$

The constant K_b is the **base dissociation constant**. At $25°C$, K_b for ammonia is 1.8×10^{-5}, and $pK_b = 4.75$.

The extent to which an acid or a base dissociates in aqueous solutions is reflected in K_a or K_b, respectively. With increasing strength of acids or bases, the extent of dissociation increases, and K_a and K_b are larger. Correspondingly, the smallest pK_a and pK_b will be associated with the strongest acids and bases.

HYDROLYSIS OF SALTS

Many salts, when placed in water, react with the solvent molecules to give acidic or basic solutions. The net reaction for the hydrolysis of a positive ion, or cation, is

$$M^+ + H_2O \rightleftharpoons MOH + H^+$$

This reaction proceeds furthest to the right when M^+ has a very high affinity for hydroxide, and therefore when MOH is a weak base. The net reaction for the hydrolysis of a negative ion, or anion, is

$$A^- + H_2O \rightleftharpoons HA + OH^-$$

This reaction proceeds furthest to the right when HA is a weak acid. From the reactions above, it can be stated that cations of weak bases hydrolyze to give acidic solutions, and anions of weak acids hydrolyze to give basic solutions.

The cations of strong bases (e.g., Na^+, K^+) and the anions of strong acids (e.g., Cl^-, NO_3^-) do not hydrolyze, and therefore do not influence $[H^+]$ or $[OH^-]$. For this reason salts composed of cations of strong bases and anions of strong acids dissolve in water without any net hydrolysis. Consequently, aqueous solutions of salts, such as NaCl, KCl, $NaNO_3$, KNO_3, and so forth, are neutral; that is, $pH = 7$.

Any salt which contains either the cation of a weak base or the anion of a weak acid will, when dissolving in water, undergo a hydrolysis reaction. This reaction will result in a solution which is not neutral. The salts of strong bases and weak acids, such as $NaCH_3CO_2$ and KCN, always give basic solutions. Salts of weak bases and strong acids, such as NH_4Cl, give acidic solutions. A salt composed of the cation of a weak base and the anion of a weak acid, such as $NH_4CH_3CO_2$ or NH_4CN, are neutral, acidic, or basic depending on the relative extents of hydrolysis of the cation and the anion.

THE COMMON ION EFFECT

The extent of dissociation of a weak acid or a weak base in aqueous solution can be altered by the addition of appropriate salts to the solution. The extent of dissociation of acetic acid can be decreased by adding sodium acetate, $NaCH_3CO_2$. This phenomenon is predicted by applying Le Chatelier's principle to the dissociation equilibrium:

$$CH_3CO_2H \rightleftharpoons H^+ + CH_3CO_2^-$$

The addition of sodium acetate to an acetic acid solution at equilibrium instantaneously increases the acetate ion concentration, resulting in a stress on the equilibrium. This stress is removed by a shift of the equilibrium to the left, that is, H^+ and $CH_3CO_2^-$ combine to form undissociated CH_3CO_2H. This effect, called **the common ion effect,** is general for all ionic equilibria, and occurs whenever the concentration of any ionic component of the equilibrium is increased. In acid dissociation equilibria, the equilibrium is also shifted to the left by addition of a strong acid such as HCl, which instantaneously increases the concentration of H^+. Thus, the pH of an acetic acid solution can be increased ($[H^+]$ decreased) by the addition of acetate ion, whereas it can be decreased ($[H^+]$ increased) by adding a strong acid.

The extent of the hydrolysis reaction which takes place when ammonia is added to water is depressed by the addition of ammonium salts, such as ammonium chloride, NH_4Cl. Strong bases, such as sodium hydroxide, NaOH, also shift the equilibrium and depress the extent of the hydrolysis reaction.

BUFFER SOLUTIONS

It is possible to prepare aqueous solutions which maintain a nearly constant pH, even when relatively large quantities of acid or base are added. Such solutions are called **buffer solutions.** Solutions containing weak acids and their conjugate bases, or weak bases and their conjugate acids, act as buffer solutions.

The properties of a buffer solution are illustrated by considering an acetic acid-sodium acetate solution and the associated equilibrium:

$$CH_3CO_2H \rightleftharpoons H^+ + CH_3CO_2^-$$

The addition of a small quantity of sodium hydroxide to this equilibrium system will result in the consumption of H^+, but the excess CH_3CO_2H present will dissociate to replace most of the H^+ consumed. The addition of a small quantity of hydrochloric acid will induce a shift in the equilibrium to consume the added H^+ and some of the excess $CH_3CO_2^-$, but will result in only a small change in pH.

These effects can be treated quantitatively in the following way. Using the K_a expression for a weak acid, HA, the following expression for H^+ is reached:

$$K_a = \frac{[H^+][A^-]}{[HA]} \qquad [H^+] = K_a \frac{[HA]}{[A^-]}$$

Taking the negative logarithm of each side of the final expression and recalling the definition of pH and pK_a give the following:

$$-\log[H^+] = -\log K_a - \log\frac{[HA]}{[A^-]}$$

$$pH = pK_a - \log\frac{[HA]}{[A^-]}$$

This expansion can be used to calculate the pH of buffer solutions of weak acids and their conjugate bases simply by considering pK_a and the ratio $[HA]/[A^-]$. These expressions are used in the analysis of titration curves of amino acids. For buffers composed of weak bases, B, and their conjugate acids, BH^+, having the equilibrium $B + H_2O \rightleftharpoons BH^+ + OH^-$, the corresponding expression is:

$$pH = 14 - pK_b - \log\frac{[BH^+]}{[B]}$$

Buffer solutions play an important role in many naturally occurring processes. All body fluids have definite pH values that must be maintained within fairly narrow ranges for proper physiological functions. The pH of the blood is normally between 7.35 and 7.45. If the pH of the blood falls below 7.0 or goes above 7.8, death occurs. Since many of the reactions that take place in our tissues form acid substances, the blood must have a mechanism to prevent such changes in pH. The equilibrium system involved in buffering the blood includes bicarbonates and carbonates, phosphates, and complex salts of proteins (p. 271).

ELECTROLYSIS

The principles of chemical equilibrium previously outlined are applicable to reversible oxidation-reduction reactions.

Oxidation-reduction reactions can be induced by passing an electric current, and it is also possible to use an oxidation-reduction reaction to produce an electric current. The former phenomenon, **electrolysis,** reported by Faraday in 1834, is performed in an **electrolytic cell,** as shown in Figure 7–3. Electric current can be generated by chemical reactions under the special conditions which exist in a **voltaic cell,** which is quite similar to an electrolytic cell. In order for an oxidation-reduction reaction to produce an electric current, it must occur spontaneously. Electrolysis is usually performed for reactions which

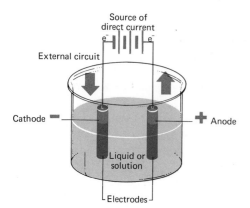

Figure 7–3   An electrolytic cell.

are not spontaneous, and so consume energy, electrical energy in this case. The study of chemical reactions involving either the consumption or generation of electrical current is called **electrochemistry.**

The reactions taking place in an electrolytic cell are not like the oxidation-reduction reactions discussed in Chapter 5 in that the reactants do not react by direct exchange of electrons between one another. The reducing agent reacts by transferring an electron to the **anode** which bears a positive charge. The electrons given to the anode travel through the external circuit to the **cathode,** which is negatively charged, where they are added to the oxidizing agent. The overall oxidation-reduction reaction can be thought of as the sum of two independent **half-reactions,** the anode half-reaction and the cathode half-reaction. In the electrolysis of molten NaCl these could be represented as follows:

anode	$2Cl^- \rightarrow Cl_2(g) + 2e^-$
cathode	$2Na^+ + 2e^- \rightarrow 2Na$
overall	$2Na^+ + 2Cl^- \rightarrow 2Na + Cl_2$

In an electrolytic cell the source of direct current in the external circuit serves as an "electron pump" which pulls electrons off the reducing agent at the anode and pushes them into the oxidizing agent at the cathode. Reactions which occur spontaneously require no "electron pumping," and in fact spontaneously generate a current in the external circuit. Reactions which do not occur spontaneously require "electron pumping."

Faraday's Law

The extent to which a chemical reaction proceeds in an electrochemical apparatus is directly proportional to how much electricity is passed. If one mole of electrons is passed through the electrochemical cell, then one gram equivalent weight of reducing agent will be oxidized at the anode, and one gram equivalent weight of oxidizing agent will be reduced at the cathode. This is a statement of **Faraday's law.** One mole of electrons, called a **faraday,** has a total negative charge of **96,500 coulombs.** If a current is passed at a known rate for a known length of time, it is possible to calculate the number of coulombs, and therefore the number of faradays, involved in an electrolysis. The rate at which a current is passed is expressed using the unit **ampere,** A, which is *one coulomb per second.* The following sample calculation illustrates the treatment of data obtained in an electrolysis experiment:

A current of 3.00 amperes is passed through an electrolytic apparatus for 10.00 minutes. How many coulombs have passed? How many faradays? What weight of Cu^{+2} from an aqueous solution would be reduced to Cu and deposited at the cathode?

The current, having units of amperes or coulombs per second, when multiplied by the time gives the number of coulombs passed:

$$10.00 \text{ min} \times 3.00 \text{ A} \times \frac{1 \text{ coul/s}}{A} \times \frac{60s}{\text{min}} = 1800 \text{ coul}$$

Since 96,500 coul is 1 faraday,

$$1800 \text{ coul} \times \frac{1 \text{ faraday}}{96,500 \text{ coul}} = 0.01865 \text{ faraday}$$

Since 1 faraday will deposit 1 gram equivalent weight of Cu at the cathode,

$$0.01865 \text{ faraday} \times \frac{1(\text{g equiv wt})\text{Cu}}{1 \text{ faraday}} \times \frac{1(\text{g atom wt})\text{Cu}}{2(\text{g equiv wt})\text{Cu}} \times \frac{63.5 \text{ g Cu}}{1(\text{g atom wt})\text{Cu}}$$

$$= 0.592 \text{ g Cu}$$

ELECTROLYSIS IN AQUEOUS SOLUTIONS

Whenever an aqueous solution is subjected to hydrolysis, H_2 and O_2 are potential products. In the electrolysis of pure water, the half-reactions are

$$\text{cathode} \quad 2H_2O + 2e^- \rightarrow H_2 + 2OH^-$$

$$\text{anode} \quad 2H_2O \rightarrow O_2 + 4e^- + 4H^+$$

The H_2 and O_2 produced bubble out of the solution. The H^+ generated at the anode migrates in the direction of the negative cathode, whereas the OH^- generated at the cathode migrates towards the anode. These ions combine to form H_2O. In order for an element to be deposited at the cathode in preference to H_2, its oxidized form, M^+, must be reduced to the reduced form, M, more easily than H^+ is reduced to H_2. At the anode O_2 will be evolved unless the reduced form of an element, X^-, is more readily oxidized to its oxidized form, X, than H_2O is oxidized to O_2. For this reason Zn^{+2}, Na^+, Al^{+3}, Mg^{+2}, and many other cations cannot be reduced at the cathode in aqueous solutions. Anions such as Cl^-, NO_3^-, and SO_4^{-2} are not oxidized at the anode in aqueous solution, since water is more easily oxidized. Thus it is seen that H_2 and O_2 will be the electrolysis products for an aqueous solution of NaCl, $Zn(NO_3)_2$, or any salt containing a cation reduced with more difficulty than H^+, and an anion oxidized with more difficulty than H_2O.

The metal ions Cu^{+2} and Ag^{+1} are more easily reduced than H^+. For this reason thin layers of these metals are often deposited on the surfaces of objects by electrolysis. This process, which uses the object to be plated as the cathode, is called **electroplating.** In the electroplating cell, the anode is composed of the metal to be plated out at the cathode, so that the anode reaction continually replenishes the metal ions in solution. A cell for the electroplating of copper and its associated half-cell reactions is shown in Figure 7–4.

VOLTAIC CELLS

Spontaneous oxidation-reduction reactions, when physically separated into half-reactions in an electrochemical cell, can produce an electric current in an external circuit

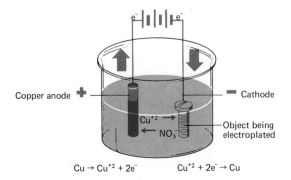

Copper anode   Cathode

Cu^{+2} →
← NO_3^-

Object being electroplated

$Cu \rightarrow Cu^{+2} + 2e^-$ $Cu^{+2} + 2e^- \rightarrow Cu$

Figure 7–4   An electrolytic cell used for electroplating copper.

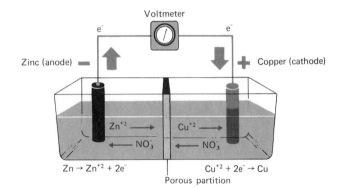

Figure 7-5   The Daniell cell.

which connects the half-reactions. An electrochemical cell in which an oxidation-reduction reaction spontaneously produces a current in an external circuit is called a **voltaic cell.**

The reaction between metallic zinc and Cu^{+2} in aqueous solution proceeds spontaneously and can be used in a voltaic cell as seen in Figure 7–5. This cell is called the **Daniell cell.** It is seen that in this cell the reaction at the anode is the oxidation of Zn to Zn^{+2}. The Zn^{+2} produced migrates toward the cathode, leaving electrons on the anode which is consequently negatively charged. At the cathode Cu^{+2} ions are reduced to Cu by obtaining electrons from the electrode, leaving it positively charged. The electron current flows from anode to cathode via the external circuit, as in electrolytic cells, but the signs on the electrodes are reversed. In order to eliminate the possibility of Cu^{+2} reacting directly with the Zn of the anode, the cell is divided into two portions. These portions are separated by a **porous partition** which allows only slow migration of the ions.

CELL POTENTIALS

The number of electrons which flow through the external circuit in a voltaic cell is related by Faraday's law to the quantity of material which reacts. The potential energy difference between electrons on the anode and electrons on the cathode is the **cell potential,** or the **electromotive force,** of the voltaic cell, and is expressed in units of **volts.** The standard cell potential, $E°$, of a given cell is dependent on the spontaneity, or driving force, with which the oxidation-reduction reaction takes place. Since the equilibrium constant, K, is related to the spontaneity of a reaction, this parameter is also related to the cell potential.

Consideration of Le Chatelier's principle indicates that the cell potential of a voltaic cell is dependent on the concentrations of any reacting ions in solution. The half-reactions and the overall reaction in the Daniell cell are

$$
\begin{array}{ll}
\text{anode} & Zn \rightleftharpoons Zn^{+2} + 2e^- \\
\text{cathode} & Cu^{+2} + 2e^- \rightleftharpoons Cu \\
\hline
\text{overall} & Cu^{+2} + Zn \rightleftharpoons Cu + Zn^{+2}
\end{array}
$$

If the half-reactions occurring at each electrode are considered to be reversible processes, then the concentrations of Zn^{+2} and Cu^{+2} will be quite influential on the spontaneity of the reaction. Considering the extreme case where $[Cu^{+2}]$ equals zero, the reaction will not proceed. As $[Cu^{+2}]$ is increased from zero, the cathode reaction proceeds with increasing spontaneity. When the $[Zn^{+2}]$ is very large, the reaction at the anode will be less spontaneous than when $[Zn^{+2}]$ is quite small. The anode reaction will proceed with the greatest spontaneity when $[Zn^{+2}] = 0$. However, as the reaction proceeds the spontaneity will decrease because of the increasing concentration of Zn^{+2}. The overall reaction, if allowed to proceed, will eventually reach equilibrium, and the cell potential is zero.

Since concentrations influence the voltage produced by a cell, comparison of different cells is made only when the same concentrations are used in each. The standard reference concentration for all ions in solution is one molar at $25°C$. For gaseous reactants the reference pressure at $25°C$ is one atmosphere. Solid and liquid reactants are considered to be in their standard reference states at $25°C$. When these conditions are imposed upon the Daniell cell, the characteristic cell potential is 1.10 volt.

COMMERCIAL VOLTAIC CELLS: BATTERIES

The **lead storage battery** used in automobiles acts as a voltaic cell when furnishing current, or discharging, and as an electrolytic cell when charging. When discharging, the anode consists of lead and the cathode is lead dioxide. The solution in which these electrodes are immersed is sulfuric acid. The reactions associated with discharging are

anode	$Pb\,(s) + SO_4^{-2} \rightarrow PbSO_4\,(s) + 2e^-$
cathode	$2e^- + PbO_2\,(s) + 4H^+ + SO_4^{-2} \rightarrow PbSO_4\,(s) + 2H_2O$
overall	$Pb\,(s) + PbO_2\,(s) + 4H^+ + 2SO_4^{-2} \rightleftharpoons 2PbSO_4\,(s) + 2H_2O$

As this reaction proceeds and reaches equilibrium, the battery discharges and the cell potential decreases. Lead sulfate is deposited on both the cathode and anode, and the concentration of H_2SO_4 decreases. As the H_2SO_4 concentration decreases, the density of the battery liquid decreases.

A lead storage battery acts as an electrolytic cell and recharges when a direct current is passed. During recharging the reactions given above are reversed. The $PbSO_4$ on the cathode and anode is consumed, and the concentration of H_2SO_4 increases, as does the density of the battery liquid. When fully charged, a lead storage battery can produce about two volts per cell. Automobile batteries use three or six cells in series to generate six or twelve volts, respectively.

Another very common voltaic cell is the **dry cell** (Fig. 7–6) or flashlight battery. In this cell a zinc metal cylindrical container, the anode, is filled with a moist paste of

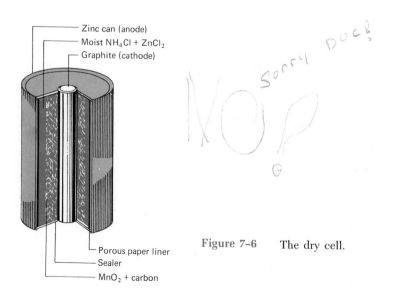

Zinc can (anode)
Moist $NH_4Cl + ZnCl_2$
Graphite (cathode)

Porous paper liner
Sealer
MnO_2 + carbon

Figure 7–6 The dry cell.

ammonium chloride and zinc chloride. A graphite rod inserted into this paste is surrounded by manganese dioxide. The container is sealed to prevent the escape of the moisture. This cell, which produces about 1.3 to 1.5 volt, has complex electrode reactions which are not well understood. It is known that zinc is oxidized to Zn^{+2} and that MnO_2 is reduced.

IMPORTANT TERMS AND CONCEPTS

activation energy
anode
buffers
catalyst
cathode
cell potential
common ion effect

electrolysis
electrolytic cell
equilibrium constant
Faraday's law
hydrolysis
K_w
law of chemical equilibrium

Le Chatelier's principle
pH
pK
pOH
rate constant
rate law
voltaic cell

QUESTIONS

1. Do low rates of chemical reactions indicate high or low activation energy?

2. Does increasing the temperature increase or decrease the frequency of collisions between reacting particles?

3. With increasing temperature the rates of chemical reactions increase by more than that predicted when considering only the increased rate of molecular collisions. Explain.

4. Why is chemical equilibrium referred to as a dynamic equilibrium?

5. For the reaction $H_2O(g) + CO(g) \rightleftharpoons H_2(g) + CO_2(g)$, which of the following would result in a change in the equilibrium composition, that is, a "shift" in the equilibrium?

   (a) increase in total pressure
   (b) addition of a catalyst
   (c) increasing the concentration of CO
   (d) decreasing volume of reaction container
   (e) increasing temperature

6. For the following reaction at equilibrium,

$$N_2(g) + 3H_2(g) \rightleftharpoons 2NH_3(g)$$

6 moles of NH_3, 3 moles of N_2, and 4 moles of H_2 are contained in a 10 liter vessel. Calculate the equilibrium constant.

7. Write the equilibrium constant expression for each of the following reactions.

   (a) $Al_2Cl_6(g) \rightleftharpoons 2AlCl_3(g)$
   (b) $CO(g) + H_2O(g) \rightleftharpoons CO_2(g) + H_2(g)$
   (c) $4NO(g) + 6H_2O(g) \rightleftharpoons 4NH_3(g) + 5O_2(g)$
   (d) $2NO_2(g) \rightleftharpoons 2NO(g) + O_2(g)$
   (e) $2Cl_2(g) + 2H_2O(g) \rightleftharpoons 4HCl(g) + O_2(g)$

8. Express the ionization equilibrium constant for the following dissociation reactions:

   (a) $KCl \rightleftharpoons K^+ + Cl^-$
   (b) $CH_3CO_2H \rightleftharpoons H^+ + CH_3CO_2^-$
   (c) $Ca(OH)_2 \rightleftharpoons Ca^{+2} + 2OH^-$
   (d) $BaSO_4 \rightleftharpoons Ba^{+2} + SO_4^{-2}$

9. Will the addition of H^+ to an aqueous equilibrium system cause an increase or decrease in $[OH^-]$? What law or principle allows this prediction?

10. In an aqueous solution which is 0.25M in NaOH, what is the $[H^+]$?

11. Calculate the pH and pOH for aqueous solutions containing the following H^+ concentrations:

(a) 10^{-6}M   (c) 0.5M   (e) 10M
(b) 2.5×10^{-3}M   (d) 4.2×10^{-8}M   (f) 0.097M

12. Which of the solutions in question 11 are acidic and which are basic?

13. As acids and bases decrease in strength, does the extent of dissociation increase or decrease? Do K_a and K_b become larger or smaller?

14. Which of the following ions will hydrolyze upon dissolving in water?

(a) anions of strong acids   (f) F^-
(b) anions of weak acids   (g) NH_4^+
(c) cations of strong bases   (h) CN^-
(d) cations of weak bases   (i) $H_2PO_4^-$
(e) Cl^- (j) HSO_4^-

15. What is meant by the common ion effect? What basic chemical principle predicts the common ion effect?

16. What is the pH of a buffer composed of 0.2M CH_3CO_2H and 0.02M $CH_3CO_2^-$?

17. What is the pH of a neutral solution at 25°C? of a basic solution? of an acidic solution?

18. In dilute aqueous solutions, what can be said about the concentrations of water molecules?

19. What can be said about the pH of NaCl and KNO_3 solutions?

20. In an electrolytic cell, the reducing agent does what?

(a) accepts electrons at the cathode
(b) transfers electrons at the anode
(c) transfers electrons directly to the oxidizing agent

21. In an electroplating apparatus, one mole of electrons is passed. What weight of Cu will be deposited at the cathode from Cu^{+2} solution?

22. What is a voltaic cell? Give an example.

23. The cell potential is dependent on the spontaneity of a reaction. Name at least two other parameters related to the cell potential.

24. An automobile battery exemplifies what kind of electrochemical cell when charging? When discharging?

25. Of what importance is the porous partition between the anode and cathode portions of the Daniell cell?

SUGGESTED READING

Kokes, Dorfman, and Mathia: Equilibria in Ionic Solutions. Journal of Chemical Education, Vol. 39, p. 93, 1962.

Lawrence and Bowman: Electrochemical Cells for Space Power. Journal of Chemical Education, Vol. 48, No. 6, p. 359, 1971.

Mogul and Schmuckler: Dilute Solutions of Strong Acids: The Effect of Water on pH. Chemistry, Vol. 42, No. 9, p. 14, 1969.

Morris: Stress, Collisions and Constants. Part II: Buffers. Chemistry, Vol. 44, No. 5, p. 15, 1971.

Murphy: New Batteries. CHEMTECH, p. 487, August 1971.

Shen: What Is the Real Value of a Faraday? Chemistry, Vol. 39, No. 2, p. 8, 1966.

Weissman: Batteries: The Workhorse of Chemical Energy Conversion. Chemistry, Vol. 45, No. 10, p. 6, 1972.

SOME
CHEMISTRY OF
NONMETALLIC
ELEMENTS

CHAPTER 8 ━━━━━━━━━━━━━━━━━━━━━━━━━━━━━━━━

The *objectives* of this chapter are to enable the student to:

1. Identify the nonmetals in the periodic table.
2. Describe the laboratory preparation of the halogens.
3. Write reactions for the preparation of the hydrogen halides.
4. Discuss the structure, bonding, and preparation of the sulfur oxides.
5. Give chemical reactions for the preparation of the oxyacids of sulfur.
6. List the oxides of nitrogen and give their structures.
7. Write the reactions involved in the Ostwald process for preparing nitric acid.
8. List the distinguishing properties of oxygen and give a method of preparation used in the laboratory.
9. Contrast the bonding and properties of ozone with those of molecular oxygen.
10. List the distinguishing properties of hydrogen and give a method of preparation used in the laboratory.

Elements in the upper right side of the periodic table are characterized by their tendency to gain electrons to complete their outer electron shell. Like the halogens, which they include, their electronegativities are relatively high, and they form many important compounds with other elements. The elements shown in Figure 8–1 are classed as **nonmetals.**

Although the nonmetals represent a relatively small group of elements, they possess a wide range of properties. About half of the group, including the inert gases, fluorine, chlorine, nitrogen, and oxygen, exist as gases, whereas bromine is a liquid and the remainder exist as solids.

THE HALOGENS

The group of elements including **fluorine, chlorine, bromine, iodine,** and **astatine** is called the **halogens,** which is derived from a Greek word meaning "salt formers."

All the halogens are too reactive chemically to be found free in nature. They are probably the most typical nonmetals of all the elements in the periodic classification.

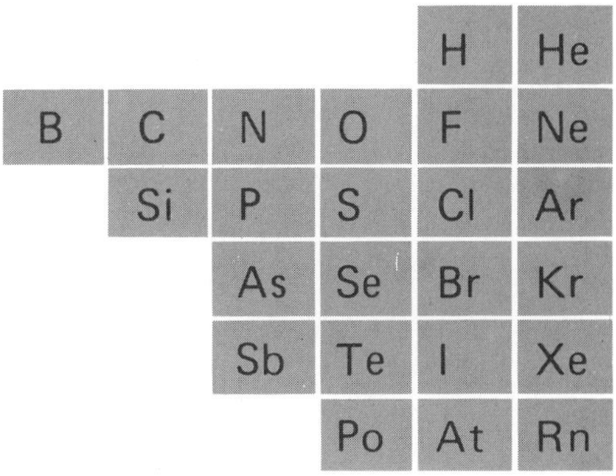

				H	He
B	C	N	O	F	Ne
	Si	P	S	Cl	Ar
		As	Se	Br	Kr
		Sb	Te	I	Xe
			Po	At	Rn

Figure 8-1 The nonmetals in the periodic table.

A comparison of their atomic structures reveals that each has seven electrons in its outer shell. They all possess a high electronegativity and readily form negative halide ions, commonly found in ionic salts. Although all the members of the group show many similar properties, these elements also illustrate graduated properties with respect to positions in their column of the periodic table. As the atomic weight increases from fluorine to iodine, the physical state of the elements changes. The first two members of the group exist as gases, whereas bromine is a liquid and iodine a solid. The melting point and boiling point temperatures and the size of the individual atoms all increase gradually from fluorine to iodine. The tendencies of the atoms to attract electrons decrease with increasing size, as reflected in the electron affinities and the electronegativity. Fluorine, which is the most electronegative halogen, exists only as the element and with an oxidation number of -1, whereas the other halogens show positive oxidation numbers in addition to 0 and -1.

Fluorine and chlorine are the most abundant of the halogens. Fluorine is found in mineral fluorides or **fluorspar,** CaF_2, and as a constituent of the aluminum ore **cryolite,** Na_3AlF_6. The most common source of chlorine in nature is $NaCl$, common table salt. Bromine occurs in nature as salts of sodium and magnesium along with chloride salts, but in much smaller concentrations than the chlorides. Iodine is found in nature as $NaIO_3$, an impurity of the $NaNO_3$ found in Chile.

All of the halogens occur as diatomic molecules in the elemental state. Fluorine and chlorine are green-yellow gases under normal conditions. Chlorine is commonly stored in steel cylinders under pressure. Bromine, the only liquid nonmetallic element under normal conditions, is a dense, reddish-brown liquid which readily vaporizes at room temperature to red vapors that have a strong pungent odor. It is moderately soluble in water, but much more soluble in alcohol, chloroform, or carbon tetrachloride. Iodine exists as a gray-black shiny solid which has violet vapors. It dissolves only slightly in water, but is very soluble in an aqueous solution of KI, as well as in organic solvents such as chloroform, ether, and alcohol.

HALIDES

The halogens combine with metals and nonmetals and among themselves. In combination with metals the halogens always have -1 as an oxidation number and are called **halides.** The halides of metals have high ionic character as indicated by their high melting points. As the metallic character of the combining element decreases, the ionic character of the binary halides gradually decreases, and the bonds become more covalent. This is illustrated by the fact that BCl_3, CCl_4, and $SiCl_4$ are all liquids at room temperature, whereas $NaCl$, $BeCl_2$, and $AlCl_3$ are all solids.

A common test for the presence of the chloride, bromide, or iodide ion is carried out by adding a solution of silver nitrate to a solution of the halogen salts. When a halogen is present, an insoluble silver salt is precipitated out of solution. The color of the precipitate helps identify the halogen, since silver chloride is white, silver bromide, pale yellow, and silver iodide, a lemon yellow.

HYDROGEN HALIDES

The hydrogen halides can each be prepared by the direct interaction of hydrogen gas and the halogen. In the laboratory where this method is impractical, HF and HCl are made by placing metal halides in warm concentrated sulfuric acid. The reaction proceeds as follows:

$$2MX(s) + H_2SO_4(l) \rightleftharpoons M_2SO_4(s) + 2HX(g)$$

This procedure cannot be used for bromides and iodides because H_2SO_4 oxidizes the halides to the free halogen:

$$2MX(s) + 2H_2SO_4(l) \rightleftharpoons X_2(g) + M_2SO_4(s) + SO_2(g) + 2H_2O$$

Hydrogen bromide and hydrogen iodide can be made using phosphoric acid in a process similar to the aforementioned:

$$3MX(s) + H_3PO_4(l) \rightleftharpoons 3HX(g) + M_3PO_4(s)$$

The hydrogen halides, all very soluble in water, form aqueous solutions in which all except HF behave as strong acids. These acids are called **hydrohalic** (i.e., hydrofluoric, etc.) **acids.**

Hydrofluoric acid cannot be kept in glass bottles because it dissolves glass by the reaction

$$4HF(aq) + SiO_2(s) \rightleftharpoons 2H_2O + SiF_4(g)$$

The SiF_4 produced is volatile and leaves the solution, so that equilibrium is never attained. Hydrofluoric acid, because of this property, is used to etch glass, and consequently solutions of HF are usually stored in special plastic or paraffin-lined bottles.

ADDITIONAL USES OF THE HALOGENS AND THEIR COMPOUNDS

In recent years the use of fluorine compounds by industry has increased considerably. Hydrogen fluoride is used as a catalyst in the petroleum industry, and hydrofluoric acid and its acid salts are used to etch glass in the manufacture of "frosted" electric light bulbs, chemical apparatus, and decorative glassware. The refrigerant Freon (CF_2Cl_2) is a fluorine compound that is commonly used in household refrigerators.

Recent research in rocket and guided-missile propellants has suggested another important use for fluorine. Compounds such as oxygen difluoride and nitrogen trifluoride can serve as high-energy oxidizers, but liquid fluorine apparently has superior properties. This use alone could result in increased production and decreased cost of fluorine. Since liquid fluorine boils at $-306°$ F, it must be kept cold with liquid nitrogen and transported in nickel alloy containers.

Large quantities of chlorine are used industrially for the bleaching of paper and cotton textiles. Considerable quantities are used in the manufacturing of bleaching powder and in the preparation of solutions of sodium hypochlorite for household bleaching agents. The water supply of almost all large cities is treated with chlorine to kill bacteria. It is an essential constituent in chlorates, which are used to manufacture matches, fireworks, and percussion caps for rifle bullets.

The second largest use of bromine is in the production of soil and seed fumigants. Organic compounds of bromine are effective fire-extinguishing and flameproofing agents. Potassium bromide is used as a mild sedative in medicine. Compounds of bromine are used in the manufacture of dyes and photographic emulsions.

Tincture of iodine, which is used as an antiseptic in medicine, is prepared by dissolving iodine in a solution of potassium iodide and alcohol. Compounds of iodine are used as drugs and medications and in the preparation of certain dyestuffs. For example, potassium iodide is used as an expectorant, and the organic compound iodoform (CHI_3) is an active disinfectant. Iodized salt contains small amounts of iodine compounds, used to prevent the occurrence of common goiter.

OXYGEN

Of all the elements in the earth's lithosphere, hydrosphere, and atmosphere, oxygen is the most abundant. One-fifth of the volume of the air, eight-ninths (by weight) of water, and approximately one-half of the earth's crust are oxygen. In the air it exists as free molecular oxygen (O_2); elsewhere it is found combined with many other elements in the form of oxides. Oxygen is an important constituent of living matter.

PREPARATION

Oxygen is usually isolated commercially from air, which is essentially a mixture of this element and nitrogen. Air is liquefied by subjecting it to a high pressure at a low temperature. When the liquid air is allowed to evaporate, the more volatile nitrogen escapes first, leaving behind fairly pure oxygen. Both the oxygen and nitrogen gases are then forced into steel cylinders under high pressure and stored for future use. Fortunately, large quantities of nitrogen gas are used in the manufacture of ammonia, and the process discussed is economically sound.

Pure oxygen is also obtained for commercial purposes by the electrolysis of water. When an electric current is passed through water, oxygen forms at the positive pole, or anode, and hydrogen forms at the negative pole, or cathode (Fig. 8–2). The oxygen and hydrogen gases thus formed are drawn off and stored under pressure.

PROPERTIES

Oxygen is a colorless, odorless, tasteless gas with a density slightly greater than air. When subjected to a high pressure at a low temperature, the gas is converted to liquid oxygen. At a temperature of $-118°C$ a pressure of 50 atmospheres is required for liquefaction. Liquid oxygen is pale blue in color, slightly more dense than water, and has a normal boiling temperature of $-183°C$.

Approximately 3 ml of gaseous oxygen will dissolve in 100 ml of water at ordinary temperatures. This slight solubility insures a supply of the gas to aquatic plants and animals, and also for the conversion of sewage and other contaminating substances in natural water into harmless material.

Oxygen is the first member of the so-called oxygen-sulfur family, which consists of oxygen, sulfur, selenium, tellurium, and the radioactive element polonium. Each member

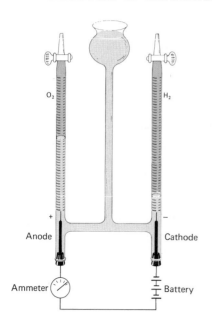

Figure 8-2   A simple apparatus for the electrolysis of water.

of this group has 6 electrons in the outer orbit of its atom. These elements, especially oxygen and sulfur, have a strong tendency to react with metals by gaining 2 electrons producing ions with a charge of -2. Oxygen and the other elements of the family also readily share 2 pairs of electrons to form covalent compounds.

When oxygen takes up 2 electrons to complete its outer shell, it assumes the stable configuration of an atom of inert gas. Its tendency to accept 2 electrons from both nonmetals and metals is responsible for its ability to form compounds called **oxides** with nearly all known elements. The formation of oxides can be readily demonstrated by burning such elements as sulfur, phosphorus, iron, and carbon in pure oxygen. For example, sulfur unites with oxygen to form sulfur dioxide:

$$S + O_2 \rightarrow SO_2\uparrow$$

Carbon burns in oxygen to form carbon dioxide:

$$C + O_2 \rightarrow CO_2\uparrow$$

The union of a substance with oxygen is called **oxidation.** The burning of wood, the rusting of iron, and the decay of plant and animal matter are examples of oxidation. Oxygen is able to unite with food and tissue substances in the body at relatively low temperatures because these reactions are hastened by catalysts called enzymes.

Ozone

The air in the neighborhood of an electrical discharge, such as in an electrical storm, has a peculiar odor caused by the formation of ozone, which is a form of oxygen represented as O_3, in contrast to ordinary molecular oxygen, O_2. When an element exists in two or more different forms possessing different physical and chemical properties, the forms are known as **allotropic modifications.** Common elements that exhibit allotropic forms are oxygen, phosphorus, and sulfur.

Molecular ozone is a bent molecule having no unpaired electrons. Since the two bond lengths are the same, ozone cannot be represented by a single Lewis structure and is best represented as a resonance hybrid of the following contributing forms:

Ozone is a colorless gas possessing a garlic-like odor and is more soluble and more dense than oxygen. Liquid ozone is dark blue in color and is capable of decomposing with explosive force, producing ordinary oxygen. Ozone is considerably more reactive than oxygen. For example, a noble metal such as silver will not combine with oxygen under ordinary conditions, but will form a film of brown silver oxide when exposed to a low concentration of ozone. The significance of the ozone layer in absorbing harmful ultraviolet radiation has been discussed in Chapter 4.

COMBUSTION

When a substance unites with oxygen so rapidly that heat and light are produced in the reaction, it is said that the substance is burning. One commonly speaks of rapid oxidation, or burning, as **combustion.**

Substances that will burn readily, such as paper, wood, and gasoline, are called **combustible;** those that will not burn, such as asbestos, stone, and clay, are called **incombustible.** The terms **flammable** and **inflammable** are both synonymous with combustible and cause confusion when incorrectly compared to combustible and incombustible. Since air is only one-fifth oxygen, substances burn less vigorously in it than in the pure gas, and many substances that are incombustible in air burn in oxygen. For example, powdered sulfur burns feebly in air but blazes up vigorously when thrust into oxygen. Although iron wire is incombustible in air, it burns brightly in pure oxygen.

SULFUR

Sulfur occurs in nature in the elemental state, as sulfide ores such as ZnS, FeS_2, and PbS, as sulfates such as $CaSO_4 \cdot 2H_2O$, and as H_2S and SO_2. Sulfur normally exists as a yellow solid that is tasteless and odorless, insoluble in water, but soluble in carbon disulfide. It is similar to carbon and oxygen in that it exists in several allotropic forms. For example, below 96°C sulfur crystallizes from solution as **rhombic** crystals, whereas if it is heated above 114°C and allowed to cool, **monoclinic** crystals are formed. An **amorphous** form, plastic sulfur (so-called), results when hot melted sulfur is poured into cold water. This allotrope is thought to consist of long chains of sulfur atoms. The plastic form is metastable and slowly transforms to the rhombic form.

Both the rhombic and monoclinic allotropes of sulfur are composed of molecules containing eight sulfur atoms in a puckered ring, as shown in Figure 8–3. The vapors of sulfur consist of an equilibrium mixture of S_8, S_4, and S_2 molecules. In CS_2 solution and in naphthalene solution, rhombic sulfur has a molecular weight of 256 amu, corresponding to S_8.

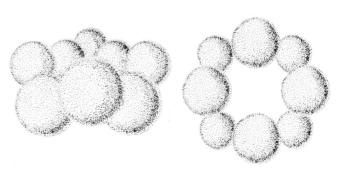

Figure 8–3 The cyclic structure of S_8.

Side view   Top view

OXIDES AND OXYACIDS

Sulfur forms the oxides SO_2 and SO_3, in which it assumes oxidation numbers of $+4$ and $+6$, respectively. The dioxide of sulfur, a gas at room temperature (b.p. $-10°C$), can be obtained by burning the element in air:

$$S(s) + O_2(g) \rightleftharpoons SO_2(g)$$

It is also formed as a by-product when metal sulfide ores are roasted in air:

$$2MS(s) + 3O_2(g) \xrightleftharpoons{\text{heat}} 2MO(s) + 2SO_2(g)$$

Molecules of sulfur dioxide are bent as is predictable by considering the total number of bonding and non-bonding pairs of electrons. The bonding is best represented as a resonance hybrid of the following two contributing forms:

Sulfur dioxide, when dissolved in water, hydrolyzes to form the weak diprotic **sulfurous acid**, $H_2SO_3(K_{a1} = 1.3 \times 10^{-2}, K_{a2} = 5.6 \times 10^{-8})$:

$$SO_2(aq) + H_2O \rightleftharpoons H_2SO_3(aq)$$

The salts of sulfurous acid, **sulfites** (M_2SO_3) and **bisulfites** $(MHSO_3)$, when placed in strongly acidic aqueous solutions, are extensively hydrolyzed, and result in the evolution of SO_2 gas from the solution. These salts also act as reducing agents being oxidized to sulfates:

$$2OH^- + SO_3^{-2} \rightleftharpoons SO_4^{-2} + H_2O + 2e^-$$

Sulfur dioxide finds many industrial uses, including use as a bleach and as a refrigerant; but by far its most important use is in the production of sulfuric acid. This very odoriferous gas is also a major atmospheric pollutant, especially in areas in which large quantities of sulfur-containing fossil fuels are burned.

Sulfur trioxide, SO_3, the anhydride of **sulfuric acid**, H_2SO_4, is prepared by the molecular oxygen oxidation of sulfur dioxide in the presence of the catalyst V_2O_5 or spongy platinum at 400 to 500°C. The product, which boils at 45°C, consists of planar triangular molecules in the gas phase considered to be resonance hybrids of the three structures as shown on p. 49.

When sulfur trioxide is placed in water, it hydrolyzes to sulfuric acid which is undoubtedly one of the most important compounds of sulfur.

$$SO_3(g) + H_2O \rightleftharpoons H^+(aq) + HSO_4^-(aq)$$

It is manufactured most commonly by the **contact process** in which sulfur dioxide is oxidized to sulfur trioxide. The sulfur trioxide is then absorbed in 98 per cent sulfuric acid, producing **fuming sulfuric acid,** or **oleum.** By the controlled addition of water, the concentration of the end product is maintained at 98 per cent and is the concentrated acid of commerce. The reactions for this process are as follows:

$$S(s) + O_2(g) \rightleftharpoons SO_2(g)$$

$$SO_2(g) + O_2(g) \xrightarrow[\text{400°-500°C}]{\text{catalyst}} SO_3(g)$$

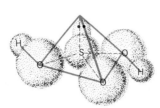

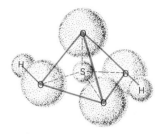

Sulfurous acid   Sulfuric acid

Figure 8-4 The structures of sulfurous acid and sulfuric acid.

$$SO_3(g) + H_2SO_4(aq)(98\%) \rightleftharpoons H_2S_2O_7(l)$$

$$H_2S_2O_7(l) + H_2O \rightleftharpoons 2H_2SO_4(aq)$$

Sulfurous acid and sulfuric acid have similar structures, as shown in Figure 8-4, in that each involves sulfur surrounded by a tetrahedral arrangement of electron pairs, either bonding or non-bonding.

The second acid dissociation step of sulfuric acid, unlike the first, does not go to completion ($K_{a2} = 1.3 \times 10^{-2}$). The acid and its salts, the **sulfates,** M_2SO_4, and **bisulfates,** $MHSO_4$, will not act as reducing agents but do have weak oxidizing powers. The acid has the best oxidizing ability and, as already mentioned, can oxidize bromides and iodides to the elements.

Alkali metal sulfates, when placed in water, undergo slight hydrolysis to produce basic solutions, whereas alkali metal bisulfates give slightly acidic solutions. The sulfates of calcium, barium, and strontium are only slightly soluble in water and are often used in qualitative testing for these cations.

NITROGEN

The elements of Group VA of the periodic table include **nitrogen, phosphorus, arsenic, antimony,** and **bismuth.** Within this group the trend from nonmetallic character to metallic on going down the group is far more obvious than in the elements of Group VIIA or Group VIA. This gradual transition is evident in physical and chemical properties. Because of the pronounced variation in properties, nitrogen and phosphorus are considered to be nonmetals; arsenic and antimony are considered intermediate between nonmetallic and metallic and are called **metalloids;** and bismuth is metallic.

Nitrogen, the only member of Group VA which occurs in nature uncombined, constitutes 78 per cent by volume of the earth's atmosphere. It is colorless, odorless, tasteless, and only slightly soluble in water. In the elemental state, nitrogen exists as diatomic molecules which have a very high bond dissociation energy of 225 kcal/mol due to a triple bond. Because of this extremely high bond dissociation energy, nitrogen is rather inert, but can be rendered very reactive by an electric discharge such as a lightning bolt. The active nitrogen thus generated reacts with oxygen and other components of the atmosphere to generate nitrogen compounds.

Nitrogen is often prepared by removing oxygen from the air by simply passing dry air over hot copper turnings. The nitrogen so produced is not pure; it contains carbon dioxide, argon, and other rare gases. Pure nitrogen may be prepared in the laboratory by heating **ammonium nitrite:**

$$NH_4NO_2(s) \xrightarrow{\text{heat}} N_2(g) + 2H_2O(g)$$

Ammonium nitrite, which is a very unstable compound, is prepared just before use from a mixture of ammonium chloride and sodium nitrite. Large quantities of nitrogen as well as oxygen are produced commercially from the distillation of liquid air.

Hydrogen Compounds

All the Group VA elements form hydrogen compounds of the general formula MH_3. These can be prepared by the hydrolysis of ionic nitrides, phosphides, and so forth:

$$Li_3M(s) + 3H_2O \rightleftharpoons MH_3(g) + 3Li^+(aq) + 3OH^-(aq)$$

Ammonia, NH_3, is also prepared in the **Haber process** by direct combination with hydrogen at high temperatures (1000°C) and pressures (up to 1000 atm) in the presence of catalysts such as iron and iron oxide:

$$N_2(g) + 3H_2(g) \rightleftharpoons 2NH_3(g)$$

Ammonia is a colorless, but very pungent, toxic gas which, when dissolved in water, hydrolyzes to produce basic solutions ($K_b = 1.8 \times 10^{-5}$). The reactivity of this pyramidal molecule is due in large part to a relatively large dipole moment and a lone pair of electrons on the nitrogen atom. Ammonia behaves as a Lewis base in combining with metal ions to form **ammine complexes** such as $[Ag(NH_3)_2]^{+1}$, $[Cu(NH_3)_4]^{+2}$, and $[Ni(NH_3)_6]^{+2}$.

Oxyacids

The important oxyacids of nitrogen are **nitrous acid, HNO_2,** and **nitric acid, HNO_3.** Nitrous acid, in which nitrogen has an oxidation number of $+3$, is unstable even in aqueous solution and decomposes to nitric acid and nitric oxide:

$$3HNO_2(aq) \rightleftharpoons HNO_3(aq) + H_2O + 2NO(g)$$

Nitrous acid is a weak acid ($K_a = 6.0 \times 10^{-6}$) and is usually prepared by acidifying nitrite salts.

Nitric acid involves nitrogen in the $+5$ oxidation state. This acid is made commercially by the **Ostwald process.** This process involves catalytic oxidation of ammonia to nitric oxide, which is oxidized by air to nitrogen dioxide and then added to water. In water nitrogen dioxide disproportionates to nitric acid and nitric oxide. The reactions involved in the preparation of nitric acid, starting with the nitrogen, are as follows:

$$N_2(g) + 3H_2(g) \rightleftharpoons 3NH_3(g)$$

$$4NH_3(g) + 5O_2(g) \xrightarrow[\text{Pt}]{900°C} 4NO(g) + 6H_2O(g)$$

$$2NO(g) + O_2(g) \rightleftharpoons 2NO_2(g)$$
$$3NO_2(g) + H_2O(l) \rightleftharpoons 2H^+(aq) + 2NO_3^-(aq) + NO(g)$$

The nitric oxide generated in the last step is recycled.

Pure nitric acid is a colorless liquid (b.p. 84.1°C) which behaves as a powerful oxidizing agent, with both H^+ and the nitrogen as potential oxidizing sites. However, the reduction products of nitric acid seldom contain hydrogen, but are composed of lower oxidation states of nitrogen, including NO_2, NO, N_2O, and N_2. The oxidizing power decreases with decreasing concentration. As shown in Figure 8–5, gaseous nitric acid is a planar molecule and the nitrate anion has a triangular planar structure.

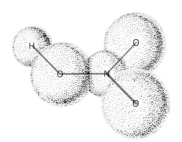
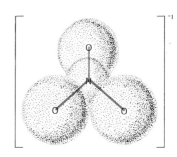

Figure 8-5   The structures of nitric acid and the nitrate ion.

PHOSPHORUS

Elementary phosphorus exists in several allotropic forms, the most important of which are white and red phosphorus.

When freshly prepared, phosphorus is a yellowish white, waxy solid which melts at 44°C and ignites in air at about 35°C. It is very poisonous and can be handled with safety only under water. It is insoluble in water, but readily dissolves in carbon disulfide and other organic solvents. This allotrope, **white phosphorus,** is composed in the solid phase, in the gas phase, and in solutions of discrete P_4 tetratomic molecules having the structure shown in Figure 8-6. If white phosphorus is heated in the presence of a trace of iodine without access to air, it is readily converted into **red phosphorus** which differs markedly from the white. It melts at 500°C, does not ignite spontaneously in air, and is not poisonous. It has a microcrystalline form and is insoluble in carbon disulfide. Red phosphorus, unlike white phosphorus, does not exist as small molecules, but rather has a crystal structure more like that of metals.

The main use of elemental phosphorus is in the manufacture of matches. The first friction matches were made of white phosphorus, which readily ignited but produced chronic poisoning of the workmen in the match industry. Red phosphorus and a sulfide, P_4S_3, are used in modern matches. For example, the head of a "strike anywhere" match is made of a mixture of P_4S_3 and a combustible substance such as sulfur and an oxidizing agent such as potassium chlorate, with glue added to bind these compounds to the matchstick. The stick is usually impregnated with ammonium phosphate to prevent after-glow when the match is extinguished. The heat of friction produced by drawing the match rapidly over most surfaces ignites the phosphorus sulfide, which sets fire to the remaining material in the head, and finally the wood in the matchstick.

Oxides and Oxyacids

The oxides P_4O_6 and P_4O_{10} are obtained by burning white phosphorus, P_4, in a limited supply of air and in an excess of air, respectively. The oxide P_4O_6 is the anhydride of **phosphorous acid,** H_3PO_3, whereas P_4O_{10} is the anhydride of **phosphoric acid,** H_3PO_4.

Figure 8-6   The structure of white phosphorus.

Probably the most important property of P_4O_{10} is its affinity for water. Because of this, it is used extensively as a thorough drying agent.

The reaction of water with P_4O_{10} can lead to the formation of several phosphoric acids. **Metaphosphoric acid, (HPO₃)ₙ,** the product of limited hydrolysis of P_4O_{10}, is a polymeric acid, which, on addition of more water, gives first **pyrophosphoric acid, $H_4P_2O_7$,** and then **orthophosphoric acid, H_3PO_4:**

$$P_4O_{10}(s) + 2H_2O \rightleftharpoons 4(HPO_3)(l)$$
$$2HPO_3(l) + H_2O \rightleftharpoons H_4P_2O_7(l)$$
$$H_4P_2O_7(l) + H_2O \rightleftharpoons 2H_3PO_4(l)$$

Orthophosphoric acid is a weak triprotic acid ($K_{a1} = 7.1 \times 10^{-3}$; $K_{a2} = 6.3 \times 10^{-8}$; and $K_{a3} = 4 \times 10^{-13}$).

The phosphoric acids and their salts play important roles in many technical processes. Orthophosphoric acid is used in the rust-proofing and preparation of sheet steel for painting. This is the "bonderizing" process employed by the automobile industry. This acid is often used as a catalyst in organic reactions, for example, in the production of the important organic compounds, alcohols. Sodium salts of the phosphoric acids find many important applications in the home and in industry. Trisodium phosphate, often called washing powder, is used to soften water and in boiler-water treatment. Monosodium phosphate is used in the manufacture of baking powders and pharmaceuticals. Hexametaphosphates, $(NaPO_3)_6$, are very useful in the treatment of hard water, since they have the property of forming a complex with calcium ions that does not precipitate. Another important sodium salt is tetrasodium pyrophosphate, $Na_4P_2O_7$, which has been used as an integral constituent of soaps and soap powders. It helps prevent the deposition of dirt particles or particles of insoluble calcium and magnesium soaps on the surface of fabrics being washed in hard water.

TOPIC OF CURRENT INTEREST

PLANT NUTRIENTS AND EUTROPHICATION

Phosphates and nitrates as major constituents of fertilizers furnish nitrogen and phosphorus necessary for plant growth. Phosphates are added to detergents and soaps to increase pH by hydrolysis and to combine with Ca^{+2} and Mg^{+2} to preclude their precipitation with soap. Unfortunately these materials are not recovered from sewage, and some fertilizer is always lost in rainfall runoff.

The growth rate of many aquatic plants is limited by the natural supply of nutrients including nitrate, NO_3^-, and phosphate, PO_4^{-3}. If these materials are furnished in large supply from sewage and agricultural runoff, the growth rate of algae and other aquatic plants can increase rapidly. As the rate of generation of decaying plant material increases, dissolved oxygen is consumed more rapidly, and anaerobic conditions are produced. Under these conditions fish and other aquatic life forms die. Increasing amounts of decaying plant material eventually convert the lake into a marsh, and the lake finally disappears. This process, called **eutrophication,** is a very slow but natural process in the absence of increased nutrients. However, it speeds to completion with man's supply of nitrates and phosphates.

CARBON

The elements of group IVA include **carbon, silicon, germanium, tin,** and **lead.** Carbon displays physical and chemical properties characteristic of a nonmetal; silicon and germanium behave as metalloids; and tin and lead are metals.

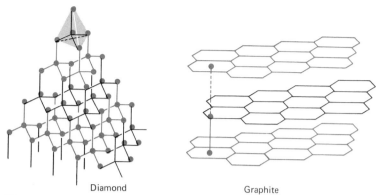

Figure 8–7   The structures of diamond and graphite. The physical properties of the two pure allotropes are strikingly different.

Silicon is the second most abundant element in the earth's crust at 26 per cent of the total. It does not occur naturally in the elemental state but rather is found as silicon dioxide or silicates. Sand, flint, and quartz are silicon dioxide, SiO_2. Metallic silicates are the primary constituents of rocks, soil, and clay.

Carbon occurs on earth both as the element and in chemical combination. Elemental carbon occurs as two pure allotropes, **diamond** and **graphite**, and in many impure **amorphous** forms such as coal. The structure of diamond involves carbon atoms each covalently bonded to four adjacent carbon atoms at the corners of a tetrahedron as shown in Figure 8–7. This bonding results in a three-dimensional structure with covalent bonds extending through the entire crystal. For this reason diamond crystals are very hard and are broken with difficulty. The graphite structure involves layer after layer of carbon atoms in a plane, each surrounded by three carbon atoms as shown in Figure 8–7. The forces between the planes are comparatively weak, so that they can "slip" relative to one another. For this reason graphite is soft and can act as a lubricant. Graphite is also an electrical conductor while diamond is an insulator.

Graphite is one of the softest and lightest of the naturally occurring minerals, whereas diamond is the hardest natural mineral known and has a much higher density than graphite. Diamond is often colorless and transparent, in contrast to graphite, which is black and opaque. Deposits of graphite are found in several parts of the world, but those in the United States do not furnish sufficient graphite for industrial requirements. Large quantities of graphite are produced artificially by heating amorphous forms of carbon in an electric furnace at extremely high temperatures. In the presence of a catalyst the carbon is gradually converted into crystalline graphite.

OXIDES AND OXYACIDS

Carbon dioxide is formed by the combination of oxygen and carbon. This gas, which has extensive biological significance, occurs in the atmosphere in a concentration of about 0.03 per cent, whereas 20 to 30 times this amount is dissolved in the water of the oceans. Gases emanating from underground pockets or accumulations in volcanic regions are often rich in carbon dioxide.

The preparation of carbon dioxide is readily achieved by burning carbon in the presence of an excess of oxygen. In the laboratory, chunks of marble or other metal carbonates are treated with dilute hydrochloric acid to produce the gas:

$$CaCO_3(s) + 2HCl(aq) \rightarrow CaCl_2(aq) + H_2O + CO_2(g)$$

Commercially, large quantities of the gas are formed as a byproduct in the preparation of alcohols by fermentation. After purification, the carbon dioxide is stored in steel cylinders under pressure.

Normally carbon dioxide exists as a colorless, odorless gas with a slightly sharp taste. When the gas is subjected to 60 or 70 atmospheres of pressure, as, for example, in the commercial steel cylinders, most of it liquefies. If this liquid carbon dioxide is allowed to escape into a container, it vaporizes so rapidly that it is quickly cooled to a temperature of $-80°C$ and part of it freezes to a white solid. Commercially, this white carbon dioxide snow is pressed into blocks of **dry ice** for sale. The extremely low temperatures and ease of handling afforded by dry ice have resulted in many applications as a refrigerant.

Carbon dioxide dissolves in water to form **carbonic acid,** H_2CO_3, which is a very weak diprotic acid ($K_{a1} = 4.2 \times 10^{-7}$; $K_{a2} = 9.7 \times 10^{-11}$):

$$CO_2(g) + H_2O \rightarrow H_2CO_3(aq)$$

If a solution of carbon dioxide and water is prepared under pressure, soda water results, which is the basis for the common carbonated beverages.

When carbon is burned in a limited supply of oxygen, carbon monoxide is formed. This compound may also be formed in a coal furnace when the carbon dioxide from combustion of the coal passes over more hot carbon:

$$CO_2(g) + C(s) \rightarrow 2CO(g)$$

Carbon monoxide is a colorless, odorless gas that differs from carbon dioxide in being combustible in air and not soluble in water. The blue flame of burning carbon monoxide can often be seen over the top layer of unburned coal in a coal furnace. One of the outstanding properties of the gas is its poisonous nature. Continuous inhalation of the gas at concentrations as low as 0.05 per cent will produce headaches, dizziness, and unconsciousness in a few hours. Breathing automobile exhaust fumes, which usually contain 7 per cent carbon monoxide, will result in death in a few minutes.

TOPIC OF CURRENT INTEREST

AIR POLLUTANTS

Three of the five major classes of primary air pollutants are nonmetal oxides. The primary air pollutants, as classified by the U.S. Environmental Protection Agency, are, **carbon monoxide, oxides of nitrogen, oxides of sulfur, hydrocarbons,** and **particulate matter.**

Carbon monoxide (CO) in the atmosphere is generated by natural processes such as forest fires and volcanic eruptions, but 90 per cent of the earth's carbon monoxide is generated by man in burning fossil fuels. The internal combustion engine generates large quantities of carbon monoxide, and urban areas, especially those near freeways and busy streets, experience carbon monoxide levels far in excess of the average atmospheric concentration.

Carbon monoxide is a primary pollutant because it is poisonous to humans. It binds to **hemoglobin** (p. 278) more strongly than does molecular oxygen and inhibits the transport of oxygen by hemoglobin from the lungs to all parts of the body.

Oxides of nitrogen—especially nitric oxide (NO) and nitrogen dioxide (NO_2)—are also produced during the combustion of fossil fuels in the internal combustion engine, particularly at high temperatures. Man-made sources of nitrogen oxides compose 99 per cent of the total found in the atmosphere. These oxides are very irritating to all animals because they form nitric acid (HNO_3) when in contact with moisture in the eyes, throat, and lungs. The increasing levels of nitrogen oxides in urban areas have resulted in federal legislation requiring all 1975 model automobiles to use catalytic converters in the exhaust system. These converters catalyze the conversion of nitrogen oxides back to N_2 and O_2.

Oxides of sulfur, SO_2 and SO_3, are produced in the combustion of coal and petroleum products containing sulfur compounds. Only 2 per cent of the sulfur oxides found in the atmosphere are from natural sources. These oxides are the major constituents of several "killer fogs" that have occurred in urban areas with large coal-burning industrial installations. The sulfur oxides form sulfurous acid (H_2SO_3) and sulfuric acid (H_2SO_4) upon contact with water and cause severe respiratory complications.

Since a marked increase in the use of coal as a primary energy source is expected in the near future (p. 3), the problem of sulfur oxides generated in coal combustion is currently under extensive investigation. Possible solutions to the problem all center around removing sulfur oxides from the gases produced by combustion or converting coal to combustible gases and removing sulfur compounds before combustion.

The term **smog** has been applied to the killer fogs generated by sulfur oxides, but the term also applies to an air pollution phenomenon which results from hydrocarbons interacting with nitrogen dioxide (NO_2) and ozone (O_3) under the influence of sunlight. Nitrogen dioxide is decomposed by sunlight into NO and oxygen atoms which react with O_2 to give O_3. The O_3 generated in this way and by lightning or other natural processes is short lived because it reacts with the NO to give back NO_2 and O_2, resulting in no net chemical change. However the presence of hydrocarbons, in large part emitted from internal combustion engines or storage tanks, interrupts this NO_2–O_3 cycle. The hydrocarbons react with atomic oxygen and then with NO and exclude O_3. This process results in oxidized hydrocarbons and dangerously high concentrations of O_3, which is very toxic. In addition O_3 oxidizes hydrocarbons so that a mixture of oxidized hydrocarbons, nitrogen oxides, and ozone tends to accumulate. Since this process is initiated by sunlight, this type of smog, called **photochemical smog**, dissipates at night. If nitrogen oxide and hydrocarbon levels of the air were decreased significantly, this type of air pollution would not prevail.

HYDROGEN

Occurrence

Atoms of hydrogen do not occur in the free form, but are combined in pairs to form a hydrogen molecule by sharing one pair of electrons. In contrast to oxygen, very little free hydrogen is found in the atmosphere. It is found to some extent in natural gas wells, although in general the quantity of hydrogen produced from natural sources is extremely limited. In comparison, the hot gases surrounding the sun are apparently rich in hydrogen, and flames of incandescent hydrogen sometimes reach thousands of miles out from the sun.

Many important compounds contain hydrogen as one of their constituents. For example, all acids and bases contain hydrogen, and water is composed of hydrogen and oxygen. Almost all organic compounds present in plant and animal tissue contain hydrogen in combined form. Like oxygen, it is an important constituent in carbohydrates, fats, and proteins, which are used as foods. It occurs combined with carbon in the important products of the petroleum industry, such as gasoline, lubricating oils, and natural gas.

Preparation

An important method for the preparation of both hydrogen and oxygen by the electrolysis of water has already been described. When an electric current is passed through water containing a small quantity of acid to increase its conductivity, hydrogen is evolved at the negative electrode while oxygen is evolved at the positive electrode. Pure hydrogen is obtained by this method and is usually stored in steel cylinders under pressure for future use. Hydrogen can also be evolved from water by the addition of **active metals.** It may be recalled that the most active metals are located in Group IA of the periodic table.

A very common method for the preparation of hydrogen in the laboratory involves the addition of dilute acids to certain metals. For example, dilute hydrochloric acid may be added to zinc granules.

$$Zn + 2HCl \rightarrow H_2\uparrow + ZnCl_2$$

Not all metals can be used effectively for the preparation of hydrogen by this method, since some metals produce only a few bubbles of hydrogen from acids, and others show no reaction.

Commercially, hydrogen is produced from **water gas,** a mixture of hydrogen and carbon monoxide. First, the water gas is prepared by passing steam over burning coke (temperature, about 1000°C). Water gas and steam are then passed over a metallic oxide at a temperature of 500°C. The metallic oxides serve as a catalyst to help in the oxidation of carbon monoxide to form carbon dioxide. These two reactions may be represented as follows:

$$\underset{\text{water gas}}{C + H_2O \rightarrow CO + H_2}$$

$$CO + H_2O \xrightarrow{\text{catalyst}} 2H_2 + CO_2$$

The mixture of hydrogen and carbon dioxide produced in the final equation is passed through water to remove the carbon dioxide, and the hydrogen may then be used immediately or stored in steel cylinders.

PHYSICAL PROPERTIES

Hydrogen, like oxygen, is a colorless, odorless, and tasteless gas. Approximately 2 ml of hydrogen will dissolve in 100 ml of water, which makes this element less soluble than oxygen. It is the lightest gas known, one liter weighing approximately 0.09 g at STP. Hydrogen can be liquefied, but only at extremely low temperatures. Certain metals can adsorb large quantities of hydrogen gas (adsorption is a surface phenomenon; in this case the hydrogen gas would adhere to the atoms on the surface of the metal). Powdered palladium, for example, can adsorb approximately one thousand times its own volume of hydrogen, and this adsorbed hydrogen is chemically very reactive.

CHEMICAL PROPERTIES

In the early forms of the periodic table, hydrogen was placed only at the top of Group I, followed by Li, Na, K, and other elements of the group. This would classify it as a metal that would readily lose one electron to form ionic compounds with non-metallic elements, or combine with them by sharing a pair of electrons. It is well known that hydrogen may also combine with metals to form ionic compounds called **hydrides,** in which the hydrogen acts as a negative ion similar to elements in Group VII. This property of hydrogen to act as a metal or a nonmetal is shown in modern periodic tables in different ways. The two most common methods are: (1) to place it at the top of Group IA and Group VIIA and (2) to center it at the top of the table with lines connecting it to the top of Groups IA and VIIA.

Hydrogen gas does not react readily with other elements at ordinary temperatures; but if the temperature is increased it will react with certain elements. When burned in the presence of oxygen, sulfur, or chlorine, hydrogen combines with these elements to form water, hydrogen sulfide, or hydrogen chloride as follows:

$$2H_2 + O_2 \rightarrow 2H_2O$$
$$H_2 + S \rightarrow H_2S$$
$$H_2 + Cl_2 \rightarrow 2HCl$$

Under proper conditions and in the presence of a catalyst, hydrogen will combine with other elements and other compounds. For example, hydrogen will combine directly with nitrogen to form the compound ammonia as shown in the following equation:

$$N_2 + 3H_2 \rightarrow 2NH_3$$

In all of the above compounds hydrogen exhibits its most common oxidation number, or oxidation state, of $+1$.

Hydrogen also has the property of combining with active metals to form a negative **hydride** ion with an oxidation state of -1. An important metal hydride, calcium hydride, is formed by passing hydrogen over hot calcium metal.

$$Ca + H_2 \rightarrow \quad CaH_2$$
Calcium hydride

Isotopes of Hydrogen

Natural hydrogen is composed of 3 isotopes, 2 of which are stable and 1 radioactive. The lightest isotope, or **protium** ($_1^1H$), is the most common and makes up 99.98 per cent of the atoms, whereas **deuterium** ($_1^2H$ or D) makes up approximately 0.02 per cent. The third isotope, which is radioactive and occurs in extremely low concentration, is **tritium** ($_1^3H$ or T). The symbols for isotopes have a left subscript designating the atomic number and a left superscript designating the mass number.

Of the approximately 1500 isotopes that are known, the three hydrogen isotopes exhibit the greatest differences in physical properties. For example, the freezing and boiling points of deuterium are higher than those of ordinary hydrogen, whereas the lighter atoms have a much higher rate of diffusion than those of deuterium.

IMPORTANT TERMS AND CONCEPTS

allotropic modification	halides	nitrogen fixation	oxyacids
combustion	hydrides	Ostwald process	ozone
contact process	metalloids	oxides	photochemical smog
eutrophication			

QUESTIONS

1. Which of the following apply to nonmetals?

   (a) high electronegativity
   (b) low electronegativity
   (c) tendency to form anions
   (d) tendency to form cations

(e) occupies upper left of periodic table
(f) occupies upper right of periodic table
(g) many of their oxides react with water to form acids
(h) many of their oxides react with water to form bases

2. In proceeding *down* the halogen column in the periodic table, which of the following are true?

(a) boiling point increases   (e) electron affinity increases
(b) boiling point decreases   (f) electron affinity decreases
(c) atom size increases   (g) abundance in nature increases
(d) atom size decreases   (h) abundance in nature decreases

3. What is the laboratory preparation of Cl_2?

4. Write the equation illustrating the testing for the presence of Br^- in a solution.

5. Write equations for the laboratory preparation of each of the four hydrogen halides.

6. How is H_2SO_4 prepared commercially? Give the reactions involved.

7. Write an equation for the commercial preparation of ammonia. What is the name of this process?

8. Active metals combine with ammonia to form what?

9. What are the names and formulas of the most important oxyacids of nitrogen?

10. Give equations for the commercial preparation of nitric acid by the Ostwald process.

11. What are the two crystalline allotropic forms of carbon? Describe their distinguishing structures.

12. Why does oxygen so readily combine with other elements to form oxides?

13. Explain the differences between oxygen and ozone.

14. Why is it difficult to show the exact position of hydrogen in the periodic table?

SUGGESTED READING

Hall: Sulfur Compounds in the Atmosphere. Chemistry, Vol. 45, No. 3, p. 16, 1972.
Keller: Hydrogen—The Simplest. Chemistry, Vol. 42, No. 10, p. 19, 1968.
Medeiros: Air Pollution: Could It Change The World's Climate? Chemistry, Vol. 47, No. 6, p. 18, 1974.
Medeiros: Carbon Monoxide: The Invisible Enemy. Chemistry, Vol. 46, No. 1, p. 18, 1973.
Medeiros: Smog Formation Simplified. Chemistry, Vol. 45, No. 1, p. 16, 1972.
Navratil: Fluorine—a Hostile Element. Chemistry, Vol. 42, No. 2, p. 11, 1969.
Rochow: The Metalloids. Lexington, Mass., D. C. Heath & Co., 1966.
Sanderson: Principles of Halogen Chemistry. Journal of Chemical Education, Vol. 41, p. 361, 1965.
Sherwin and Weston: Chemistry of the Non-Metallic Elements. New York, Pergamon Press, 1966.

SOME CHEMISTRY OF METALLIC ELEMENTS

The *objectives* of this chapter are to enable the student to:

1. Identify the representative metals, the transition metals, and the innertransition metals in the periodic table.
2. Discuss general techniques used in recovery of metals from ores.
3. Use ionization potential data to explain commonly occurring oxidation states of metals.
4. Write an equation illustrating the hydrolysis of a metal oxide.
5. Define the term amphoterism.
6. Define and explain the "lanthanide contraction."
7. Discuss general properties of the transition metals.
8. List the most important oxidation states for the first row transition metal elements.
9. Define coordination compound, ligand, coordination sphere, and coordination number.
10. Illustrate with a drawing *cis-trans* isomerism in square planar or octahedral complexes.

    The elements which have the properties of metals are shown in Figure 9–1. These are found in every group of the periodic table except those of the oxygen family, the halogens, and the inert gases. The metals are subdivided into the **representative metals,** the **transition metals,** and the **innertransition metals.** The representative metals found in the A groups of the periodic table have either empty or completely filled inner *d*-orbitals. For example, potassium and calcium have empty 3*d*-orbitals, whereas gallium has a completely filled set of 3*d*-orbitals. The transition metals have from one to ten electrons in inner *d*-orbitals and no electrons in the valence shell *p*-orbitals. The elements scandium through nickel contain partially filled 3*d*-orbitals, whereas copper and zinc contain ten electrons in 3*d*-orbitals. None of these elements has electrons in 4*p*-orbitals. The innertransition metals have a partially filled or a completely filled set of *f*-orbitals. For example, the elements cerium, Ce, through thulium, Tm, have a partially filled set of 4*f*-orbitals, whereas ytterbium, Yb, and lutetium, Lu, have filled sets of 5*f*-orbitals.

    It is difficult to give an all-inclusive definition for metals, although there are many

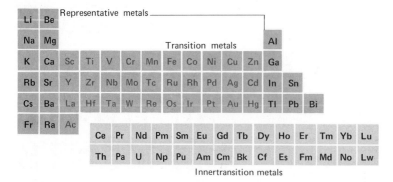

Figure 9–1   The metals in the periodic table.

properties that are common to the majority of these elements. For example, most metals exhibit a metallic luster, are good conductors of electricity and heat, and are solids at room temperature. When combined with other elements they always exist in positive oxidation states. Metals usually possess a degree of hardness greater than the nonmetals, although sodium and potassium are very soft metals, and mercury is a liquid. The common metals possess to a varying degree the properties of malleability and ductility. A metal is malleable if it can be hammered or rolled into very thin sheets, whereas it is ductile if it can be drawn into a wire without breaking. Gold and silver are the most malleable and ductile of metals. Sheets of gold can be prepared that are less than 0.00025 mm thick, and a gram of gold can be drawn into a wire over 7 kilometers long.

Since many of the compounds of metals are ionic, the chemical behavior of the metals is often largely determined by properties associated with the metallic ion. These include ionization potential of the element, atomic radius of the element, ionic radius, and charge on the ion.

THE OCCURRENCE AND RECOVERY OF METALS

Most metals are found in chemical combination in nature. Only the relatively nonreactive metals, such as copper, silver, platinum, and gold, are found in nature in their elemental state. Mineral deposits containing chemically combined metals are called **ores.** The most common ores contain metal oxides, sulfides, halides, carbonates, silicates, and sulfates. Although many ores contain only one compound of a metal, they are often mixtures.

The many processes of recovering metals from their ores compose the field of **metallurgy.** Although unique metallurgical processes are used for almost every metallic element, these processes do have some common features. These include concentration of the ore, reduction to the elemental state, and refinement to the desired purity.

The initial processes in the concentration of an ore usually involve removal of undesirable minerals called **gangue.** This is usually done by crushing and grinding the mineral to a finely divided state and then separating the ore and gangue by taking advantage of some difference in physical property, such as density or the ability to wet.

Sulfide and carbonate ores are often converted to oxides by roasting in air:

$$2MS(s) + 3O_2(g) \xrightarrow{\text{heat}} 2MO(s) + 2SO_2(g)$$

$$MCO_3(s) \xrightarrow{\text{heat}} MO(s) + CO_2(g)$$

Reduction of metal ions in ores to elemental metal is performed in a number of ways. Chemical reductions are performed using carbon, carbon monoxide, hydrogen, active metals, and electrolytic processes. Some metallic sulfides, when heated, undergo a self-reduction as follows:

$$MS(s) + O_2(g) \xrightarrow{\text{heat}} M(s) + SO_2(g)$$

Impure metals obtained in the reduction of ores are purified most commonly by electrolytic processes. Occasionally, low-melting metals are separated from impurities by melting and recrystallization, or by distillation.

ALLOYS

Although special uses are found for many metals in a pure state, most metals are used as mixtures. These mixtures of metals, usually designed to generate desired physical or chemical properties, are called **alloys.** The preparation of an alloy is most often performed to adjust hardness, melting point, or chemical reactivity, such as corrosive resistance. Alloys are almost always harder than their major component. Alloys are prepared by melting metals together to form a molten homogeneous state and then allowing this state to crystallize. Nonmetals are often included in metals to give special properties.

Alloys are similar to metals in that they are good thermal and electrical conductors. They usually display other physical properties of metals such as malleability, ductility, and metallic luster.

THE REPRESENTATIVE METALS

The representative metals include the alkali metals, the alkaline earth metals, the Group III elements except boron, the Group IV elements tin and lead, and bismuth.

THE ALKALI METALS

The **alkali metals, lithium, sodium, potassium, rubidium, cesium,** and **francium,** are the members of Group IA. The atoms of these elements are the largest found in any period. Because of the relatively large distance of the outermost electrons from the nucleus, these elements have the smallest ionization potentials in each period. As expected, none of the alkali metals is found in nature because of their very high reactivity.

Freshly prepared alkali metals have a highly lustrous surface, very good electrical and thermal conductivity, and relatively low melting and boiling points. The melting points and the boiling points decrease with increasing atomic number. Because of their low boiling points, these metals are often purified by distillation.

The loss of one electron by an alkali metal atom results in the electron configuration of an inert gas. For each of the alkali metals, the first ionization potential is much smaller than the second ionization potential (energy required to remove an electron from M^{+1}). Because of this, these elements always occur in chemical combination as $+1$ ions.

The alkali metals combine directly with most nonmetals to form binary compounds, virtually all of which are ionic and exist as crystalline solids. On combining with oxygen, lithium forms the oxide Li_2O, sodium forms the peroxide Na_2O_2, and potassium, rubidium, and cesium give superoxides of formula MO_2. Each of these compounds hydrolyzes when added to water to give basic solutions. Crystalline alkali metal hydroxides, when dissolved in water, behave as strong electrolytes.

Compounds of alkali metals, including salts of oxyacids, are highly soluble in water. However, many compounds of lithium show much lower solubility than corresponding compounds of the other elements.

The alkali metals react violently with water to give hydrogen and basic solutions:

$$2M(s) + 2H_2O \rightleftharpoons H_2(g) + 2M^+(aq) + OH^-(aq)$$

THE ALKALINE EARTH METALS

The **alkaline earth metals, beryllium, magnesium, calcium, strontium, barium,** and **radium,** are the members of Group IIA. These elements are slightly less reactive than the alkali metals, but are too reactive to occur naturally in the elemental state. They do occur in many minerals, primarily as halides, carbonates, and sulfates. The elements are usually recovered from these minerals by converting to halides and electrolyzing the fused halide.

The chemical and physical properties of the alkaline earth metals can usually be predicted by considering the properties of the neighboring alkali metals and how they change with decreased atomic and ionic radii. Each of the alkaline earth metals has a highly lustrous surface and is a good electrical conductor.

When reacting, the alkaline earth metals tend to lose two electrons to generate M^{+2} ions. These ions have the same electronic configurations as inert gases and the M^{+1} ions of the neighboring alkali metal. There are no stable compounds in which a Group IIA element exhibits an oxidation number of $+1$.

Binary and ternary compounds of calcium, strontium, and barium are highly ionic and exist as typical crystalline ionic solids which conduct electricity when in the molten state. Beryllium and, to a much smaller extent, magnesium have a pronounced tendency to form compounds with covalent bond characteristics. These are rather poor conductors in the molten state.

Group IIA elements are quite reactive and combine with most nonmetals to form binary compounds. Each of the elements combines with oxygen to give the oxide MO. All the oxides, except beryllium oxide, BeO, hydrolyze on addition to water to form the hydroxides, $M(OH)_2$, which are relatively insoluble but are strong bases. Beryllium oxide is insoluble in water but dissolves slowly in concentrated acid solution to give $[Be(H_2O)_4]^{+2}$. It also dissolves in concentrated basic solution to give $[Be(OH)_4]^{-2}$. These reactions are represented as follows:

acidic solution $\quad BeO(s) + 3H_2O + 2H^+(aq) \rightleftharpoons [Be(H_2O)_4]^{+2}(aq)$

basic solution $\quad BeO(s) + H_2O + 2OH^-(aq) \rightleftharpoons [Be(OH)_4]^{-2}(aq)$

Thus, beryllium oxide acts as a base when added to an acid, and as an acid when added to a base. Such behavior is referred to as **amphoterism,** and beryllium oxide is called **amphoteric.**

THE GROUP IIIA METALS

All the Group IIIA elements except boron behave as metals. These include **aluminum, gallium, indium,** and **thallium.** Aluminum, the third most abundant element in the earth's crust, at 7.5 per cent, is the most abundant metal.

The trends of various properties, such as hardness and reactivity, observed for the Group IIA metals are even more pronounced in the case of the Group IIIA metals.

Aluminum is a very hard, strong, lustrous metal, which in air always has a surface film of aluminum oxide, Al_2O_3. This film protects the metal from corrosion.

Although the +3 oxidation state is most common in compounds of the Group IIIA metals, the +1 oxidation state is known in each case. The stability of the +1 oxidation state increases going down the family and is quite important in the chemistry of thallium. The Group III metals react completely with most nonmetals, including the halogens, oxygen, and sulfur.

The Group IVA Metals

Tin and **lead** are the only Group IVA elements which behave as metals. Tin, which is known to exist as three temperature-dependent allotropes, is at room temperature a malleable, lustrous metal, but a poor electrical conductor. Elemental tin is used extensively in alloys. Some of the important alloys of tin are **solder** (33% Sn, 67% Pb), **babbitt metal** (90% Sn, 7% Sb, 3% Cu), **pewter** (85% Sn, 7% Cu, 6% Bi, 2% Sb), and several **bronzes** which contain from 10 to 20% tin.

Lead, which appears as dull gray, does have a lustrous surface when freshly exposed. It is very soft and has a relatively low melting point. Elemental lead is used extensively in the pure state in storage battery plates, lead pipes, bullets, and cable covering. Many alloys of lead are also used, including **solder** and **type metal** (82% Pb, 15% Sb, 3% Sn).

In the chemistry of the Group IVA metals, the +2 and +4 oxidation states are both prevalent. The +2 oxidation state is more important in lead chemistry.

THE TRANSITION METALS

The **transition metals** are those elements which have partially filled sets of d-orbitals or filled sets of d-orbitals with empty outer p-orbitals. The **innertransition metals** are really a subgroup of the transition metals. These elements have a partially filled set of f-orbitals or a filled set of f-orbitals.

The Innertransition Series

The innertransition metal series, **cerium,** Ce, through **lutetium,** Lu, which is referred to as the **lanthanides,** has a partially or totally filled set of 4f-orbitals. These orbitals are located considerably inside of the 6s-orbital. The physical and chemical behavior of the members of this series is consequently not significantly related to the 4f-orbital electron configurations. The major differences in the properties observed within the series are due to a smooth decrease in atomic or ionic radius with increasing atomic number. This effect is called the **"lanthanide contraction."**

The second row of innertransition metals, called the **actinides,** contains only six elements found in nature. The other elements of the series have all been synthesized by nuclear processes.

General Properties of the Transition Metals

Proceeding across the first transition metal series, many of the trends within these properties can be attributed to the effect of increasing nuclear charge. Trends similar to those observed in this series are also observed in the second and third transition metal series.

With the exception of the Group IIB metals **zinc, cadmium,** and **mercury,** all of the transition metals have relatively high melting and boiling points. **Mercury** is a liquid which freezes at $-38.9°C$, and cadmium and zinc are solids which melt at $321°C$ and $420°C$, respectively. All of the transition metals are good thermal and electrical conduc-

TABLE 9-1   THE OXIDATION STATES OF THE FIRST
TRANSITION METAL SERIES

Sc	Ti	V	Cr	Mn	Fe	Co	Ni	Cu	Zn
+3	+2	+2	+2	+2	+2	+1	+2	+1	+2
	+3	+3	+3	+3	+3	+2	+3	+2	
	+4	+4	+4	+4	+4	+3		+3	
		+5	+5	+6	+6	+4			
			+6	+7					

tors. The Group IB elements, **copper, silver,** and **gold,** are outstanding electrical conductors.

The elements of the first transition metal series, as well as the second and third series, typically display several oxidation states. These are shown in Table 9–1 for the first transition metal series. Unlike the representative metals, many of the oxidation states of the transition metals leave an odd number of electrons on the metal ion. As a result of this, many transition metal compounds are **paramagnetic** (i.e., attracted to a magnetic field). The transition metals form many simple binary or ternary compounds. However, the chemistry of transition metal ions is dominated by the tendency of the ion to surround itself with the maximum possible number of Lewis bases, whether molecules, atoms, or ions. This behavior results in the formation of **coordination compounds** which are discussed in a later section.

RECOVERY OF IRON

Iron is the second most abundant metal, constituting 4.7 per cent of the earth's crust. The major ores are **hematite,** Fe_2O_3, **limonite,** $FeO(OH)$, **siderite,** $FeCO_3$, and **magnetite,** Fe_3O_4.

To recover metallic iron from its ores, the oxides are reduced by heating in the presence of coke and limestone. The burning coke or carbon combines with oxygen to form carbon dioxide and carbon monoxide. The carbon monoxide formed reacts with the iron oxide in the ore to form iron and carbon dioxide:

$$3CO(g) + Fe_2O_3(s) \rightarrow 3CO_2(g) + 2Fe(l)$$

The iron formed in this way, called **pig iron,** may be converted into several other types of iron, such as cast iron, malleable iron, wrought iron, and sponge iron. For the preparation of cast iron, pig iron is melted with scrap iron in the foundry and is poured into metal or sand molds. If it is rapidly cooled in the metal molds, it is very hard and brittle and is known as white cast iron. When allowed to cool slowly in sand molds, gray cast iron is formed, which is not as hard or brittle as the white variety and is readily machineable. Small amounts of other metals, such as nickel, copper, or chromium, are added to cast iron to produce alloys whose properties are superior to the original cast iron. If cast iron is remelted and kept at a faint red heat for several days and then slowly cooled, malleable iron is produced. This type of iron is stronger than gray cast iron and has a greater resistance to shock.

A special type of iron that is ductile and malleable and can be forged and welded is called wrought iron. This iron is produced by gradually oxidizing the impurities out of melted pig iron until the product becomes soft and pasty. Large balls of this semisolid mass are removed from the furnace and rolled into sheets. The resultant product is nearly pure iron containing approximately 1 per cent slag. The slag is not dissolved in the iron but forms very thin films that separate the iron fibers. This content of slag apparently imparts strength, toughness, flexibility, and resistance to corrosion and oxidation.

Sponge iron is made by the reduction of iron ore by natural gas. The iron does not melt, but remains as a soft spongy mass. This type of iron is in great demand for magnets in motors and in electric appliances.

Steel is an alloy of iron and carbon containing very small amounts of the same impurities present in cast iron. The carbon content varies from 0.1 to 1.5 per cent and the degree of hardness increases with increasing carbon content.

COORDINATION COMPOUNDS

A **coordination compound** consists of a central metal ion bonded to surrounding atoms, molecules, or ions by coordinate covalent bonds. Transition metal ions have a high tendency to form coordination compounds, because they have partially filled d-orbitals which can accept electron pairs. The representative metal ions also form coordination compounds.

The atoms or ions surrounding the central metal ion in a coordination compound are called **ligands,** and they constitute the **primary coordination sphere.** The number of ligands in the primary coordination sphere is the **coordination number.** In the complex ion $[PtCl_4]^{-2}$, Pt^{+2} is the central metal ion, and the primary coordination sphere consists of four ligands which are chloride ions. The coordination number is four. In the complex compound $[Pt(NH_3)_3Cl]Cl$, the primary coordination sphere contains three ammonia molecules and a chloride ion. A second chloride ion is part of the compound, but it is outside of the primary coordination sphere. In writing the formulas of complex compounds, the central metal ion and the contents of the primary coordination sphere are enclosed in brackets.

The most common coordination numbers found in coordination compounds of the transition metal ions are two, four, and six. The geometries of typical complex ions are shown in Figure 9–2. Species of coordination number two, such as $[Ag(NH_3)_2]^+$, are linear. Species of coordination number four are either tetrahedral or square planar, and six coordinate compounds are octahedral.

Molecules, ions, and atoms which have lone pairs of electrons available for donation can act as ligands. Ligands which bond through one lone pair of electrons are called **monodentates.** Ligands having more than a single site which can furnish a lone pair of electrons are called **chelates** (Greek, *chele*—claw) or **polydentates.** Those which bond to a metal ion through two sites, called **bidentate ligands,** include **ethylenediamine** (en), **oxalate** (ox), and **acetylacetonate** (acac), as shown in Figure 9–3. Polydentate ligands having up to six and more donor sites are known. When chelates coordinate to metal ions, they form closed rings which include the metal ion. In the compounds $[Pt(en)_2]^{+2}$ and $[Co(en)_3]^{+2}$, two and three rings are formed as seen in Figure 9–4.

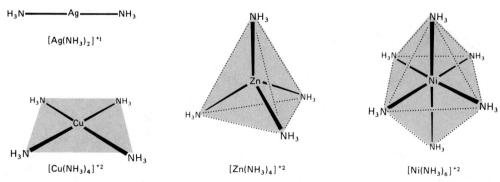

Figure 9–2   The common geometries of two, four, and six coordinate complex ions.

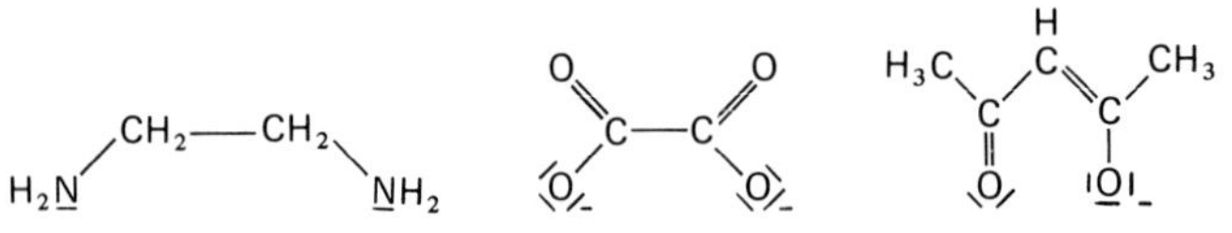

Figure 9-3   Some bidentate ligands.

(en)   (ox)$^{-2}$ (acac)$^{-1}$

ISOMERISM

When a coordination compound has two or more of each different ligand on the central metal ion, it is possible to have more than one structural formula. This phenomenon is referred to as **stereoisomerism**. This is illustrated by the two possible structures for $[Pt(NH_3)_2Cl_2]$ shown in Figure 9–5.

These two structures are called **isomers** of the compound. The isomer with like ligands occupying adjacent positions in the coordination sphere is called the **cis-isomer**. The other isomer is the **trans-isomer**. In the tetrahedral complex $[Zn(NH_3)_2Cl_2]$, isomerism of the type mentioned cannot prevail because there is only one possible structure. In the octahedral complex $[Co(en)_2Cl_2]$, cis- and trans-isomers prevail as seen in Figure 9–6. In addition there are two types of cis-isomers called **optical isomers** which are mirror images as shown in Figure 9–6.

AQUO COMPLEXES

Metal ions in aqueous solution, in the absence of other strongly coordinating ligands, have water molecules in their primary coordination spheres. The coordination number of a metal ion in water is dependent upon the charge on the ion and the ionic radius, but most aquo complexes have a coordination number of six. Some of those which have been established include:

$$[Al(H_2O)_6]^{+3}; \ [Cr(H_2O)_6]^{+3}; \ [Fe(H_2O)_6]^{+3}; \ [Cu(H_2O)_6]^{+2}$$

In aqueous solution the formation of coordination compounds other than aquo complexes is best visualized as a metathesis reaction in which a ligand replaces water in the primary coordination sphere. This is exemplified by the reaction of Cu^{+2} ion in water on the addition of ammonia:

$$[Cu(H_2O)_6]^{+2}(aq) + 4NH_3(aq) \rightleftharpoons [Cu(NH_3)_4(H_2O)_2]^{+2}(aq) + 4H_2O$$

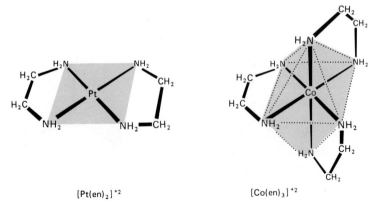

[Pt(en)$_2$]$^{+2}$ [Co(en)$_3$]$^{+2}$

Figure 9–4   Closed rings in chelates.

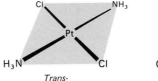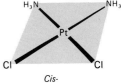

Figure 9-5   *Cis-* and *trans-*[Pt(NH₃)Cl₂].

Trans-   Cis-

THE IMPORTANCE OF COORDINATION COMPOUNDS

Coordination compounds and complex ions play an important role in many processes. Various chelating ligands are often used to remove unwanted ions from aqueous solution. This is a common procedure in water softening. In the photographic process, **thiosulfate ion,** $S_2O_3^{-2}$, is used as a ligand to remove silver ions from the film by the formation of $[Ag(S_2O_3)_2]^{-3}$ in the developing process. Many complex ions play an important role in the catalysis of organic reactions, including hydrogenations.

Undoubtedly, one of the most striking roles of a coordination compound is the oxygen transport behavior of hemoglobin (p. 278), an iron coordination compound. There are many other important biochemical processes in which coordination compounds play an important role, including chemistry of vitamins such as vitamin B_{12} (p. 331), which contains cobalt, and the hydrolysis of proteins. Other biologically important coordination compounds, including chlorophyll which contains magnesium (p. 357), cytochrome C (p. 341), and ferredoxin (p. 357), both containing iron, will be treated more extensively in the biochemistry section of this book.

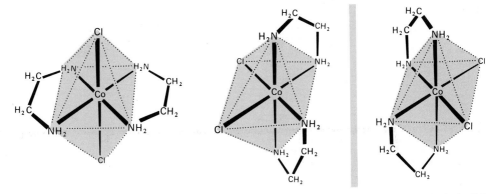

Figure 9-6   The *trans-*isomer of [Co(en)₂Cl₂] (left), and the optical isomers of *cis-*[Co(en)₂Cl₂] (right).

TOPIC OF CURRENT INTEREST

METAL IONS—NECESSARY AND TREACHEROUS

"A little bit goes a long way" adequately describes the behavior of metal ions in the chemical reactions of life. It is known that many metal ions are essential to man if he is to maintain a healthy body. These include ions of calcium, cobalt, copper, iron, magnesium, potassium, and sodium. A deficiency or excess of these ions serves as a perturbing influence on body chemistry and results in symptoms of poor health. Extreme excesses result in toxic effects.

Some metals which are prevalent in the environment have no known beneficial metabolic function and behave as toxic agents even in relatively low concentrations. Such is the case with mercury and lead. These metals behave similarly in the body in that they are not rapidly expelled as are most metal ions. Consequently if ingestion continues even slowly at very low concentrations, the cumulative effects lead to toxic levels.

Although most mercury compounds are insoluble in water and cannot be transported in natural systems, it has been discovered that soluble forms such as CH_3Hg^+ can be generated from metallic mercury and insoluble mercury salts by the action of anaerobic bacteria in the mud and sediment in lakes and streams. This conversion permits mercury to be transported into the food chains of various life forms, including fish. This can be a hazard to man, since its ingestion, even in relatively small amounts, gives toxic reactions.

Lead, which is used more extensively by man than is mercury, is more mobile, since it forms water-soluble and volatile compounds. Although lead poisoning has been traced to the earliest use of lead utensils by man, more recently detected sources include house paint pigments and "moonshine" whiskey which has been distilled from homemade stills which use lead solder in pipe joints. It is indeed alarming that a drastic increase of lead in the environment has been detected since 1940. This date marks the first use of tetraethyl lead as a gasoline additive to reduce "knock" resulting from preignition. Lead escapes the engine as $PbClBr$ and other relatively volatile compounds which constitute 98 per cent of the lead pollution in air. This problem should be decreased to some extent by the increased use of unleaded gasolines required by automobiles with catalytic converters.

IMPORTANT TERMS AND CONCEPTS

alkali metal
alkaline earth metal
alloy
chelate
cis-trans isomers
coordination compound

innertransition metal
ligand
metallurgy
monodentate
polydentate

primary coordination sphere
representative metal
steel
stereoisomerism
transition metal

QUESTIONS

1. Name the three subdivisions of the metals, indicating the differences according to electron configuration.

2. What groups of the periodic table do *not* contain metals?

3. How do most metals occur in nature? Give exceptions.

4. What is metallurgy?

5. Name three methods for purifying elemental metals.

6. What are alloys? How are they made?

7. Write an equation illustrating the hydrolysis of an alkaline earth metal oxide in water.

8. Which of the following properties are generally true for the transition metals?

   (a) high boiling points
   (b) low boiling points
   (c) good thermal and electrical conductors
   (d) poor thermal and electrical conductors
   (e) paramagnetic
   (f) not paramagnetic

(g) usually display only one oxidation state
(h) usually display several oxidation states

9. Match the following columns:

   (a) coordination compound
   (b) ligand
   (c) coordination number
   (d) primary coordination sphere

   (w) ions or atoms surrounding central metal ion
   (x) all ligands directly bonded to the metal ion
   (y) central metal with atoms, ions, or molecules bonded by coordinate covalent bonds
   (z) number of ions or atoms surrounding central metal ion

10. Write chemical reactions which represent dissolving an amphoteric metal oxide, MO_2, in acidic solution and in basic solution.

11. What is the difference between *cis-* and *trans-*isomers? Illustrate your answer by drawing the *cis-* and *trans-*isomers of square planar $[Pd(H_2O)_2I_2]$.

SUGGESTED READING

Battista: Chromium, The Metal That Glitters. Chemistry, Vol. 42, No. 1, p. 19, 1969; Aluminum, Featherweight Champion of Metals. Vol. 42, No. 3, p. 14, 1969; Titanium, The Cinderella of Metals, Vol. 42, No. 5, p. 13, 1969.
Evans: Biological Function of Copper. Chemistry, Vol. 44, No. 6, p. 10, 1971.
Frieden: The Chemical Elements of Life. Scientific American, Vol. 227, No. 1, p. 52, 1972.
House: Beryllium. Chemistry, Vol. 44, No. 11, p. 10, 1971.
Weand: Lithium—The Lightest Metal. Chemistry, Vol. 44, No. 7, p. 10, 1971.

INTRODUCTION AND GENERAL PRINCIPLES OF ORGANIC CHEMISTRY

CHAPTER 10

The *objectives* of this chapter are to enable the student to:

1. Give a definition of organic chemistry.
2. Compare the properties of organic and inorganic compounds.
3. Define structural isomers.
4. Recognize the basic functional groups in organic chemistry.
5. Understand the unique role of carbon in chemistry.
6. Recognize the relationship between bonding and the shapes of organic molecules.

At the beginning of the nineteenth century considerable evidence had been accumulated concerning the nature, properties, and reactions of inorganic compounds. In contrast to this large body of knowledge about inorganic compounds, relatively little was known about organic compounds. It was known at that time that organic compounds were composed of only a few elements, such as carbon, hydrogen, oxygen, nitrogen, and sulfur; and that, in contrast to inorganic materials, the organic compounds were easily combustible and many of them were sensitive to heat and strong acids and bases.

The compounds investigated by these early chemists, however, were all obtained from either vegetable or animal sources and are what are commonly known as "natural products." Since these initial organic compounds had been produced by living organisms, the early theory of organic chemistry postulated that organic compounds could only arise through the operation of a "vital force" inherent in living cells. Consequently, the chemical synthesis of organic compounds in the laboratory seemed impossible to these early chemists.

In 1828, however, a German chemist, Friedrich Wöhler, discovered that by heating an aqueous solution of the inorganic salt ammonium cyanate, urea was produced. This material isolated by Wöhler was identical to urea isolated from urine. Wöhler correctly concluded that ammonium cyanate rearranged under the influence of heat to give urea. Urea and ammonium cyanate both contain the same number and kinds of atoms, but these atoms are arranged differently.

$$NH_4OCN \rightarrow CO(NH_2)_2$$

Ammonium cyanate   Urea

This discovery, of course, dealt a severe blow to the "vital force" theory for the synthesis of organic compounds. Wöhler's experiments marked the beginning of the end of this theory and set the stage for a new era in organic chemistry. Since that time the organic compounds which have been synthesized in the laboratory by organic chemists far outnumber those isolated and identified from natural sources, although the branch of organic chemistry known as "natural products" is still an important area of research in organic chemistry.

Although Wöhler's conversion of ammonium cyanate to urea dispelled the "vital force" idea and established a link between inorganic and organic chemistry, the designation "organic" has persisted as a convenient means of classifying groups of compounds having some features in common. Today, it is recognized that the common feature among compounds classified as "organic" is that they all contain the element carbon. Hence, the present-day definition of organic chemistry is **the chemistry of carbon compounds.** Conversely, compounds not containing carbon are designated as inorganic compounds. [*]

The importance of organic chemistry to modern society is evident if we consider our surroundings and environment. Most of the clothes that we wear are made of rayon, dacron, nylon, orlon, or some other synthetic fiber developed in the organic chemistry laboratory. Not only the fiber used in making clothing but even the dyes employed to color these fabrics are the result of organic research. Although the past forty years have seen startling developments in man-made fibers, certainly the future will deliver even greater developments in this area.

Modern plastics not only have revolutionized many of our commonplace utensils, such as cups, plates, glasses, and so forth, but have replaced even wood and metal in many of our household furnishings, building materials, and even toys. In today's world plastic materials are as commonplace as wood materials were a hundred years ago.

Even common, naturally occurring organic materials such as petroleum have felt the improving hand of the organic chemist. The chemist and chemical engineer have designed catalytic cracking methods for transforming raw petroleum into improved gasoline for automotive and aviation fuels, and have developed methods for utilizing the nongasoline fractions of petroleum to make raw materials for synthetic rubber and plastics. In fact, much of our modern chemical technology has been developed from by-products of the petroleum industry.

COMPARISON OF ORGANIC AND INORGANIC COMPOUNDS

Although inorganic chemistry preceded the study of organic chemistry by many years, the number of known organic compounds now far exceeds the number of known inorganic compounds. Today, there are well over 3,000,000 known organic compounds, and it should be obvious that some sort of logical and consistent theory of reactions is necessary if one is to study such a multitude of chemical structures.

In contrast to inorganic compounds, which can contain any of the elements in various

---

[*] The student will recognize that several carbon-containing compounds such as CO_2, CO, H_2CO_3, and other carbonates have been previously classified as inorganic. These compounds were classified as inorganic before Wöhler's experiments and before carbon was recognized as the common feature of organic compounds. For convenience, their classification has remained the same and they are studied as part of inorganic chemistry.

combinations, organic compounds must contain the element carbon. In addition to carbon, other elements such as hydrogen, oxygen, fluorine, chlorine, bromine, iodine, nitrogen, phosphorus, and sulfur are commonly found in organic compounds. The number of atoms present in an organic compound is often large (for example, $C_{20}H_{40}$) in contrast to most inorganic compounds (for example H_3PO_4 and $K_2Cr_2O_7$) which generally contain few atoms. In addition, the structures of organic molecules are often complex, such as choles- terol, tetrahydrocannabinol (marijuana), and lysergic acid diethylamide (LSD), and may contain ring structures as well as carbon-carbon chains (see pp. 321 and 391).

Carbon structures, such as C_2H_6O, may exist as more than one structure. The two compounds dimethyl ether and ethyl alcohol are known as **isomers.** Isomers can be defined as compounds which have the same molecular formula, but which have different atomic arrangements (refer also to the previous example of ammonium cyanate and urea). The physical properties and reactions of isomers can be quite similar or quite different

Dimethyl ether   Ethyl alcohol

depending on how the atoms are arranged in the isomeric structures. In the previous example, dimethyl ether and ethyl alcohol exhibit quite different physical properties. Dimethyl ether is a gas at room temperature, whereas ethyl alcohol is a liquid (b.p. 78°C). These two compounds are also quite different chemically, the alcohol being quite reactive with many reagents such as HCl and H_2SO_4, whereas dimethyl ether is unreactive with these reagents.[*]

If the natures of the bonds and the organic groups in the isomers are quite similar, similar properties and reactivities can be expected. For example, C_4H_{10} can exist as the following isomers:

Butane   2-Methylpropane
(b.p. 0°C)   (b.p. −10°C)

These two isomeric hydrocarbons have similar boiling points and react with similar chemical reagents. As the number of carbon atoms in the molecule increases, the number of possible isomeric structures also increases, and it is the ability of carbon to form many isomeric structures that accounts in part for the large number of known organic com- pounds. Most inorganic compounds, with the exception of complex ions, do not form isomeric structures. Hence, for a particular combination of atoms in an inorganic molecule only one structure is possible.

The types of bonding in organic and inorganic compounds also differ and account for the large difference in some of the physical properties of organic and inorganic compounds. Whereas most inorganic compounds are composed of ions and held together by strong electrostatic forces, most organic compounds are composed of covalently bonded

---

[*] As will be obvious from later discussions of alcohols and ethers, the presence of the —OH group in the alcohol will account for its extensive reactivity compared to the ether which has no —OH groups.

atoms and are relatively nonpolar materials. This difference in bonding is reflected in the physical properties such as boiling point, melting point, and solubility. Most inorganic compounds have high melting points and high boiling points (generally $>1000°C$), whereas most organic compounds melt at temperatures less than $300°C$ and boil at temperatures less than $500°C$ (refer to p. 61 for review of polarity).

$$Na^+Cl^-$$

Sodium chloride
(m.p. 801°C)
(b.p. 1413°C)

$$CH_3\overset{\displaystyle O}{\overset{\|}{C}}-NH_2$$

Acetamide
(m.p. 81°C)
(b.p. 222°C)

Since most inorganic compounds are made up of ions held together electrostatically, it would be expected that inorganic compounds should be soluble in polar solvents and, as expected, most inorganic compounds are soluble in the polar solvent water. Water breaks the bond between the ions in the inorganic crystal and hydrates the individual ions. It is also found that these hydrated ions conduct an electric current and behave as good electrolytes. On the other hand, most organic compounds are insoluble in a polar solvent like water but are quite soluble in nonpolar solvents like ether, benzene, and hydrocarbons. Since dissolution of an organic compound into an organic solvent does not produce ions, most solutions of organic compounds do not conduct an electric current and are classified as nonelectrolytes.°

THE ROLE OF CARBON IN ORGANIC CHEMISTRY

Why define and separate a branch of chemistry for one element, such as carbon, and classify the chemistry of the other hundred or so elements as another branch of chemistry, namely, inorganic chemistry? The unique character and emphasis on the atom carbon can be summarized as follows:

1. Its position in the periodic table: Carbon is in the middle of the second period in the periodic table and has an atomic number of six. Consequently, it has six orbital electrons. Two of these orbital electrons make up the first ($1s^2$) shell of electrons, leaving four electrons in the outer valence shell available for bonding purposes. Carbon, being in the middle of the periodic table, is neither strongly electronegative nor strongly electropositive, and therefore has little tendency to form either cations (C^{+4}) or anions (C^{-4}). In fact, carbon forms bonds with other elements by sharing electrons (covalent bonds) and attains the inert gas configuration in this manner.

   Methane:†

$$CH_4 \equiv H\cdot\cdot\overset{\displaystyle \overset{H}{\cdot\cdot}}{\underset{\displaystyle \underset{H}{\cdot\cdot}}{C}}\cdot\cdot H$$

° Although an organic compound such as trimethyl amine, $(CH_3)_3N$, does not conduct an electric current, treatment of this amine with HCl produces a salt, $[(CH_3)_3\overset{+}{N}H]Cl^-$, which does conduct an electric current. Consequently, some organic compounds, which can be converted into ions by the appropriate acid or base reaction, can behave as conductors, but this is not the normal behavior of most organic compounds.

† In this type of representation, the symbol of the element (C) represents the nucleus and the inner shell (non-bonding electrons). The dots represent the valence electrons involved in covalent bond formation. This type of formula is called a Lewis formula. It shows only the number of valence electrons involved in covalent bond formation and gives no indication of the spatial orientation of the atoms in the molecule.

In a molecule like methane, each hydrogen shares an electron (forms a covalent bond) with carbon, and carbon in turn shares an electron with each hydrogen atom. In order to complete the bonding capacity of carbon, four hydrogen atoms are necessary. In addition, hydrogen, by sharing an electron, has attained the electronic configuration of He, and carbon has attained the electronic configuration of Ne.

2. Ability to bond with itself and to form multiple bonds with itself: Carbon, because of its small atomic radius and because of the strength of carbon-carbon bonds, has the striking property of being able to form bonds with itself. Although other atoms, such as silicon, are able to do this to some extent, carbon possesses this property more than any other element. This property accounts for the many organic compounds known in this branch of chemistry. For example, if two carbon atoms share an electron pair and form a covalent bond, and if the other valences of the carbon atoms are filled by forming covalent bonds to hydrogen, the molecule ethane, C_2H_6, results:

$$\cdot \overset{..}{\underset{..}{C}} - \overset{..}{\underset{..}{C}} \cdot {}^{\circ} + 6H \cdot \rightarrow H - \overset{\overset{\displaystyle H}{|}}{\underset{\underset{\displaystyle H}{|}}{C}} - \overset{\overset{\displaystyle H}{|}}{\underset{\underset{\displaystyle H}{|}}{C}} - H \quad \text{or} \quad H_3CCH_3\dagger$$

Ethane

Of course, if carbon were to share two electron pairs between each carbon atom, and if hydrogen atoms are used to complete the bonding capacity of carbon in this system, the molecule ethylene, C_2H_4, results. In ethylene, there is a double (or multiple) bond between the carbon

$$\cdot \overset{..}{C} = \overset{.}{C} \cdot + 4H \cdot \rightarrow \quad \overset{H}{\underset{H}{>}} C = C \overset{H}{\underset{H}{<}} \quad \text{or} \quad H_2C = CH_2$$

Ethylene

atoms.[δ] If this same sort of process is used by two carbons to share three electron pairs between each carbon, and then to use hydrogen atoms to complete any unused bonding capacity, the molecule acetylene, C_2H_2, results, as follows:

$$H - C \equiv C - H \qquad \text{or} \qquad HC \equiv CH$$

Acetylene

Acetylene again contains a multiple bond between the carbon atoms, and in this case we have a triple bond between carbon atoms. Applying this same process further, the molecule C_2, $C \equiv C$, could be formed by sharing all four valence electrons between two carbon atoms. However, four bonds between two carbon atoms has not been observed, and only single-, double-, and triple-bonded carbon atoms have been found in organic compounds.

[°] A covalent bond formed by sharing an electron between two atoms can be represented by a dash for the sake of convenience.

[†] The structural formula in which all the covalent bonds are indicated by a dash is known as an *expanded structural formula*. For convenience, the dashes (covalent bonds), especially to hydrogen, are often omitted and the *condensed* structural formula is used.

[δ] As will become obvious later in this chapter and in the next chapter, the type of bonding, either single or multiple, between carbon atoms has a dramatic effect on the shape of the molecule and on the chemical reactivity of the organic compound.

The process used in the preceding paragraphs could be repeated again and again using additional carbon and hydrogen atoms to give even longer carbon chains. In addition, carbon can share electrons, not only with itself and with hydrogen but with many other simple elements to form cyclic organic compounds as well as linear-chain compounds. Some examples of these various types of compounds are as follows:

H_3CCl $H_3C—O—CH_3$ H_3CCH_2I

Methyl chloride   Dimethyl ether   Ethyl iodide

$$\begin{array}{c} O \\ \parallel \end{array}$$

H_3CNH_2 H_3CCCH_3 $\begin{array}{c} H_2C—CH_2 \\ \diagdown \diagup \\ CH_2 \end{array}$

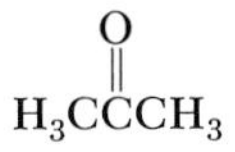

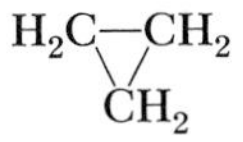

Methyl amine   Acetone   Cyclopropane

THE SHAPES OF ORGANIC MOLECULES

Ionic compounds, such as those commonly found among inorganic molecules, are held together by electrostatic forces between positive and negative ions. Electrostatic forces of this type, such as in Na^+Cl^-, are exerted symmetrically in all directions, and the ions can be thought of as a point charge, or a sphere of unit charge on which the charge is distributed equally over the surface of the sphere.

In contrast to the nondirectional nature of electrostatic forces, covalent bonds are directional in nature and give a definite shape to the molecule which depends on the type of covalent bond. In the simple examples methane, ethane, ethylene, and acetylene, which were considered in the previous section, different shapes and bond angles are found in each case. In methane, the carbon atom is considered to be at the center of a regular tetrahedron and the four bonds to hydrogen are directed to the corners of the tetrahedron (see Fig. 10–1). The molecule can be pictured as follows:

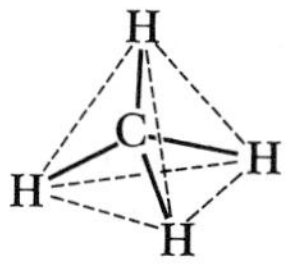

Methane

Figure 10–1

Other methods of determining chemical shapes and geometry, such as x-ray and electron-diffraction, have confirmed the regular tetrahedron shape of molecules such as CH_4 and CCl_4. The bond angles in molecules of this shape are 109.5°.

When carbon-carbon bonds are linked together in the formation of more complex molecules, such as ethane, propane, and so forth, the shape of the molecules is a series of tetrahedrons which share a common corner. The normal carbon-carbon single bond distance in molecules such as this is 1.54 Å.° Longer linear-chain molecules can be assembled by adding on additional tetrahedrons which share a common corner.

In a compound, such as ethylene, C_2H_4, the formation of the carbon-carbon double bond in the molecule imposes certain geometric requirements on the shape of the

---

° 1 Å = 10^{-8} cm = 10^{-1} nm

molecule. First, the introduction of the double bond limits rotation around the carbon-carbon bond. In compounds, such as ethane, which share a corner of a tetrahedron, there is free rotation° around the carbon-carbon single bond. However, introduction of the carbon-carbon double bond restricts any free rotation (360°), and for all practical purposes no rotation is allowed in this molecule unless the carbon-carbon double bond is broken. Secondly in a molecule which contains a carbon-carbon double bond the atoms attached to the carbon atoms with the double bond are also coplanar (all the carbon and hydrogen atoms in ethylene lie in the same plane) with bond angles of 120° and a carbon-carbon bond length of 1.34 Å. The bond angles and bond distances again may vary slightly depending upon what atoms are attached to carbon, but the gross overall features of the molecule will not change.

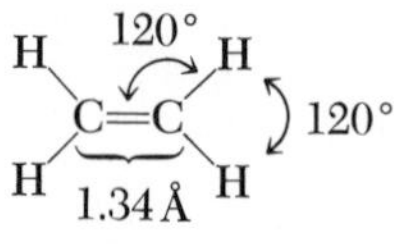

$$\begin{array}{c} \text{H} \quad \overset{120°}{\curvearrowright} \quad \text{H} \\ \underset{\text{H}}{\diagdown} \text{C} = \text{C} \underset{\text{H}}{\diagup} \quad \Big) \, 120° \\ \text{H} \quad 1.34\,\text{Å} \quad \text{H} \end{array}$$

Ethylene

In acetylene, C_2H_2, and other molecules containing a carbon-carbon triple bond, even greater deviations from the simple tetrahedral structures occur. X-ray and electron-diffraction methods have shown that compounds containing a $-C\equiv C-$ linkage are linear molecules with a carbon-carbon bond length of 1.21 Å, as illustrated below for acetylene:

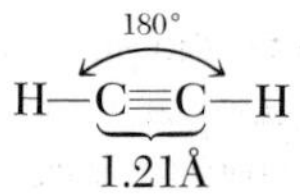

$$\overset{180°}{\overset{\curvearrowright}{\text{H}-\text{C}\equiv\text{C}-\text{H}}}$$
$$1.21\text{Å}$$

Acetylene

In cyclic organic compounds some deviation from the normal bond angles illustrated above may be expected, as the constraining of the carbon atoms into rings of certain sizes will force the atoms into unusual and strained shapes. For example, cyclopropane must be a planar molecule, since three points (the three carbon atoms) define a plane. The bond angles in cyclopropane must necessarily be equal, since all the atoms are identical, and have been shown to be 60°. Since the normal bond angle of a single

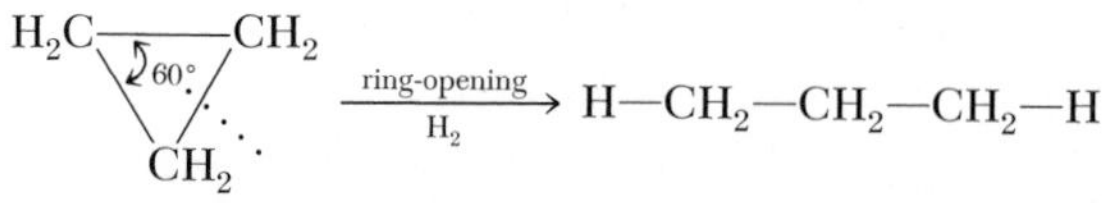

$$\begin{array}{c} \text{H}_2\text{C} \underline{\qquad} \text{CH}_2 \\ \diagdown \, 60° \diagup \\ \diagdown \diagup \\ \text{CH}_2 \end{array} \quad \xrightarrow[\text{H}_2]{\text{ring-opening}} \quad \text{H}-\text{CH}_2-\text{CH}_2-\text{CH}_2-\text{H}$$

Cyclopropane   Propane

carbon-carbon bond is 109.5° (see Fig. 10–1), to constrain or compress these bond angles from 109.5° to 60° will introduce strain into the molecule. Consequently, we might expect cyclopropane and any other highly strained compound to be particularly susceptible to ring-opening reactions, since after ring-opening the bond angles become approximately 109.5° again.

Experimentally, it is commonly found that the chemical susceptibility to ring-opening reactions does increase with increasing amount of ring strain, and cyclopropane does

---

° In actuality there is a small energy barrier to rotation, since the atoms do occupy space and must pass each other on rotation. However, in most compounds containing carbon–carbon single bonds this barrier is very small.

undergo ring-opening reactions with many chemical reagents. As the size of the ring increases, the bond angles increase, and the amount of strain decreases. Consequently, ring systems higher than cyclopropane are less prone to undergo ring-opening reactions. Ring systems higher than cyclopropane can be either planar or "puckered" (nonplanar) systems. In four- and five-membered rings there is some "puckering" of the ring, and these ring systems are not completely planar.° In higher ring systems, such as cyclohexane, bond angles of 120° would be expected if the molecule were a planar hexagon, and extensive strain and ring-opening properties would be anticipated for cyclohexane. However, cyclohexane has been found, both experimentally and by x-ray analysis, to be a nonstrained and nonplanar molecule, and, hence, is not susceptible to easy ring-opening reactions owing to the presence of bond angles of 109.5°. The cyclohexane molecule generally exists in two arrangements known as a "chair" and a "boat" form (see p. 163). The chair form in most cases is the preferred form, since it again minimizes steric repulsions of the hydrogen atoms in the molecule. The chair and boat forms are not isolable.

ORGANIC FUNCTIONAL GROUPS

When the chemistry of organic compounds is considered in detail in the following chapters, it will become apparent that only certain bonds and organic groups participate in the chemical reaction, and that most of the carbon chain structure of the molecule remains unchanged in going from reactants to products. The following reaction of t-butyl alcohol is illustrative:

$$(CH_3)_3COH + \quad HBr \quad \rightarrow (CH_3)_3CBr + HOH$$

| t-Butyl alcohol | Hydrogen bromide | t-Butyl bromide | Water |

In this reaction the —OH group of the alcohol is lost, and its place on the carbon chain is taken by the bromine atom of the hydrogen bromide. In turn, the —OH lost by the alcohol combines with the hydrogen of the hydrogen bromide to produce water as the other product of this reaction.† The important point to consider here is that only a small portion of the organic molecule undergoes change. The bond between the carbon atom and the —OH group is broken, and a new bond between carbon and bromine is formed. Other than these simple changes, all the remaining bonds (all carbon and hydrogen in this case) in the organic compound remain unchanged. This is a characteristic feature of organic molecules, that only certain atoms or groups of atoms in an organic molecule determine the chemistry of the class of compounds containing that particular atom or group of atoms. *The atom or group of atoms that defines the structure of a particular class of organic compounds and determines its properties is called the **functional group**.* In the particular example above, the functional group in the alcohol is the —OH group, and the functional group in the product is the halogen atom, Br.

A large portion of organic chemistry is concerned with the transformation of one functional group into another. A basic understanding of organic chemistry is a mastery of the properties and reactivities of each type of functional group, and how one functional group can be transformed into another functional group of different properties and

---

° In the planar compounds, there are unfavorable steric repulsions between the hydrogen atoms caused by compression of the hydrogen atoms trying to occupy the same space. In the "puckered" form, these steric repulsions are minimized, since the hydrogen atoms are staggered in space and this form is more stable (this can be seen more clearly by building a model of these compounds).

† The student should be aware that this explanation is very simplified, and that the details in converting the alcohol to the bromide are more involved than merely exchanging groups or atoms. This will become apparent when we look at a mechanism for this type of reaction.

TABLE 10-1   SIMPLE FUNCTIONAL
CLASSES OF ORGANIC COMPOUNDS

ILLUSTRATIVE EXAMPLE	NAME OF FUNCTIONAL CLASS	FUNCTIONAL GROUP
$CH_3CH_2CH_3$	Alkanes	—
$CH_3CH\!=\!CH_2$	Alkenes	$C\!=\!C$
$CH_3C\!\equiv\!CH$	Alkynes	$C\!\equiv\!C$
$CH_3CH_2CH_2OH$	Alcohols	$-OH$
$\overset{\text{H}}{CH_3CH_2C\!=\!O}$	Aldehydes	$\overset{\text{H}}{-C\!=\!O}$
$\overset{O}{\overset{\|}{CH_3CCH_3}}$	Ketones	$\overset{O}{\overset{\|}{-C-}}$
$\overset{O}{\overset{\|}{CH_3CH_2COH}}$	Acids	$\overset{O}{\overset{\|}{-COH}}$
$CH_3CH_2CH_2NH_2$	Amines	$-NH_2$
$CH_3OCH_2CH_3$	Ethers	$-C-O-C-$
$CH_3CH_2CH_2Br$	Halides	$-Br$

reactivity. In later chapters, using simple molecules, we shall learn to associate a particular set of properties with a particular functional group. When encountering a more complicated molecule, we may expect the properties of this molecule to roughly approximate the properties of the various functional groups contained in the molecule.

A summary of the typical functional groups to be taken up in later chapters is presented in Table 10-1. Although it is not assumed that the student understands the chemistry of these functional groups, it is extremely helpful at this stage to learn the names of some of the simple functional classes, and especially to learn to associate a particular atom or group of atoms with a particular functional class.

IMPORTANT TERMS AND CONCEPTS

condensed structural formula   expanded structural formula   isomer   structural formula
covalent bond   functional group   single bond   triple bond
double bond

QUESTIONS

1. Which of the following compounds are isomers?

(a) $CH_3CH_2CH_2OH$ (d) H_3CCHCH_3
(b) $CH_3CHClCH_3$ $|$
(c) $CH_3CH_2CH_3$ OH
                        (e) $CH_3CH_2CH_2Cl$

2. Which of the following compounds are identical?

(a) 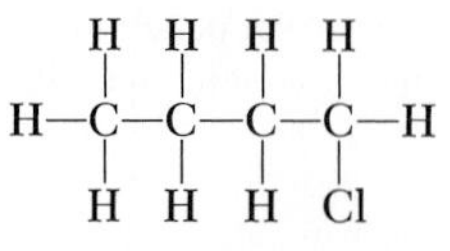
$$
\begin{array}{ccccc}
 & H & H & H & H \\
 & | & | & | & | \\
H- & C- & C- & C- & C-H \\
 & | & | & | & | \\
 & H & H & H & Cl \\
\end{array}
$$

(b)
$$
\begin{array}{ccccc}
 & H & H & Cl & H \\
 & | & | & | & | \\
H- & C- & C- & C- & C-H \\
 & | & | & | & | \\
 & H & H & H & H \\
\end{array}
$$

(c)
$$
\begin{array}{ccccc}
 & H & H & H & H \\
 & | & | & | & | \\
H- & C- & C- & C- & C-H \\
 & | & | & | & | \\
 & H & Cl & H & H \\
\end{array}
$$

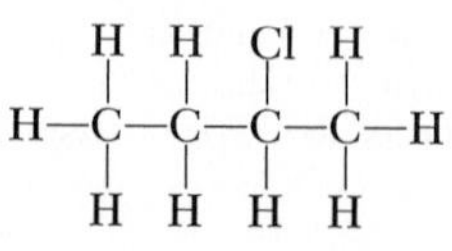

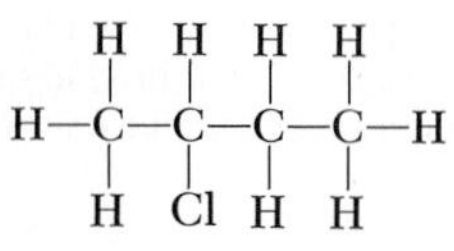

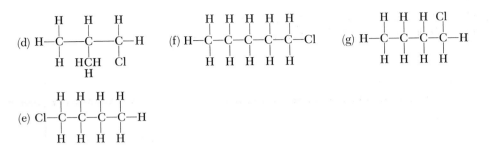

(d)
$$H-\underset{\underset{H}{|}}{\overset{\overset{H}{|}}{C}}-\underset{\underset{H}{\overset{|}{C}H}}{\overset{\overset{H}{|}}{C}}-\underset{\underset{Cl}{|}}{\overset{\overset{H}{|}}{C}}-H$$

(f)
$$H-\underset{\underset{H}{|}}{\overset{\overset{H}{|}}{C}}-\underset{\underset{H}{|}}{\overset{\overset{H}{|}}{C}}-\underset{\underset{H}{|}}{\overset{\overset{H}{|}}{C}}-\underset{\underset{H}{|}}{\overset{\overset{H}{|}}{C}}-\underset{\underset{H}{|}}{\overset{\overset{H}{|}}{C}}-Cl$$

(g)
$$H-\underset{\underset{H}{|}}{\overset{\overset{H}{|}}{C}}-\underset{\underset{H}{|}}{\overset{\overset{H}{|}}{C}}-\underset{\underset{H}{|}}{\overset{\overset{H}{|}}{C}}-\underset{}{\overset{\overset{Cl}{|}}{C}}-H$$

(e)
$$Cl-\underset{\underset{H}{|}}{\overset{\overset{H}{|}}{C}}-\underset{\underset{H}{|}}{\overset{\overset{H}{|}}{C}}-\underset{\underset{H}{|}}{\overset{\overset{H}{|}}{C}}-\underset{\underset{H}{|}}{\overset{\overset{H}{|}}{C}}-H$$

3. Represent the structure of $CH_3CH_2CH_2CH_3$, using a Lewis formula.

4. Write expanded structural formulas for *all* the different isomers of:

   (a) C_3H_8O (d) C_3H_5Cl
   (b) $C_2H_3Br_3$ (e) C_3H_8
   (c) $C_2H_4Br_2$

5. Which of the following pairs of compounds are isomers?

   (a) $CH_3CH_2CH_2CH_3$ and $(CH_3)_2CHCH_3$
   (b) $CH_3CH=CH_2$ and $CH_3C\equiv CH$
   (c) $H_2C\overset{\displaystyle CH_2}{\underset{\displaystyle \overset{|}{C}}{\diagup\diagdown}}$ and $\underset{H_2C-CH_2}{\overset{H_2C-CH_2}{|\qquad|}}$

   (d) $CH_3CH_2CH_2\overset{\overset{\displaystyle H}{|}}{C}=O$ and $CH_3\overset{\overset{\displaystyle O}{\|}}{C}CH_2CH_3$

   (e) $CH_3CH_2CH_2\overset{\overset{\displaystyle O}{\|}}{C}OH$ and $CH_3CH_2\overset{\overset{\displaystyle O}{\|}}{C}OCH_3$

6. Classify each of the following compounds with respect to the shape of the molecule.

   (a) C_2Cl_4 (d) CH_2Cl_2
   (b) CBr_4 (e) C_2F_6
   (c) C_2Cl_2

7. In each of the following compounds, pick out the functional group and classify each compound into a functional class.

   (a) $CH_3CH_2\overset{\overset{\displaystyle O}{\|}}{C}CH_3$

   (f) $CH_3CH_2CH_2\overset{\overset{\displaystyle OH}{|}}{C}HCH_3$

   (b) $(CH_3)_2CHCH_2C(CH_3)_3$

   (g) $\triangleright\!\!-\overset{\underset{\displaystyle O}{\|}}{C}-OH$

   (c) $CH_3CH_2OCH_2CH_3$

   (h) (cyclohexanone structure with O)

   (d) $CH_3CH=CHCH_3$

   (i) $CH_3CHBrCH_3$

   (e) $CH_3CH_2NH_2$

   (j) $CH_3C\equiv CCH_3$

SUGGESTED READING

Dence: Conformational Analysis or How Some Molecules Wiggle. Chemistry, Vol. 43, No. 6, p. 6, 1970.

Kurzer, and Sanderson: Urea in the History of Organic Chemistry. Journal of Chemical Education, Vol. 33, p. 452, 1956.

HYDROCARBONS

The *objectives* of this chapter are to enable the student to:

1. Define the class of organic compounds known as hydrocarbons.
2. Describe and give examples of the class of hydrocarbons called alkanes.
3. Draw a structural formula of an alkane given the IUPAC name.
4. Write the IUPAC name of an alkane given the structural formula.
5. Recognize and name the most common alkyl groups.
6. Understand the concept of homologous series.
7. Appreciate the role of petroleum in modern society.
8. Understand the conversion of other functional groups to alkanes.
9. Understand the role of combustion as a source of useful energy.
10. Distinguish between substitution and addition types of reactions.

The simplest organic compounds are hydrocarbons, which contain only the elements carbon and hydrogen. Replacement of a carbon-hydrogen bond by a functional group, such as those given in Chapter 10, gives rise to the various classes of organic compounds, and these functional classes may be thought of as derivatives of hydrocarbons.

Although hydrocarbons contain only two elements, these elements may be combined in several ways. For example, in the following compounds the carbon atoms may be linked together to form a linear chain or a ring. Also, the molecule may contain only carbon-

$$CH_3CH_2CH_2CH_2CH_2CH_3$$

n-Hexane
C_6H_{14}

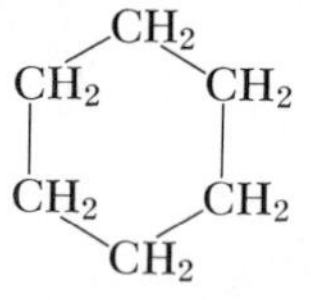

Cyclohexane
C_6H_{12}

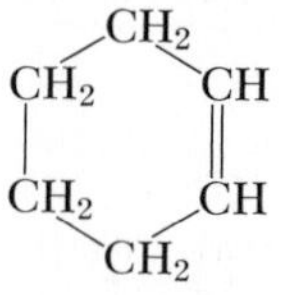

Cyclohexene
C_6H_{10}

$$CH_3CH_2CH_2CH_2CH=CH_2$$

1-Hexene
C_6H_{12}

$$CH\equiv CH$$

Acetylene
C_2H_2

carbon single bonds or may contain carbon-carbon multiple bonds. Even between members of the same class of compounds, for example *n*-hexane and cyclohexane, neither of which contains double bonds, the ratio of carbon to hydrogen is not constant, and hence the molecular formula is not an indication of the type of hydrocarbon structure. For example, the molecular formula C_6H_{12} as shown previously could refer to either cyclohexane or 1-hexene. The chemistry of these two compounds is quite different even though they have the same ratio of carbon to hydrogen atoms.

TABLE 11–1   HYDROCARBONS

CLASS	DISTINGUISHING FEATURE	SIMPLEST EXAMPLE
Alkanes	$-\overset{\displaystyle |}{\underset{\displaystyle |}{C}}-\overset{\displaystyle |}{\underset{\displaystyle |}{C}}-$	CH_4, Methane
Alkenes	$\overset{\diagdown}{\diagup}C=C\overset{\diagup}{\diagdown}$	C_2H_4, Ethylene
Alkynes	$-C{\equiv}C-$	C_2H_2, Acetylene

Within the class of compounds known as hydrocarbons there are different degrees and types of chemical reactivity, and in order to classify the properties and chemical reactions of hydrocarbons, it is convenient to divide hydrocarbons into several subclasses. The basis for classification is the number of covalent bonds formed between the carbon atoms in the compounds. If only carbon-carbon single bonds are involved in the compound, the class is known as **alkanes,** or **saturated hydrocarbons.** The term "saturated" means that only one pair of electrons is shared covalently between any two bonded atoms in the molecule. Therefore, *n*-hexane and cyclohexane are alkanes by this definition. If the molecule contains multiple carbon-carbon bonds (more than one pair of electrons is shared covalently between any two bonded atoms), the compounds are classified on the basis of the number of multiple bonds between any two bonded atoms in the molecules. For example, cyclohexene and 1-hexene both contain a carbon-carbon double bond and would therefore be classed in the same category. However, acetylene contains a carbon-carbon triple bond and would not be classed with cyclohexene or 1-hexene. Compounds containing carbon-carbon double bonds are known as **alkenes,** and compounds containing carbon-carbon triple bonds are known as **alkynes.** Hence, cyclohexene and 1-hexene are alkenes and acetylene is an alkyne. These classifications of hydrocarbons are summarized in Table 11–1.

ALKANES

The simplest saturated hydrocarbon is methane. Higher members of this series can be developed by replacing one or more of the hydrogens by one or more carbon atoms and by completing the valence requirements of the added carbon atom with hydrogen atoms. Several examples of this stepwise development of a carbon chain are illustrated below:

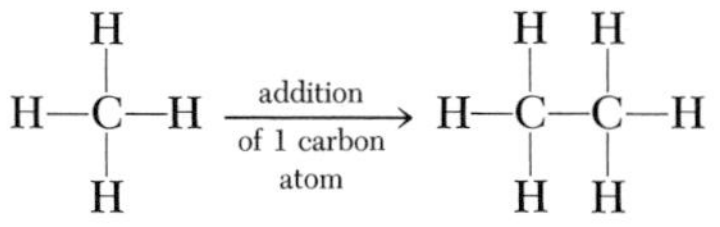

Ethane

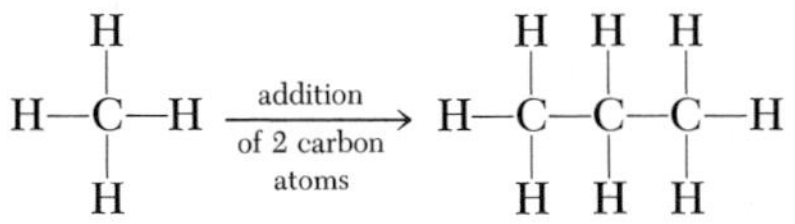

Propane

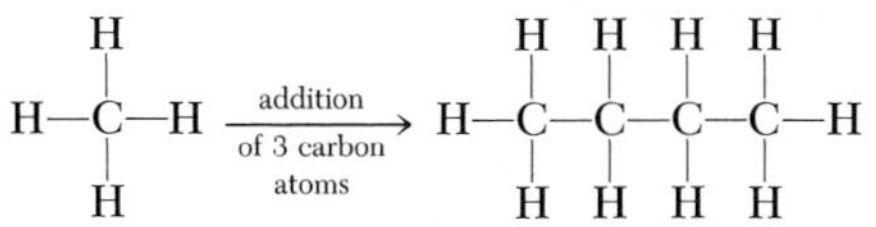

Normal butane

Another and easier way to view this process of building a carbon chain structure is that the propane carbon chain is developed from ethane by replacing a hydrogen of ethane by a —CH$_3$ group. Similarly, normal° butane is developed from propane by replacement of the terminal hydrogen with a —CH$_3$ group. Higher members of this series are built up in an analogous manner as shown below:

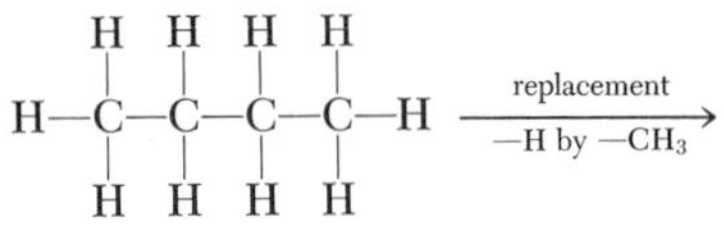

H H H H
| | | |
H—C—C—C—C—H $\xrightarrow[\text{—H by —CH}_3]{\text{replacement}}$
| | | |
H H H H
Normal butane

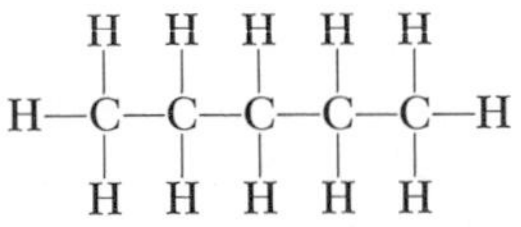

H H H H H
| | | | |
H—C—C—C—C—C—H   or   CH$_3$CH$_2$CH$_2$CH$_2$CH$_3$ or   C$_5$H$_{12}$
| | | | |
H H H H H
Normal pentane

In the compounds illustrated above, the molecular formulas conform to the general formula C$_n$H$_{2n+2}$, in which n is the number of carbon atoms in the molecule. This general formula fits all linear (straight chain) alkanes and can be used to predict the molecular formula of any linear alkane. For example, an alkane containing seven carbon atoms (heptane) would have a molecular formula C$_7$H$_{16}$; one with eight carbons (octane), C$_8$H$_{18}$; one with nine carbons (nonane), C$_9$H$_{20}$; one with ten carbons (decane), C$_{10}$H$_{22}$, and so on. Inspection of each of these structures shows that propane differs from butane by a —CH$_2$— unit; pentane differs from hexane by a —CH$_2$— unit; hexane differs from heptane by a —CH$_2$— unit, and so on. This —CH$_2$— unit is known as a methylene group. Each higher member in this series differs from the next lower or higher one by one more or one less methylene unit. A series of compounds of this type in which each member differs from the next higher or lower member by a constant increment is known as a **homologous series.**

A family of compounds in a homologous series exhibit characteristic features: (a) they all contain the same elements and can be represented by a single general formula; (b) each homolog differs from the one above and below it in the series by a —CH$_2$— unit; and (c) all the homologs show similar and closely related physical and chemical properties. Consequently, it is not necessary to investigate the properties of every single organic compound. Rather, the properties of each homologous class are studied with representative members of that series, and these properties are used to predict the behavior of the other members of this series.

MOLECULAR MODELS

One drawback to the representation of structure illustrated in the preceding section is that the structures are shown only in two dimensions. Sometimes it is advantageous to depict the three-dimensional features of the molecule. The chemist does this by the use of molecular models. Two types of molecular models are frequently used, and these are illustrated on the next page.

The ball and stick model (Tinkertoy) shows clearly the bond angles in the molecule. However, the real bonded atoms are not spherical in shape as shown by this model and

---

° **Normal** refers to the fact that all of the carbon atoms in the chain are arranged in a linear manner. This is in contrast to other isomers having the same molecular formula, but in which the carbon atoms are arranged in other than a linear chain.

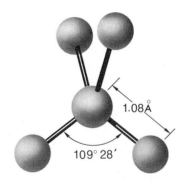

Figure 11-1   Methane.

Ball and Stick Model   Space-Filling Model

are not separated in space by rigid bonds. The space-filling models (called Van der Waals models) are exact scale models which show the relative size of each atom. However, the bond angles and bond distances are more difficult to see. Each type of model has its own utility, and the chemist uses the appropriate model depending upon the type of information desired.

The ball and stick model and the space-filling model for butane are illustrated below:

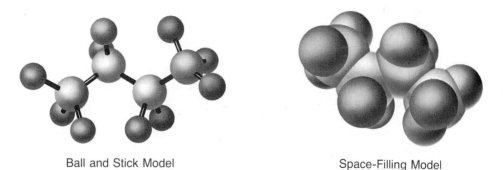

Ball and Stick Model   Space-Filling Model

Figure 11-2   Butane.

The cycloalkanes are best shown with the ball and stick type model, and the boat and chair form of cyclohexane are shown below:

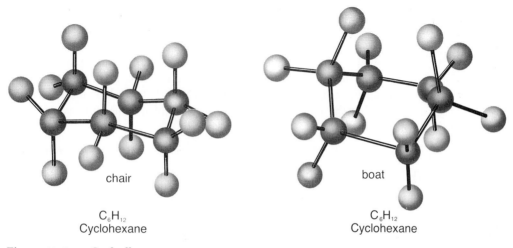

chair   boat

C_6H_{12} C_6H_{12}
Cyclohexane   Cyclohexane

Figure 11-3   Cycloalkanes.

NOMENCLATURE

The nomenclature of the simple straight chain alkanes is straightforward. The first four members have common names (methane, ethane, propane, and butane), but the stem names of the higher members are derived from the number of carbon atoms in the chain, and the -ane ending is added to the stem name. Therefore, the names hexane, heptane, octane, and so forth are used for alkanes containing six, seven, and eight carbons in the chain.

In the preceding section normal butane was developed by replacing a terminal hydrogen in propane by a $—CH_3$ group. Since both ends of the propane molecule are equivalent, it doesn't matter which terminal hydrogen is replaced by the $—CH_3$ group. However, another possibility still exists for developing the carbon chain by this process. If the hydrogen of the $—CH_2—$ group in propane is replaced by a $—CH_3$ group, an isomer of normal butane is formed, namely, isobutane.

Isobutane

In the case of propane, only two isomers of butane can be developed by replacing hydrogen by a $—CH_3$ group. With higher members of this series, many isomers are possible by replacement of the different hydrogens. For example, in the octane series, C_8H_{18}, eighteen structural isomers are possible. Of course, when the number of isomers of a particular carbon chain structure becomes high, the naming of these isomers becomes difficult, and an unambiguous nomenclature is necessary to avoid confusion. The naming system selected should be simple, but must provide a name that fits one and only one structure. The International Union of Pure and Applied Chemistry (IUPAC) has recommended a system which is used universally by organic chemists to name alkanes. The rules of this system are as follows:

1. The characteristic ending -ane is applied to the stem name to obtain the name of a linear saturated hydrocarbon.
2. For branched chain alkanes, the compound is named as a derivative of the hydrocarbon corresponding to the *longest continuous carbon chain* in the molecule.
3. Substituents (atoms or groups of atoms) are indicated by a suitable prefix and a number to indicate their position on the carbon chain.
4. Numbering of the longest continuous carbon chain (Rule 2) must be done in such a way that the numbers giving the position of the substituents are kept as low as possible.

For example, the following compound can be named in two different ways:

$$\overset{1}{C}H_3\overset{2}{C}H\overset{3}{C}H_2\overset{4}{C}H_2\overset{5}{C}H_3 \quad or \quad \overset{5}{C}H_3\overset{4}{C}H\overset{3}{C}H_2\overset{2}{C}H_2\overset{1}{C}H_3$$
$$\underset{CH_3}{|} \qquad\qquad\qquad \underset{CH_3}{|}$$

(A) 2-Methylpentane   (B) 4-Methylpentane

In both structures, (A) and (B), the longest continuous carbon chain contains five carbons; hence, the compound will be named as a derivative of pentane. The —CH_3 group (methyl group) is attached at position 2 in structure (A) or at position (4) in structure (B). Rule 4 demands that the lowest number be used in numbering substituents; hence, 2-methylpentane is the correct name for this compound—not 4-methylpentane. Similarly, the following compounds are correctly named as indicated following the IUPAC rules:

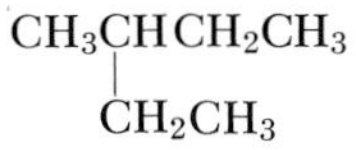

$$CH_3CHCH_2CH_3 \qquad \text{3-Methylpentane, not 2-ethylbutane}$$ 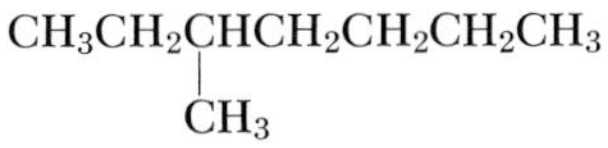
$$|$$
$$CH_2CH_3$$

$$CH_3CH_2CHCH_2CH_2CH_2CH_3 \qquad \text{3-Methylheptane, not 5-methylheptane}$$
$$|$$
$$CH_3$$

When more than one substituent is present on the carbon chain, each substituent is given a number. For example, structure (C) is correctly named as 2,2-dimethylbutane—

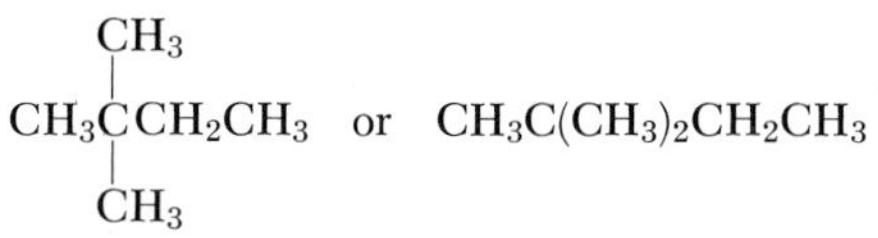

$$CH_3$$
$$|$$
$$CH_3CCH_2CH_3 \quad \text{or} \quad CH_3C(CH_3)_2CH_2CH_3$$
$$|$$
$$CH_3$$
$$\text{(C)}$$
2,2-dimethylbutane

not 2-dimethylbutane.° Even though the substituents are identical, each must be given a number, and the number of identical substituents is also indicated by the prefix di-.

In the preceding examples, the substituents (other than hydrogen atoms) were designated as methyl or ethyl. The names of these substituents were derived from the alkane containing the same number of carbon atoms by changing the **-ane** ending to **-yl.** These groups are derived from the parent alkane by removing one of the hydrogen atoms and are known as **alkyl** groups (from alk**ane** → alk**yl**). Therefore, methyl and ethyl groups can be formulated from methane and ethane as follows:

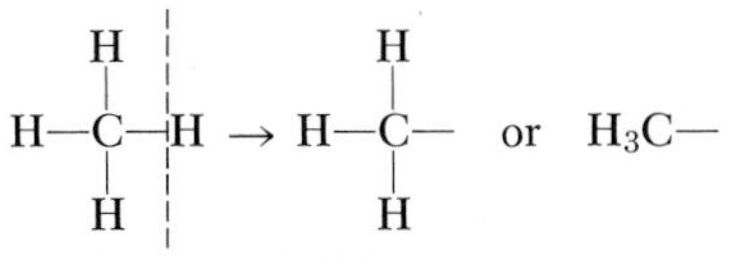

$$\begin{array}{ccc} H & & H \\ | & & | \\ H-C-H & \rightarrow & H-C- \quad \text{or} \quad H_3C- \\ | & & | \\ H & & H \\ \text{Methane} & & \text{Methyl group} \end{array}$$

$$\begin{array}{ccc} H \; H & & H \; H \\ | \; | & & | \; | \\ H-C-C-H & \rightarrow & H-C-C- \quad \text{or} \quad CH_3CH_2- \quad \text{or} \quad C_2H_5- \\ | \; | & & | \; | \\ H \; H & & H \; H \\ \text{Ethane} & & \text{Ethyl group} \end{array}$$

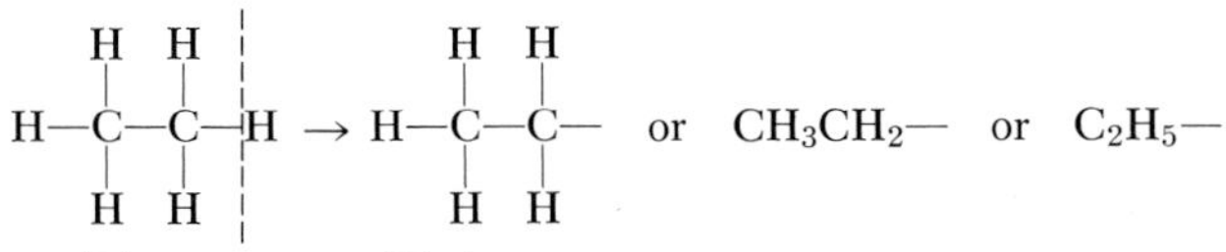

In compounds containing more than two carbon atoms, the number of alkyl groups will

---

° It is evident that by counting the number of carbon atoms indicated in the name given to the longest continuous chain and adding to this the number of carbons indicated in the substituent names, the total number of carbons in the compound is obtained. This is a good way to check that no carbons have been omitted in naming the compound.

depend on the number of different types of hydrogens in the molecule. From propane, for example, two different alkyl groups are possible:

$$
\begin{array}{c}
\underset{\text{Propane}}{
\begin{array}{c}
\text{H} \quad \text{H} \quad \text{H} \\
| \quad\; | \quad\; | \\
\text{H}-\text{C}-\text{C}-\text{C}-\text{H} \\
| \quad\; | \quad\; | \\
\text{H} \quad \text{H} \quad \text{H}
\end{array}}
\rightarrow
\underset{\substack{\text{Normal propyl} \\ (\text{n-propyl})}}{
\begin{array}{c}
\text{H} \quad \text{H} \quad \text{H} \\
| \quad\; | \quad\; | \\
\text{H}-\text{C}-\text{C}-\text{C}- \\
| \quad\; | \quad\; | \\
\text{H} \quad \text{H} \quad \text{H}
\end{array}}
\quad \text{or} \quad CH_3CH_2CH_2- \quad \text{or} \quad nC_3H_7-
\end{array}
$$

$$
\begin{array}{c}
\begin{array}{c}
\text{H} \quad \text{H} \quad \text{H} \\
| \quad\; | \quad\; | \\
\text{H}-\text{C}-\text{C}-\text{C}-\text{H} \\
| \quad\; | \quad\; | \\
\text{H} \quad \text{H} \quad \text{H}
\end{array}
\rightarrow
\underset{\substack{\text{2-Propyl or} \\ \text{isopropyl}}}{
\begin{array}{c}
\text{H} \quad \text{H} \quad \text{H} \\
| \quad\; | \quad\; | \\
\text{H}-\text{C}-\text{C}-\text{C}-\text{H} \\
| \qquad\quad | \\
\text{H} \qquad\quad \text{H}
\end{array}}
\quad \text{or} \quad CH_3CHCH_3 \quad \text{or} \quad isoC_3H_7-
\end{array}
$$

From higher homologues, even more possibilities can be formulated. In general, naming of alkyl groups by this method is only reasonable for groups containing a small number of carbons. This nomenclature is usually employed for compounds containing 1 to 4 carbon atoms. Table 11–2 summarizes some of the more important alkyl groups.

Cycloalkanes are analogous to straight chain alkanes, except that the ends of the carbon chain are joined together in a ring. This process of linking the ends of the chain into a ring requires the use of one additional valence from each terminal carbon atom. Consequently, two less C—H bonds are formed, and the general formula for these compounds is C_nH_{2n}, where n equals the number of carbon atoms. The parent name of the cyclic hydrocarbon is derived by adding the prefix **cyclo-** to the linear alkane having the same number of carbon atoms.

TABLE 11–2   ALKYL GROUPS

ALKYL GROUP	IUPAC NAME	COMMON NAME
CH_3-	Methyl	Methyl
CH_3CH_2-	Ethyl	Ethyl
$CH_3CH_2CH_2-$	Propyl	n-Propyl
CH_3CHCH_3	Methylethyl	Isopropyl°
$CH_3CH_2CH_2CH_2-$	Butyl	n-Butyl
$CH_3CH_2CHCH_3$	1-Methylpropyl	s-Butyl°°
CH_3CHCH_2- CH_3	2-Methylpropyl	Isobutyl°
CH_3CCH_3 CH_3	Dimethylethyl	t-Butyl°°
$CH_3CH_2CH_2CH_2CH_2-$	Pentyl	n-Pentyl (Amyl)

° iso- refers to any structure having a terminal CH_3CHCH_3 grouping.
°° s- and t- refer to secondary and tertiary. In this system of nomenclature, the carbon having the unsatisfied valence is the focal point. Whether this carbon is referred to as primary, secondary, or tertiary depends upon whether it is attached to 1, 2, or 3 additional carbon atoms.

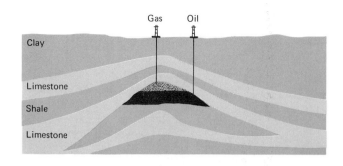

Figure 11-4 An illustration of an oil dome, or anticline, showing deposits of petroleum and natural gas.

SOURCES OF HYDROCARBONS

Natural gas and petroleum are the most important natural sources of hydrocarbons. Large deposits of these substances have been formed over the years by the gradual decomposition of marine life and other biological materials. These deposits usually accumulate under a dome-shaped layer of rock several thousand feet under the earth's surface (Fig. 11-4). When a hole is drilled through the rock layer, the pressure under the dome forces the gas or oil to the surface. After the pressure is released, pumps are required to bring the remaining oil to the surface.

NATURAL GAS

Natural gas is an excellent source of low molecular weight alkanes. Natural gas occurs in most parts of the United States, but most of it is produced in the southwest. In recent years a vast network of pipelines has been installed to carry natural gas from Texas to other parts of the United States. The propane and butane are removed by liquefaction before the gaseous fuel is introduced into the pipelines for distribution. The liquid propane and butane are stored under pressure in steel cylinders from which they are released as a gaseous fuel to be used in rural areas and in locations that are not supplied by natural gas mains.

PETROLEUM

The crude oil, or petroleum, obtained from oil wells is another rich source of hydrocarbons. The hydrocarbons in petroleum, in contrast to those of natural gas, are of higher molecular weight. Petroleum has been known for several centuries and has been used for many purposes, particularly as a fuel. It was not until recent years, however, that petroleum was separated into its hydrocarbon components. It is a very complex mixture of hydrocarbons, and its composition varies with the location of the oil field from which it is obtained. It contains mainly a mixture of alkanes, cycloalkanes, and aromatic hydrocarbons. In addition to the hydrocarbons, petroleum contains about 10 per cent by weight of sulfur, nitrogen, and oxygen compounds.

Petroleum is separated into its hydrocarbon fractions by the process of fractional distillation. A fractional distillation column consists of a tall column containing perforated plates or irregularly shaped glass or ceramic pieces designed to promote intimate contact between the distilling vapors and the refluxing liquid that condenses and runs back down the column. The effect of such a column is to concentrate the lower-boiling constituents in the vapors as they rise, and to enrich the reflux with the higher-boiling constituents. By proper construction and operation, the various petroleum fractions can be removed

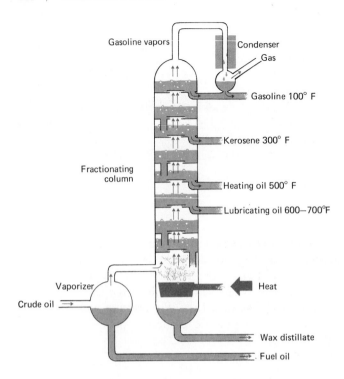

Figure 11–5   A diagram of a fractionating column, showing the various levels from which the petroleum fractions are removed.

from different levels in the fractionating column (Fig. 11–5). In the petroleum industry, these distillation procedures are called the refining process.

Gases from Petroleum.   The first products given off during the distillation of petroleum are the gaseous hydrocarbons containing from 1 to 5 carbon atoms. This hydrocarbon mixture contains both saturated and unsaturated hydrocarbons which can be separated from each other by chemical methods. The unsaturated gases are used in the production of aviation fuel, synthetic rubber, and other organic compounds. The saturated hydrocarbons, especially propane and butane, are liquefied and sold as **bottled gas.** Over two billion gallons of these liquefied gases from both petroleum and natural gas are used in the United States every year.

Petroleum Ether.   The second fraction that is distilled from petroleum is called petroleum ether. This consists mainly of pentanes, hexanes, and heptanes, and is used extensively as fat solvents; paint, varnish, and enamel thinner; and dry-cleaning agents.

Gasolines.   In the early days of the automobile industry simple distillation of petroleum gave more than enough gasoline to supply the demands. This type of gasoline is called **straight-run gasoline** and is composed essentially of alkanes plus minor amounts of cycloalkanes and aromatic hydrocarbons. The composition of gasoline varied considerably and depended on the source of petroleum from which it was distilled.

TOPIC OF CURRENT INTEREST

OCTANE NUMBERS AND ANTI-KNOCK AGENTS

The average motorist, if he is observant, has probably noticed that the service station where he buys his gasoline has posted on the gas pumps the octane number of the fuel that he has purchased. What are these numbers, what do they mean, and what relationship do they have to the performance of his vehicle?

Generally the gasoline fraction that is obtained from petroleum is not entirely satisfactory for use in today's high compression automobile engines. It causes the engine to knock or to develop the familiar "ping" when the engine is accelerated too rapidly or when the automobile goes up a steep grade with subsequent loss of power. It has been found that the knocking characteristics of a gasoline depend on the structure of the constituent hydrocarbons. The higher the proportion of straight chain hydrocarbons in the gasoline, the more knock. On the other hand, as the percentage of branched chain hydrocarbons or aromatic hydrocarbons in the gasoline is increased, the knock in the engine decreases and the engine operation becomes smoother. The performance of a gasoline is expressed as its **octane number,** and the octane rating is a measure of its performance in the internal combustion engine.

The straight chain hydrocarbon n-heptane is a poor fuel and causes excessive knock in a gasoline engine. It is **arbitrarily** assigned an octane rating of **zero.** Isooctane, a good fuel with little knocking characteristics, is assigned a rating of **100.** The octane rating of an unknown gasoline mixture is determined by comparing its performance in a standard engine to mixtures of n-heptane and isooctane in the same engine. The numerical value of the octane rating of the unknown gasoline is equal to the percentage of isooctane in the heptane/isooctane mixture of the same performance characteristics. Thus, a gasoline of the same knock as a mixture of 90 per cent isooctane and 10 per cent n-heptane is said to have an octane rating of 90. Most regular grades of gasoline have octane ratings of 92 to 94, and premium grades of gasoline average 98 to 100.

$$CH_3CH_2CH_2CH_2CH_2CH_2CH_3 \qquad \underset{\overset{|}{CH_3}}{\overset{\overset{CH_3}{|}}{CH_3-C}}-CH_2-\underset{\overset{|}{CH_3}}{\overset{\overset{H}{|}}{C}}-CH_3$$

n-Heptane   Isooctane

Consequently, an obvious way to increase the octane rating and the performance of a gasoline is to increase its percentage of branched chain hydrocarbons. This is done in petroleum refineries by a process called thermo-cracking. In this process the lower molecular weight hydrocarbons from petroleum are treated with various catalysts at temperatures of 400 to 500°C. This process increases the percentage of branched chain hydrocarbons and aromatics in a given fraction. Unfortunately, this additional processing increases the cost of gasoline.

Another way to increase the octane rating (other than thermo-cracking) is to add small quantities of a so-called antiknock agent to the gasoline. The most common antiknock agent used has been tetraethyl lead, $(CH_3CH_2)_4Pb$, an organometallic compound. Other additives such as TCP (tricresyl phosphate) and boron hydrides have also been used in recent years. Usually, the tetraethyl lead is used in combination with ethylene dibromide (CH_2BrCH_2Br) to form a mixture referred to as **ethyl fluid.** From one to three milliliters of this ethyl fluid is added per gallon of gasoline to improve the antiknock performance. However, the by-products of this ethyl fluid after combustion of the fuel in the engine are lead bromide and metallic lead, which are emitted with the exhaust gases. The toxicity of lead by-products from auto exhausts has caused much public concern in recent years. Consequently, several "lead-free" gasolines have appeared on the market, and government legislation has been enacted to eventually ban lead additives from gasoline products. Undoubtedly, other additives will come under scrutiny, and the petroleum industry is attempting to develop other suitable methods of increasing octane rating without the use of harmful additives.

As the number of automobiles increased, the supply of straight-run gasoline was insufficient to meet the demands. In addition to the shortage of gasoline, the petroleum industry was faced with an excess of other fractions distilled from petroleum. It was found as early as 1912 that lower molecular weight hydrocarbons could be produced from the higher molecular weight fractions by heating them to a temperature of from 400 to 500°C.

At this temperature some of the bonds of the larger hydrocarbon molecules break to form lower boiling range fractions. If the temperature is increased beyond 500°C, small molecular weight gaseous hydrocarbons are produced. This process, called **thermo-cracking,** not only produces smaller hydrocarbons, but also yields unsaturated and aromatic hydrocarbons. More gasoline with a higher octane number was therefore produced by the cracking process. Research by petroleum engineers has resulted in the development of several cracking processes that employ petroleum fractions of lower molecular weight and higher molecular weight than those in gasoline. Catalysts have also been found that increase the yield of high octane gasoline from these petroleum fractions.

Kerosene. In the early days of the petroleum industry the most important fraction from petroleum was kerosene. It was used for lighting purposes, for cooking, and for heating. With the advent of the electric light and the automobile, the demand for kerosene decreased and that for gasoline increased. **Kerosene** is composed of a mixture of saturated, unsaturated, and aromatic hydrocarbons containing from 12 to 15 carbon atoms.

Gas Oil and Fuel Oil. The next higher boiling fraction after kerosene contains a mixture of hydrocarbons whose smallest members have 15 carbon atoms. This fraction contains gas oil, fuel oil, and diesel oil. The name **gas oil** is derived from the fact that this fraction was used originally to enrich water gas for use as a fuel. Large quantities of **fuel oil** are used in furnaces that burn oil, whereas **diesel oil** is used in diesel engines. This fraction may also be cracked to produce gasoline.

Lubricating Oils. Lubricating oils are produced from the fraction of petroleum that distills at the highest temperature, usually over 300°C. This fraction consists of hydrocarbons with 20 or more carbon atoms and can be separated into oils of different viscosity by fractional distillation. An example of the importance of viscosity in oils is the switch from an oil of relatively high viscosity to one of low viscosity when preparing an automobile for winter driving. Recently, by the proper combination of oils and substances such as detergents, a superior multirange viscosity automobile oil has been produced. Such oils remain fluid at low temperatures and possess a greater viscosity at high temperatures than regular motor oil.

The Residual Fraction. The residual material that is left after the removal of the distillable fractions of petroleum usually contains either asphalt or paraffin types of hydrocarbons. **Paraffin wax** is prepared from the residue of paraffin base oil and consists of straight chain alkanes with 26 to 30 carbon atoms. This residue also yields petrolatum, which is commonly known as petroleum jelly or Vaseline. **Petrolatum** is a semisolid substance that is used as a pharmaceutical base for many salves and ointments. The asphalt type crude oil produces a residue containing pitch or asphalt that is used in roofing material, protective coatings, paving, and asphalt tiles for floors.

PHYSICAL PROPERTIES

The lower molecular weight alkanes, methane through butane, are gases at ordinary temperatures and pressures. The C_5 to C_{17} alkanes (m.p. $< 20°C$) are liquids at room temperature, and compounds containing more than C_{17} are solids at room temperature. The alkanes in the C_{26} and C_{36} range make up the substance known as paraffin wax, and these alkanes are sometimes referred to as paraffin hydrocarbons. Table 11–3 summarizes some of the physical properties of the more common alkanes.

The boiling points of the alkanes show a regular increase of approximately 20 to 30°C with the introduction of each new methylene group. Branching of the carbon chain lowers the boiling point of the alkane, and, with compounds that can exist as isomers, the straight chain isomer is always the highest boiling isomer.

The saturated hydrocarbons are almost completely insoluble in water, but are soluble in organic solvents.

TABLE 11-3   ALKANES

NAME	M.P. °C	B.P. °C	MOL. FORMULA	SP. GRAVITY (as liquids)
Methane	−183	−162	CH_4	—
Ethane	−172	−89	C_2H_6	—
Propane	−187	−42	C_3H_8	—
n-Butane	−135	−0.5	C_4H_{10}	—
n-Pentane	−130	+36	C_5H_{12}	0.626
n-Hexane	−94	69	C_6H_{14}	0.659
n-Heptane	−90	98	C_7H_{16}	0.683
n-Octane	−57	126	C_8H_{18}	0.703
n-Nonane	−54	151	C_9H_{20}	0.718
n-Decane	−30	174	$C_{10}H_{22}$	0.729
n-Undecane	−26	196	$C_{11}H_{24}$	0.740
n-Dodecane	−10	216	$C_{12}H_{26}$	0.749
n-Tridecane	−6	235	$C_{13}H_{28}$	0.757
n-Tetradecane	+6	251	$C_{14}H_{30}$	0.764
n-Pentadecane	10	268	$C_{15}H_{32}$	0.769
n-Hexadecane	18	280	$C_{16}H_{34}$	0.775
n-Heptadecane	22	303	$C_{17}H_{36}$	0.777
n-Octadecane	28	308	$C_{18}H_{38}$	0.777
n-Nonadecane	32	330	$C_{19}H_{40}$	0.778
n-Eicosane	36	343	$C_{20}H_{42}$	0.778

PREPARATION OF ALKANES

Many of the alkanes are actually obtained from petroleum and natural gas, either directly or by fractional distillation of petroleum products. In many cases, however, a particular compound or isomer may be required that is not available or easily obtained commercially from petroleum. Then the compound must be synthesized in the laboratory by the organic chemist. The following are some of the more common methods used to prepare alkanes.

REDUCTION OF ALKENES (HYDROGENATION)

The carbon-carbon double bond present in alkenes can add a mole of hydrogen (H_2) in the presence of a catalyst to give an alkane as the final product.

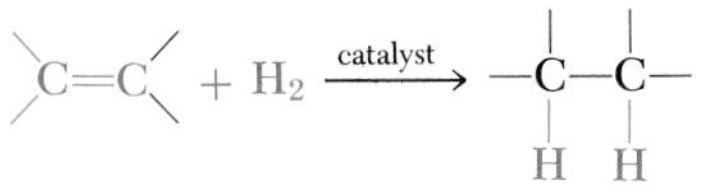

$$\text{C=C} + H_2 \xrightarrow{\text{catalyst}} \underset{H\ \ \ H}{-C-C-}$$

The catalysts most commonly used are palladium (Pd), platinum (Pt), and nickel (Ni).

$$CH_3CH{=}CH_2 + H_2 \xrightarrow{\text{Pt}} CH_3CH_2CH_3$$
$$\text{Propene} \qquad\qquad \text{Propane}$$

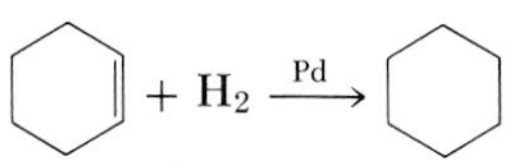

Cyclohexene   Cyclohexane

The addition of hydrogen (reduction) to biologically important compounds generally occurs enzymatically. For example, the important enzyme lactic dehydrogenase catalyzes the reduction of pyruvic acid to lactic acid (see p. 329).

Electrolytic Decarboxylation of Carboxylic Acids. The decarboxylation of a solution of the acid salt may also be carried out electrolytically. This type of reaction is known as a **Kolbe reaction.** The reaction differs from other decarboxylation methods in that the main product obtained is not a hydrocarbon containing one less carbon atom than the salt, but a hydrocarbon containing two less carbon atoms than twice the number of carbons in the original salt, as illustrated for sodium acetate (the product is a dimer of the two alkyl groups in the salt):

$$2\ CH_3\overset{\displaystyle O}{\overset{\displaystyle \|}{C}}-O^-Na^+ + 2H_2O \xrightarrow[\text{current}]{\text{electric}} CH_3CH_3 + 2CO_2 + 2NaOH + H_2$$

Sodium acetate   Ethane

In biologic systems the coenzyme, thiamine pyrophosphate (TPP), functions as the catalyst in the oxidative decarboxylation of pyruvic acid to carbon dioxide and acetaldehyde (see p. 327). This type of biochemical decarboxylation prevents the accumulation of pyruvic acid in the blood and tissues.

CHEMICAL REACTIONS

The alkanes are the least reactive organic compounds. They are resistant to strong acids and alkalies, and under ordinary conditions are resistant to oxidizing agents. For these reasons they are commonly employed as inert solvents in chemical reactions of the other functional groups. They do, however, react with halogens and oxygen under special conditions.

COMBUSTION

Hydrocarbons will react with oxygen when ignited in the presence of excess oxygen. The products of the combustion reaction are carbon dioxide and water. This reaction may be expressed as:

$$C_nH_{2n+2} + \frac{3n+1}{2}\ O_2 \rightarrow nCO_2 + (n+1)\ H_2O + \text{heat}$$

The heat given off in this reaction is called the heat of combustion. It is the utilization of this heat that accounts for much of the commercial use of hydrocarbons as heating fuels. Gasoline, which is composed of hydrocarbons containing 6 to 12 carbon atoms per molecule, burns similarly with oxygen in an automobile engine. The gases CO_2 and H_2O power the pistons of the engine, and the heat evolved is the heat carried away by the cooling system. Some specific examples are shown below:

$$CH_4 + 2O_2 \rightarrow CO_2 + 2H_2O + 213\ \text{kcal/mole}$$
$$C_2H_6 + 3\tfrac{1}{2}O_2 \rightarrow 2CO_2 + 3H_2O + 373\ \text{kcal/mole}$$

If combustion occurs in the absence of sufficient oxygen, carbon monoxide is formed. Carbon monoxide is the toxic gas found in the exhaust fumes of a car.

$$C_2H_6 + 2\tfrac{1}{2}O_2 \rightarrow 2CO + 3H_2O$$

The oxidation of organic compounds in biochemical systems occurs enzymatically to also give carbon dioxide, water, and cellular energy. Since the human cells could not adequately handle the large amounts of energy released in a combustion reaction, biologic oxidation is not completely analogous to combustion. It occurs in a stepwise sequence to release energy in small amounts useful to the cells. For example, the biologic oxidation of glucose (p. 353) releases large amounts of energy to be used in muscular work. Similarly, the enzymatic oxidation of fatty acids provides useful available energy for the human body (p. 362).

HALOGENATION

The halogens, fluorine, chlorine, and bromine, will react under the proper conditions with alkanes. Iodine does not react with alkanes. Fluorine undergoes a controlled reaction with alkanes only under very carefully controlled conditions, and because of its extreme reactivity it is generally not too useful for ordinary laboratory reactions.

Consequently, chlorine and bromine are the only practical halogens useful in the laboratory for reaction with alkanes. At higher temperatures or in the presence of sunlight, a chlorine or bromine atom can substitute for the hydrogen of an alkane. These reactions are known as **halogenation** reactions, or more specifically as **chlorination** or **bromination**, depending on whether a chlorine or a bromine atom substitutes for the hydrogen. The overall reaction can be summarized as:

$$Cl_2 + CH_4 \xrightarrow[\text{sunlight}]{\text{heat or}} CH_3Cl + HCl$$

Only half of the halogen ends up in the organic product; the other half ends up as hydrogen chloride. Substitution reactions always give two products, whereas addition reactions (such as the hydrogenation of alkenes to give alkanes) give only one product.

Substitution reactions in many instances give rise to more than two products. In the case of the chlorination of methane to give methyl chloride, the product, methyl chloride, may react further (via the same type of halogenation reaction) to give methylene chloride (CH_2Cl_2), which can react further to give chloroform ($CHCl_3$), which reacts further to give carbon tetrachloride (CCl_4). In actual practice, chlorination of methane gives all four products. The amounts of each depend on the amount of chlorine used and the reaction conditions.

Higher homologues increase the complexity of even the simple monohalogenation step. For example, in the chlorination of propane, two possible monochlorinated products are possible, depending upon which of the two different types of hydrogens chlorine substitutes for. In practice, both products are obtained. Further halogenation of these monochlorinated products produces an even more complex mixture. Because of this

$$Cl_2 + CH_3CH_2CH_3 \xrightarrow[\text{sunlight}]{\text{heat or}} CH_3CH_2CH_2Cl + CH_3CHClCH_3 + HCl$$
Propane   *n*-Propyl chloride   Isopropyl chloride
(1-chloropropane) (2-chloropropane)

multiplicity of products, halogenation of alkanes is not a particularly useful laboratory reaction. However, in many commercial products where ultra pure products are not always required, the mixtures obtained are useful, and this type of reaction is commercially important for preparing halogenated alkanes used in such products as dry-cleaning agents and fire extinguishers.

PYROLYSIS (CRACKING) OF ALKANES

Although alkanes are generally some of the most stable organic compounds, they can be broken down (cracked into smaller fragments) by heating to high temperatures

$(400\text{--}700°C)$ in the absence of air (to avoid combustion). This is an important reaction in the petroleum industry for converting high molecular weight alkanes into fragments in the gasoline range, thereby increasing the amount of gasoline obtainable from crude petroleum. In the petroleum industry many catalysts have been developed which affect the "cracking" reaction at much lower temperatures than simple pyrolysis. A simple illustrative example of pyrolytic cracking is shown below:

$$CH_3CH_2CH_2CH_3 \xrightarrow{700°C} CH_2{=}CHCH_2CH_3 + CH_3CH{=}CHCH_3 + CH_3CH_3 +$$

$$CH_2{=}CH_2 + CH_3CH{=}CH_2 + CH_4$$

Both carbon-carbon and carbon-hydrogen bonds are broken at these temperatures, and a complex mixture of products is obtained.

IMPORTANT TERMS AND CONCEPTS

addition reaction	cycloalkane	hydrocarbon	methylene group
alkane	halogenation	hydrogenation	petroleum
combustion	homologous series	IUPAC rules	substitution reaction
cracking reaction			

QUESTIONS

1. Draw out all the possible isomers for the compound having a molecular formula, C_6H_{14}. Give the correct IUPAC name to each of the compounds.

2. Which of the isomeric compounds in Question 1

   (a) contains a tertiary butyl group?
   (b) is isohexane?
   (c) contains an isopropyl group?
   (d) has the lowest boiling point?
   (e) has the highest boiling point?

3. Draw a correct structure for each of the following compounds:

   (a) 1,2-dimethylcyclohexane
   (b) 4-isopropyl-5-methyldecane
   (c) *n*-heptane
   (d) tertiary butyl bromide
   (e) 3-methyl-4-ethyl-5-isopropyloctane

4. Name the following compounds by the IUPAC system.

   (a) $(CH_3)_2CHC(CH_3)_2C(C_2H_5)_3$
   (b) $(CH_3)_2CHC(C_2H_5)_3$
   (c) $(CH_3)_3CCH_2CH(CH_3)CH_2CH(CH_3)_2$

   (d) 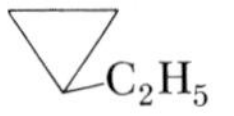 $\triangle\!\!-C_2H_5$

(e)

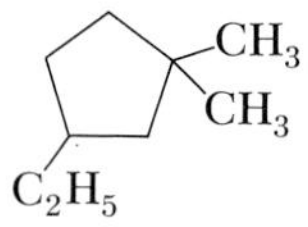

5. Explain what is wrong with *each* of the following names. Give the *correct* name for each compound.

    (a) 1,1-diethylbutane
    (b) 1,1,2-trimethylbutane
    (c) 4-isopropyl-5-methylhexane
    (d) 2-dimethylbutane
    (e) 1,3,3-trimethylcyclobutane

6. Write a chemical equation for each of the following chemical reactions. Indicate any necessary catalysts and reaction conditions. Name each of the organic products of the reaction.

    (a) complete combustion of propane
    (b) electrolytic decarboxylation of sodium propionate ($CH_3CH_2CO_2Na$)

SUGGESTED READING

Be Sure to Understand Octane Numbers. Chemistry, Vol. 47, p. 3, 1974.

Conrad and Sabin: Motor Fuel Quality as Related to Refinery Processing and Antiknock Compounds. Journal of Chemical Education, Vol. 34, p. 262, 1957.

Evienx: The Geneva Congress on Organic Nomenclature, 1892. Journal of Chemical Education, Vol. 31, p. 326, 1954.

Hurd: The General Philosophy of Organic Nomenclature. Journal of Chemical Education, Vol. 38, p. 43, 1961.

Kimberlin: Chemistry in the Manufacture of Modern Gasoline. Journal of Chemical Education, Vol. 34, p. 569, 1957.

Medeiros: Lead From Automobile Exhaust. Chemistry, Vol. 44, No. 10, p. 7, 1971.

Orchin: Determining the Number of Isomers from a Structural Formula. Chemistry, Vol. 42, No. 5, p. 8, 1969.

Reti: Leonardo Da Vinci's Experiments on Combustion. Journal of Chemical Education, Vol. 29, p. 590, 1952.

ALKENES AND ALKYNES

CHAPTER 12

The *objectives* of this chapter are to enable the student to:

1. Understand the concept of pi bonds is unsaturated hydrocarbons.
2. Define and explain geometrical isomerism.
3. Write structural formulas and IUPAC names of alkenes and alkynes.
4. Describe and understand the concept of elimination reactions.
5. Distinguish between primary, secondary, and tertiary alcohols.
6. Predict the direction of an elimination reaction.
7. Describe Markownikoff's rule.
8. Describe the generalized concept of an electrophilic addition reaction.
9. Describe the basic chemistry of alkynes.

Alkenes are distinguished from alkanes by the presence of the carbon-carbon double bond in the molecule. Other terms used to denote alkenes are **olefins** (from olefiant gas—the old name for ethylene) and **unsaturated hydrocarbons** (to denote the presence of a multiple bond in the molecules). The presence of the carbon-carbon double bond in the molecule makes these kinds of compounds highly reactive compared to alkanes.

In Chapter 10 the differences in bonding between alkanes and alkenes were briefly noted. In alkanes, which contain only carbon-carbon single bonds, essentially free rotation is allowed around the axis of the bond between the two carbon atoms. The orbitals used in the bonding of alkanes were sp^3 and gave a tetrahedral structure. However, in alkenes three sp^2 orbitals are used to form the carbon-hydrogen bonds and the carbon-carbon single bond. The use of sp^2 orbitals in bonding gives a planar arrangement of the carbon and hydrogen atoms involved° (see also Fig. 3–9). The remaining carbon-carbon bond is formed by overlap of the $2p$ orbital remaining on each carbon atom as illustrated in Figure 12–1. Hence, the first carbon-carbon bond is formed by the overlap of two sp^2 orbitals (called a sigma [σ] bond) and the second carbon-carbon bond is formed by overlap of two $2p$ orbitals (called a pi [π] bond). The π-bond is much weaker than the σ-bond and is the first bond broken in the chemical reactions of most alkenes.

Resultant features of this multiple bonding in alkenes are: (1) The bond distance between the two carbon atoms of the multiple bond is 1.34 Å compared to 1.54 Å in an alkane; (2) Rotation about the carbon-carbon double bond is restricted—no free rotation is allowed, since the π-bond would have to be broken to allow this rotation; (3)

° The student is encouraged to use models to confirm this coplanar arrangement of the atoms and the restricted rotation in alkenes.

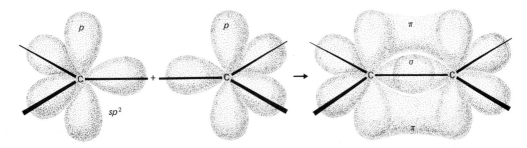

Figure 12-1   Bonding orbitals in a carbon-carbon double bond.

Because of restricted rotation around the multiple bond, a new type of isomerism is possible—namely, *geometrical isomerism* in which the two isomers differ only in the arrangement of the four atoms or groups of atoms that are attached to the multiple bond.

NOMENCLATURE AND STEREOCHEMISTRY

Alkenes have the general formula C_nH_{2n}, and cycloalkenes have the general formula C_nH_{2n-2}. Two systems again are most commonly used to name these compounds. The correct nomenclature system is the IUPAC system. In this system the characteristic ending **-ene** is added to the stem name (formed by dropping the **-ane** ending from the alkane having the same number of carbon atoms); the carbon atoms in the *longest continuous chain containing the double bond* are numbered so that the carbon atoms of the double bond have the lowest possible numbers. Therefore the location of the double bond is indicated by the lowest possible number, and this number is placed in front of the name. Substituents are listed in front of the alkene name and are preceded by the appropriate number to indicate their position. The other system of nomenclature is the use of common names for the lower members of this series. The common names are formed by dropping the **-ane** ending of the parent alkane and adding the ending **-ylene** to the remaining root stem.

$$\underset{\substack{\text{Ethene}\\ \text{(ethylene)}}}{\overset{H}{\underset{H}{>}}C=C\overset{H}{\underset{H}{<}} \quad \text{or} \quad H_2C{=}CH_2}$$

$$\underset{\substack{\text{Propene}\\ \text{(propylene)}}}{H{-}\overset{H}{\underset{H}{C}}{-}C{=}C\overset{H}{\underset{H}{<}} \quad \text{or} \quad CH_3CH{=}CH_2}$$

In olefins containing four carbon atoms or more, not only are isomers possible by varying the arrangement of the carbon atoms, but isomerism can also result from a shift in the position of the double bond without changing the carbon skeleton itself, as illustrated below for the pentenes:

$$CH_2{=}CHCH_2CH_2CH_3 \qquad \text{1-Pentene}$$

$$CH_3CH{=}CHCH_2CH_3 \qquad \text{2-Pentene}$$

$$\underset{CH_3}{\overset{CH_3}{>}}CHCH{=}CH_2 \qquad \text{3-Methyl-1-butene}$$

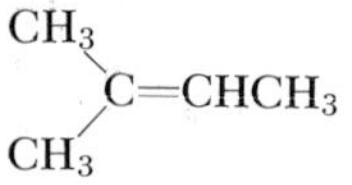

CH₃
  \
   C=CHCH₃   2-Methyl-2-butene 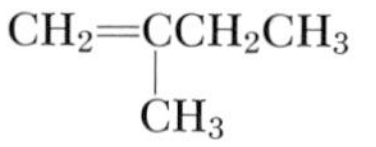
  /
CH₃

CH₂=CCH₂CH₃   2-Methyl-1-butene
       |
      CH₃

The only difference between 1-pentene and 2-pentene is the position of the double bond. Further examination of the pentene isomers has revealed that there are two different pentenes which have the structure $CH_3CH=CHCH_2CH_3$. The structure of these two pentenes has been shown to be:

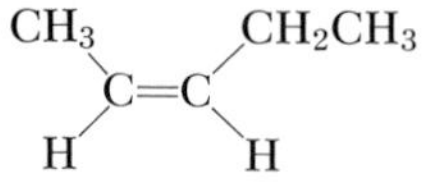

CH₃   CH₂CH₃   CH₃   H
  \   /   \   /
   C=C   and   C=C
  /   \   /   \
 H   H   H   CH₂CH₃

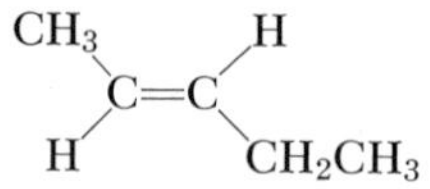

       cis-2-Pentene   *trans*-2-Pentene
           (A)   (B)

The only difference between these two isomers is that in one structure (A) the two hydrogens attached to the double bond are on the same side of the double bond (this is called a *cis* arrangement) and the two alkyl groups are also on the same (but opposite to the hydrogen) side of the double bond. In structure (B), the two hydrogens attached to the double bond are on opposite sides of the double bond (this is called a *trans*-arrangement), and the two alkyl groups are also *trans* across the double bond. Because of the restricted rotation around the double bond, these two compounds are fixed in respect to the arrangement of the atoms and alkyl groups around the double bond. Both the *cis*- and *trans*- compounds do exist, and both have been isolated and their structures unambiguously proved.

The necessary requirements for *geometrical isomerism* are: (1) that restricted rotation of some kind be present in the molecule. This may be either a double bond in olefins or the presence of a ring system which also prevents free rotation. Consequently, cyclic alkanes also exhibit geometric isomerism; (2) that neither carbon atom involved in the restricted rotation may hold identical groups. Some examples of geometric isomerism are shown below:

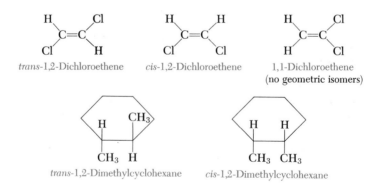

H   Cl   H   H   H   Cl
 \   /   \   /   \   /
  C=C   C=C   C=C
 /   \   /   \   /   \
Cl   H   Cl   Cl   H   Cl

trans-1,2-Dichloroethene   *cis*-1,2-Dichloroethene   1,1-Dichloroethene
                                                        (no geometric isomers)

      H   CH₃   H   H

    CH₃ H   CH₃ CH₃
trans-1,2-Dimethylcyclohexane   *cis*-1,2-Dimethylcyclohexane

Many naturally occurring biologic systems also exhibit geometrical isomerism. For example, a fatty acid, such as oleic acid [$CH_3(CH_2)_7CH=CH(CH_2)_7COOH$] is found in nature only in the cis form. *Cis*- and *trans*- vitamin A (p. 333) are also important in the human visual cycle. Steroid systems (p. 320) are fused cyclic rings which also exhibit *cis-trans* isomerism with the *cis* configuration being the most common form.

TABLE 12-1   PHYSICAL PROPERTIES OF ALKENES AND DIENES

COMPOUND	STRUCTURE	M.P. °C	B.P. °C
Ethylene	$CH_2{=}CH_2$	−169	−104
Propylene	$CH_3CH{=}CH_2$	−185	−48
1-Butene	$CH_3CH_2CH{=}CH_2$	−185	−6
2-Butene (cis)	$CH_3CH{=}CHCH_3$	−139	+4
2-Butene (trans)	$CH_3CH{=}CHCH_3$	−106	+1
1-Pentene	$CH_3CH_2CH_2CH{=}CH_2$	−165	+30
1-Hexene	$CH_3(CH_2)_3CH{=}CH_2$	−140	+64
Cyclobutene		—	+2
Cyclopentene	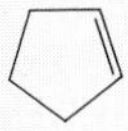	−135	+44
Cyclohexene	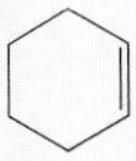	−104	+83
Allene	$CH_2{=}C{=}CH_2$	−136	−35
1,3-Butadiene	$CH_2{=}CH{-}CH{=}CH_2$	−109	−4

In **cycloalkenes,** the possibility of *cis-trans* isomerism is limited by the constraints of the ring. Cycloalkenes of less than seven carbons in the ring exist only in the *cis*-arrangement. The corresponding *trans*-isomer in the smaller rings is too highly strained and cannot be isolated.

PHYSICAL PROPERTIES OF ALKENES

The alkenes are quite similar to the alkanes in physical properties. Alkenes containing 2 to 4 carbon atoms are gases; those containing 5 to 18 carbon atoms are liquids; and those containing more than 18 carbons are solids. They are relatively insoluble in water but are soluble in concentrated sulfuric acid. The physical properties of some of the more common alkenes are summarized in Table 12–1.

PREPARATION OF ALKENES

The most general laboratory synthesis of alkenes involves a reaction known as an **elimination** reaction in which a molecule of water or hydrogen halide is removed from adjacent carbon atoms in a saturated compound. The generalized reaction may be summarized as:

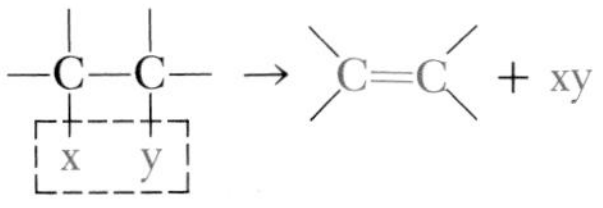

$$-\overset{|}{\underset{\underset{x}{|}}{C}}-\overset{|}{\underset{\underset{y}{|}}{C}}- \rightarrow {\large >}C{=}C{\large <} + xy$$

x = H
y = OH or halogen

In this process the group —x is removed in such a manner that it leaves its electron pair which then forms an additional bond (the double bond) between the carbon atoms involved.

DEHYDRATION OF ALCOHOLS

When the molecule (xy) lost in an elimination reaction is water (HOH), the reaction is called a **dehydration reaction.** The starting material which is dehydrated is an alcohol, and the alcohol can be either a straight chain or a cyclic alcohol. The catalysts employed for dehydration reactions are acidic, and all have a strong affinity for water. Acids such as concentrated sulfuric acid and phosphoric acid, or aluminum oxide and phosphorus pentoxide at high temperatures, are generally employed. For example cyclohexane can be prepared from cyclohexanol by this method. In some alcohols the elements of H_2O

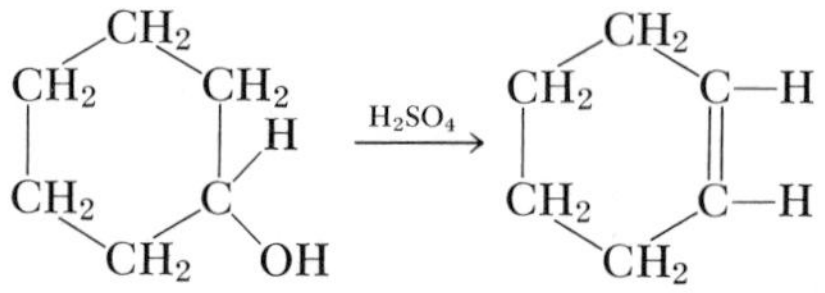

Cyclohexanol   Cyclohexene

can be lost in more than one way. For example, in 2-butanol either 1-butene or 2-butene may be formed by the elimination of water.

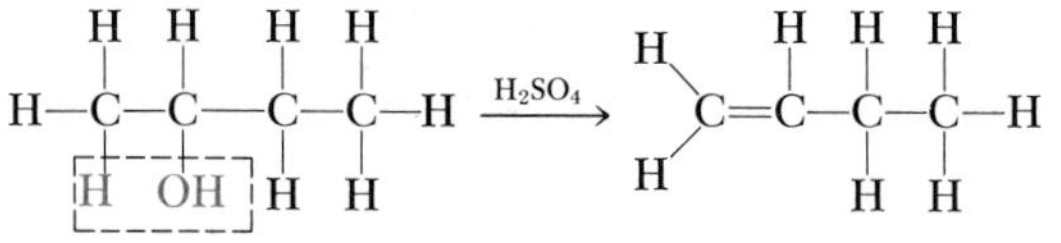

1-Butene

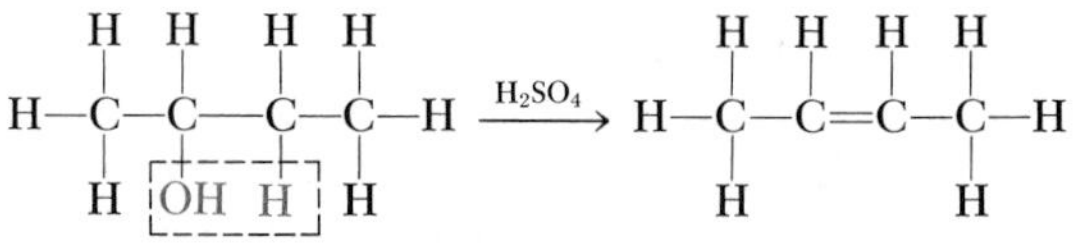

2-Butene

In alcohols of this type, mixtures of both possible olefins are generally obtained. The ratio of 2-butene to 1-butene in this reaction is $4:1$.

The ease of dehydration of alcohols depends upon the type of alcohol used. Alcohols are classified as primary, secondary, and tertiary by counting the number of alkyl groups bonded to the carbon atom bearing the —OH group, as follows:

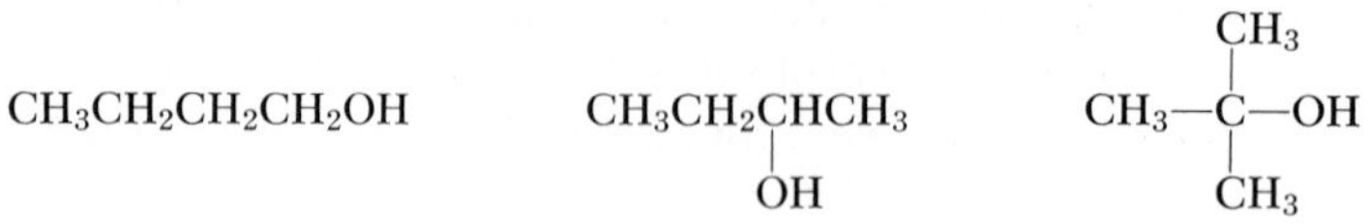

Primary alcohol   Secondary alcohol   Tertiary alcohol
1-Butanol   2-Butanol   2-Methyl-2-propanol

The ease of dehydration has been proved to be: tertiary $>$ secondary $>$ primary. Thus, it is easier to dehydrate 2-methyl-2propanol than it is to dehydrate 1-butanol.

Dehydration (loss of H_2O) is not unique to the formation of a carbon-carbon double bond. Other functional groups, such as ether linkages, in the dehydration of sugars to form disaccharides, polysaccharides, glycogen, starch, and cellulose (see Chapter 24), or amide linkages, in the formation of protein molecules by the loss of water between amino acids (see p. 273), are important biochemical dehydrations that will be encountered later.

DEHYDROHALOGENATION OF ALKYL HALIDES

An elimination reaction, similar to the dehydration of alcohols, can occur with **alkyl halides** (compounds containing a halogen bonded to an alkyl group) in the presence of base to eliminate the elements of —HX (where X = halogen) and form an alkene. For example, if isopropyl chloride is treated with potassium hydroxide in alcohol, propene, potassium chloride, and water are formed as products.

$$CH_3CHClCH_3 + KOH \xrightarrow{\text{alcohol}} CH_3CH{=}CH_2 + KCl + H_2O$$

Isopropyl chloride   Propene
(2-chloropropane)

The ease of removal of halogen in reactions of this type depends on the type of halogen eliminated and the structure of the alkyl halide. For example, tertiary butyl chloride undergoes elimination much easier than n-butyl chloride, and the ease of dehydrohalogenation is in the following order: tertiary alkyl halides > secondary > primary. Also, the ease of dehydrohalogenation for a particular type of carbon structure increases as the halogen is varied from fluorine through iodine. Consequently, it is easier to dehydrohalogenate isopropyl iodide than it is to dehydrohalogenate isopropyl chloride.

The generally accepted scheme for dehydrohalogenation is as follows:

$$HO^- + CH_2{-}CH{-}CH_3 \rightarrow CH_2{=}CHCH_3 + H_2O + Cl^-$$

The base, OH^- in the example, abstracts a proton from the adjacent carbon atom to form water. The electron pair left behind by the proton being abstracted then forms the carbon-carbon double bond with the ejection of a chloride ion. This type of reaction is known as a β-elimination, since the proton is being abstracted from the carbon atom which is *beta* to the group being eliminated (Cl^-). In compounds which are not symmetrical, two different olefins can be produced. In 2-chlorobutane, both 1-butene and *cis*- and *trans*-2-butene can be formed. In this particular compound, 80% of the 2-butene and 20% of the 1-butene are produced.

REACTIONS OF ALKENES

In the previous chapter the reactions of alkanes were shown to be reactions of a substitution type in which the carbon-hydrogen bond is broken and a new bond is formed between carbon and a new atom or group of atoms. Two products are formed in reactions of this type; for example, methyl chloride and hydrogen chloride are formed in the chlorination of methane (cf. p. 173).

In contrast to this behavior of alkanes, *the most characteristic reaction of alkenes is addition to the double bond*. The π-bond of the olefin is broken and two new single bonds are formed in this process. *Only one product is formed in addition reactions of this type*. A generalized reaction scheme for an addition reaction can be represented by the following equation:

$$\begin{array}{c}\diagup\\ \diagdown\end{array}C=C\begin{array}{c}\diagup\\ \diagdown\end{array} + xy \rightarrow \begin{array}{c}|\quad|\\ -C-C-\\ |\quad|\\ X\quad y\end{array}$$

Addition reaction

The π-bond (the weaker of the two bonds) is broken and a new bond is formed between carbon and atom or group —x, and between carbon and atom or group —y, to yield the addition product, which now contains only single bonds between all atoms or groups of atoms. The molecule or compound xy can be one of many types of materials added to olefins. We shall only consider several simple types of compounds which have been added to olefins, but it should be kept in mind that this is a very general and widely used reaction by the organic chemist and one of the more important types of reactions in organic chemistry. In later chapters we shall again encounter addition reactions to other multiple bonds, such as $>C=O$, or $—C\equiv N$. The necessary requirement for an addition reaction is the presence of a multiple bond, and addition reactions are not necessarily limited to carbon-carbon multiple bonds.

HYDROGENATION OF ALKENES

In Chapter 11, the preparation of alkanes was shown to be possible by the catalytic hydrogenation of alkenes. This is a general reaction of alkenes and is important both in the laboratory and commercially in the preparation of many organic compounds.

$$\begin{array}{c}\diagup\\ \diagdown\end{array}C=C\begin{array}{c}\diagup\\ \diagdown\end{array} + H_2 \xrightarrow{\text{catalyst}} \begin{array}{c}|\quad|\\ -C-C-\\ |\quad|\\ H\quad H\end{array}$$

Catalytic hydrogenation is also of considerable commercial importance. Margarine and cooking shortenings are prepared by this type of reaction from vegetable oils. Vegetable oils are long chain unsaturated acids which contain one or more $>C=C<$ linkage. For example, oleic acid, $CH_3(CH_2)_7CH=CH(CH_2)_7COOH$, m.p. $+16°C$, linoleic acid, $CH_3(CH_2)_4CH=CH—CH_2CH=CH(CH_2)_7COOH$, m.p. $-5°C$, and linolenic acid, $CH_3CH_2CH=CHCH_2CH=CHCH_2CH=CH—(CH_2)_7COOH$, m.p. $-11°C$, are all converted to stearic acid, $CH_3(CH_2)_{16}COOH$, m.p. $+71°C$ via the addition of one, two, or three moles of hydrogen, respectively. Note the increase in the melting point as the number of double bonds is decreased. Consequently, by controlling the extent of hydrogenation, a fat of the desired melting point can be obtained. Commercial fats such as "Crisco" and "Spry" contain approximately 20 to 25 per cent saturated fatty acids, 65 to 75 per cent oleic acid, and 5 to 10 per cent linoleic acid.

TOPIC OF CURRENT INTEREST

POLYUNSATURATES AND HEART DISEASE

Fats and oils are the fatty acid esters of glycerol and are commonly called **triglycerides.** The distinction between a fat and an oil is purely arbitrary. An oil is a fat that is liquid at room temperature. They are found in large quantities as fat deposits in animals and as oils in nuts and seeds. Most of the fats found in food

$$CH_2-O-\overset{\displaystyle O}{\overset{\|}{C}}-R_1$$

$$CH-O-\overset{\displaystyle O}{\overset{\|}{C}}-R_2$$

$$CH_2-O-\overset{\displaystyle O}{\overset{\|}{C}}-R_3$$

Triglyceride (R_1, R_2, R_3—long chain unsaturated groups)

are mixtures of a number of esters of glycerol whose composition varies depending upon the environmental conditions under which they are formed. The oils (liquid fats) are richer in unsaturates (that is, R_1, R_2, R_3, contain more than one double bond). The relative unsaturations of some common edible fats are shown in the following tabulation.

Fat	Relative Unsaturation
Coconut oil	0.3
Butter	1.0
Lard	2.2
Olive oil	3.1
Corn oil	4.6
Soybean oil	4.9
Hard margarine (average)	2.7
Soft margarine (average)	3.2

The lipid cholesterol is present in the blood and in all body tissues. In the disease atherosclerosis, commonly called hardening of the arteries, excessive amounts of cholesterol are deposited on the arterial walls of the blood vessels and constrict them. Such action results in a rise in the blood pressure and a subsequent increase in the incidence of heart attacks. Recent research has suggested that the presence of polyunsaturated fats (in contrast to polysaturated fats) in the diet reduces the level of cholesterol in the blood, thereby decreasing the possibility of atherosclerosis and the risk of heart attack. Although such a cause-and-effect relationship has not been unambiguously proven, it has become popular for manufacturers of vegetable short-enings and margarines (oils that have been partially hydrogenated to have the consistency of butter) to stress in their advertising campaigns that these products are produced from 100 per cent polyunsaturated oils. Perusal of the above tabulation indeed does show that these oils are "more" polyunsaturated relative to butter. Whether this difference of a factor of two to five in relative unsaturation from natural saturated fats is enough to significantly decrease the rate of atherosclerosis is a question still under active investigation.

In relation to the question of eating too much saturated fat, some interesting nonmedical research has been recently reported. Since beef and lamb have relatively high concentrations of saturated fatty acids, many doctors have advised their patients with high serum cholesterol levels to substitute other forms of protein in their diet—to the dismay of the patients. However, it has been recently found that lambs raised on a milk diet high in linoleic acid show up to 10 times as much polyunsaturated fatty acid in their body fat as control lambs not raised on this diet. Consequently, it may be possible to produce polyunsaturated veal, and experimental studies on animals are currently in progress. Perhaps in the future, the diet of the average person will include a variety of polyunsaturated dietary products (including meat) as a means of reducing cholesterol levels in the blood.

ADDITION OF HALOGEN

Halogens behave similar to hydrogen and undergo 1,2-addition across the double bond. In practice, chlorine and bromine are the halogens generally employed. Fluorine can be added to olefins under special conditions but is too reactive for general laboratory use. Iodine adds reversibly to olefins, and most 1,2-vicinal iodides are unstable.

$$CH_3-\underset{\underset{CH_3}{|}}{C}=CH_2 + Br_2 \rightarrow CH_3-\underset{\underset{CH_3}{|}}{\overset{\overset{Br}{|}}{C}}-CH_2Br$$

Isobutylene
(2-methylpropene)

1,2-Dibromo-2-methylpropane

The addition of bromine to a $>C=C<$ bond is the basis for a simple chemical test for this functional group. Since a bromine solution in carbon tetrachloride is red, disappearance of this red color upon addition of an unknown organic compound is a strong indication that the unknown compound contains the $>C=C<$ linkage.

The addition of iodine to a $>C=C<$ linkage is also the basis of a test to determine the content of unsaturated fatty acids in lipids (p. 315). Since the amount of unsaturation of a fat or oil is proportional to the number of $>C=C<$ linkages in the fatty acid, the amount of iodine absorbed by a lipid can be employed to determine the degree of unsaturation. The degree of unsaturation is called the *iodine number,* and is defined as the grams of iodine absorbed by 100 grams of fat or oil. For example, the iodine number for oleic acid can be calculated as shown below.

$$CH_3(CH_2)_7CH=CH(CH_2)_7COOH + I_2 \rightarrow CH_3(CH_2)_7CHICHI(CH_2)_7COOH$$

oleic acid

$$\text{Iodine Number} = \frac{\text{Mol. Wt. } (I_2) \times \text{no. of } C=C \text{ linkages} \times 100}{\text{Mol. Wt. (fatty acid)}}$$

$$= \frac{(254)(1)(100)}{282.5} = 90$$

For linoleic acid, the iodine number is 181. In general, a high iodine number indicates a high degree of unsaturation.

ADDITION OF ACIDS

Acids such as sulfuric acid and the hydrogen halides (HF, HCl, HBr, HI) can be added across a carbon-carbon double bond to give either alkyl hydrogen sulfates or alkyl halides, as illustrated below for ethylene:

$$CH_2=CH_2 + H_2SO_4(HOSO_2OH) \rightarrow CH_3CH_2OSO_2OH$$

Ethylene   Sulfuric acid   Ethyl hydrogen sulfate

$$CH_2=CH_2 + HBr \rightarrow CH_3CH_2Br$$

Ethyl bromide

In unsymmetrical olefins, two possible products could result. For example, in the addition of hydrogen iodide to propene, either isopropyl iodide or n-propyl iodide could be produced. An empirical rule, called **Markownikoff's rule,** generalizes the addition of unsymmetrical reagents to unsymmetrical olefins as follows: *When an unsymmetrical reagent adds to an unsymmetrical olefin, the positive part of the unsymmetrical reagent becomes attached to the carbon atom of the double bond which bears the greatest number of hydrogen atoms.* This rule predicts isopropyl iodide (which is the actual product of the reaction) as the product in the following reaction.

$$CH_3CH=CH_2 + HI \rightarrow CH_3CH_2CH_2I \text{ or } CH_3CHICH_3$$

Propene   n-Propyl iodide   Isopropyl iodide

Other unsymmetrical reagents add similarly.

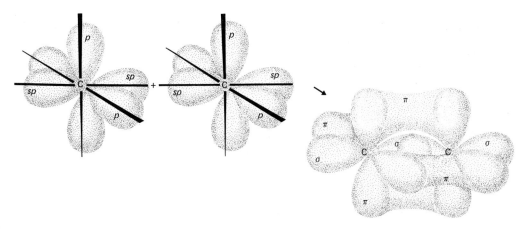

Figure 12-2   Schematic representation of bonding orbitals in a carbon-carbon triple bond.

ALKYNES

Alkynes are organic compounds that contain a carbon-carbon triple bond. The general molecular formula which describes these compounds is C_nH_{2n-2}. This class of compounds is also referred to as "acetylenes," after the first member of this series, acetylene itself, $HC{\equiv}CH$.

The carbon-carbon triple bond in alkynes is composed of one sigma bond formed by using an sp-orbital (see Fig. 3–9) from each carbon, and two π-bonds formed from the remaining p-orbitals on the respective carbon atoms. The two π-bonds are perpendicular to one another and enclose the carbon-carbon sigma bond in a cylinder of electron density (Fig. 12–2).

The alkynes are similar to alkanes and alkenes in physical properties. They are insoluble in water and soluble in organic solvents. Their boiling points are similar to those of the corresponding alkanes and alkenes, and the physical properties of some of the more common alkynes are listed in Table 12–2. Since the alkynes involve bonding using sp-orbitals, they are linear molecules. Consequently, *cis-trans* isomers are not possible for compounds such as 2-butyne ($CH_3C{\equiv}CCH_3$) as was found for the 2-butenes.

NOMENCLATURE

In the IUPAC nomenclature system alkynes are named by adding the ending -yne to the stem name for the corresponding hydrocarbon, and the position of the triple bond is given the lowest number. Other substituents are named as before. Some typical examples are shown in Table 12–2.

METHODS OF PREPARATION

Acetylene is prepared industrially by the pyrolysis of methane.

$$2CH_4 \xrightarrow{1500°C} HC{\equiv}CH + H_2$$

Methane   Acetylene

TABLE 12-2   PHYSICAL PROPERTIES
OF ALKYNES

COMPOUND	STRUCTURE	M.P. °C	B.P. °C
Acetylene	$HC{\equiv}CH$	−82	−84 (subl.)
Propyne	$CH_3C{\equiv}CH$	−103	−23
1-Butyne	$CH_3CH_2C{\equiv}CH$	−126	+8
2-Butyne	$CH_3C{\equiv}CCH_3$	−32	+27
1-Pentyne	$CH_3CH_2CH_2C{\equiv}CH$	−106	+40
2-Pentyne	$CH_3C{\equiv}CCH_2CH_3$	−109	+56
1-Hexyne	$CH_3CH_2CH_2CH_2C{\equiv}CH$	−132	+71
2-Hexyne	$CH_3C{\equiv}CCH_2CH_2CH_3$	—	+84

Laboratory preparations of acetylenes are similar to those used for the preparation of alkenes.

DEHYDROHALOGENATION

The vicinal dihalides required in this reaction are conveniently prepared via the addition of halogens to alkenes. Thus, the synthesis of propyne may be envisioned as a two-step process starting from propene:

$$CH_3CH{=}CH_2 + Br_2 \longrightarrow CH_3CHBrCH_2Br$$
Propene   1,2-Dibromopropane

$$CH_3CHBrCH_2Br + 2KOH \xrightarrow{alcohol} CH_3C{\equiv}CH$$
1,2-Dibromopropane   Propyne

REACTIONS OF ALKYNES

Since acetylenes contain two π-bonds, it might be expected that the characteristic reaction of acetylenes would also be an addition reaction. In fact, this prediction is realized experimentally, and alkynes undergo addition reactions like olefins, except that two moles of reagent are required to saturate the two π-bonds. In many cases, addition reactions to acetylenes can be controlled to give an olefin derivative, which is formed via addition of one mole of reagent to a triple bond. Some of the more characteristic reactions of alkynes are summarized in the following section.

HYDROGENATION

Exhaustive catalytic (Pt, Pd, Ni) hydrogenation of alkynes gives alkanes as the final product. Partial hydrogenation gives an alkene. In compounds which can exhibit *cis-trans* isomerism, the *cis*-isomer is the predominant isomer formed on partial catalytic hydrogenation of an acetylene.

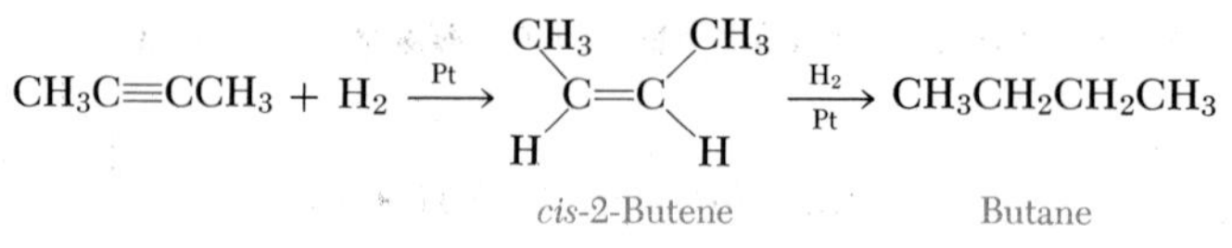

$$CH_3C{\equiv}CCH_3 + H_2 \xrightarrow{Pt} \underset{H \quad\quad H}{\overset{CH_3 \quad CH_3}{C{=}C}} \xrightarrow[Pt]{H_2} CH_3CH_2CH_2CH_3$$

cis-2-Butene   Butane

HALOGENATION

As with the alkenes, chlorine and bromine add easily to a triple bond. Fluorine is generally too vigorous, and iodine generally does not form stable addition products.

$$HC{\equiv}CH + Cl_2 \longrightarrow ClCH{=}CHCl \xrightarrow{Cl_2} CHCl_2CHCl_2$$

Acetylene   1,2-Dichloroethylene   1,1,2,2-Tetrachloroethane

ADDITION OF HYDROGEN HALIDES

The addition of hydrogen halides (HF, HCl, HBr, and HI) leads first to vinyl* halides and then to 1,1-dihalides. These addition reactions follow Markownikoff's rule.

$$HC{\equiv}CH + HBr \longrightarrow CH_2{=}CHBr \xrightarrow{HBr} CH_3CHBr_2$$

Acetylene   Vinyl bromide   1,1-Dibromoethane
                          (1-bromoethene)

TOPIC OF CURRENT INTEREST

VINYL CHLORIDE—CARCINOGEN

Vinyl chloride ($H_2C{=}CHCl$) has been used in the plastics industry for over forty years and has provided a host of commercial applications by forming polyvinyl-chloride (PVC) polymers. In addition, the monomer has been used as a propellant in aerosol consumer products such as hair sprays and pesticides.

However, it has recently been disclosed that several workers in vinyl chloride plants died of an unusual form of liver cancer called angiosarcoma. All of these workers had been exposed to vinyl chloride over a period of 14 to 27 years before the cancers appeared. Consequently, the number of people exposed to vinyl chloride before the recognition of this hazard runs into the thousands, and the possibility exists that more cases may soon be found. In addition to the workers exposed to this olefin, many millions of others have been unknowingly exposed to this material as a propellant in aerosol products. To avoid any further exposure to the general population, manufacture of vinyl chloride for use as an aerosol has been stopped, and the sale of vinyl chloride aerosol sprays has been banned by the Environmental Protection Agency (EPA). In addition, the safety standard of 500 parts per million (ppm) was revised downward to 50 ppm in vinyl chloride manufacturing plants, although continued criticism to lower the limit to zero ppm still abounds. The discovery of a carcinogen (cancer-causing agent) such as vinyl chloride, after such long-term use and exposure, obviously raises the crucial issue of other potential carcinogens to which we are unknowingly exposed.

IMPORTANT TERMS AND CONCEPTS

addition reactions	cycloalkene	geometrical isomerism	σ bond
alkyl halide	dehydration	Markownikoff's rule	*trans* isomer
alkynes	dehydrohalogenation	olefin	vinyl group
cis isomer	elimination reaction	π bond	

---

*The $CH_2{=}CH-$ is known as the vinyl group. $CH_2{=}CHCl$ (vinyl chloride) is an important olefin in preparing the commercial polyvinyl chloride.

QUESTIONS

1. Draw a structural formula for each of the following compounds:

   (a) *trans*-2-hexene
   (b) *cis*-2,3-dichloro-2-butene
   (c) 1-methylcyclopentene
   (d) *trans*-1,2-dibromocyclohexane
   (e) 4-ethyl-1-octene
   (f) 3-hexyne
   (g) *cis*-diiodoethylene
   (h) 2-methyl-2-butene

2. Name each of the following compounds by the IUPAC system:

   (a) $CH_3CH_2C \equiv CH$
   (b) $CH_2 = CHCBr = CHCH_3$
   (c) $(CH_3)_2C = CHCH_3$
   (d) $CH_3CH = CCl_2$
   (e) $CH_2 = CF_2$
   (f) 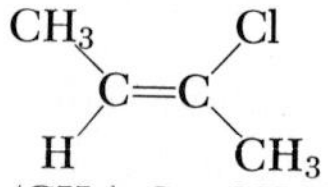 $CH_3 \quad Cl$
           $C = C$
         $H \quad CH_3$
   (g) $(CH_3)_2C = CHCH_2CH(CH_3)_2$
   (h) $CH_3CHClC \equiv CCH_3$

3. Draw out all the possible structural isomers for the compound having a molecular formula, C_5H_{10}. Name each isomer according to the IUPAC system.

4. Which of the isomeric compounds in Question 3

   (a) contain(s) no geometric isomers?
   (b) has(ve) no double bonds?
   (c) is (are) symmetrical?
   (d) has the highest boiling point?
   (e) contain(s) geometric isomers, but no double bonds?

5. Indicate what is wrong with each of the following names. Give the *correct* name for each of the following compounds:

   (a) 4-methyl-3-pentene
   (b) 6-methylcyclohexene
   (c) 2-isopropyl-1-propene
   (d) 4,6-dimethylcyclohexene
   (e) *cis*-4-heptene

6. Write equations for the reactions of 1-pentene with the following reagents:

   (a) Br_2 (b) H_2SO_4 (c) HI   (d) H_2/Ni

7. Write equations for the reactions of 1-butyne with the following reagents:

   (a) excess Cl_2
   (b) 1 mole Br_2
   (c) excess HI
   (d) 1 mole HCl
   (e) excess H_2/Ni

8. Using reactions discussed in this chapter and the preceding chapter, show how each of the following compounds could be prepared from isopropyl iodide:

   (a) propene
   (b) propyne
   (c) isopropyl chloride
   (d) 2,2-dibromopropane

9. Starting from propyne, show how each of the following compounds could be prepared:

   (a) propane
   (b) 2-chloropropene
   (c) isopropyl iodide

SUGGESTED READING

Cohen: The Shape of the 2p and Related Orbitals. Journal of Chemical Education, Vol. 38, p. 20, 1961.

Jones: The Markownikoff Rule. Journal of Chemical Education, Vol. 38, p. 297, 1961.

Polyunsaturated Lamb Chops. Research Reporter, Chemistry, Vol. 46, No. 9, p. 33, 1973.

Polyunsaturated In Food Products. A. Mancott and J. Tietjen, Chemistry, Vol. 47, No. 10, p. 29, 1974.

Tucker: Catalytic Hydrogenation Using Raney Nickel. Journal of Chemical Education, Vol. 27, p. 489, 1958.

Vinyl Chloride and Aerosol Sprays. Chemistry, Vol. 47, No. 7, p. 4, 1974.

Vinyl Chloride Hazard. Research Reporter, Chemistry, Vol. 47, No. 10, p. 27, 1974.

The Vinyl Chloride Story. J. W. Moore, Chemistry, Vol. 48, No. 6, p. 12, 1975.

AROMATIC HYDROCARBONS

CHAPTER 13

The *objectives* of this chapter are to enable the student to:

1. Distinguish the class of hydrocarbons known as aromatics.
2. Understand the stability of aromatic hydrocarbons relative to other polyunsaturated hydrocarbons, such as alkenes and alkynes.
3. Describe the concept of resonance, resonance hybrids, and delocalization of electrons.
4. Write the structural formulas and names of aromatic hydrocarbons.
5. Give a definition of ortho, meta, and para isomers.
6. Understand the generalized concept of aromatic substitution reactions.
7. Recognize an ortho-para directing group and a meta-directing group.
8. Predict the orientation in aromatic substitution reactions.
9. Distinguish between reactions which occur on the aromatic ring *versus* reactions which occur in the alkyl side chain.
10. Define a heterocyclic compound.
11. Recognize some of the basic five- and six-membered heterocyclic structures.

Early in the nineteenth century a class of organic compounds was isolated from aromatic substances such as oils of cloves, vanilla, wintergreen, cinnamon, bitter almonds, and benzoin. These compounds were pleasant smelling substances, and the term **aromatic** was given to this class of compounds to denote their aroma. In fact, many of these "aromatics" are still used in the perfumery and flavor extract industries because of their distinctive and pleasant odors.

The investigation of the chemistry of this class of compounds soon made it evident that the aromatic compounds were not related in an obvious manner to alkanes, alkenes, or alkynes and constituted a new class of hydrocarbons. Further chemical investigation also made it evident that all of the members of this class of hydrocarbons were structurally related to a cyclic hydrocarbon, **benzene,** which has the molecular formula C_6H_6. Benzene is not a pleasant smelling substance like many of its derivatives, and the original meaning of the term aromatic can only be loosely applied to benzene. However, the term has been carried over to include the chemistry of benzene and benzene derivatives; hence, this class of hydrocarbons is still called **aromatic hydrocarbons.** Benzene, itself, was first isolated by Michael Faraday in 1825.

Density, combustion analysis, and molecular weight studies established the molecular formula as C_6H_6. If a straight chain structure is written for a compound of this molecular formula, several double bonds or triple bonds must be included in the carbon chain to attain the molecular formula C_6H_6. Surprisingly, however, it was found that benzene did not decolorize a solution of bromine in carbon tetrachloride (therefore, did not undergo addition of bromine), and it was not oxidized by potassium permanganate, a reagent which easily oxidizes alkenes and alkynes. Therefore, structures containing several double and/or triple bonds were ruled out for the structure of benzene.

Further investigation of benzene established the following facts: (1) Catalytic hydrogenation of benzene indicates that benzene absorbs three moles of hydrogen and gives cyclohexane, C_6H_{12}, as the final product. Therefore, benzene must contain a cyclic six-membered ring of carbon atoms. The addition reaction also suggests that benzene contains an unsaturation which is equivalent to three double bonds. (2) Benzene can be chlorinated or brominated in the presence of a catalyst to give only the organic compound (C_6H_5X). Experimental evidence indicates that all six hydrogens in benzene are equivalent. If they were not equivalent, more than one isomer would be possible from this kind of

$$C_6H_6 + X_2 \xrightarrow{Fe} C_6H_5X + HX$$
$$(X = Cl, Br)$$

halogenation experiment. Furthermore, the halogenation reaction gives two products, indicating that this reaction is a **substitution** reaction and not an **addition** reaction.

The modern structure of benzene is depicted as the formation of a six-membered ring of atoms using the sp^2 orbitals of carbon. The remaining p orbitals (one on each carbon) then overlap to form a π-bond (similar to a π-bond in an alkene, except that the π-bond can be formed with either one of two adjacent neighbors). This π-bond is formed by the overlap of six p orbitals which results in a region of π-electron density above and below the hexagon structure, such that the hexagonal carbon skeleton is encased in a donut shape π-electron cloud. Thus, benzene can be depicted as shown in Figure 13–1.

Another representation commonly used is a double headed arrow between the two resonance structures.

The double headed arrow indicates that these structures are resonance forms. Remember that resonance forms involve only a change in the distribution of electrons (see p. 49) and that the *real structure* is a composite of all the possible resonance forms. Thus, benzene is neither of the resonance structures, nor is it simply an equal mixture of the two, but it is a structure that lies between these two extremes and is commonly represented by the symbol . The circle indicates that the π-electron density is distributed evenly over the six carbon atoms of the ring.

The delocalization° of electrons stabilizes the molecule to such an extent that benzene does not react like a simple alkene.

It has been demonstrated that compounds like benzene, which have a continuous

Resonance hybrids

π-electron cloud encompassing all the carbon atoms of the ring system, exhibit unusual stability compared to analogous systems in which the electrons would be fixed in double bonds. Therefore, oxidation or addition reactions which would disrupt the continuous π-cloud are unfavorable energetically, and do not occur under ordinary conditions.

In addition to benzene and substituted benzenes, other aromatic hydrocarbons are

---

° Electrons that occupy an orbital around more than two nuclei are called *delocalized electrons*.

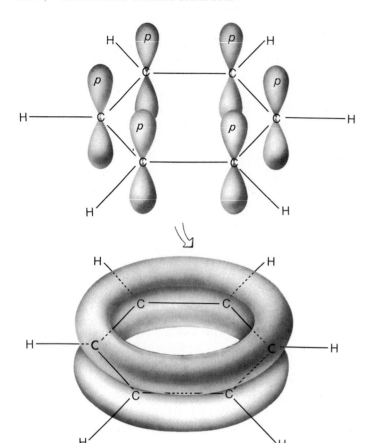

Figure 13-1 Representation of a π electron cloud in benzene.

known which have two or more rings **fused** or **condensed.** Again, the π-electron system is continuous over the entire carbon skeleton, and these compounds are more stable than would be expected for similar compounds containing fixed double bonds. Some of the more common condensed ring systems and their physical properties are given in Table 13-1.

NOMENCLATURE

Many aromatic compounds are named by common names, or as derivatives of the parent hydrocarbon by naming the substituent attached to the ring followed by the name of the aromatic hydrocarbon.

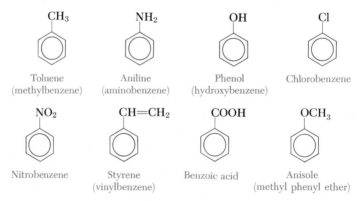

The C_6H_5 group is known as the **phenyl** group.

TABLE 13–1   PHYSICAL PROPERTIES OF
AROMATIC HYDROCARBONS

Compound	Structure	M.P. °C	B.P. °C
Benzene	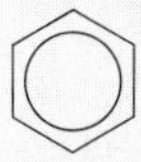	+6	+80
Toluene	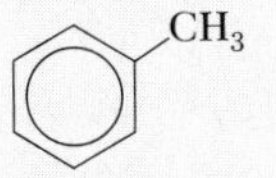 CH₃	−95	+111
Ethylbenzene	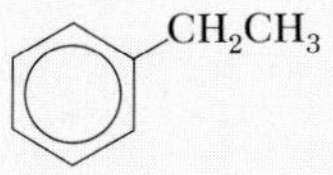 CH₂CH₃	−95	+136
Isopropylbenzene	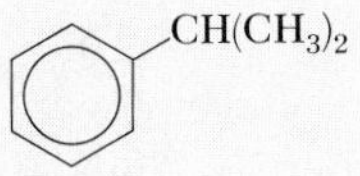 CH(CH₃)₂	−96	+152
Naphthalene	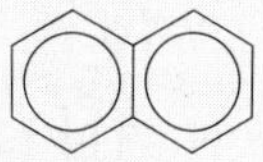	+80	+218
Anthracene	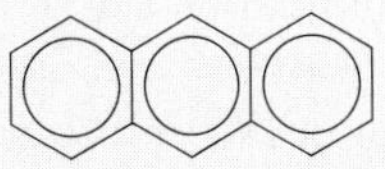	+217	+355
Phenanthrene	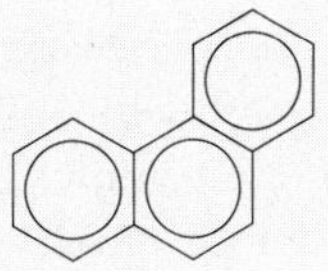	+100	+340
o-Xylene	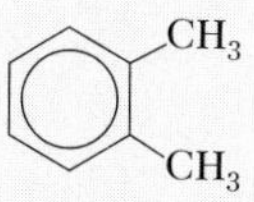 CH₃ CH₃	−25	+144
m-Xylene	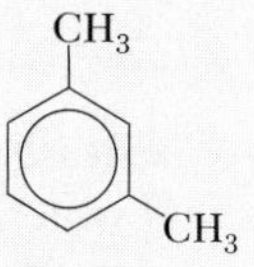 CH₃ CH₃	−48	+139
p-Xylene	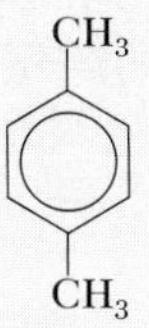 CH₃ CH₃	+13	+138

When two substituents are attached to the benzene ring, two systems of nomenclature are used. The position and number of each substituent can be indicated by the appropriate number and prefix. Alternatively, the relative positions of the two substituents can be indicated by the prefixes *ortho-* (*o-*), *meta-* (*m-*), and *para-* (*p-*) to indicate either a 1,2-; 1,3; or 1,4- position of the substituents relative to each other. For aromatic hydrocarbons containing more than two substituents, the numbering system is used. The manner in which the numbers are applied is not always consistent. Sometimes it is done alphabetically, and sometimes it is done by assigning the most important substituent the lowest number, and numbering the other substituents accordingly. Some examples of polysubstituted benzenes are shown below:

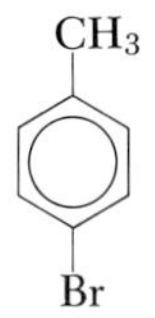

$$
\underset{\substack{p\text{-Bromotoluene}\\ (4\text{-bromo-1-methylbenzene})}}{\text{CH}_3\text{-C}_6\text{H}_4\text{-Br}}
\qquad
\underset{2,4\text{-Dichlorotoluene}}{\text{CH}_3\text{-C}_6\text{H}_3\text{-Cl}_2}
\qquad
\underset{\substack{o\text{-Nitrotoluene}\\ (2\text{-nitrotoluene})}}{\text{CH}_3\text{-C}_6\text{H}_4\text{-NO}_2}
$$

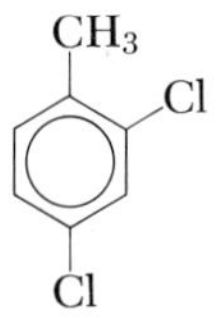

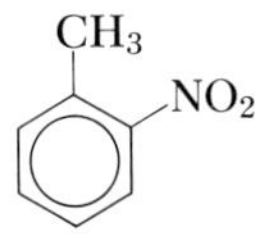

PHYSICAL PROPERTIES OF AROMATIC HYDROCARBONS

Benzene and its homologues are similar to other types of hydrocarbons with respect to their physical properties. They are insoluble in water but soluble in organic solvents. The boiling points of the aromatic hydrocarbons are slightly higher than those of the alkanes of similar carbon content. For example, n-hexane, C_6H_{14}, boils at 69°C, whereas benzene, C_6H_6, boils at 80°C. The physical properties of some of the common aromatic hydrocarbons are summarized in Table 13–1.

Aromatic hydrocarbons are quite flammable and should be handled with caution. Benzene is toxic when taken internally and must be used with proper precautions in any commercial process. Prolonged inhalation of its vapors results in a decreased production of red and white corpuscles in the blood which may prove fatal. Consequently, compounds of this class should only be handled under well-ventilated conditions. In addition, some of the more complex polynuclear aromatic hydrocarbons are carcinogenic and should be handled accordingly.

METHOD OF PREPARATION

FRIEDEL-CRAFTS REACTION

The French chemist Charles Friedel and the American chemist James Crafts discovered an alkylation reaction which could be used to prepare benzene derivatives. They found that when benzene is treated with an alkyl halide in the presence of a Lewis acid like aluminum chloride, an alkylated benzene is produced. For example, ethylbenzene could be similarly prepared from benzene, ethyl bromide, and aluminum bromide.[*] The major limitation on this kind of reaction is that in many cases the aromatic hydrocarbon

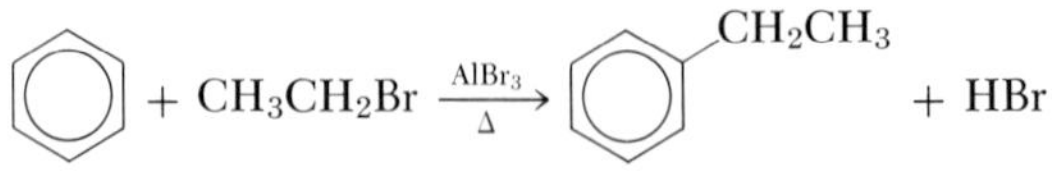

$$
\underset{\text{Benzene}}{C_6H_6} + \underset{\text{Ethyl bromide}}{CH_3CH_2Br} \xrightarrow[\Delta]{AlBr_3} \underset{\text{Ethylbenzene}}{C_6H_5CH_2CH_3} + HBr
$$

obtained contains an alkyl substituent that has a rearranged carbon skeleton. For example, the reaction of benzene, n-propyl chloride, and aluminum chloride would be expected to yield n-propylbenzene. However, cumene is the major product of this reaction.

---

[*] Although the aluminum halides are generally used, other acid catalysts such as H_2SO_4, BF_3, HF, and $SnCl_4$ have been used, depending on the type of starting material and its reactivity.

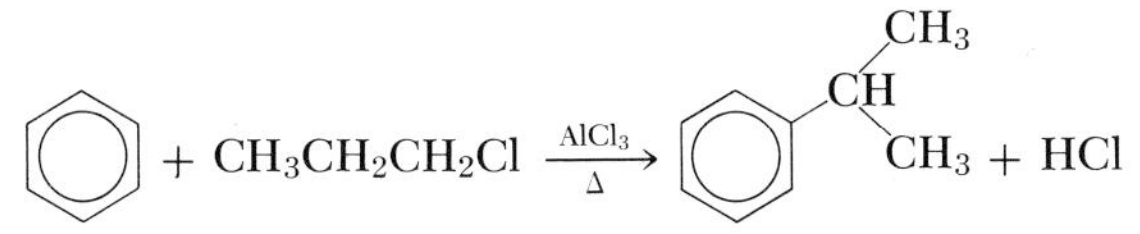

$$\text{Benzene} + CH_3CH_2CH_2Cl \xrightarrow[\Delta]{AlCl_3} \text{Cumene} + HCl$$

Benzene   n-Propyl chloride   Cumene
(Isopropyl benzene)

Rearrangement of the carbon skeleton of the alkyl halide is usually observed with longer chain alkyl halides, and hence, this reaction is of limited use for the preparation of aromatic hydrocarbons containing a straight chain alkyl substituent.

REACTIONS OF AROMATIC HYDROCARBONS

The most characteristic reactions of aromatic hydrocarbons are substitution reactions rather than addition reactions. Consequently, two products are formed in the reaction. One is an organic product in which an atom or group of atoms has substituted for the hydrogen atom of the aromatic ring. The second product is usually an inorganic acid or water.

HALOGENATION

With chlorine or bromine, halogenation occurs readily in the presence of a Lewis acid catalyst. Usually, iron or an iron halide containing the same halide atom as the halogenating agent is used as the catalyst. Fluorine is generally too reactive to be used

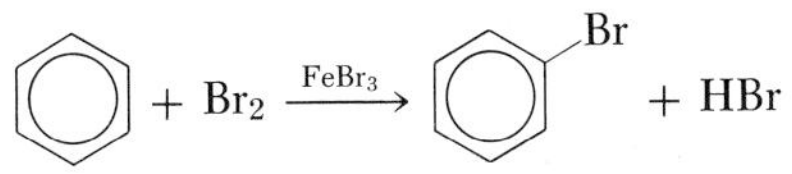

$$\text{Benzene} + Br_2 \xrightarrow{FeBr_3} \text{Bromobenzene} + HBr$$

Benzene   Bromobenzene

in substitution reactions of this type, and iodine is generally too unreactive to be of use in this type of reaction. The introduction of the first halogen (chlorine or bromine) into the aromatic ring makes the ring less susceptible to further attack and deactivates it, and the introduction of a second or third halogen atom becomes progressively more difficult.

NITRATION

The introduction of a nitro group ($-NO_2$) into an aromatic ring can be readily carried out using a mixture of concentrated nitric and sulfuric acids. The introduction

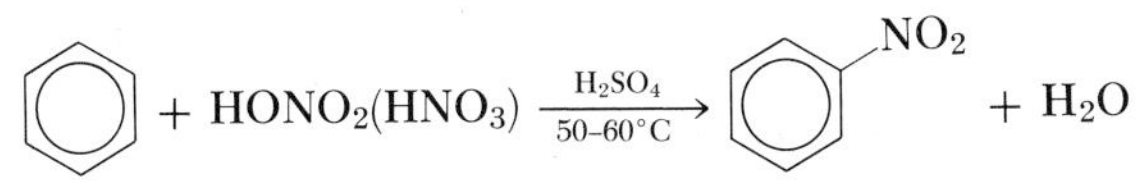

$$\text{Benzene} + HONO_2(HNO_3) \xrightarrow[50-60°C]{H_2SO_4} \text{Nitrobenzene} + H_2O$$

Benzene   Nitric acid   Nitrobenzene

of a nitro group into the ring also deactivates the aromatic ring to further substitution, and more vigorous conditions must be used to introduce a second or third nitro group. Nitrobenzene has a harmful physiological effect on the red blood corpuscles and on the liver. Therefore, caution should be used in handling this material, and inhalation of its vapor should be avoided.

SULFONATION

The introduction of a sulfonic acid ($-SO_3H$) group can be accomplished by treating benzene with concentrated sulfuric acid at elevated temperatures, or with fuming sulfuric acid (sulfuric acid containing SO_3) at moderate temperatures.

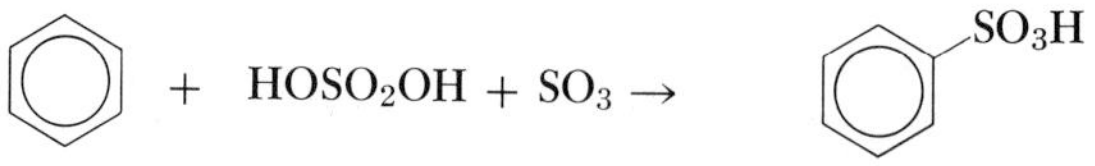

Benzene   Fuming sulfuric acid   Benzenesulfonic acid

Sulfonic acids are strong acids and are usually water soluble. Their acidity is generally comparable to that of the mineral acids.

Although sulfonic acids themselves are important, derivatives of sulfonic acids have received extensive treatment, since many of them were found to exhibit physiological action. Sulfanilamide, the forerunner of the sulfa drugs, is a sulfonic acid derivative (see p. 249). Sulfanilamide was found to be effective in the treatment of streptococcus infections, pneumonia, puerperal fever, gonorrhea, and gas gangrene. Unfortunately, sulfanilamide is only slightly soluble in water and tends to crystallize from aqueous solutions. When administered orally, the drug is absorbed and eventually carried to the kidney for excretion. When the dose is large, or under prolonged therapy, the kidneys are damaged by the accumulated sulfanilamide. Other toxic reactions, including methemoglobinemia, caused a search for derivatives that were less toxic. Sulfathiazole, sulfapyridine, sulfaguanidine, and sulfadiazine are among these derivatives (see p. 393).

Sulfonic acids dissolve readily in water and ionize completely to the sulfonate, and the introduction of a sulfonic acid group

$$C_6H_5SO_3H + H_2O \rightarrow C_6H_5SO_3^- + H_3O^+$$

Sulfonic acid   Sulfonate

in the molecule is frequently used to increase the water solubility of many dyes, medicinals, and synthetic detergents.

Early synthetic detergents were prepared from alkylbenzenes by the following scheme:

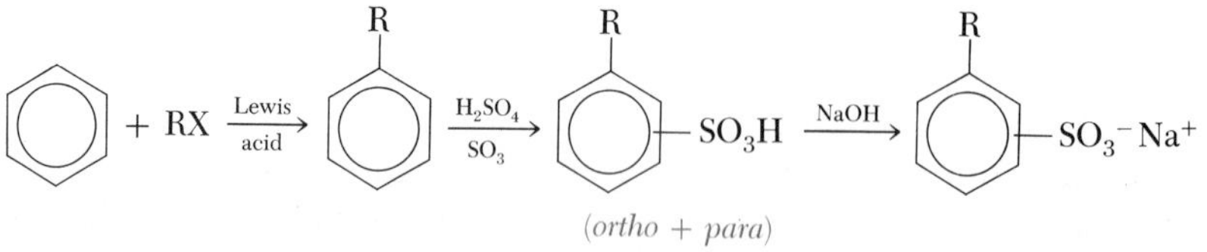

(ortho + para)

The alkyl portion (R) was generally a highly branched C_{12}—C_{18} alkyl group. These synthetic detergents exhibited good detergent properties and were widely used until the late 1960's. By this time the increasing amount of "foam" which was appearing in the nation's rivers was recognized as due to the nonbiodegradability of these synthetic detergents. Bacteria which normally degrade hydrocarbons via oxidation could not degrade the highly branched side chains of the detergents. Eventually these types of detergents were banned by Federal law.

However, it was soon found that when the alkyl portion (R) of the molecule was a straight chain alkyl group that the synthetic detergent became biodegradable, and today a variety of the biodegradable type of synthetics are available on the market.

ORIENTATION IN AROMATIC SUBSTITUTION REACTIONS

The introduction of an atom or a group of atoms into an unsubstituted benzene ring presents no problem as to the position taken by the new atom or groups of atoms. Since all the carbon-hydrogen bonds in benzene are equivalent, substitution of a halogen atom, a nitro group, a sulfonic acid group, an alkyl group, or an acyl group can give only one product. However, if a second group is substituted on the ring, more than one isomer is possible, and the position of the new atom or group of atoms becomes important. For example, nitration of toluene could give any or all of the following products:

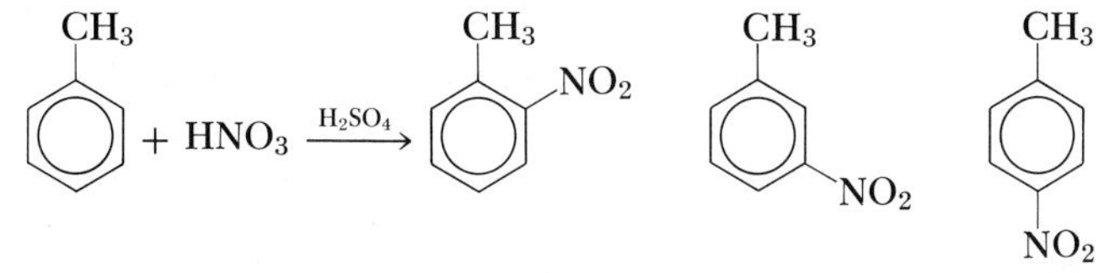

Toluene   o-Nitrotoluene,   m-Nitrotoluene,   p-Nitrotoluene

From an extensive study of many reactions of substituted benzene derivatives, a set of simple rules has been deduced which can be used to predict the expected product in reactions of this type.

ORIENTATION RULES

The predominant products are predicted by the following orientation rules:

1. The position of the second substituent is determined by the group already present on the ring.
2. The atom or group of atoms already present on the ring may be divided into two classes.

Class A: Atoms or groups of atoms which orient the new group predominantly into the *ortho-* and *para-* positions. The members of this class are called *ortho-para* directors and include atoms or groups such as —NH$_2$, —OH, —OCH$_3$, alkyl groups (CH$_3$, C$_2$H$_5$, and so forth), Cl, Br, and I. Except for the halogens, atoms and groups of this class activate the aromatic ring (relative to benzene) toward substitution. Thus, toluene is nitrated more easily than benzene, whereas nitration of chlorobenzene will require more vigorous conditions than the nitration of benzene.

Class B: Atoms or groups of atoms which orient the new group predominantly into the *meta-* position. The members of this class are called *meta* directors and include groups such as NO$_2$, COOH, CHO, CN, CO$_2$CH$_3$, and SO$_3$H. Atoms and groups of this class all deactivate the aromatic ring (relative to benzene) toward substitution.

Though these rules predict the major products, in most cases small amounts of the other isomers are also formed, but generally only to a minor extent.

Using these orientation rules, the nitration reaction of toluene would be expected to give mainly *ortho* and *para* nitrotoluenes, since the methyl group is an *ortho-para* director.

Since the product which is formed in the aromatic substitution reaction of a substituted benzene is determined by the incumbent substituent, the order in which the substituents are introduced becomes important. For example, bromination followed by nitration produces the *ortho* and *para* isomers. However, nitration followed by bromi-

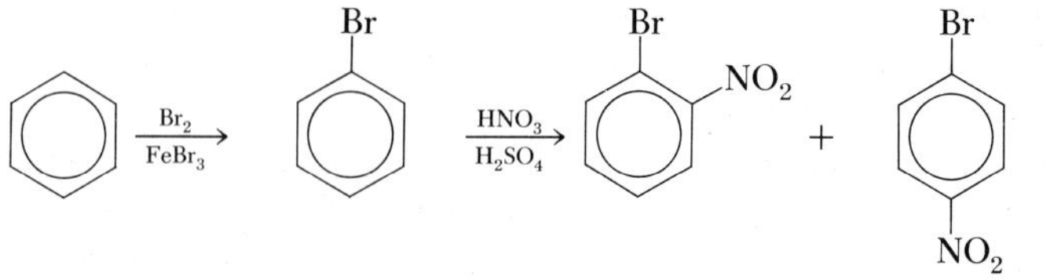

Benzene   Bromobenzene   *o*-Nitro-   *p*-Nitro-
                           bromobenzene bromobenzene

nation gives mainly the *meta* isomer. Consequently, when designing an organic synthe-

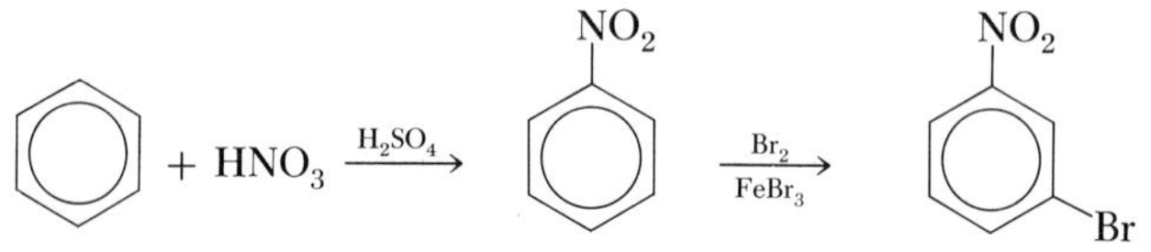

Benzene   Nitrobenzene   *m*-Nitrobromobenzene

sis which involves substituted benzene derivatives, the organic chemist must plan carefully the order in which the substituents are introduced, otherwise undesired isomeric products can be formed.

SIDE-CHAIN REACTIONS OF AROMATIC HYDROCARBONS

A substituted aromatic hydrocarbon can undergo two possible modes of chemical attack. It can undergo aromatic substitution on the aromatic ring itself, or reaction can take place in the group (side chain) attached to the aromatic ring. The mode of chemical reaction will be dependent on the type of side chain, the kind of chemical reagent, and the conditions used for carrying out the reaction. For example, toluene undergoes brom-

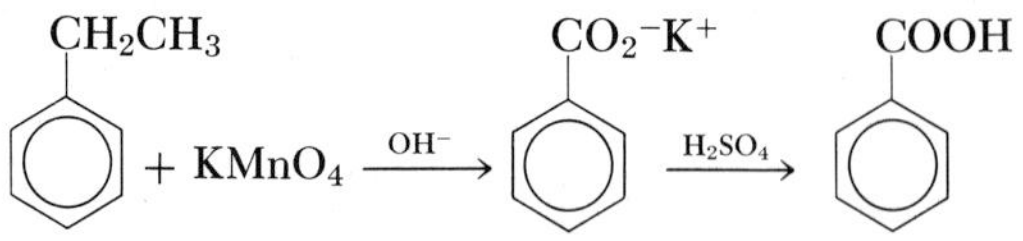

Ethylbenzene   Potassium benzoate   Benzoic acid

ination in the ring in the presence of Lewis acids (see p. 195). In many cases, however, the reagent and conditions used will affect only the side chain, and the aromatic ring will be unaffected by the chemical reaction. For example, aromatic hydrocarbons are resistant to oxidizing agents. Consequently, an alkyl group side chain can be oxidized to a carboxylic acid without oxidizing the aromatic ring. The alkyl group, regardless of its length, is degraded to the —COOH group. If more than one alkyl group is attached to the ring, a polyfunctional acid results on oxidation. Thus, *para*-xylene gives terephthalic acid, an important compound in the synthesis of Dacron (see Chapter 19).

Similarly, the aromatic ring is stable to chemical reduction. Thus, nitro groups can be reduced by tin and hydrochloric acid to give the amino group. Nitrobenzene gives aniline as a reduction product.

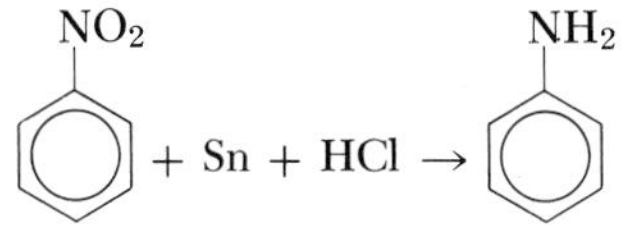

$$\text{Nitrobenzene} \quad + \text{Sn} + \text{HCl} \rightarrow \quad \text{Aniline}$$

Derivatives of aromatic amines have important medicinal properties and are used as drugs. Nitration of an aromatic hydrocarbon followed by chemical reduction of the nitro group to the amino group provides a facile route to the intermediates used in many of these medicinal compounds.

HETEROCYCLIC COMPOUNDS

As has been stated previously, the two major types of organic compounds are the aliphatic and the cyclic. If the cyclic compounds are composed of rings of carbon atoms only, as in benzene and its derivatives, they are called **carbocyclic.** When atoms other than carbon are also included in the ring the compounds are termed **heterocyclic.** The most commonly occurring elements other than carbon in these ring structures are oxygen, nitrogen, and sulfur.

The basic ring structure of a heterocyclic compound is called the **heterocyclic nucleus.** In the majority of heterocyclic compounds the nucleus consists of five- or six-membered rings that contain either one or two elements other than carbon. Examples of the important heterocyclic nuclei and their derivatives will be considered in the following sections.

FIVE-MEMBERED RINGS

Several important heterocyclic compounds are derived from a heterocyclic ring made up of 1 oxygen and 4 carbon atoms. This ring is known as **furan,** and one of its most important derivatives is the α-aldehyde, **furfural.** For purposes of nomenclature, the rings are numbered counterclockwise starting with the element other than carbon. In the furan nucleus illustrated here, the carbon atoms adjacent to the oxygen are α, the next ones are called β carbons.

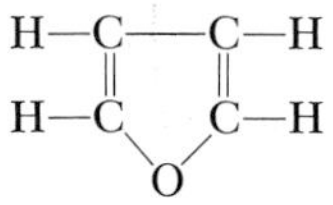

Furan 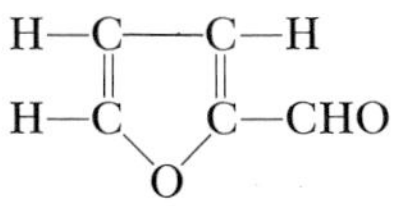   Furfural 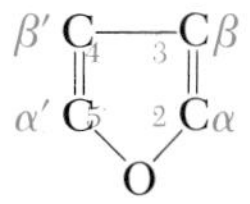   Pyrrole

Another important five-membered heterocyclic ring, containing nitrogen in place of oxygen, is **pyrrole.** It is a constituent of many important naturally occurring substances such as hemoglobin, chlorophyll, amino acids, and alkaloid drugs.

If pyrrole is condensed with a benzene ring, another type of heterocyclic nucleus called **indole** is produced. Indole and 3-methyl indole, which is called **skatole,** are formed during the putrefaction of proteins in the large intestine. They are responsible for the characteristic odor of feces. One of the most important derivatives of the indole nucleus is the amino acid **tryptophan.** Tryptophan is present in most proteins and is an essential constituent of the diet of growing animals (see p. 370).

The other example of a five-membered heterocyclic ring, in which sulfur takes the place of oxygen or nitrogen, is **thiophene.**

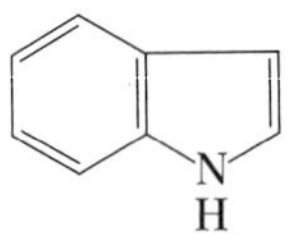

Indole 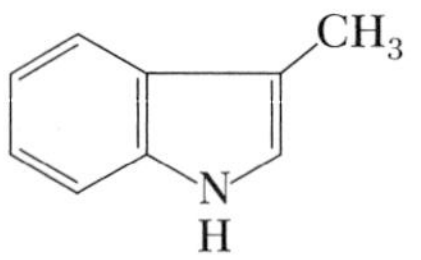   Skatole 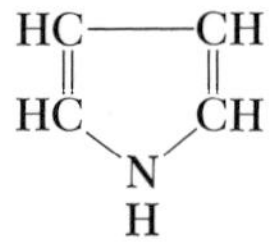   Pyrrole 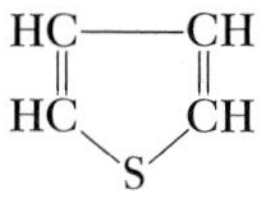   Thiophene

SIX-MEMBERED RINGS

The most common six-membered rings containing one element other than carbon are pyran and pyridine.

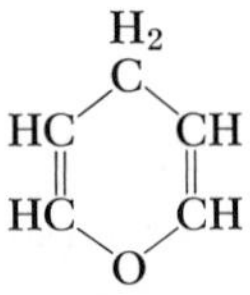

Pyran 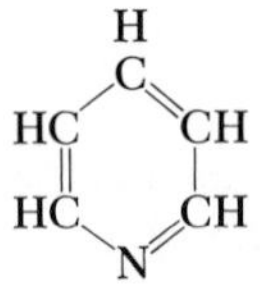   Pyridine

The pyran ring is present in anthocyanin, which is responsible for the color of flowers, and in rotenone, a plant material that is used as an insecticide.

Pyridine is a common heterocyclic compound obtained from coal tar. Pyridine is a liquid with a characteristically disagreeable odor. It behaves as a weak base and is a good solvent for both organic and inorganic compounds. It is used to manufacture pharmaceuticals such as sulfa drugs, antihistamines, and steroids. The methyl pyridines are known as picolines and may be oxidized to the corresponding picolinic acids. The acid obtained from the oxidation of β-picoline, or 3-methyl pyridine, is known as **nicotinic acid** (see p. 328).

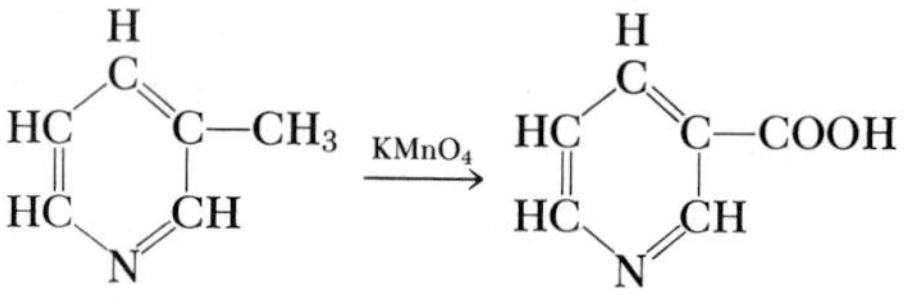

β-Picoline   Nicotinic acid

Nicotinic acid and its amide are members of the vitamin B complex. Consumption of a diet lacking in these compounds results in a deficiency disease called pellagra. Nicotinamide is used by the body for the manufacture of coenzymes NAD and NADP, which are essential for the proper functioning of certain dehydrogenating enzymes (see p. 328).

IMPORTANT TERMS AND CONCEPTS

alkylation	heterocyclic compound	pyridine
aromatic hydrocarbon	*meta*-directing groups	pyrrole
delocalized electrons	nitration	resonance hybrid
aromatic substitution	*ortho-para* directing groups	sulfa drug
Friedel-Crafts reaction	phenyl group	sulfonation
furan	pyran	

QUESTIONS

1. Draw a structural formula for each of the following compounds:

(a) toluene
(b) naphthalene
(c) *p*-nitrotoluene
(d) *m*-bromotoluene
(e) *p*-chlorotoluene
(f) *m*-diethylbenzene
(g) *o*-dinitrobenzene
(h) 2,4-dinitrotoluene
(i) 1,3,5-trimethylbenzene
(j) 1,4-dichloro-2,5-dibromobenzene

2. Draw out all the possible structural isomers for aromatic compounds with the following molecular formulas and name each of the isomers:

(a) C_8H_{10}
(b) C_8H_9Cl
(c) $C_7H_6Br_2$
(d) $C_6H_3Cl_2Br$
(e) C_7H_6ClBr

3. Write equations for each of the following reactions, and name each of the organic products obtained:

(a) benzene + isopropyl chloride ($AlCl_3$ catalyst)
(b) toluene + Br_2 (Fe catalyst)
(c) toluene + H_2SO_4 + SO_3
(d) nitrobenzene + concentrated nitric acid + sulfuric acid (heat)
(e) chlorobenzene + bromine (Fe catalyst)
(f) benzenesulfonic acid + Br_2 (Fe catalyst)
(g) toluene + nitric acid + sulfuric acid
(h) ethylbenzene + H_2SO_4 + SO_3

4. Show how the following conversions may be carried out in the laboratory. More than one step may be necessary. Give any necessary catalysts.

(a) benzene to *m*-chloronitrobenzene
(b) benzene to benzoic acid

5. How many tetrachlorobenzenes are there? Name them.

SUGGESTED READING

Duewell: Aromatic Substitution. Journal of Chemical Education, Vol. 43, p. 138, 1966.
Gero: Kekule's Theory of Aromaticity. Journal of Chemical Education, Vol. 31, p. 201, 1954.
Marsi and Wilen: Friedel-Crafts Alkylation. Journal of Chemical Education, Vol. 40, p. 214, 1963.
Varshni: Directive Influence of Substituents in the Benzene Ring. Journal of Chemical Education, Vol. 30, p. 342, 1953.
Waack: The Stability of the Aromatic Sextet. Journal of Chemical Education, Vol. 39, p. 469, 1962.

ALCOHOLS
AND ETHERS

CHAPTER 14

The *objectives* of this chapter are to enable the student to:

1. Recognize an alcohol and a phenol.
2. Write the structure and names of alcohols and phenols.
3. Explain the anomalous physical properties of alcohols.
4. Describe the conversion of other functional groups into alcohols or phenols.
5. Distinguish the different modes of hydration of olefins used in alcohol formation.
6. Describe the basic functional group reactions of alcohols and phenols.
7. Recognize some of the more important alcohols and phenols and their application in modern society.
8. Recognize an ether or epoxide.
9. Write the structural formulas and names of the common ethers and epoxides.
10. Describe the preparation of simple ethers and epoxides.
11. Recognize the difference in reactivity between ethers and epoxides.
12. Describe the basic chemical reactions of ethers.
13. Explain the ring opening reactions of epoxides.
14. Recognize some of the important ethers and epoxides and their current uses.

ALCOHOLS

The basic elements of the compounds contained in the preceding chapter were carbon and hydrogen. The addition of a third element, oxygen, greatly extends the number of types of possible organic compounds. Oxygen has six electrons in its valence shell and must therefore share two electrons to attain the stable inert gas configuration. The manner in which oxygen shares electrons with carbon and hydrogen determines the type of organic compound. For example, when oxygen shares one electron with carbon and one electron with hydrogen, the type of organic compound called an **alcohol** results. If each valence electron is shared with a separate carbon atom, a class of compounds called **ethers** is produced. If both valence electrons are shared with the same carbon atom to form a carbon-oxygen double bond, an aldehyde or ketone results, depending upon the other groups or atoms bonded to carbon. The various types of covalent bonding possible for oxygen are illustrated below:

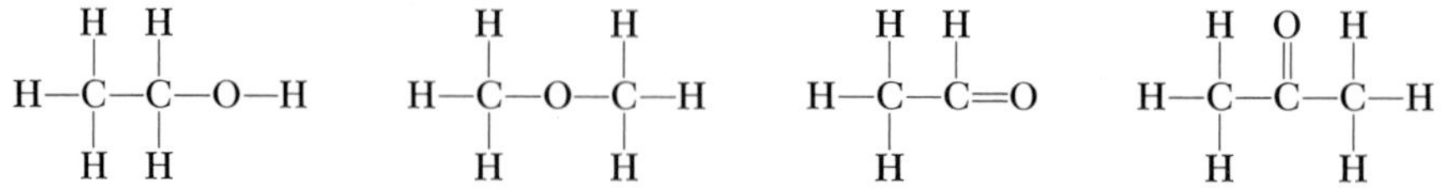

| Ethyl alcohol | Dimethyl ether | Acetaldehyde | Acetone |

Of all the classes of organic compounds, alcohols are probably the best known. For centuries it has been recognized that alcoholic beverages contain ethyl, or grain, alcohol.

Also, most temporary automobile antifreezes contain methyl, or wood alcohol. Most permanent type antifreezes contain ethylene glycol (CH_2OHCH_2OH), an alcohol containing two hydroxyl groups.

Alcohols are also very important biologically, since the alcohol group occurs in a variety of compounds associated with biologic systems. For example, most sugars contain several hydroxyl (—OH) groups (see Chapter 24), and starch, cellulose, and glycogen contain thousands of hydroxyl groups. Cholesterol, hormones, and other related steroids also contain the alcohol functional group (see Chapter 25), as well as fat soluble vitamins such as vitamin A and vitamin D (Chapter 26). In fact, the reaction of the alcohol functional group in vitamin A (p. 333) is important in the human visual process. Not only alcohols themselves, but also derivatives of alcohols are important biologically. For example, fats (p. 315), which are simple derivatives of the important trihydroxy alcohol, glycerol, $CH_2OHCHOHCH_2OH$, and phosphatides (p. 319), which are complex derivatives of glycerols, are also important in biologic processes.

Alcohols may be considered derivatives of hydrocarbons in which a hydrogen of the hydrocarbon has been replaced (substituted) by a hydroxyl group. Another way of considering alcohols is to view them as the organic analogs of water, in which one of the hydrogen atoms of water has been replaced by an alkyl or aryl group. If the latter way of viewing alcohols is accepted, the chemical and physical behavior of alcohols might be anticipated to be similar in some respects to water.

$$H \!-\! OH \qquad R \!-\! OH \qquad Ar \!-\! OH$$

Water   Aliphatic alcohol   Aromatic° alcohol

NOMENCLATURE OF ALCOHOLS

Alcohols are named in the IUPAC system by replacing the **-e** ending of the corresponding alkane with the characteristic **-ol** ending of the alcohols. Other substituents are named and their positions on the carbon chain indicated by the appropriate number and prefix. Since the hydroxyl group may appear at more than one position on the carbon chain, its position must also be indicated by the number of the carbon atom to which it is attached. Common names are used for the lower members of this series and are formulated by naming the alkyl group attached to the hydroxyl group, followed by the term alcohol. Some typical examples illustrating these nomenclature systems are illustrated below and in Table 14–1.

Compound	Name	Classification
CH_3OH	Methanol Methyl alcohol	Primary
CH_3CH_2OH	Ethanol Ethyl alcohol	Primary
$CH_3CH_2CH_2OH$	1-Propanol n-Propyl alcohol	Primary
$CH_3CHOHCH_3$	2-Propanol Isopropyl alcohol	Secondary
$CH_3CH_2CH_2CH_2OH$	1-Butanol n-Butyl alcohol	Primary
$(CH_3)_3COH$	2-Methyl-2-propanol t-Butyl alcohol	Tertiary

° The symbol Ar— is used to denote an aromatic ring. In this case, if Ar— = C_6H_5, the aromatic alcohol would be phenol, C_6H_5OH.

TABLE 14-1   PHYSICAL PROPERTIES OF
ALCOHOLS AND PHENOLS

COMPOUND	STRUCTURE	M.P.(°C)	B.P.(°C)
Methyl alcohol	CH_3OH	−97	+65
Ethyl alcohol	CH_3CH_2OH	−114	+78
n-Propyl alcohol	$CH_3CH_2CH_2OH$	−126	+97
Isopropyl alcohol	$(CH_3)_2CHOH$	−89	+82
n-Butyl alcohol	$CH_3CH_2CH_2CH_2OH$	−90	+118
Isobutyl alcohol	$(CH_3)_2CHCH_2OH$	−108	+108
t-Butyl alcohol	$(CH_3)_3COH$	+25	+83
Phenol	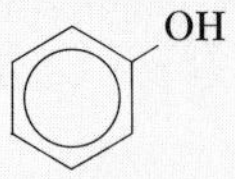 ⬡—OH	+41	+182

The aromatic alcohols are named as derivatives of the parent compound phenol (or carbolic acid). Some typical examples are represented below:

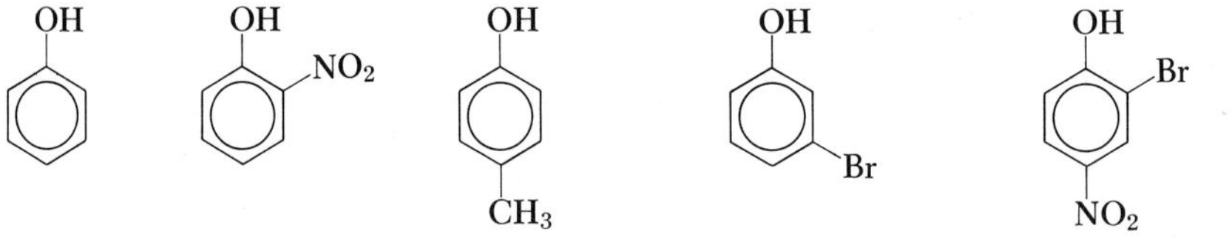

| Phenol | o-Nitrophenol | p-Methylphenol (p-cresol) | m-Bromophenol | 2-Bromo-4-nitrophenol |

The methyl derivatives are also known by the common name, **cresols.**

Phenols also play an important biologic role. The amino acid phenylalanine is biologically oxidized to a phenol derivative, dopa. Dopa is currently being used in the

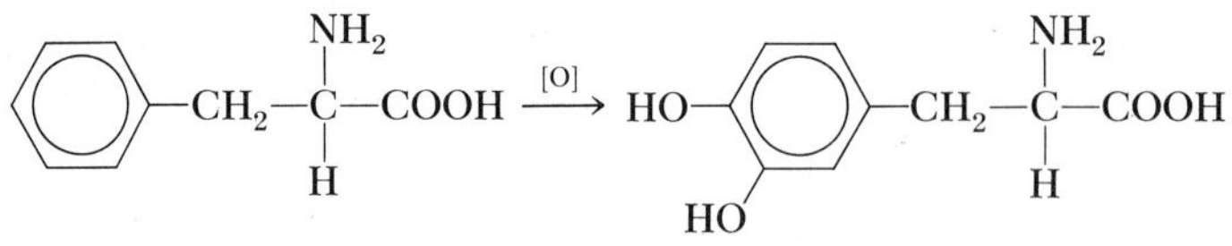

Phenylalanine   Dopa

control of Parkinson's disease and has shown great success in restoring normal life to persons affected with this disease.

PHYSICAL PROPERTIES OF ALCOHOLS

The introduction of a hydroxyl group into a hydrocarbon has a pronounced effect on the physical properties of the compound. In contrast to hydrocarbons, which are insoluble in water, the short chain alcohols (methanol through the butanols) are soluble in water. As the number of carbons in the alcohol increases, the solubility in water decreases, and the physical properties approach those of the saturated hydrocarbons.

The boiling points of alcohols are also abnormally high compared to the saturated hydrocarbons of comparable molecular weight. The large increase in boiling point is due to the presence of the hydroxyl group. The effect of this group has been attributed to the presence of hydrogen bonds formed between alcohol molecules in the liquid state, similar to the type of hydrogen bonding encountered in water. This type of association can also be used to explain the solubility of the short chain alcohols in water. When these alcohols are dissolved in water, hydrogen bonding occurs between the hydroxyl group of the alcohol and the hydroxyl group of the water.

The aromatic alcohol, **phenol,** is slightly soluble in water, but very soluble in alcohol, ether, and other organic solvents. In contrast to aliphatic alcohols, which are weaker acids $(Ka \sim 10^{-16})$ than water $(Ka = 10^{-14})$, phenols are much stronger acids than water.

METHODS OF PREPARATION OF ALCOHOLS

HYDROLYSIS OF ALKYL HALIDES

Alkaline hydrolysis of alkyl halides, in which the halogen atom of the alkyl halide is displaced by a hydroxyl group, is a useful method for preparing alcohols. A generalized representation of this reaction is shown below:

$$R\text{—}X + OH^- \xrightarrow{H_2O} R\text{—}OH + X^-$$

$$X = Cl, Br, I$$
$$R = Alkyl\ group$$

Aqueous sodium or potassium hydroxide, or aqueous silver oxide° (Ag_2O), is generally used as the hydrolyzing agent. The reactivity of the halides in this reaction is $RI > RBr > RCl$. Aromatic halides, in which the halogen is directly attached to the ring, are inert to this type of displacement reaction. A competing side reaction in this method of preparation is the elimination of hydrogen halide from the alkyl halide to give an olefin (see p. 181). Therefore, the best yields of the alcohols are obtained from primary alkyl halides using silver oxide at moderate temperatures. A typical example is outlined below:

$$CH_3CH_2CH_2I + Ag_2O \xrightarrow{H_2O} CH_3CH_2CH_2OH + AgI\downarrow$$
n-Propyl iodide ⟶ n-Propyl alcohol

HYDRATION OF OLEFINS

Alkenes undergo hydration in the presence of strong acids such as sulfuric acid to give alcohols as the final product. The hydration reaction follows Markownikoff's rule; therefore, secondary and tertiary alcohols are formed except in the case of ethylene. A typical example is shown below:

$$CH_3CH{=}CH_2 + H_2O \xrightarrow{H_2SO_4} CH_3CHOHCH_3$$
Propylene ⟶ Isopropyl alcohol

Addition of water also plays an important role biologically in the metabolism of carbohydrates. For example, as part of the Krebs cycle (p. 355), fumaric acid is converted

---

° The aqueous solution of silver oxide may be represented as AgOH (silver hydroxide). The silver oxide method promotes less elimination than the aqueous alkali method.

to malic acid and *cis*-aconitic acid is converted to isocitric acid. Both of these conversions involve the enzymatic hydration of a carbon-carbon double bond.

Until 1955 the preparation of primary alcohols such as 1-butanol *via* the hydration of alkenes was an unfulfilled goal of the organic chemist. The anti-Markownikoff addition of water to olefins was successfully developed by Professor Herbert C. Brown at Purdue University by the alkaline hydrogen peroxide oxidation of organoboranes. The organoboranes could be easily formed by the reaction of diborane° with olefins.

$$CH_3CH_2CH{=}CH_2 + B_2H_6 \rightarrow 2\ (CH_3CH_2CH_2CH_2)_3B$$
$$\text{1-Butene} \qquad \text{Diborane} \qquad \text{Tri-}n\text{-butylborane}$$

$$2(CH_3CH_2CH_2CH_2)_3B \xrightarrow[\text{OH}^-]{H_2O_2} 6CH_3CH_2CH_2CH_2OH + H_3BO_3$$
$$\text{1-Butanol} \qquad \text{Boric acid}$$

As outlined in this reaction sequence, the boron always adds to the least substituted carbon of the olefinic bond. Consequently, with terminal olefins, primary alcohols always result. This type of anti-Markownikoff hydration is commonly called *hydroboration*.

METHOD OF PREPARATION OF PHENOLS

The methods of preparation of aliphatic alcohols outlined previously are generally unsatisfactory for preparing phenols in the laboratory. The hydrolysis reaction, however, has been used commercially to prepare phenol. The reaction requires high temperatures and pressure. The initially formed salt is subsequently converted to phenol.

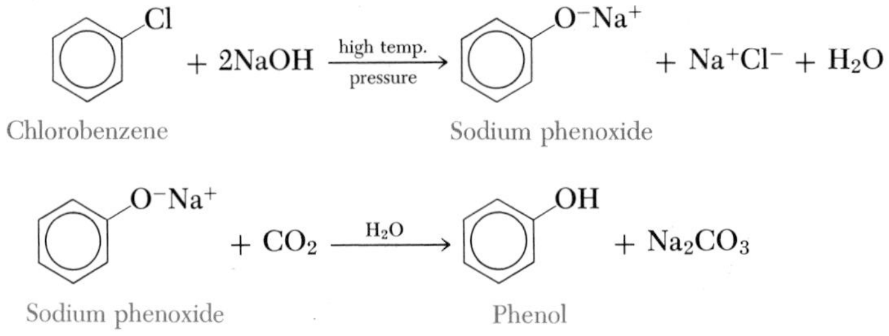

Chlorobenzene   Sodium phenoxide

Sodium phenoxide   Phenol

REACTIONS OF ALCOHOLS AND PHENOLS

Alcohols and phenols can undergo two types of reactions which involve the hydroxyl group. A reaction can occur to cleave the oxygen-hydrogen bond, or a reaction can occur to cleave the carbon-oxygen bond resulting in loss of the hydroxyl group. Reactions which involve oxygen-hydrogen bond cleavage are considered first.

SALT FORMATION

Because of their increased acidity, phenols will form salts with aqueous alkalies.

° Diborane can be conveniently generated by the reaction of sodium borohydride and boron trifluoride: $3NaBH_4 + BF_3 \rightarrow 2B_2H_6 + 3NaF$.

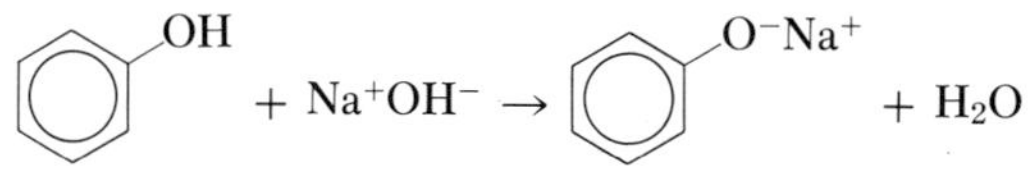

$$\text{Phenol} + Na^+OH^- \rightarrow \text{Sodium phenoxide} + H_2O$$

Aliphatic alcohols will not form salts with aqueous alkali. The free phenol can be regenerated by treatment of the salt with acids such as hydrogen chloride. Although aliphatic alcohols do not form salts with aqueous alkalies, both alcohols and phenols will form salts with active metals such as sodium, potassium, magnesium, and so forth. These reactions are similar to the reaction of water with active metals to give an alkali metal hydroxide and hydrogen.

$$CH_3CH_2OH + Na \rightarrow CH_3CH_2O^-Na^+ + \tfrac{1}{2}H_2\uparrow$$

Ethyl alcohol   Sodium ethoxide

It is interesting to note that the salts of alcohols (called alkoxides) are strong bases when used in a nonaqueous solvent. Just as HO^- is a strong inorganic base, alkoxide ions (RO^-, where R is an alkyl group) are strong organic bases. In aqueous solution, the alkoxides are hydrolyzed back to the alcohol; therefore, they must be used in a nonaqueous medium.

ESTER FORMATION

In the presence of an acid catalyst, alcohols will react with a carboxylic acid to form an ester. Experiments carried out using an alcohol containing oxygen-18 instead of the normal oxygen-16 have shown that the oxygen of the alcohol ends up as one of the oxygen atoms in the ester. Therefore, only the oxygen-hydrogen bond of the alcohol is broken, and the hydroxyl group used in forming the by-product water comes from the carboxylic acid and not the alcohol.

A typical example of ester formation is illustrated below:

$$CH_3CH_2\overset{18}{O}H + CH_3\overset{O}{\overset{\|}{C}}{-}OH \xrightarrow{H^+} CH_3\overset{O}{\overset{\|}{C}}{-}\overset{18}{O}CH_2CH_3 + H_2O$$

Ethyl alcohol   Acetic acid   Ethyl acetate

Inorganic acids also form esters with alcohols. For example, glyceryl trinitrate (nitroglycerine), which is the explosive ingredient in dynamite, is produced by the esterification of glycerol with nitric acid.

$$HOCH_2CHCH_2OH + 3\,HONO_2 \longrightarrow \underset{\overset{|}{ONO_2}\ \ \overset{|}{ONO_2}}{\overset{\overset{|}{ONO_2}}{CH_2CHCH_2}} + 3\,H_2O$$

Glycerol   Nitric acid   Glyceryl
                                          trinitrate

Similarly, phosphate esters, which are of extreme importance in biochemistry, can also be produced by the esterification of alcohols with phosphoric acid.[*] In biochemical systems, a more complex phosphoric acid derivative will be used to form complex phosphate esters.

---

[*] Either 1, 2, or 3 of the acidic hydrogens of phosphoric acid can be esterified.

$$\text{HO}-\overset{\overset{\displaystyle O}{\|}}{\underset{\underset{\displaystyle OH}{|}}{P}}-\text{OH} + \text{R}-\text{O H} \overset{H^+}{\rightleftharpoons} \text{HO}-\overset{\overset{\displaystyle O}{\|}}{\underset{\underset{\displaystyle OH}{|}}{P}}-\text{OR} + \text{H}_2\text{O}$$

Phosphoric   Alcohol   Phosphate ester
acid

REPLACEMENT OF THE HYDROXYL GROUP

DEHYDRATION

Alcohols undergo loss of water (dehydration) in the presence of acid catalysts under the proper conditions. These reactions have been considered earlier under the preparation of olefins (see p. 180).

CONVERSION TO ALKYL HALIDES

The hydroxyl group of alcohols can be replaced with a halogen atom by several types of reagents. With hydrogen chloride, the reaction is generally carried out using zinc chloride ($ZnCl_2$) as a catalyst. The order of reactivity is tertiary > secondary > primary alcohols.

$$\text{ROH} + \text{HCl} \xrightarrow{ZnCl_2} \text{RCl} + \text{H}_2\text{O}$$

The order of reactivity of the hydrogen halides in this type of reaction is HI > HBr > HCl. The reaction sequence is dependent on the type of alcohol undergoing displacement of the hydroxyl group. With primary and secondary halides the sequence involves protonation of the hydroxyl group and subsequent displacement of water by a halide ion, as outlined below:

$$\text{R } \overset{..}{\underset{..}{O}}\text{H} + \text{HX} \rightarrow \text{R } \overset{+}{O}\text{H}_2 + \text{X}^-$$

X = halogen
R = primary or secondary

$$\text{X}\!\longrightarrow\!\text{R} \overset{+}{O}\text{H}_2 \rightarrow \text{ X}\!-\!\text{R} + \text{H}_2\text{O}$$

It is much easier to displace water than hydroxide ion (HO^-), hence the function of the proton catalyst is to facilitate the displacement reaction. With most tertiary alcohols, the protonated alcohol undergoes ionization to form an ion, which then picks up a halide ion to form the alkyl halide.

$$\text{R}-\overset{\overset{\displaystyle R'}{|}}{\underset{\underset{\displaystyle R''}{|}}{C}}-\overset{..}{\underset{..}{O}}\text{H} + \text{HX} \rightarrow \text{R}-\overset{\overset{\displaystyle R'}{|}}{\underset{\underset{\displaystyle R''}{|}}{C}}-\overset{+}{O}\text{H}_2 \xrightarrow{-H_2O} \text{R}-\overset{\overset{\displaystyle R'}{|}}{\underset{\underset{\displaystyle R''}{|}}{C}}{}^+ \xrightarrow{X^-} \text{R}-\overset{\overset{\displaystyle R'}{|}}{\underset{\underset{\displaystyle R''}{|}}{C}}-\text{X}$$

X = halogen   + X$^-$
R,R',R'' = Alkyl groups

OTHER REACTIONS OF ALCOHOLS AND PHENOLS

In addition to the reactions involving cleavage of the oxygen-hydrogen bond of the hydroxyl group and displacement of the hydroxyl group by carbon-oxygen bond cleavage, alcohols and phenols can undergo other kinds of reactions which involve the carbon-hydrogen bonds of the alcohol or the phenol. In phenols most of these reactions involve aromatic substitution reactions similar to those observed earlier for benzene. The hydroxyl group attached to the aromatic ring activates the aromatic ring toward attack, and phenol undergoes substitution much more easily than either benzene or toluene. The hydroxyl group is an *ortho-para* director. A typical reaction of phenol involving substitution on the aromatic ring is illustrated below:

HALOGENATION

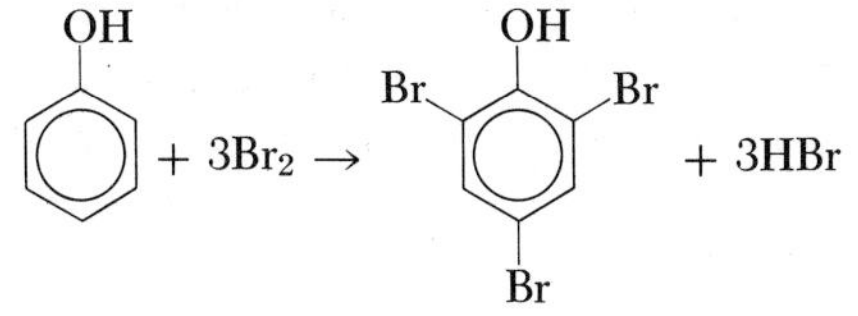

Phenol   2,4,6-Tribromophenol

The hydroxyl group so activates the ring that substitution by halogen occurs at all the possible *ortho* and *para* positions. Special precautions must be used to obtain the monohalogenated product. Nitration and sulfonation occur similarly.

OXIDATION REACTIONS

Oxidizing agents such as alkaline potassium permanganate or potassium dichromate and sulfuric acid oxidize primary and secondary alcohols to carboxylic acids and ketones, respectively. This reaction will be considered in more detail in Chapters 16 and 17. Under the normal conditions of this reaction, tertiary alcohols are not oxidized. A typical example is given below (see also Chapters 16 and 17).

$$CH_3CH_2OH + K_2Cr_2O_7 \xrightarrow{H_2SO_4} CH_3CHO \rightarrow CH_3\overset{\overset{\displaystyle O}{\|}}{C}OH$$

Ethyl alcohol   Acetaldehyde   Acetic acid

IMPORTANT ALCOHOLS AND PHENOLS

Some of the more important alcohols and phenols and their uses are discussed in the following pages of this chapter.

IMPORTANT ALCOHOLS

Methyl Alcohol. Methyl alcohol is commonly called wood alcohol because it was once exclusively produced by the destructive distillation of wood. A synthetic process which has largely supplanted the wood distillation method involves the reaction of carbon monoxide and hydrogen under a high pressure and a temperature of 350°C in the presence of zinc and chromium oxide catalysts.

$$CO + 2H_2 \xrightarrow[Cr_2O_3]{ZnO} CH_3OH$$

Carbon   Methyl
monoxide   alcohol

Methyl alcohol is a colorless, volatile liquid with a characteristic odor. It is used as a denaturant for ethyl alcohol, as an antifreeze for automobile radiators, and as the raw material for the synthesis of other organic compounds. When taken internally, methyl alcohol is poisonous, small doses producing blindness by degeneration of the optic nerve, whereas large doses are fatal. The taste, odor, and poisonous properties of wood alcohol make it a desirable denaturing agent to be added to ethyl alcohol to prevent its use in beverages. During the prohibition era in the United States, many persons were blinded and others died after drinking ethyl alcohol denatured in this fashion. An individual who was blinded temporarily after drinking a small amount of methyl alcohol was said to be "blind drunk."

Ethyl Alcohol.   Ethyl alcohol is commonly known as alcohol, or as grain alcohol since it may be made by fermentation of various grains. It is prepared commercially by the fermentation of the sugars and starch of common grains, potatoes, or black-strap molasses. The yeast used in fermentation contains enzymes that catalyze the transformation of more complex sugars into simple sugars, and then into alcohol and carbon dioxide, as shown in the following reactions:

$$C_{12}H_{22}O_{11} + H_2O \xrightarrow{\text{enzymes}} 2\ C_6H_{12}O_6 \xrightarrow{\text{enzymes}} 4\ C_2H_5OH + 4\ CO_2$$

| Complex sugar | | Simple sugar | | Ethyl alcohol | Carbon dioxide |

Enzymes and fermentation will be studied more completely in a later chapter.

Ethyl alcohol is a colorless, volatile liquid with a characteristic pleasant odor. Industrial ethyl alcohol contains approximately 95 per cent alcohol and 5 per cent water. It is difficult to remove all the water from alcohol, since in simple distillation processes a constant boiling mixture of 95 per cent alcohol and 5 per cent water is formed. It is an excellent solvent for many substances and is used in the preparation of medicines, flavoring extracts, and perfumes. Alcohol is widely used in the hospital as an antiseptic, a vehicle for medications, and as a rubbing compound to cleanse the skin and lower a patient's temperature.

The concentration of alcohol in beverages is usually expressed as per cent or "proof." The relationship between proof and per cent alcohol concentration may be shown in the following examples. The common 100 proof whiskey is 50 per cent alcohol, whereas the standard laboratory 95 per cent alcohol is 190 proof. Beer and wine contain from 3 to 20 per cent, whereas whiskey, rum, vodka, and gin contain from 35 to 45 per cent alcohol.

Ethylene Glycol.   All the alcohols so far considered have been monohydroxy alcohols. Since it is possible to replace a hydrogen atom on more than one carbon atom in a hydrocarbon with a hydroxyl group, it is possible to have polyhydroxy (polyhydric) alcohols. The simplest polyhydric alcohol would be one formed by replacing a hydrogen on each of the two carbons of ethane by a hydroxyl group. This compound is called **ethylene glycol** ($HOCH_2CH_2OH$). It is prepared by oxidizing ethylene to ethylene oxide (see pp. 213 and 214) and hydrolyzing the oxide to ethylene glycol. Ethylene glycol is water soluble and has a very high boiling point compared to those of methyl and ethyl alcohols. These properties make it an excellent permanent, or nonvolatile, type of antifreeze for automobile radiators. Antifreeze preparations such as **Prestone** and **Zerex** consist of ethylene glycol plus a small amount of a dye.

Glycerol.   The most important trihydric alcohol is glycerol, which is sometimes called glycerin. It is an essential constituent of fat (an ester of glycerol and fatty acids) and may be prepared by the hydrolysis of fat as represented in the following equation:

$$\text{Fat} + \text{Hydrolysis} \longrightarrow \underset{\underset{\displaystyle OH}{|}}{H_2C}\!-\!\underset{\underset{\displaystyle OH}{|}}{CH}\!-\!\underset{\underset{\displaystyle OH}{|}}{CH_2} + \text{Fatty acids}$$

Glycerol

Glycerol is obtained commercially as a by-product of the manufacture of soap, and from a synthetic process that uses propylene from the catalytic cracking of petroleum as a starting material. Glycerol is a syrupy, sweet-tasting substance that is soluble in all proportions of water and alcohol. It is nontoxic and is often used for the preparation of liquid medications. Since it has the ability to take up moisture from the air, it tends to keep the skin soft and moist when applied in the form of cosmetics and lotions.

Phenol. Phenol has strong antiseptic properties, and as a class of compounds phenols are active germicides. However, because of its extreme toxicity, phenol itself is rarely used as an antiseptic. It is caustic, causes blistering of the skin, and is a violent poison when taken internally. Dilute solutions of the cresols (Lysol), however, are used in hospitals as disinfectants.

Phenol is used as a standard for comparison of other germicides, and the efficiency of other antiseptics is measured in arbitrary units called the *phenol coefficient.* For example, a 1 per cent solution of a germicide that is as effective as a 5 per cent solution of phenol in destroying an organism is assigned a phenol coefficient of 5.

Other important commercial phenol derivatives are hexyl resorcinol, used in mouthwashes, and hexachlorophene, used in soaps, deodorants, and toothpastes.

Hexyl resorcinol

Hexachlorophene

ETHERS

Ethers are closely related to the alcohols and may be considered a derivative of an alcohol in which the hydrogen of the hydroxyl group has been replaced by an alkyl or aryl group. Consequently, ethers may be considered to be organic derivatives of water (HOH) in which both hydrogens have been replaced by alkyl or aryl groups. In addition, cyclic ethers are possible in which the oxygen atom is part of the cyclic structure. The three-membered ring structures containing an oxygen atom as part of the ring are also called **epoxides.** As noted earlier, ethers are isomeric with the alcohols containing the same number of carbon atoms.

Ethers are named either by common names or by naming the two alkyl or aryl groups linked to the oxygen atom, followed by the word "ether." If one of the alkyl or aryl groups has no simple name, the compound is named as an **alkoxy** derivative.° Some typical examples of ethers are given in Table 14–2.

Since ethers are essentially hydrocarbons with a single oxygen atom, their physical properties would be expected to parallel those of hydrocarbons. They are colorless, insoluble in water, soluble in strong concentrated acids (whereas alkanes are not), soluble in organic solvents, and in general have densities and boiling points similar to hydrocarbons of corresponding molecular weight. It is interesting to note that dimethyl ether, which is isomeric with ethanol, is a gas at room temperature, whereas ethanol is a liquid at room temperature. Ethers, which have no hydroxyl group, cannot form hydrogen bonds as can alcohols; consequently, the boiling points are not abnormally high like the alcohols.

° The —OR group is known as an alkoxy group. The name of this type of group is formed from the hydrocarbon name of the —R group by dropping the **-ane** and adding **-oxy.**

TABLE 14–2   PHYSICAL PROPERTIES
OF ETHERS

COMPOUND	STRUCTURE	M.P. °C	B.P. °C
Dimethyl ether	CH_3OCH_3	-140	-25
Diethyl ether	$CH_3CH_2OCH_2CH_3$	$-116\ (-123)$	$+35$
Methyl ethyl ether	$CH_3OCH_2CH_3$	—	$+8$
Di-n-propyl ether	$CH_3CH_2CH_2OCH_2CH_2CH_3$	-122	$+91$
Diisopropyl ether	$(CH_3)_2CHOCH(CH_3)_2$	-60	$+68$
Tetrahydrofuran	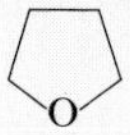	-108	$+66$
Anisole	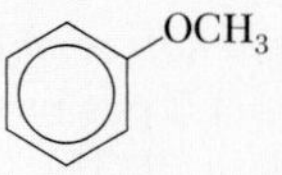	-37	$+154$
Diphenyl ether	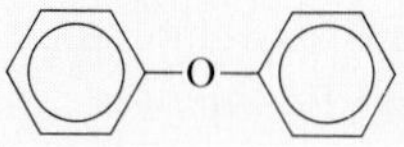	$+27$	$+259$

The physical properties of some of the more common ethers are summarized in Table 14–2.

METHODS OF PREPARATION OF ETHERS

WILLIAMSON SYNTHESIS OF ETHERS

A general preparation of both symmetrical and unsymmetrical ethers is a reaction discovered by the British chemist Alexander Williamson. The reaction involves a displacement of halide ion from an alkyl halide by an alkoxide ion (obtained from an alcohol or phenol). The generalized scheme of this reaction, and several specific examples, are illustrated below:

$$RO^-M^+\ +\ RX\ \rightarrow\ ROR'+\ M^+X^-$$
Metal alkoxide   Alkyl halide   Ether   Salt

$M^+ = Na, K$ (usually)
$R =$ alkyl or aryl
$R' =$ alkyl

$$CH_3O^-Na^+\ +\ CH_3CH_2I \rightarrow CH_3OCH_2CH_3\ +\ Na^+I^-$$
Sodium methoxide   Ethyl iodide   Methyl ethyl ether

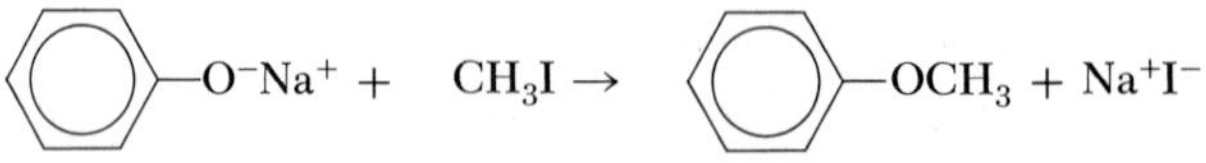

Sodium phenoxide   Methyl iodide   Anisole

Compared to the ease of displacement of halide ion from an alkyl halide by alkoxide, it is difficult to displace halide ion from an aromatic ring by alkoxide ion. Therefore, in preparing mixed aromatic-aliphatic ethers, such as anisole, the reaction must be carried out using sodium phenoxide displacement on methyl iodide. Reversal of the types of reagents does not give anisole.

The cyclic three-member ethers (epoxides) can also be prepared by an intramolecular type of Williamson synthesis. Addition of hypohalous acids to alkenes gives halohydrins, which undergo intramolecular cyclization in the presence of base to give epoxides, as outlined below:

$$\underset{\substack{\text{Ethylene}\\\text{chlorohydrin}}}{\overset{\text{Cl}}{\underset{\text{OH}}{CH_2-CH_2}}} + KOH \rightarrow \left[\overset{\text{Cl}}{\underset{O^-}{CH_2-CH_2}} \right] \rightarrow \underset{\substack{\text{Ethylene oxide}}}{H_2C\overset{}{\underset{O}{\diagdown\diagup}}CH_2} + K^+Cl^- + H_2O$$

Commercially, ethylene oxide is prepared from ethylene and oxygen in the presence of a silver gauze catalyst.

$$\underset{\text{Ethylene}}{CH_2\!=\!CH_2} + \tfrac{1}{2}O_2 \xrightarrow[\Delta,\text{ pressure}]{\text{Ag catalyst}} \underset{\substack{\text{Ethylene oxide}}}{H_2C\overset{}{\underset{O}{\diagdown\diagup}}CH_2}$$

REACTIONS OF ETHERS AND EPOXIDES

Except for the saturated hydrocarbons, ethers are the most unreactive of any of the simple functional groups. They are stable to dilute acids and bases and are also resistant to many oxidizing and reducing agents. They are similar to alkanes in their lack of chemical reactivity. This lack of chemical reactivity, however, makes ethers quite suitable as solvents for many chemical reactions. However, under more vigorous reaction conditions, ethers do undergo cleavage reactions with concentrated mineral acids, and the aromatic ethers do undergo ring substitution reactions.

REACTION WITH ACIDS

With the hydrogen halides, ethers can be cleaved at high temperatures. Hydrogen iodide and hydrogen bromide are usually used for this purpose. With one equivalent of the hydrogen halide, an alcohol and an alkyl halide are produced. With excess hydrogen halide, two moles of alkyl halide are produced if both groups attached to the oxygen atom of the ether are aliphatic. If one of the groups is aromatic, a mole of the corresponding phenol is formed, since phenols are not converted to aryl halides by hydrogen halides. Some typical cleavage reactions are illustrated in the following examples:

$$CH_3CH_2OCH_2CH_3 + HI \text{ (excess)} \xrightarrow{\Delta} 2CH_3CH_2I$$

$$\underset{\text{Anisole}}{\bigcirc\!\!-OCH_3} + HI \xrightarrow{\Delta} \underset{\text{Methyl iodide}}{CH_3I} + \underset{\text{Phenol}}{\bigcirc\!\!-OH}$$

AROMATIC SUBSTITUTION REACTIONS OF AROMATIC ETHERS

If one of the groups attached to the ether oxygen atom is an aromatic group, the usual halogenation, nitration, sulfonation, and alkylation reactions can be carried out on the aromatic ring without affecting any cleavage of the ether linkage. The alkoxy group of such a mixed ether is an *ortho-para* directing group and also activates the ring towards substitution. A typical bromination of phenyl ethyl ether (phenetole) is shown on p. 214, but it should be kept in mind that similar reactions occur for nitration, sulfonation, and Friedel-Crafts alkylation.

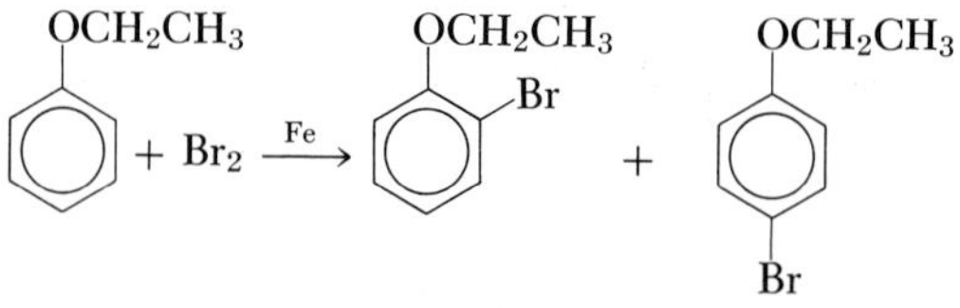

Phenetole   o-Bromophenetole   p-Bromophenetole

REACTIONS OF EPOXIDES

In contrast to the chemical inertness of simple ethers, epoxides contain a strained three-membered ring and are generally much more chemically reactive than ordinary ethers. As might be expected, the vast majority of reactions of epoxides are ring-opening reactions to relieve the strain in the three-membered ring. Both acid-catalyzed and base-catalyzed ring-opening reactions are known, and several such reactions are illustrated for ethylene oxide:

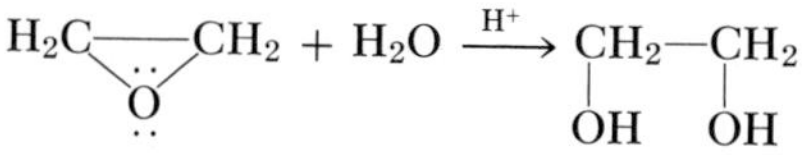

Ethylene oxide   Ethylene glycol

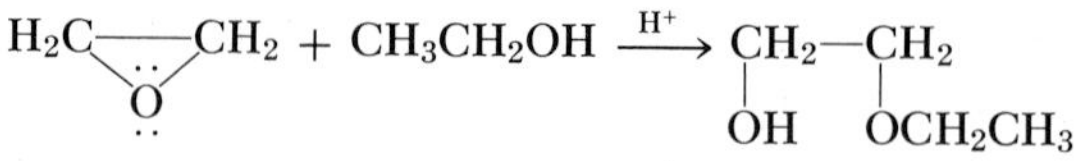

Ethylene oxide   Ethanol   2-Ethoxyethanol (cellosolves)

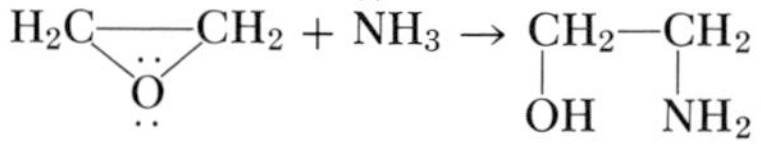

Ethylene oxide   Ethanolamine

SOME IMPORTANT ETHERS AND THEIR USES

Diethyl ether, which is often called ethyl ether or simply ether, is extensively used as a general anesthetic. It is easy to administer and causes excellent relaxation of the muscles. Blood pressure, pulse rate, and rate of respiration as a rule are only slightly affected. The main disadvantages are its irritating effect on the respiratory passages and its aftereffect of nausea. More recently, methyl propyl ether has been used as a general anesthetic. It has been claimed that this substance, called Neothyl, is less irritating and more potent than ethyl ether.

Diethyl ether is also an excellent solvent for fats and is often used in the laboratory for the extraction of fat from foods and animal tissue. In general, ethers are good solvents for fats, oils, gums, resins, and most functional derivatives of hydrocarbons. Ethylene oxide is used as a fumigating agent for seeds and grains and as the starting material in the preparation of the antifreeze ethylene glycol and the cellosolve solvents ($ROCH_2CH_2OH$, where R is an alkyl group) used in varnishes and lacquers.

Ethers occur also in many biologically active molecules. Tetrahydrocannabinol, mescaline, and morphine all contain the ether linkage, and this functionality is also present in DNA and RNA nucleic acids as a furan ring oxygen in the sugar portion of the molecule.

IMPORTANT TERMS AND CONCEPTS

alcohol	epoxide	grain alcohol	phenol
anisole	ester	hydration of olefins	phosphate esters
cellosolve	ethylene glycol	hydroboration	Williamson synthesis
cresol	ethylene oxide	hydrogen bonding	wood alcohol
diethyl ether	glycerol		

QUESTIONS

1. Draw out all the possible structural isomers of an alcohol which has a molecular formula $C_5H_{12}O$. Name each of these isomers according to the IUPAC system.

2. Which of the structural isomers in Question 1

   (a) are primary alcohols?
   (b) are secondary alcohols?
   (c) are tertiary alcohols?
   (d) on dehydration will give 2-pentene?

3. Give each of the following compounds an appropriate name:

   (a) $(CH_3)_2CHCH_2CH_2CH_2OH$
   (b) $(CH_3CH_2)_3COH$

   (c)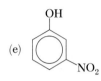

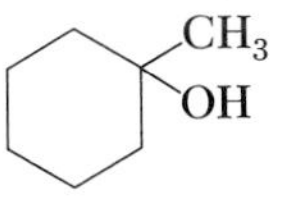

   (e)

   (d) $CH_3CH_2CH(CH_3)CH(OH)CH(C_2H_5)CH_2CH_3$

   (f) $CH_3CHOHCHOHCH_2OH$
   (g) $(CH_3)_2CHCHBrCH_2CH_2OH$

4. What is wrong with each of the following names? Give the correct name for each compound.

   (a) 1,1-dimethyl-1-butanol
   (b) 4-methyl-4-pentanol
   (c) 5-nitrophenol
   (d) 2-ethyl-2-propanol
   (e) 4-pentene-2-ol

5. Write structural formulas for each of the following compounds.

   (a) isobutyl alcohol
   (b) isopropyl alcohol
   (c) *t*-butyl alcohol
   (d) 3-pentanol
   (e) 3,3-dimethylcyclohexanol
   (f) 2,4,4-trimethyl-2-heptanol
   (g) *meta*-bromophenol
   (h) 1-bromo-2-methyl-3-hexanol
   (i) *cis*-3-ethylcyclopentanol
   (j) *trans*-2-pentene-1-ol

6. Which of the following alcohols are easily oxidized?

   (a) $CH_3CH_2CH_2OH$
   (b) $CH_3CHOHCH_2CH_3$
   (c) $(CH_3)_3COH$
   (d) $(CH_3)_3CCH_2OH$
   What is the functional class of compounds formed when *each* of the above alcohols is oxidized?

7. Write equations for the reaction of isopropyl alcohol with the following reagents:

(a) hydrogen iodide   (d) Al_2O_3/heat
(b) potassium   (e) $K_2Cr_2O_7/H_2SO_4$
(c) acetic acid

8. Draw a structural formula for each of the following compounds:

(a) methyl isopropyl ether   (d) ethyl ether
(b) propylene oxide   (e) 2-bromo-4-ethoxyhexane
(c) *p*-nitroanisole

9. Name each of the following compounds:

(a)

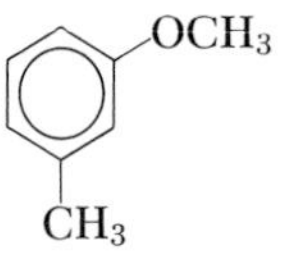

(d) $HOCH_2CHOHCH_2OH$

(b) $(CH_3)_2CH-O-CH(CH_3)_2$ (e) $CH_3OCH_2CH_2OCH_3$

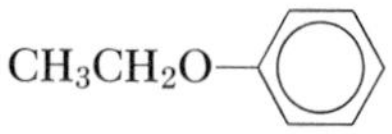

(c) CH_3CH_2O-⟨◯⟩

10. Write equations for the reactions of anisole with the following reagents:

(a) excess HI/heat   (c) sodium
(b) $KMnO_4$ (d) HNO_3/H_2SO_4

11. Show by equations how the following conversions may be carried out. More than one step may be necessary.

(a) methyl *n*-propyl ether from *n*-propyl iodide
(b) phenyl ethyl ether from phenol
(c) methyl isopropyl ether from propene

12. Write equations for the reactions of ethylene oxide with the following reagents:

(a) *n*-propyl alcohol/H^+ (d) ethylene glycol/H^+
(b) sodium methoxide/CH_3OH (e) ethanolamine ($HOCH_2CH_2NH_2$)
(c) methylamine (CH_3NH_2)

SUGGESTED READING

Beecher: Anesthesia. Scientific American, Vol. 196, No. 1, p. 70, 1957.
Ferguson: Hydrogen Bonding and Physical Properties of Substances. Journal of Chemical Education, Vol. 33, p. 267, 1956.
Krantz: Volatile Anesthetics and Analgesics. Journal of Chemical Education, Vol. 37, p. 169, 1960.
Weaver: Glycerol. Journal of Chemical Education, Vol. 29, p. 524, 1952.
Webb: Hydrogen Bond, "Special Agent." Chemistry, Vol. 41, No. 6, p. 16, 1968.

HALOGEN DERIVATIVES OF HYDROCARBONS

CHAPTER 15

The *objectives* of this chapter are to enable the student to:

1. Recognize an alkyl or aryl halide.
2. Write the structures and names of the halogen derivatives of hydrocarbons.
3. Describe the methods for the introduction of halogen into organic compounds.
4. Explain the concept of anionic displacement reactions.
5. Define an organometallic compound.
6. Recognize some simple fluorocarbon compounds and their current applications.
7. Recognize some important fungicides and insecticides.

In previous chapters compounds have been discussed that contain a carbon-halogen bond. They were found to be the basic starting materials in some of the preparations of olefins, alcohols, and ethers. The utility of this class of compounds should now be evident, and the purpose of this chapter is to correlate, extend, and amplify somewhat, the properties, reactions, and uses of the halogen derivatives of hydrocarbons.

Halogen derivatives of hydrocarbons can be simply defined as *compounds in which a hydrogen of a hydrocarbon has been replaced by a halogen,* in which the halogen can be fluorine, chlorine, bromine, or iodine. The carbon atom bonded to the halogen atom may be a carbon that is part of an alkyl, vinyl, acetylenic, aromatic, or heterocyclic system.

The most widely used types of halogen derivatives are the ones containing the halogen bonded either to an alkyl group (called **alkyl halides**) or to an aromatic ring (called **aryl halides**), and most of the emphasis in this chapter will be on these two types of halides. Although many reactions of alkyl and aryl halides may appear quite similar, in the most important reaction (displacement of the halogen atom by another atom or group of atoms) of halides these two types of halides behave quite differently. Aryl halides are generally quite resistant to displacement reactions under ordinary conditions, whereas most alkyl halides undergo displacement reactions very easily. This contrasting behavior will be discussed in more detail later in this chapter.

PROPERTIES OF HALIDES

Within any series of alkyl or aryl halides, the boiling points increase with increasing molecular weight; consequently, the boiling points increase in the order: fluorides < chlo-

TABLE 15-1. PHYSICAL PROPERTIES
OF HALIDES

COMPOUND	STRUCTURE	M.P. °C	B.P. °C
Methyl fluoride	CH_3F	-142	-78
Methyl chloride	CH_3Cl	-98	-24
Methyl bromide	CH_3Br	-94	$+4$
Methyl iodide	CH_3I	-66	$+42$
Methylene chloride	CH_2Cl_2	-95	$+40$
Chloroform	$CHCl_3$	-64	$+62$
Carbon tetrachloride	CCl_4	-23	$+77$
Vinyl chloride	$CH_2{=}CHCl$	-154	-13
Tetrachloroethylene	$CCl_2{=}CCl_2$	-22	$+121$

rides $<$ bromides $<$ iodides, as illustrated in Table 15-1 for the methyl halides. The physical properties of some of the more common halides are summarized in Table 15-1.

The organic halides are insoluble in water. The monofluoro and monochloro derivatives are less dense than water, and the monobromo and monoiodo compounds are more dense than water.

METHODS OF PREPARATION OF ORGANIC HALIDES

Several of the preparations of halides have been discussed in earlier chapters and will be noted here to correlate this material, but these previous methods will not be discussed in detail again, and the student is urged to go back and review these preparations in the previous chapters.

DIRECT HALOGENATION

Direct halogenation of alkanes was shown earlier (p. 173) to lead to a mixture of halogenated hydrocarbons, and, except for a few special cases, is not generally applicable to preparing alkyl halides. Indirect methods are used to prepare the commercially important halides **carbon tetrachloride** and **chloroform.** Carbon tetrachloride is prepared industrially by the chlorination of carbon disulfide. It is a colorless liquid, insoluble in water, soluble in organic solvents, and more dense than water. It is extensively used in

$$CS_2 + 3Cl_2 \xrightarrow{\text{SbCl}_5} CCl_4 \quad + \quad S_2Cl_2$$

Carbon   Carbon   Sulfur
disulfide   tetrachloride monochloride

the laboratory as an extraction solvent and is used commercially as a solvent for oils and greases. Because of its high solvent power, it has also been used as a dry-cleaning agent and as a household cleaning agent. In contrast to the hydrocarbons, alcohols, and ethers, carbon tetrachloride is nonflammable and has found wide use as a fire-extinguishing agent. However, the oxidation of carbon tetrachloride at the temperatures of a fire gives phosgene ($COCl_2$), a toxic gas, and fire extinguishers containing carbon tetrachloride should only be used where adequate ventilation is available.

Chloroform ($CHCl_3$) is obtained commercially by the reduction of carbon tetrachloride using iron and steam as the reducing agent. Chloroform is a sweet-smelling volatile liquid (b.p. 62°C) that once was widely used as an anesthetic. Toxic effects sometimes result from its use, however, and it has been replaced by other anesthetics.

$$CCl_4 + H_2O \xrightarrow{Fe} CHCl_3 + HCl$$

Carbon   Chloroform
tetrachloride

Chloroform undergoes photochemical (sunlight) oxidation to give phosgene unless stabilized, and commercially available chloroform contains $3/4$% ethanol as a stabilizer to prevent air oxidation.

Although direct halogenation of alkanes is unsuitable for preparing alkyl halides, it is the most widely used method for preparing aryl halides (see reactions of aromatic hydrocarbons) containing chlorine or bromine. Substituted aryl halides can also be prepared satisfactorily by this method.

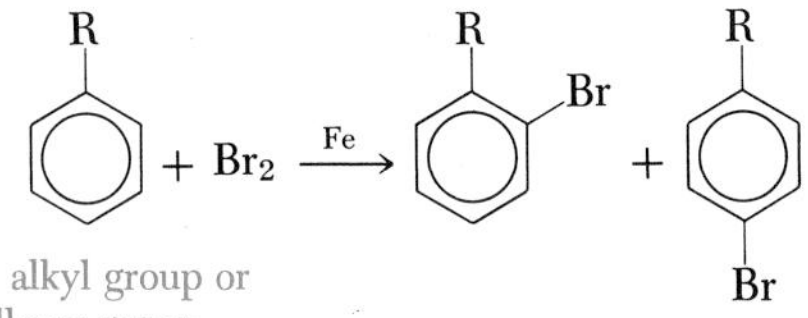

R = alkyl group or
alkoxy group

ADDITION OF HYDROGEN HALIDES TO OLEFINS

The addition of hydrogen halides (HF, HCl, HBr, and HI) to olefins has been shown previously (see reactions of alkenes) to give alkyl halides. The addition reaction follows Markownikoff's rule.

$$CH_3CH{=}CH_2 + HCl \rightarrow CH_3CHClCH_3$$

Propene   Isopropyl chloride

Replacement of Other Functional Groups

HYDROXYL GROUP OF ALCOHOLS

The hydroxyl group of alcohols can be replaced by a halide atom using the hydrogen halides (see reactions of alcohols). Fluorides are generally not prepared by this method, but chlorides, bromides, and iodides are attainable. Phenols do not usually undergo this type of reaction.

$$ROH + HX \rightarrow RX + H_2O$$
$$X = Cl, Br, I$$

REACTIONS OF HALIDES

The most common type of reaction of alkyl halides involves displacement of the halogen atom. This type of reaction may be rationalized as the attack of a negative ion on an alkyl halide to bring about displacement of a halide ion. Because of the great variety of anions that will cause displacement of a halide ion from an alkyl halide, numerous types of functional derivatives can be prepared by this kind of reaction. Consequently, alkyl halides are one of the basic starting materials for many of the general classes of hydrocarbon derivatives. Some of the more important of such displacement reactions are summarized in the following equations. The general reaction may be summarized as follows:

$$N^- + R\!-\!X \rightarrow N\!-\!R + X^-$$

N⁻ = anion
X = halogen

ANION DISPLACEMENT OF HALIDE ION

Alcohol formation:	$R\!-\!X + \bar{O}H \rightarrow R\!-\!OH + X^-$
Ether formation:	$R'\!-\!X + \bar{O}R \rightarrow R'\!-\!O\!-\!R + X^-$
Amine formation:	$R\!-\!X + \bar{N}H_2 \rightarrow R\!-\!NH_2 + X^-$

$$\text{Ester formation:} \quad R'\!-\!X + RCO_2^- \rightarrow R'\!-\!O\!-\!\overset{\overset{O}{\|}}{C}\!-\!R + X^-$$

Acetylene formation:	$R'\!-\!X + RC\!\equiv\!C^- \rightarrow R'\!-\!C\!\equiv\!CR + X^-$
Nitrile formation:	$R\!-\!X + \bar{C}N \rightarrow R\!-\!CN + X^-$
Mercaptan formation:	$R\!-\!X + \bar{S}H \rightarrow R\!-\!SH + X^-$

In this type of displacement reaction, the reactivity is dependent on the type of anion and the kind of alkyl group. For any given anion, the tendency toward displacement decreases in the order: primary > secondary > tertiary. Aromatic halides do not generally undergo this type of displacement reaction under normal conditions. Vinyl halides are likewise inert to this kind of displacement reaction under normal conditions.

Let us consider a specific example of this type of reaction, namely, the conversion of methyl bromide to methyl alcohol by hydroxide.

$$CH_3Br + \ ^-OH \rightarrow CH_3OH + Br^-$$

The sequence of this reaction has been shown to involve simultaneous formation of the $-\!\overset{|}{\underset{|}{C}}\!-\!OH$ bond and cleavage of the $-\!\overset{|}{\underset{|}{C}}\!-\!Br$ bond as shown below in B. The hydroxide attacks the carbon bearing the bromine from the backside (A) and the bromine

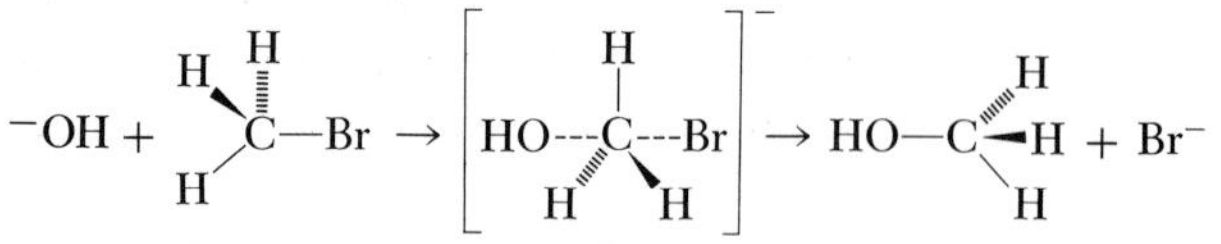

leaves from the front side (C). In this situation (B), the two electronegative Br and OH groups assume positions as far apart as possible. After departure of the bromide ion, the final product (C) has been inverted; i.e., the molecule has been essentially turned inside out and the hydroxyl group is on the opposite face of the molecule (relative to the other groups attached to carbon). For most molecules inversion at carbon by this type of reaction does not cause any change in the molecule; however, if the carbon atom is an asymmetrical° carbon, the resultant product will have the opposite configuration.

Since attack by the anion involves backside attack on carbon, the other groups attached to this carbon play a significant role in the ease of this reaction. As the size of these groups attached to carbon increases, the ease of approach of the anion decreases,

---

°An asymmetric carbon is a carbon atom which has four different groups attached to it.

since it encounters increased resistance to approach at the backside of the carbon being attacked. Consequently, the ease of attack is primary > secondary > tertiary, since this order is also the order of increasing size of these groups.

FORMATION OF ORGANOMETALLIC REAGENTS

Compounds in which a metal is bonded to carbon are called organometallic compounds. Compounds containing most of the known metals have been prepared and studied. Several of these organometallic compounds are well known to most people. For example, tetraethyl lead, $(C_2H_5)_4Pb$, is an important antiknock ingredient added to most gasolines. Mercurochrome and Merthiolate are two mercury derivatives used as antiseptics.

Three important biologic compounds which contain metals as part of the molecule are chlorophyll, which is important in photosynthesis and contains magnesium (see p. 357), heme, which is important as an oxygen carrier in the blood and contains iron (see p. 280), and vitamin B_{12}, which contains cobalt (see p. 331).

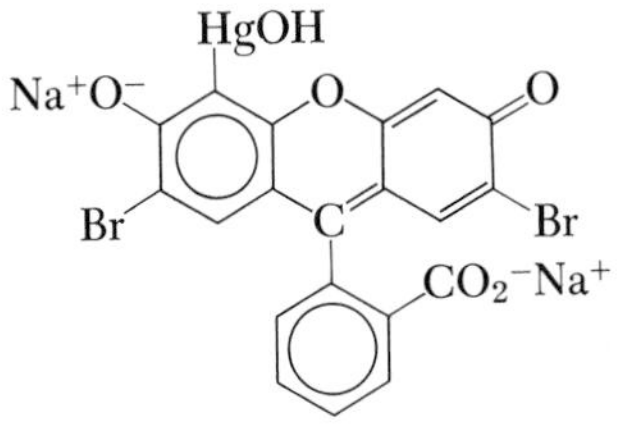

Mercurochrome

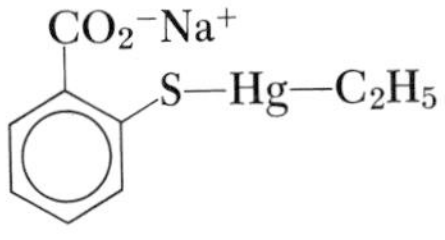

Merthiolate

Of the numerous and diversified kinds of organometallic compounds, the magnesium compounds have been the most useful to the organic chemist in the laboratory.

The preparation and use of organomagnesium compounds were developed by the French chemist, Victor Grignard, and this type of compound is known as a **Grignard reagent.** Both alkyl and aryl halides form a Grignard reagent upon reaction with metallic magnesium in anhydrous ether.

$$CH_3CH_2I + Mg \xrightarrow{\text{ether}} [CH_3CH_2]^- \overset{+}{Mg}I$$

Ethyl iodide   Ethyl magnesium iodide

Grignard reagents are highly reactive compounds and undergo reaction with many functional groups (particularly carbonyl-containing functional groups) to give useful products. Applications of these organometallic reagents will be introduced in subsequent chapters.

AROMATIC SUBSTITUTION REACTIONS OF ARYL HALIDES

Although aromatic halides undergo anion displacement reactions only with extreme difficulty, they undergo the normal type of aromatic substitution reactions without too much difficulty. The halogen atoms are *ortho-para* directing; therefore, bromobenzene gives *ortho-* and *para*-dibromobenzene on bromination. Nitration, sulfonation, and alkylation proceed similarly.

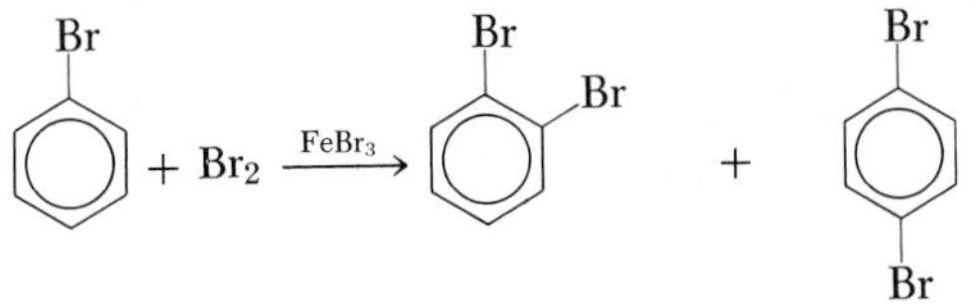

Bromobenzene   *o*-Dibromobenzene   *p*-Dibromobenzene

FLUORINE COMPOUNDS

Because of the great reactivity of fluorine, it is not normally useful as a general purpose fluorinating agent except under special conditions. Consequently, organic compounds containing fluorine are usually prepared by special methods. The most important fluorine compounds are those in which all the hydrogen atoms have either been replaced by fluorine (this class is known as fluorocarbons) or by a combination of fluorine and other halogens. Substances containing many fluorine atoms are generally inert to oxidizing and reducing agents and to most acids. Consequently, they find extensive use in refrigeration, aerosols, lubricants, electrical insulators, and in the plastics industry.

The most widely used fluorine compounds are the methane and ethane derivatives sold under the trade names Freon, Genetron, and Ucon. Some typical examples are shown below:

$$CF_2Cl_2 \qquad CFCl_3 \qquad CFCl_2CF_2Cl \qquad \begin{array}{c} CF_2{-}CF_2 \\ | \qquad | \\ CF_2{-}CF_2 \end{array}$$

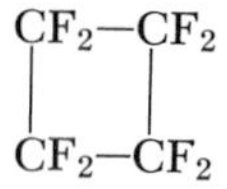

Freon-12	Freon-11	Freon-113	Freon-C318
Dichlorodifluoromethane	Trichlorofluoromethane	1,1,2-Trifluoro-trichloroethane	Octafluorocyclobutane

Freon-12 is volatile (b.p. $-28°C$) and nontoxic, and is used as a refrigerant in the majority of household and commercial refrigeration units and air conditioners. Freon-C318 (octafluorocyclobutane) is odorless, tasteless, nontoxic, and extremely stable to hydrolysis, and has application in the aerosol industry.

Probably the best known of the fluorine containing compounds is the polymer known as **Teflon,** which is made by polymerization of tetrafluoroethylene, $CF_2{=}CF_2$. Teflon contains the repeating unit $({-}CF_2{-}CF_2{-})_n$. This fluorocarbon polymer is inert to most chemical reagents, has excellent electrical insulating properties, maintains its lubricating properties over a wide range (-50 to $+300°C$), and has "non-sticking" properties for most materials. This latter characteristic has been exploited in making Teflon coated frying pans, cookie sheets, pie tins, and so forth, which can be used for cooking and baking without grease, as foodstuffs do not adhere to Teflon.

Another important property of fluorocarbons is their nonwettability. Materials and fabrics coated with a fluorocarbon become water and oil repellant. Fabrics treated with fluorochemicals become stain resistant, and this particular property has been put to practical use in the manufacture of "Scotch Guard" coated fabrics.

TOPIC OF CURRENT INTEREST

FLUOROCARBONS—BLOOD SUBSTITUTES

In some startling recent research, scientists at several universities have perfected a stand-in for erythrocytes, the blood cells which transport oxygen and carbon dioxide. The most promising erythrocyte replacements are fluorocarbons, such as $C_8F_{17}Br$, perfluorodecalin, and fluorinated ethers. Chemically and biologically stable, these fluids can absorb and release fifteen times as much oxygen and carbon dioxide as water can. Thus, they mimic blood's most important function—gas transport.

The most dramatic experiment to date is the work by Dr. Leland Clark, who submerged mice for one hour in perfluorobutyl tetrahydrofuran containing dissolved oxygen. The mice survived and showed no ill effects. Similarly, dogs have lived more than four years after a fluorocarbon transfusion.

Thus far, the only drawback to use in humans has been the retention of fluorocarbon in the liver. Current research is directed at finding a fluorocarbon which has the required properties but is also excreted from the liver in a short time. The fluorocarbons shown below exhibit promise in this direction.

$CF_3CF_2CF_2CF_2CF_2CF_2CF_2CF_2Br$

$C_8F_{17}Br$

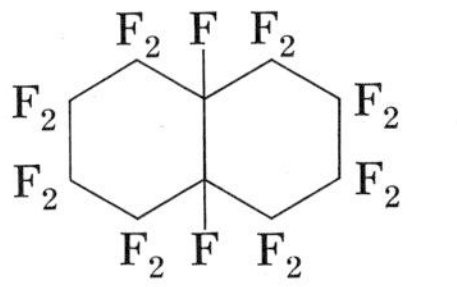

Perfluorodecalin

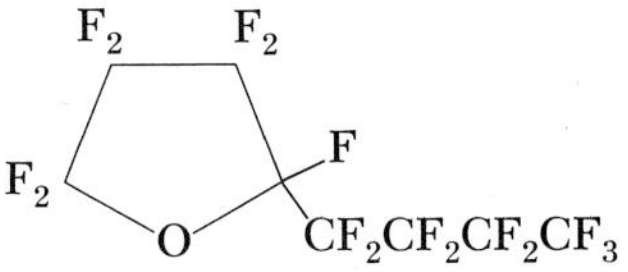

Perfluorobutyl tetrahydrofuran

A potential immediate application of "synthetic" blood would be the use of fluorocarbons to save a person's life when blood for transfusion is not readily available. In such diseases as aplastic anemia, for which long-term transfusion treatments are impractical, fluorocarbon solutions could conceivably prolong lives for many many months, if not indefinitely.

Also, it has been found that certain oxygenated fluorocarbons actually carry more oxygen than does blood, and a blood-starved limb threatened with gangrene, for example, might be saved by emergency treatment with oxygen-saturated fluorocarbons.

Another potential application is the use of oxygenated fluorocarbons in underwater oxygen support facilities to reduce effects of rapid decompression (the bends) by eliminating the breathing atmosphere containing nitrogen which, under pressure, dissolves in the blood and forms bubbles if pressure is released too quickly.

MISCELLANEOUS HALOGEN COMPOUNDS

Halogen compounds play an important role in the chemical industry as synthetic starting materials. In addition, many halogen-containing compounds find application as fumigants (CH_3Br) for controlling various insects and rodents, and as fungicides and pesticides. Some of the more common halogen-containing pesticides are shown below:

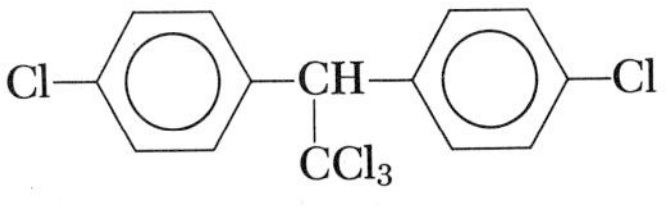

DDT

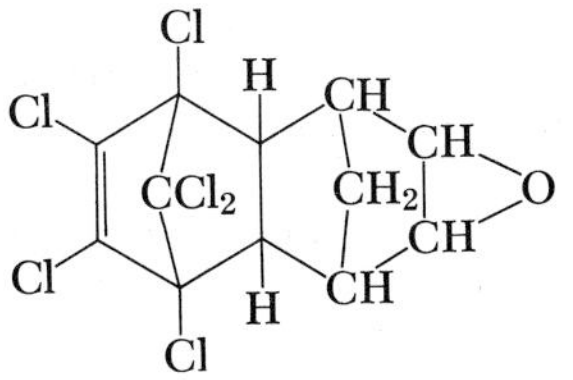

Dieldrin

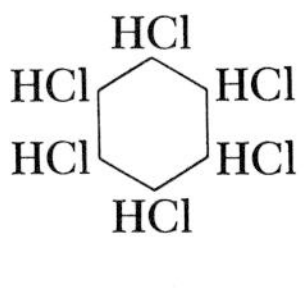

Lindane

In recent years the use of these chlorinated hydrocarbons has aroused public interest and criticism. Although successful control of malaria and the boll weevil have been the virtues of DDT and related compounds, more recent data has shown that these substances are accumulated in high concentration in fish and fowl. This increase in DDT concentration in the animal food chain has threatened the survival of several species of birds and fishes by affecting their reproductive cycle. Consequently, much controversy has arisen in recent years over the continued use of these chlorinated hydrocarbons, and renewed interest in biodegradable insecticides, chemosterilants, and insect attractants has arisen in the search for other means to control insects.

TOPIC OF CURRENT INTEREST

CHEMICAL COMMUNICATION

All species of animals communicate vital information to each other via chemical messages. When used among members of the same species, chemical compounds which are secreted by an animal and elicit a specific kind of behavior are called **pheromones.** Such chemical substances emitted by an animal may pass on information for selecting mates, finding food, avoiding enemies, or raising the alarm in case of danger. They may also trigger physiological changes, such as molting and new growth.

Insect sex attractants are probably the most potent and widely used of the pheromones. For example, the male silk moth has two finely tuned antennae that can detect a certain chemical compound emitted by the female moth. When this pheromone emitted by the female moth triggers a nerve impulse in the male moth, a message is received in the moth's brain and it moves upwind to claim a mate. So sensitive are the male moths to this type of chemical message that it has been claimed that male silkworm moths have been attracted to a female moth a quarter of a mile away, and that male gypsy moths are sexually excited by as little as 1×10^{-12} grams of attractant. Many of these sex attractant pheromones are long-chain epoxides, unsaturated alcohols, esters, and carboxylic acids.

$$(CH_3)_2CH(CH_2)_4\overset{\displaystyle H}{\underset{\displaystyle O}{C}}\!\!-\!\!\overset{\displaystyle H}{C}(CH_2)_9CH_3 \qquad CH_3(CH_2)_2CH\!=\!CH\!-\!CH\!=\!CH(CH_2)_9OH$$

Disparlure

Female gypsy moth attractant

Female silkworm moth attractant

$$CH_3(CH_2)_7CH\!=\!CH(CH_2)_6CO_2CH_3 \qquad CH_3(CH_2)_7CH\!=\!CH\!-\!CH\!=\!CH\!-\!CH_2\!-\!CO_2H$$

Hexalure

Female black carpet beetle attractant

In the past 20 years interest in pheromones has grown because the ultimate goal of this research is to put pheromones to work as lures in the control of insect pests. The objective in the use of sex attractant pheromones is to lure the unsuspecting male into a trap. This trap may contain a chemical which will eradicate the insect, or it may contain a chemosterilant which sterilizes the insect. Alternatively, the trap may be baited with a pheromone to confuse the insect so that it cannot locate a proper mate.

For example, traps containing disparlure were recently utilized in Pennsylvania forests infested with gypsy moths. Each trap was baited with disparlure; and the forest area was saturated with so much sex lure that the excited male would look in vain for an authentic female until he gave up mating altogether. Although the full results of these "saturation tests" are not in yet, the early results seem promising.

Similarly, in California the pink bollworm moth, a destructive cotton pest which had become resistant to DDT, was effectively controlled by the use of hexalure. Traps baited with hexalure were used to reveal the presence of the destructive bollworm moth. The infested area was then saturated with sterile moths which carried out unproductive mating with the female moths.

A recent (1974) report by medical workers at Emory University suggests that pheromones may have played a significant role in mating patterns of early humans. These workers have found a combination of volatile aliphatic acids in human females that are proven olfactory sexual stimulants in lower primate males. Significantly, the production of these pheromones appears to coincide directly with human ovulation, lending credence to the theory that pheromones play a role, or perhaps played a role, in human reproduction. Researchers in this area generally believe that there was probably a link between the pheromones and human sexuality in the early days of the development of the human species. At the same time they are skeptical about effects of pheromones in contemporary humans, because it is likely that many women who lead normal sex lives probably do not allow their accumulation in any significant quantities because of the scrupulous personal hygiene currently in vogue.

IMPORTANT TERMS AND CONCEPTS

alkyl Halide	DDT	Grignard reagent	organometallic
aryl Halide	dieldrin	lindane	Teflon
chloroform	freon	anion displacement reaction	

QUESTIONS

1. Draw a structural formula for each of the following compounds:

   (a) tertiary butyl bromide   (d) 1,1,2-trichlorotrifluoroethane
   (b) *p*-iodotoluene   (e) carbon tetrabromide
   (c) cyclohexyl bromide

2. Name each of the following compounds:

   (a) $CHCl_3$ (c)

   (b) $CHCl=CCl_2$ (d) CH_3CHICH_3

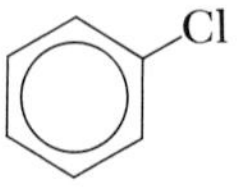

3. Indicate what is wrong with each of the following names. Give the correct name for each compound.

   (a) 2,4,4-trichloropentane
   (b) 2-ethyl-4-chloropentane
   (c) 1,4,5-tribromobenzene
   (d) *trans*-3,4-dichloro-3-pentene
   (e) 5-chlorotoluene

4. Write equations for the reactions of *n*-propyl bromide with the following reagents:

   (a) sodium ethoxide   (d) silver oxide/H_2O
   (b) potassium cyanide   (e) NaSH
   (c) Mg/ether

SUGGESTED READING

C. & E. News Special Report: Fluorocarbons. Chemical and Engineering News, July 18, p. 92, 1960.
Fluorocarbons: Possible Blood Substitute. Research Reporter, Chemistry, Vol. 46, No. 8, p. 20, 1973.
Jacobson and Beroza: Insect Attractants. Scientific American, Vol. 211, No. 8, p. 20, 1964.
Keller: The DDT Story. Chemistry, Vol. 43, p. 8, 1970.

Lykken: Chemical Control of Pests. Chemistry, Vol. 44, p. 18, 1971.

Preparing Fluorocarbons—A New Method. Chemistry, Vol. 43, p. 30, 1970.

Rheinboldt: Fifty Years of the Grignard Reaction. Journal of Chemical Education, Vol. 27, p. 476, 1950.

Solomon: A Bloodless Coup. The Sciences, No. 12, p. 11, 1972.

Vollmer and Gordon: Chemical Communication. Chemistry, Vol. 47, No. 10, p. 6, 1974.

Wilson: Pheromones. Scientific American, Vol. 208, No. 5, p. 100, 1963.

ALDEHYDES AND KETONES

CHAPTER 16

The *objectives* of this chapter are to enable the student to:

1. Recognize an aldehyde and a ketone.
2. Write the structures and names of the more common aldehydes and ketones.
3. Describe the conversion of other functional groups into the carbonyl functional group.
4. Describe the reactions of carbonyl compounds.
5. Understand the addition reactions of aldehydes and ketones.
6. Recognize an acetal and a hemiacetal.
7. Recognize the important aldehydes and ketones and their current applications.

Compounds containing the carbonyl group (—C̈—) are known as **aldehydes** or **ketones.** If one of the atoms linked to the carbonyl group is a hydrogen atom, the compound is an aldehyde (—C̈—H). The other atom or group attached to the carbonyl group of an aldehyde may be hydrogen, alkyl, or aryl. In the case of ketones, both of the groups attached to the carbonyl group are either alkyl or aryl. Cyclic ketones also exist in which the carbonyl group is part of the ring. Cyclic aldehydes are not possible.

NOMENCLATURE

In the IUPAC system, the characteristic ending for aldehydes is **-al,** and the characteristic ending for ketones is **-one.** These endings are added to the stem name of the hydrocarbon having the same number of carbon atoms. As usual, the compound is named as a derivative of the longest continuous chain of the carbon atoms, including the carbonyl functional group. In the case of aldehydes, the —**CHO** group must always appear at the end of the chain and is always indicated by the number 1 (the lowest number), although this number does not appear in the name. In the case of ketones, however, the carbonyl group may appear at various positions in a carbon chain, and its position must be designated by the lowest possible number. All other substituents are indicated by the appropriate number and prefix to indicate their positions on the carbon chain.

Common names are also used to name the aldehydes and ketones. Aldehydes are generally named as derivatives of the corresponding acid *to which they can be oxidized.* The **-ic** or **-oic** ending of the acid is dropped and replaced by the term **aldehyde.** Ketones, with the exception of acetone, are named according to the alkyl or aryl groups attached

TABLE 16–1   PHYSICAL PROPERTIES
OF ALDEHYDES AND KETONES

COMPOUND	STRUCTURE	M.P. °C	B.P. °C
Formaldehyde	HCHO	−92	−21
Acetaldehyde	CH_3CHO	−123	+21
Propionaldehyde	CH_3CH_2CHO	−81	+49
Chloral	CCl_3CHO	−58	+98
Acetone	CH_3COCH_3	−95	+56
Methyl ethyl ketone	$CH_3COCH_2CH_3$	−86	+80
Cyclohexanone	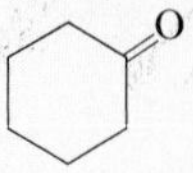	−45	+157
Acetophenone	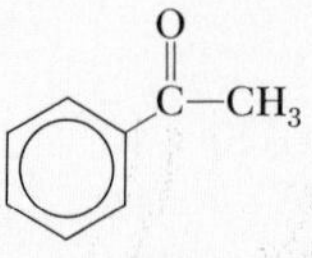	+20	+202
Benzophenone	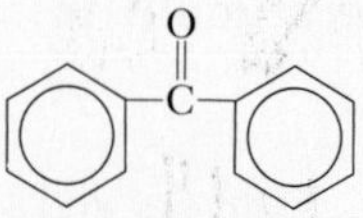	+48	+306
Acrolein	$CH_2{=}CHCHO$	−88	+53
Crotonaldehyde	$CH_3CH{=}CHCHO$	−77	+104
Benzaldehyde	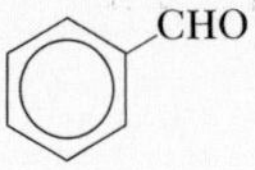	−26	+179
Salicylaldehyde	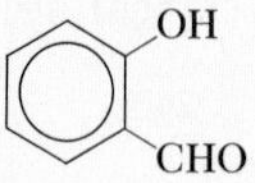	−7	+197

to the carbonyl function, followed by the word **ketone.** Some typical examples are illustrated below (the common name is in parentheses) (cf. also Table 16–1).

$$CH_3\overset{O}{\overset{\|}{C}}CH_3 \qquad CH_3CH_2\overset{O}{\overset{\|}{C}}CH(CH_3)_2 \qquad CH_3\underset{CH_3}{\overset{|}{C}}HCH_2CH_2CHO$$

Propanone   2-Methyl-3-pentanone   4-Methylpentanal
(acetone)   (ethyl isopropyl ketone)

PHYSICAL PROPERTIES

Formaldehyde is a gas at room temperature (20°C). All other simple aliphatic and aromatic aldehydes are colorless liquids. Most of the lower molecular weight aldehydes have a sharp odor, but the odor becomes more fragrant as the molecular weight increases. The aromatic aldehydes have been used as flavoring agents and perfumes. Benzaldehyde, for example, is a constituent of the seeds of bitter almonds and was once called "oil of bitter almond." It is a colorless liquid with a pleasant almondlike odor.

All of the more common lower molecular weight ketones are liquids. Acetone, the simplest ketone, is a moderately low boiling liquid (b.p. 56°C). With few exceptions all other ketones are either liquids or solids (Table 16–1).

The lower molecular weight aldehydes and ketones are soluble in water. Limited solubility occurs around 5 to 6 carbon atoms.

ACETALDEHYDE SYNDROME

The metabolism of ethyl alcohol, from alcoholic beverages, in the body involves a series of complex chemical reactions. The ethyl alcohol, CH_3CH_2OH, is first oxidized to acetaldehyde which is further oxidized to acetyl coenzyme A (CoA) and eventually to carbon dioxide and water.

$$CH_3CH_2OH \xrightarrow{\text{enzyme}} \underset{\text{Acetaldehyde}}{CH_3\overset{\overset{\displaystyle O}{\|}}{C}-H} \xrightarrow{\text{enzyme}} \text{Acetyl CoA} \xrightarrow{\text{enzyme}} CO_2 + H_2O$$

Ethyl alcohol

It has been found that when the concentration of acetaldehyde in the blood is significantly increased, the individual experiences a sharp decrease in blood pressure, a reddening of the face, a more rapid heart beat, and a total uncomfortable feeling called the "acetaldehyde syndrome." The chemical compound disulfiram has been found to compete with the enzyme that oxidizes the acetaldehyde to CoA.

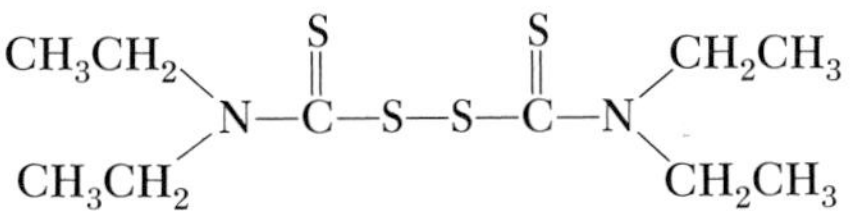

$$\underset{CH_3CH_2}{\overset{CH_3CH_2}{\diagdown}}N-\overset{\overset{\displaystyle S}{\|}}{C}-S-S-\overset{\overset{\displaystyle S}{\|}}{C}-N\underset{CH_2CH_3}{\overset{CH_2CH_3}{\diagup}}$$

Disulfiram

Therefore, when this compound is used, the acetaldehyde oxidation is interrupted, the acetaldehyde builds up in the blood, and the symptoms noted above are experienced. Medical testing of this chemical with alcoholics has shown that the uncomfortable feelings experienced when taking this material help the alcoholic to refrain from additional alcohol consumption. Although disulfiram does not cure alcoholism, it is hoped that the abstinence caused by its use will build up the self-confidence of the alcoholic and allow him to overcome his problem by total abstinence.

METHOD OF PREPARATION

OXIDATION

Under the proper conditions, primary alcohols can be oxidized to give aldehydes, and secondary alcohols can be oxidized to give ketones. These reactions can be viewed as a "dehydrogenation" process, since the elements of H_2 are lost in going from reactant to product. In the case of aldehydes, the process is somewhat complicated by the fact that aldehydes are more easily oxidized than alcohols. Consequently, as the aldehyde is formed, it can (without proper precautions) undergo further oxidation to give a carboxylic acid. The preparation of acetaldehyde by this method is illustrated in the following equation:

$$CH_3CH_2OH + K_2Cr_2O_7 \xrightarrow{H_2SO_4} CH_3\overset{\overset{\displaystyle H}{|}}{C}=O$$

Ethyl alcohol   Acetaldehyde
(B.P. 78°C)   (B.P. 21°C)

Formaldehyde is prepared industrially by a variation of this type of oxidation reaction.

$$CH_3OH + \frac{1}{2}O_2 \xrightarrow[\text{high temps.}]{Cu} \overset{\displaystyle O}{\underset{\displaystyle \|}{H-C-H}} + H_2O$$

Methyl alcohol   Formaldehyde

Oxygen is used in this process to convert the hydrogen into water. The formaldehyde-water mixture obtained is passed into water, and the solution formed (40% formaldehyde) is sold under the trade name **Formalin.**

The problem of further oxidation in the preparation of ketones from secondary alcohols is nonexistent under the normal conditions of the oxidation reactions. Ketones are quite resistant to further oxidation, except under extreme conditions, and the oxidation of the secondary alcohol stops very cleanly at the ketone stage. A typical example of this kind of reaction is shown below:

$$CH_3CHOHCH_3 + K_2Cr_2O_7 \xrightarrow{H_2SO_4} CH_3\overset{\displaystyle O}{\overset{\displaystyle \|}{C}}CH_3$$

Isopropyl alcohol   Acetone

REACTIONS OF CARBONYL COMPOUNDS

OXIDATION

As noted in the preparation of aldehydes, oxidation of aldehydes occurs very readily to give a carboxylic acid. Even mild oxidizing agents bring about the oxidation of aldehydes. Two such mild oxidizing reagents are **Tollen's reagent,** an ammoniacal solution of silver nitrate, and **Fehling's** or **Benedict's solution,** alkaline solutions of copper sulfate.° Oxidation of an aldehyde with Tollen's reagent produces metallic silver (usually as a silver mirror), whereas Fehling's or Benedict's solution gives a red precipitate of cuprous oxide. These simple visual tests make it easy to distinguish aldehydes from ketones, since ketones are not oxidized by these reagents. Hence, no silver mirror or red precipitate is obtained on treatment of a ketone with either Tollen's reagent or Fehling's and Benedict's solutions. The overall reaction of these reagents with aldehydes is summarized below:

$$RCHO + 2Ag(NH_3)_2^+ + 2OH^- \longrightarrow RCO_2^-NH_4^+ + 2Ag{\downarrow} + 3NH_3 + H_2O$$

Aldehyde   Tollen's reagent   Acid salt   Silver
           (colorless)   metal

$$RCHO + 2Cu^{++}\text{ (complex)} \xrightarrow{NaOH} RCO_2^-Na^+ + Cu_2O{\downarrow}$$

Aldehyde   Fehling's or Benedict's   Acid   Red precipitate
           solution (blue)   salt

This reaction is the basis of the Fehling and Benedict tests for determining the presence of sugar in urine.

The metabolic oxidation of sugars, which are the most important biologic aldehydes, allows the body to convert the sugar into energy. An important step in the Embden-Meyerhof pathway (p. 354) oxidation of glucose is the oxidation of glyceraldehyde-3-phosphate.

° Fehling's solution also contains sodium potassium tartrate, and Benedict's solution contains sodium citrate. These salts form tartrate and citrate complexes with the cupric ion and help keep the cupric ion in solution.

Addition Reactions

The most characteristic type of reaction of aldehydes and ketones is an addition reaction across the carbon-oxygen double bond. A similar type of reaction is characteristic of compounds containing carbon-carbon multiple bonds (see reactions of alkenes and alkynes). With carbonyl groups, however, the polarization of the bond dipole is such that the carbon end of the dipole is the positive end. Therefore, it is expected that the carbon atom of the carbonyl group would be attacked by the negative end of the adding

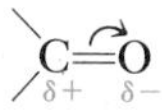

$$\underset{\delta+ \quad \delta-}{\overset{\frown}{\diagdown C = O}}$$

reagent, and that the addition reactions of aldehydes and ketones would be anion addition reactions. Overall reaction:

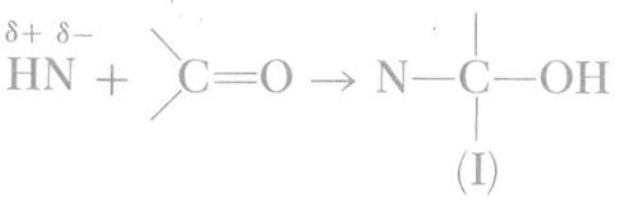

$$\underset{\delta+ \ \delta-}{HN} + \diagup\overset{\diagdown}{C}=O \rightarrow N-\overset{|}{\underset{|}{C}}-OH$$
(I)

In some cases, the initially formed addition product, (I), can undergo a dehydration reaction (loss of H_2O), so that the actual product isolated may not always appear exactly like (I). Some typical addition reactions of aldehydes and ketones are outlined below.

Addition of Hydrogen Cyanide

The addition of hydrogen cyanide gives an α-hydroxy nitrile,[*] called a cyanohydrin. These types of compounds are valuable in the preparation of hydroxy acids, amino acids, and sugars.

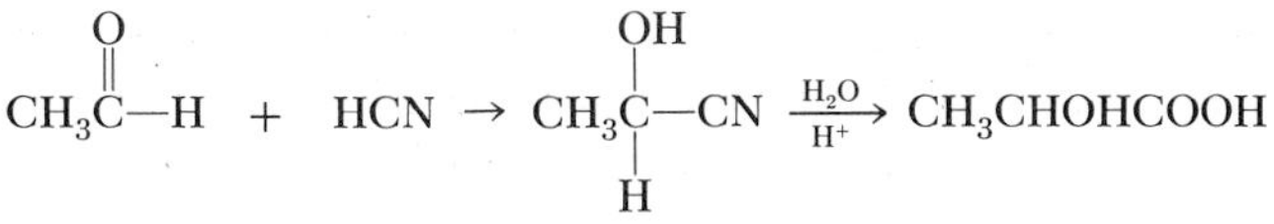

$$\underset{\text{Acetaldehyde}}{\overset{O}{\overset{\|}{CH_3C}}-H} \ + \ \underset{\substack{\text{Hydrogen} \\ \text{cyanide}}}{HCN} \ \rightarrow \ \underset{\text{Cyanohydrin}}{\overset{OH}{\overset{|}{CH_3\underset{|}{\overset{|}{C}}-CN}}} \ \xrightarrow[H^+]{H_2O} \ \underset{\text{Lactic acid}}{CH_3CHOHCOOH}$$

ADDITION OF ORGANOMETALLIC REAGENTS

Organometallic reagents, particularly Grignard reagents, also add across the carbonyl group. Hydrolysis of the initially formed complex gives alcohols as the final product. Aldehydes, except for formaldehyde, give secondary alcohols, and ketones give tertiary alcohols. This type of reaction provides one of the best laboratory methods for preparing secondary and tertiary alcohols. The generalized scheme of this kind of reaction is outlined below.

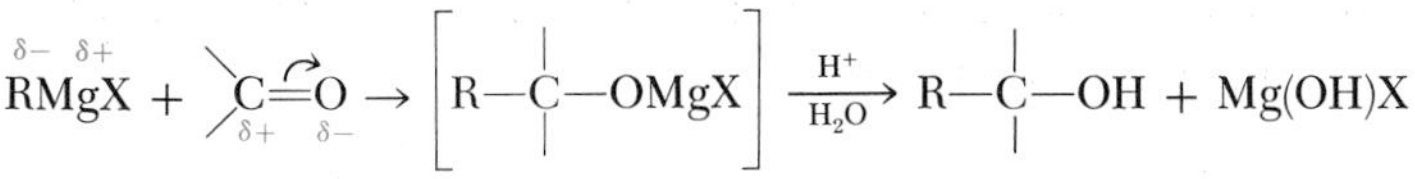

$$\underset{\delta- \ \delta+}{RMgX} + \underset{\delta+ \ \delta-}{\diagup\overset{\diagdown}{C}\overset{\frown}{=}O} \rightarrow \left[R-\overset{|}{\underset{|}{C}}-OMgX \right] \xrightarrow[H_2O]{H^+} R-\overset{|}{\underset{|}{C}}-OH + Mg(OH)X$$

(not isolated)

---

[*] Organic compounds containing the $-C \equiv N$ grouping are called either cyanides (relating them to the inorganic analogues) or nitriles.

ADDITION OF ALCOHOLS

Compounds called **hemiacetals** and **acetals** are formed by reacting aldehydes with alcohols in the presence of anhydrous HCl. One molecule of the alcohol adds to the carbonyl group to form a **hemiacetal,** which, in turn, can form an **acetal** by reaction with another molecule of alcohol.

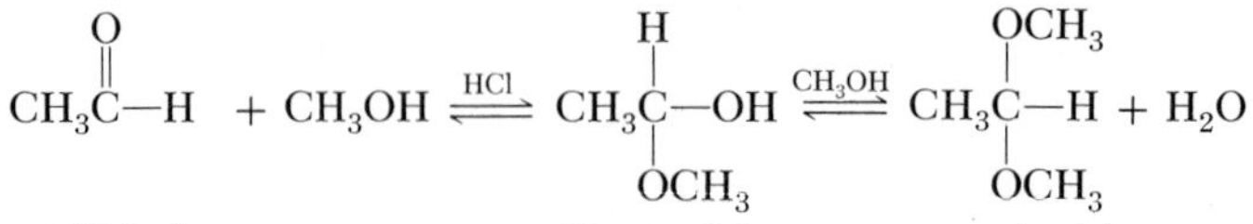

$$
\underset{\text{Acetaldehyde}}{CH_3\overset{\overset{O}{\|}}{C}-H} + CH_3OH \underset{}{\overset{HCl}{\rightleftharpoons}} \underset{\text{Hemiacetal}}{CH_3\underset{\underset{OCH_3}{|}}{\overset{\overset{H}{|}}{C}}-OH} \overset{CH_3OH}{\rightleftharpoons} \underset{\text{Acetal}}{CH_3\underset{\underset{OCH_3}{|}}{\overset{\overset{OCH_3}{|}}{C}}-H} + H_2O
$$

Acetals behave like ethers in their reactivity. They are stable to base but react with dilute aqueous acid to regenerate the alcohol and aldehyde from which they were prepared.

The acetal structure occurs in a number of important biologic compounds. Some sugars, such as glucose, ribose, and fructose, exist in nature as cyclic hemiacetals (see also p. 305).

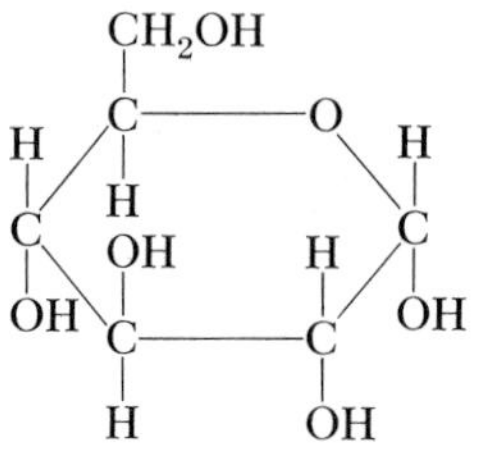

α-D-Glucopyranose

ADDITION OF AMMONIA AND AMMONIA DERIVATIVES

Except for the reaction of ammonia and formaldehyde, most aldehydes and ketones do not give stable addition products with ammonia. However, ammonia derivatives, in which a hydrogen of an —NH bond has been replaced by another group, do undergo addition to aldehydes and ketones, followed by loss of water, as shown in the following reaction:

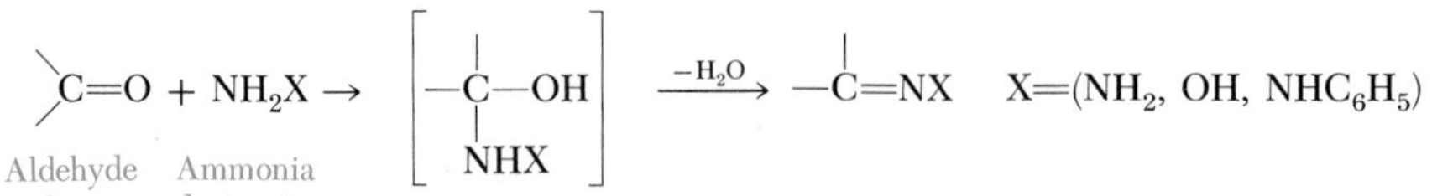

$$
\underset{\substack{\text{Aldehyde} \\ \text{or ketone}}}{\overset{}{\diagdown}C=O} + \underset{\substack{\text{Ammonia} \\ \text{derivative}}}{NH_2X} \rightarrow \underset{\text{Addition product}}{\left[-\underset{\underset{NHX}{|}}{\overset{|}{C}}-OH \right]} \xrightarrow{-H_2O} -\overset{|}{C}=NX \quad X=(NH_2, \ OH, \ NHC_6H_5)
$$

AROMATIC SUBSTITUTION REACTIONS

The carbonyl group is a *meta*-directing group and is also a deactivating group, and more vigorous reaction conditions must be used to bring about aromatic substitution. The halogenation of a typical aromatic aldehyde is shown below. Similar reactions occur on nitration, sulfonation, and alkylation.

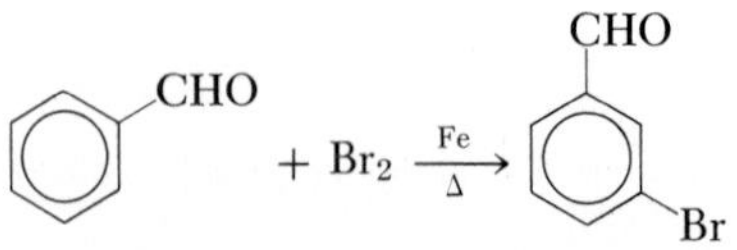

Benzaldehyde   *m*-Bromobenzaldehyde

USES OF IMPORTANT ALDEHYDES AND KETONES

As noted earlier, formaldehyde is a colorless gas that readily dissolves in water; a 40% solution is known as **formalin.** Formalin acts as a disinfectant and is used in embalming fluid and as a preservative of various tissues.

Acetaldehyde is used in the production of acetic acid, ethyl acetate, synthetic rubber, and other organic compounds. Paraldehyde, a cyclic trimer of acetaldehyde, is more stable than acetaldehyde and serves as a source of the latter compound when heated. It is an effective hypnotic, or sleep producer, but has been replaced by other drugs since it has an irritating odor and an unpleasant taste.

Benzaldehyde is an important intermediate in the preparation of drugs, dyes, and other organic compounds. As noted earlier, it is also used in flavoring agents and perfumes. Cinnamaldehyde, ⟨O⟩—CH=CHCHO, is a constituent of the oil of cinnamon obtained from cinnamon bark.

Acetone, the most important ketone, is prepared by the oxidation of isopropyl alcohol, or by the butyl alcohol fermentation process from cornstarch. It is used as a solvent for cellulose derivatives, varnishes, lacquers, resins, and plastics.

Methyl ethyl ketone $(CH_3COCH_2CH_3)$ is used by the petroleum industry in the dewaxing of lubricating oils. It is an excellent solvent for fingernail polish and is used as a polish remover.

Acetophenone is used as an intermediate in the preparation of other compounds and in the preparation of perfumes. If one of the hydrogen atoms on the methyl group is replaced by a chlorine atom, chloroacetophenone $(C_6H_5COCH_2Cl)$ is formed. This compound is a lachrymator and is commonly used as tear gas.

IMPORTANT TERMS AND CONCEPTS

acetal	Benedict's solution	cyanohydrin	hemiacetal
acetaldehyde	benzaldehyde	formaldehyde	Tollen's reagent
acetone	carbonyl group	formalin	

QUESTIONS

1. Draw a structural formula for each of the following compounds:

   (a) heptanal
   (b) 2-hexanone
   (c) *m*-nitrobenzaldehyde
   (d) cyclopentanone

   (e) 5-bromo-3-methylpentanal
   (f) *p*-chloroacetophenone
   (g) methyl cyclohexyl ketone
   (h) 3-methylcyclohexanone

2. Name each of the following compounds.

   (a) $(CH_3)_2CHCHO$

   (b)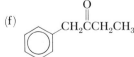

   (c) $CH_3CH_2\overset{O}{\overset{\|}{C}}CH(CH_3)_2$

   (d) CH_2BrCH_2CHO

   (e) CCl_3CH_2CHO

   (f) ⟨O⟩$CH_2\overset{O}{\overset{\|}{C}}CH_2CH_3$

   (g) $CH_3CH=CHCHO$

   (h) $CH_3CH_2\overset{O}{\overset{\|}{C}}C(CH_3)_3$

3. What is the oxidizing agent in: (a) Fehling's solution; (b) Benedict's solution; (c) Tollen's reagent?

4. Write equations for the reactions of butyraldehyde with the following reagents:

(a) $Ag(NH_3)_2^+$ (c) phenylhydrazine ($C_6H_5NHNH_2$)
(b) CH_3MgI/hydrolysis

SUGGESTED READING

Labianca: Acetaldehyde Syndrome and Alcoholism. Chemistry, Vol. 47, No. 10, p. 21, 1974.
Moore: The Art and the Science of Perfumery. Journal of Chemical Education, Vol. 37, p. 434, 1960.

CARBOXYLIC ACIDS AND ACID DERIVATIVES

== CHAPTER 17

The *objectives* of this chapter are to enable the student to:

1. Recognize a carboxylic acid or a carboxylic acid derivative.
2. Write the names and structures of carboxylic acids and acid derivatives.
3. Explain the acidity of carboxylic acids relative to water and alcohols which also contain the hydroxyl group.
4. Deduce the effect of substituents on the acidity of carboxylic acids.
5. Describe the preparation of carboxylic acids and acid derivatives.
6. Explain the reactions of carboxylic acids.
7. Recognize the important carboxylic acids, acid derivatives, and their current uses.

Organic compounds composed of the three elements carbon, hydrogen, and oxygen have been discussed in the previous chapters. These compounds included alcohols, ethers, aldehydes, and ketones. Further oxidation of primary alcohols or aldehydes eventually leads to the formation of another group of important organic compounds, **carboxylic acids,** which contain the carboxyl functional group. These acids are widely distributed in nature, especially in foodstuffs.

Important derivatives of organic acids that will be considered in this chapter include salts, esters, acid halides, acid anhydrides, and amides.

NOMENCLATURE

Carboxylic acids are defined as organic compounds which contain one or more carboxyl groups (—COOH) in the molecule. Some typical acids are illustrated in Table 17–1. Hydrogen, alkyl, and aryl groups may be attached to the carboxyl group, and more than one carboxyl group may be present in the molecule.

Since organic acids* occur so widely in nature either as the free acid or an acid derivative, they were among the earliest organic compounds investigated. Common names

---

* The carboxyl group may be written as $-\overset{O}{\underset{\|}{C}}-OH$, —COOH, or —CO$_2$H.

==================== 235

TABLE 17–1   IONIZATION CONSTANTS OF
CARBOXYLIC ACIDS

ACID	STRUCTURE	K_a
Acetic acid	CH_3COOH	1.8×10^{-5}
Propionic acid	CH_3CH_2COOH	1.0×10^{-5}
n-Butyric acid	$CH_3CH_2CH_2COOH$	1.5×10^{-5}
Chloroacetic acid	CH_2ClCO_2H	1.5×10^{-3}
Dichloroacetic acid	$CHCl_2CO_2H$	5.0×10^{-2}
Trichloroacetic acid	CCl_3CO_2H	1×10^{-1}
α-Chloropropionic acid	$CH_3CHClCO_2H$	1.6×10^{-3}
β-Chloropropionic acid	$CH_2ClCH_2CO_2H$	8×10^{-5}
Benzoic acid	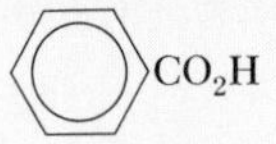 ⬡CO_2H	6×10^{-5}
p-Chlorobenzoic acid	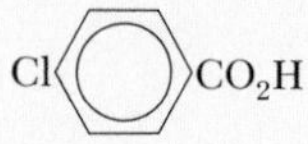 Cl⬡CO_2H	1×10^{-4}

were given to many of these acids, and these names are still used frequently today. For the aliphatic acids, the IUPAC nomenclature is also used. This system adds the ending **-oic acid** to the stem name of the hydrocarbon with the corresponding number of carbon atoms. The carboxyl group is given the number **1,** and all the other usual IUPAC rules are then applied. The aromatic acids are named either by common names or as derivatives of the parent aromatic acid, benzoic acid. Some typical illustrations of the use of these nomenclature systems are shown below. The common name (for the aliphatic acids) is given in parentheses.

$$
\underset{\substack{\text{Methanoic acid}\\ \text{(formic acid)}}}{\overset{\overset{\textstyle O}{\|}}{HCOH}}
\qquad
\underset{\substack{\text{Ethanoic acid}\\ \text{(acetic acid)}}}{\overset{\overset{\textstyle O}{\|}}{CH_3COH}}
\qquad
\underset{\substack{\text{Propanoic acid}\\ \text{(propionic acid)}}}{\overset{\overset{\textstyle O}{\|}}{CH_3CH_2COH}}
\qquad
\underset{\substack{\text{Butanoic acid}\\ \text{(butyric acid)}}}{\overset{\overset{\textstyle O}{\|}}{CH_3CH_2CH_2COH}}
$$

In some cases the position of substituents on the aliphatic chain is indicated by the Greek letters α, β, γ, δ, and so forth (cf. Table 17–1).

Di-acids, tri-acids, and so forth, are named in the IUPAC system by adding the ending **-dioic acid, -trioic acid,** and so forth, to the name of the hydrocarbon containing the same number of carbon atoms. In addition, common names are used for some of these compounds (cf. p. 241).

PHYSICAL PROPERTIES OF CARBOXYLIC ACIDS

Carboxylic acids are, as the name indicates, acidic compounds. When dissolved in water, they can ionize slightly to donate a proton to a more basic substance, such as water. Most simple carboxylic acids are only slightly ionized in water, and these acids

$$
\underset{\substack{\text{Carboxylic}\\ \text{acid}}}{\overset{\overset{\textstyle O}{\|}}{R-C-OH}} + H_2O \rightleftharpoons \underset{\substack{\text{Carboxylate}\\ \text{anion}}}{\overset{\overset{\textstyle O}{\|}}{R-C-O^-}} + H_3O^+
$$

TABLE 17-2   PHYSICAL PROPERTIES OF
ACIDS AND ACID DERIVATIVES

COMPOUND	STRUCTURE	M.P. °C	B.P. °C
Formic acid	$\overset{\text{O}}{\overset{\|}{\text{HCOH}}}$	+8	+101
Acetic acid	CH_3CO_2H	+17	+118
Propionic acid	$CH_3CH_2CO_2H$	−21	+141
n-Butyric acid	$CH_3CH_2CH_2CO_2H$	−4	+164
Isobutyric acid	$(CH_3)_2CHCO_2H$	−46	+153
n-Valeric acid	$CH_3(CH_2)_3CO_2H$	−34	+186
Methyl acetate	$CH_3CO_2CH_3$	−98	+57
Ethyl acetate	$CH_3CO_2CH_2CH_3$	−84	+77
Acetyl chloride	CH_3COCl	−112	+52
Acetamide	CH_3CONH_2	+82	+222
Acetic anhydride	$(CH_3CO)_2O$	−73	+140
Benzoic acid	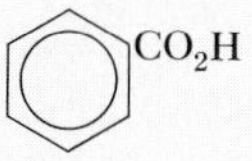 $\bigcirc\!\!-CO_2H$	+122	+249
Benzoyl chloride	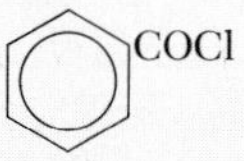 $\bigcirc\!\!-COCl$	−1	+197
Benzamide	 $\bigcirc\!\!-CONH_2$	+130	+290
Phthalic anhydride	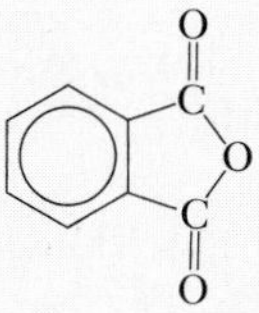 (phthalic anhydride structure)	+130	+285
Methyl benzoate	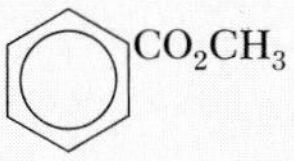 $\bigcirc\!\!-CO_2CH_3$	−12	+199

are fairly weak. The ionization constant, K_a, is a measure of the acid strength of these compounds. For the unsubstituted aliphatic acids, the acidity varies with changes in the alkyl group. Substitution of the α-hydrogens by highly electronegative groups, such as the halogens, has a significant effect on the acidity. The effect is an additive one, and the acidity increases as the number of electron-attracting groups in the α-position increases. This effect is only large when the electron-attracting group is in the α-position. When this type of group is further removed from the carboxyl group, the effect on the acidity of the acid is small.

Aromatic acids are very slightly more acidic than the aliphatic acids. Strongly electron-attracting substituents on the ring also increase the acidity of the aromatic acid, and the effect is generally about the same as in the aliphatic acids. Some typical acids and substituted acids are tabulated in Tables 17–1 and 17–2 with their (approximate) K_a to illustrate the preceding points.

Like alcohols, the carboxyl group contains a hydroxyl group which is capable of forming hydrogen bonds with other acid molecules, or with other similar kinds of molecules, such as water. Consequently, the lower molecular weight carboxylic acids are water soluble, the borderline solubility being about 4 to 5 carbon atoms.

PREPARATION OF CARBOXYLIC ACIDS

OXIDATION

Oxidation of primary alcohols and aldehydes (see reactions of aldehydes) gives carboxylic acids. This type of oxidation can be used to prepare both aliphatic and aromatic acids. Oxidizing agents, such as potassium dichromate and sulfuric acid or alkaline potassium permanganate, are generally used for these reactions.

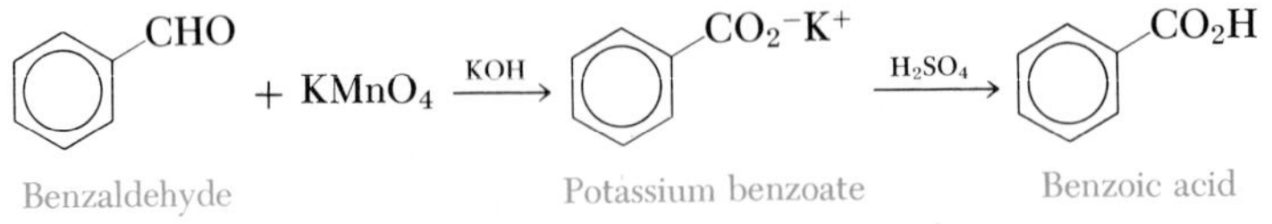

Benzaldehyde   Potassium benzoate   Benzoic acid

Di-acids can be prepared similarly by oxidation of the corresponding diols or di-aldehydes. This type of oxidation method gives a carboxylic acid with the same number of carbon atoms as the original alcohol or aldehyde.

Benzoic acid and other aromatic acids can also be prepared by the oxidation of alkyl side chains with either nitric acid or alkaline potassium permanganate (cf. p. 198). Since the alkylbenzene can be easily prepared by an alkylation reaction (see reactions of aromatic hydrocarbons), this is a useful preparative method for some aromatic acids. Similar types of oxidations of aliphatic hydrocarbons require more vigorous conditions, are not easily controlled, and are not of extensive preparative value.

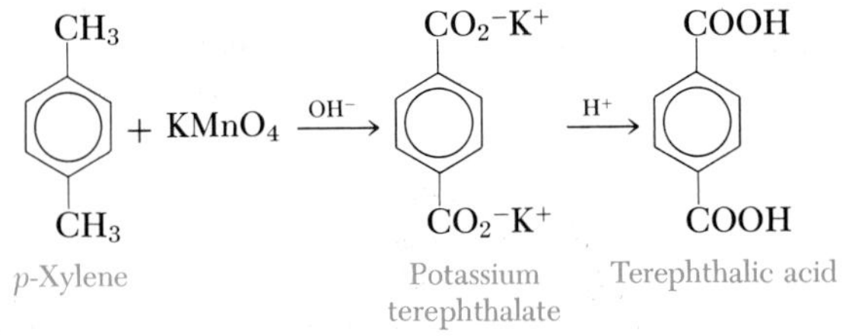

p-Xylene   Potassium terephthalate   Terephthalic acid

CARBONATION OF ORGANOMETALLIC REAGENTS

When organometallic reagents, such as Grignard reagents, are treated with carbon dioxide (CO_2), a complex is formed that gives a carboxylic acid on hydrolysis. This reaction may be rationalized as the addition of the organometallic reagent across the carbonyl group (similar to the addition of organometallics to aldehydes and ketones), as shown below:

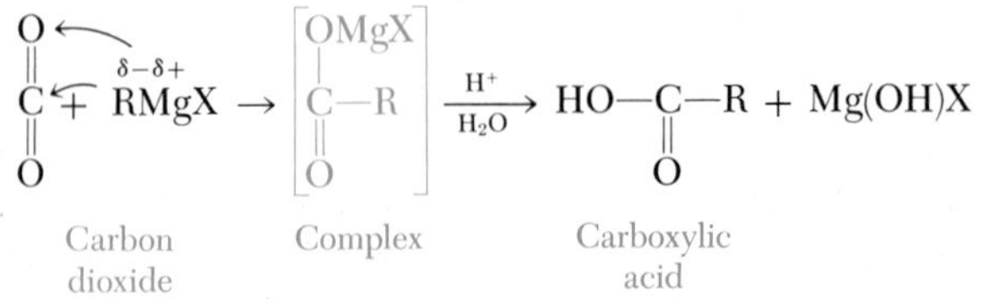

Carbon dioxide   Complex   Carboxylic acid

Both aliphatic and aromatic carboxylic acids can be prepared from the appropriate alkyl or aryl halides.

HYDROLYSIS OF NITRILES

The acid or base hydrolysis of nitriles affords another convenient method of preparing certain classes of carboxylic acids. The nitriles are conveniently prepared from alkyl halides via an anion displacement reaction (cf. p. 220).

$$R\,C\equiv N \xrightarrow[\substack{\text{or}\\ \text{base}}]{\text{acid}} R\overset{\displaystyle O}{\overset{\|}{C}}OH$$

REACTIONS OF CARBOXYLIC ACIDS

SALT FORMATION

Like their inorganic analogues, carboxylic acids react with bases to form salts. The name of the salt is formed by naming the cation, followed by the name of the acid in which the **-ic** ending (of either the common or IUPAC name) has been changed to **-ate**. Many of these salts are water-soluble. Since these salts are the salts of weak acids, they are readily hydrolyzed, and the acid can be regenerated by treatment of the acid salt with a stronger acid such as HCl or H_2SO_4. A typical example is shown below:

$$CH_3CO_2H + NaOH \rightarrow CH_3CO_2^-Na^+ + H_2O$$
Acetic acid   Sodium acetate

CONVERSION INTO FUNCTIONAL DERIVATIVES

Acid Halides (Acyl Halides).[*] The —OH group of a carboxylic acid can be replaced by a halogen atom to give a class of compounds called **acid halides,** RCOX (where X = halogen). This reaction is analogous to the replacement of the —OH group of alcohols by halogen to give alkyl or aryl halides. The hydrogen halides cannot be used to prepare the acid halides. Thionyl chloride is the most widely used reagent for this conversion.

$$CH_3\overset{\displaystyle O}{\overset{\|}{C}}OH + SOCl_2 \rightarrow CH_3\overset{\displaystyle O}{\overset{\|}{C}}Cl + SO_2\uparrow + HCl\uparrow$$
Acetic acid   Thionyl   Acetyl
chloride   chloride

Esters. Carboxylic acids react with alcohols in the presence of a strong acid catalyst such as HCl or H_2SO_4 to give a class of compounds called **esters,** RCO_2R' (see reactions of alcohols). The name of the ester is formed by changing the **-ic** ending of the acid (either the common or IUPAC name) to **-ate,** preceded by the name of the alkyl or aryl group derived from the alcohol. A typical example is shown on page 240.

As noted in Chapter 14 phosphate esters are associated with almost all biologic reactions.

---

[*] The $R-\overset{\displaystyle O}{\overset{\|}{C}}-$ group is known as the acyl group. In many reactions of acids, this portion of the acid molecule remains unaffected in the chemical reaction. Many acid derivatives, particularly the halide derivatives, are named according to this group. The name is formed by changing the **-ic** ending of the common acid name to **-yl.**

$$CH_3\overset{O}{\overset{\|}{C}}OH + CH_3CH_2OH \underset{}{\overset{H^+}{\rightleftharpoons}} CH_3\overset{O}{\overset{\|}{C}}OCH_2CH_3 + H_2O$$

Acetic acid   Ethyl alcohol   Ethyl acetate

Acid Anhydrides. Carboxylic acids undergo dehydration (loss of a mole of water) when treated with dehydrating agents to give a class of compounds called **acid anhydrides.** Anhydrides are named by adding the name anhydride after the acid from which the anhydride is derived. Cyclic anhydrides, containing five- or six-membered rings, are formed from the appropriate di-acids on dehydration. Some representative examples are shown below:

$$2CH_3\overset{O}{\overset{\|}{C}}-OH \xrightarrow{H_2SO_4} CH_3\overset{O}{\overset{\|}{C}}-O-\overset{O}{\overset{\|}{C}}-CH_3$$

Acetic acid   Acetic anhydride

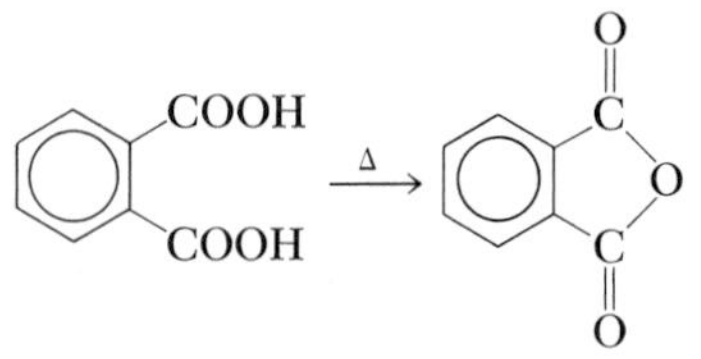

Phthalic acid   Phthalic anhydride

Amides. Carboxylic acids react with ammonia to give salts, which on heating give a class of compounds called **amides, $RCONH_2$.** Substituted amides, $RCONHR'$ or

$$CH_3\overset{O}{\overset{\|}{C}}OH + NH_3 \rightarrow CH_3CO_2^-NH_4^+ \xrightarrow{\Delta} CH_3\overset{O}{\overset{\|}{C}}NH_2 + H_2O$$

Acetic acid   Ammonium acetate   Acetamide

$RCONR_2'$, can be prepared by using amines (RNH_2 or R_2NH) in place of ammonia. The unsubstituted amides ($RCONH_2$) are named by dropping the **-ic** and **-oic** endings of the common and IUPAC names of the acid and adding the term **-amide.** When groups other than hydrogen are present on the nitrogen atom, their position and type is indicated by adding the prefix **N-** to indicate their position, followed by the name of the group attached to the nitrogen atom.

The formation of the amide linkage catalyzed by an enzyme is of great biologic importance. For example, when two amino acids react to form a substituted amide, the carbonyl group of one amino acid reacts with the α-amino group of the other amino acid to form a **peptide.** Proteins are high-molecular weight compounds that contain many

$$\underset{NH_2}{\overset{O}{\overset{\|}{RCHCOH}}} + \underset{R}{\overset{H}{\overset{|}{HNCHCOOH}}} \underset{}{\overset{Enzyme}{\rightleftharpoons}} \underset{NH_2 \quad R}{RCHC\overset{O}{\overset{\|}{}}\overset{H}{\overset{|}{N}}CHCOOH}$$

Peptide

amino acids linked by peptide (amide) bonds. These types of compounds will be discussed in more detail in Chapter 21.

IMPORTANT ACIDS, ACID DERIVATIVES, AND THEIR USES

Formic acid occurs free in nature in small amounts, and its presence is made known in an unpleasant fashion. Anyone who has been bitten by an ant or brushed against stinging nettles has felt the irritating effect of formic acid injected under the skin. Formic acid is a colorless liquid with a sharp, irritating odor. It is a slightly stronger acid than most carboxylic acids and produces blisters when it comes in contact with the skin. It is used in the manufacture of esters, salts, plastics, and oxalic acid.

Acetic acid has been known for centuries as an essential component of vinegar. It is formed from the oxidation of ethyl alcohol that is produced from the fermentation of fruit juices in the preparation of the vinegar. Cider vinegar from fruit juices contains about 4% acetic acid in addition to flavoring and coloring agents from the fruit. White vinegar is prepared by diluting acetic acid to the proper concentration with water.

Acetic acid is produced commercially by the catalytic oxidation of ethylene. Commercially produced acetic acid is about 99.5% pure and is called "glacial acetic acid," since at temperatures below 17°C it freezes to an icelike solid.

Lactic acid, $CH_3CHOHCOOH$, is an important hydroxy acid. It is formed when lactose (milk sugar) is fermented by lactobacillus bacteria. The taste of sour milk and buttermilk is due to the presence of lactic acid. Sour milk is often used in baking for its leavening effect on the dough. The lactic acid reacts with the sodium bicarbonate of baking soda to produce carbon dioxide throughout the dough.

In the process of muscular contraction, lactic acid is formed by the muscle tissues and released into the blood stream. In many of the so-called "cycles" involved in the oxidation of carbohydrates, lipids, and proteins to produce energy in the body, lactic acid is an essential component.

Oxalic acid, HO_2CCO_2H, occurs in the leaves of vegetables such as rhubarb and is one of the strongest naturally occurring acids. It is used to remove iron stain from fabrics and from porcelain ware, and to bleach straw and leather goods.

Citric acid (cf. p. 355) is a normal constituent of citrus fruits. Large quantities are produced by a fermentation process from starch or molasses. It is commonly employed to impart a sour taste to food products and beverages. In fact, most of the citric acid produced annually is used by the food and soft drink industries.

Benzoic acid is a colorless solid, slightly soluble in hot water. It is used in the synthesis of organic compounds, and its sodium salt (sodium benzoate) is used as an antiseptic and food preservative. Several substituted benzoic acids are also important commercial materials. *Para*-**aminobenzoic acid** is classified as a vitamin. It is apparently necessary for the proper physiological functioning of chickens, mice, and bacteria, but as yet has not been proved essential in the diet of humans. Certain derivatives of this acid are used as local anesthetics. Procaine, or Novocain, is probably the most important local anesthetic derived from *p*-aminobenzoic acid.

p-Aminobenzoic acid

Procaine

ACID DERIVATIVES

Organic acid derivatives, such as salts, are often used in place of the acids for many commercial uses. **Sodium acetate** is often used for its buffering effect in reducing the acidity of inorganic acids. It is also used in the preparation of soaps and pharmaceutical agents. Calcium propionate is added to bread to prevent molding. The calcium salt of lactic acid is sometimes used to supplement the calcium of the diet. Many infants receive additional iron in their diets from the salt, ferric ammonium citrate. Magnesium citrate has long been used as a saline cathartic. A solution of sodium citrate is employed to prevent the clotting of blood used in transfusions, and potassium oxalate will prevent the clotting of blood specimens drawn for analysis in the clinical laboratory. The sodium salt of salicylic acid is used extensively as an antipyretic or fever-lowering agent, and as an analgesic in the treatment of rheumatism and arthritis.

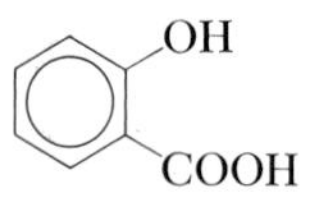

Salicylic acid

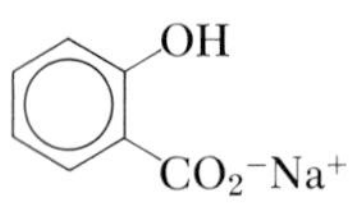

Sodium salicylate

Many esters occur in free form in nature and are responsible for the odor of most flowers and fruits. The characteristic taste and odors of different esters find application in the manufacture of artificial flavoring extracts and perfumes. Synthetic esters that are commonly used as food flavors are amyl acetate for banana, octyl acetate for orange, ethyl butyrate for pineapple, amyl butyrate for apricot, isobutyl formate for raspberry, and ethyl formate for rum.

TOPIC OF CURRENT INTEREST

SWEETNESS AND TASTE

Taste response can be classified into four basic reactions: sour, salty, bitter, and sweet. Although sweetness is probably the most studied and best understood taste effect from a molecular standpoint, the variation of relative sweetness with chemical structure is still not completely known. For example, β-D-mannose is bitter whereas α-D-mannose is sweet. A small change in chemical structure can have a pronounced effect on taste.

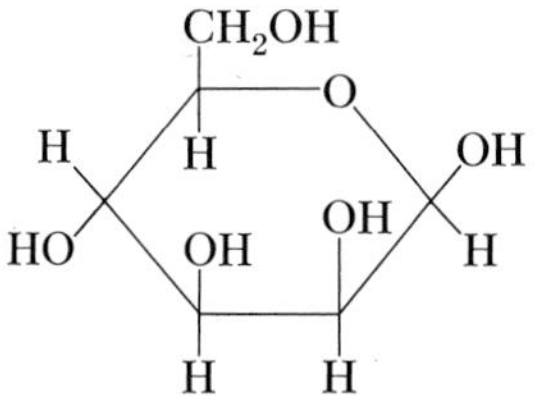

β-D-mannose

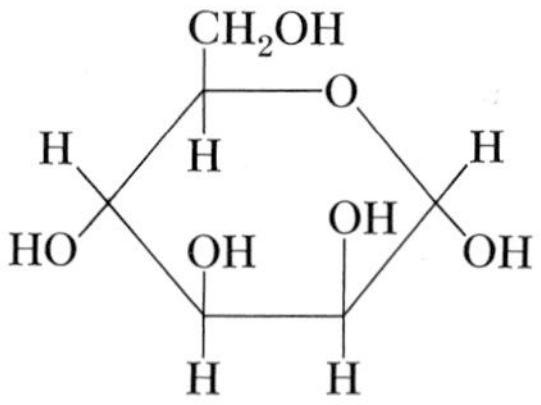

α-D-mannose

In the past, various sugars such as sucrose, glucose, and fructose were commonly used sweeteners in foodstuffs. In recent years, a wave of diet consciousness swept the country, and a demand for a noncaloric sugar substitute arose. The first such substitute discovered was saccharin, which has been found to be 300 times sweeter than sucrose. Until 1970, another group of widely used sweeteners were the cyclamates. This type of sweetener is only 30 times sweeter

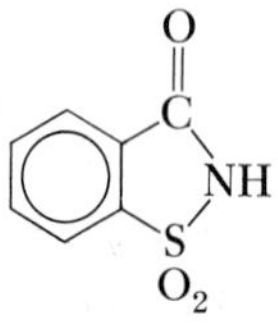

Saccharin

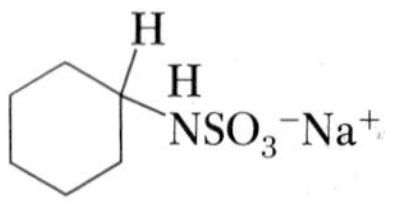

Sodium cyclamate

than sucrose, but it leaves less aftertaste than saccharin and was widely used in a variety of "low-calorie" foods and drinks. However, cyclamates were reported to cause bladder cancer in rats and in 1970 were banned by the Food and Drug Administration. More recent evaluation tests suggest that the earlier cancer danger may have been overstated, and cyclamates may eventually be cleared for use. Meanwhile, several other sweeteners have been developed and appear promising in early evaluation trials. Most of these new low-calorie sweeteners are amino acids,

$$HO_2CCH_2CHCNCHCCH_3$$

L-Aspartyl-L-phenylalanine methyl ester   6-Chloro-D-tryptophan

peptides, or proteins. Dipeptides such as L-aspartyl-L-phenylalanine methyl ester are nearly 200 times sweeter than sucrose, leave no unpleasant aftertaste, and are low in calories. Enzymatic hydrolysis leads to easily metabolized natural amino acids.

Similarly, an amino acid derivative, 6-chloro-D-tryptophan, has been developed by Eli Lily and Co. This compound is more than 1000 times sweeter than sugar and is currently being tested for toxicity. Results thus far have been encouraging.

Perhaps one of the most intriguing of the protein sweeteners is a glycoprotein which was isolated from a berry grown on the west coast of Africa and is marketed under the trade name Miraculin. It is currently marketed in the form of candy fruit drops, which, if eaten before an unsweetened acid-based dessert, make the dessert taste sweet. Miraculin has no effect on nonacid foods such as meat, bread, and coffee.

These protein-type sweeteners appear to offer many advantages over cyclamates and saccharin. Much of the current research in this area is directed at finding natural proteinlike sweeteners of even greater sweetness and increased heat stability, since the major drawback of all the current protein sweeteners is that heat destroys their sweetening ability. Consequently, sugar substitutes of this type are of limited value in any process which involves heating as part of the packaging procedure. Nevertheless, these types of compounds show great promise of providing a safe, low-calorie sugar substitute for dieters.

Other esters are commonly used as therapeutic agents in medicine. Ethyl acetate is employed as a stimulant and antispasmodic in colic and bronchial irritations. It is also applied externally in the treatment of skin diseases caused by parasites. Ethyl nitrite, when mixed with alcohol, is called elixir of niter and is used as a diuretic and antispasmodic. Amyl nitrite is used to lower blood pressure temporarily and causes relaxation of muscular spasms in asthma and in the heart condition known as angina pectoris. Glyceryl trinitrate, or nitroglycerin, is a vasodilator that has physiological action similar to amyl nitrate.

Methyl benzoate and methyl salicylate are two of the aromatic esters used in the manufacture of perfumes and flavoring agents. Methyl benzoate has the odor of new-mown hay, whereas methyl salicylate smells and tastes like wintergreen. One of the most important esters of salicylic acid is that formed with acetic acid. **Acetylsalicylic acid,** or aspirin, is the common antipyretic and analgesic drug. An increasing number of medications for the relief of simple headaches and the pain of rheumatism and arthritis contain aspirin combined with other pharmaceutical agents. For example, aspirin combined with antacid and buffering agents is absorbed more rapidly from the intestinal tract and should therefore afford more rapid relief of aches and pains.

Polyesters make up some of the most important commercial polymers. One of the

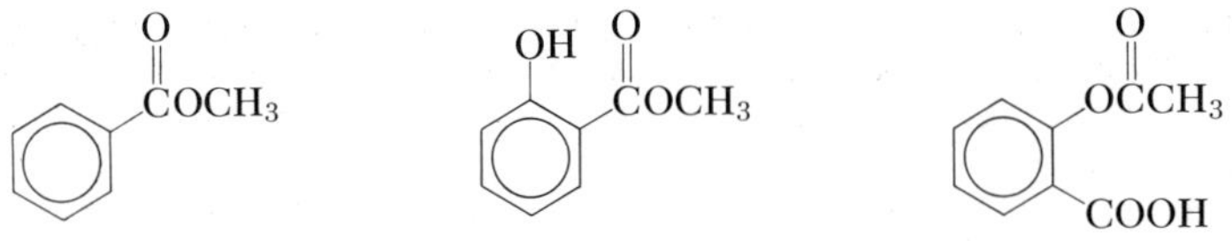

Methyl benzoate   Methyl salicylate   Acetylsalicylic acid

most interesting and important types of vinyl plastics is the acrylic resins. They are polymers of acrylic acid esters, such as methyl acrylate and methyl methacrylate.

$$H_2C=CHCOOH \qquad H_2C=CHCO_2CH_3 \qquad H_2C=C(CH_3)CO_2CH_3$$

Acrylic acid   Methyl acrylate   Methyl methacrylate

Polymerization of methyl methacrylate results in a clear, transparent plastic that can be used in place of glass. Lucite, Plexiglas, and Perspex are some of the trade names given this polymer. It can be formed into strong, flexible sheets that are highly transparent and lighter than glass, or it can be molded into transparent articles.

Another important polyester, Dacron, is prepared from terephthalic acid and ethylene glycol. The initial condensation produces an ester linkage, but the initial condensation product still has two functional groups left which can react further. This type of condensation continues until thousands of the ester units are contained in the chain. Polymers of this type will be considered in more detail in Chapter 19.

Phthalic anhydride is another very important organic compound. Large quantities of this material are used as plasticizers of synthetic resins and in the manufacture of the glyptal type of weather-resistant protective coating.

The most important amide is **urea,** which is a naturally occurring diamide of carbonic acid. The synthesis of this compound by Wöhler in 1828 was the keystone in the development of organic chemistry. Urea is prepared by the action of heat and pressure on carbon dioxide and ammonia.

$$CO_2 + 2NH_3 \xrightarrow[\text{pressure}]{\Delta} H_2NCNH_2 + H_2O$$

Urea

On heating, urea decomposes to ammonia and isocyanic acid. Alcohols react with isocyanic acid to form carbamates.

$$H_2NCNH_2 \xrightarrow{\Delta} NH_3 + HN=C=O \xrightarrow{CH_3CH_2OH} H_2NCOCH_2CH_3$$

Urea   Ammonia   Isocyanic   Ethyl carbamate
                                     acid

One of the most important commercial uses of urea is the base catalyzed condensation of urea with substituted diethyl malonates to produce the sedatives known as the **barbiturates.**

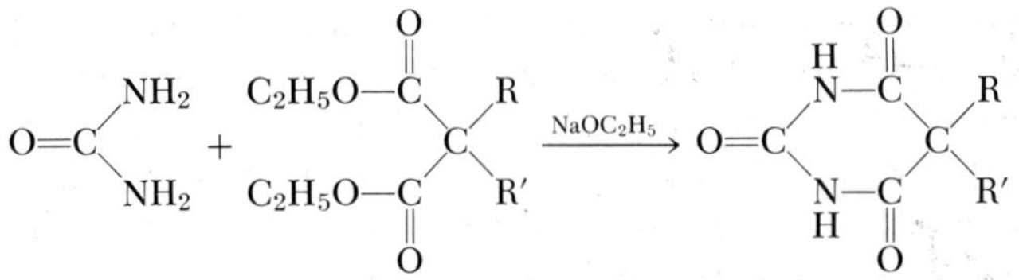

Urea   Substituted diethyl   A barbiturate
                     malonates

An important polyamide, nylon, is prepared by the condensation of adipic acid and hexamethylenediamine (see Chapter 19).

IMPORTANT TERMS AND CONCEPTS

acetic acid	amino acid	citric acid	oxalic acid
acid anhydride	aspirin	ester	peptide
acid halide (acyl halide)	benzoic acid	lactic acid	urea
amide	carboxyl group	nitrile	

QUESTIONS

1. Draw a structural formula for each of the following compounds:

(a) potassium propionate (e) butyramide
(b) isobutyryl chloride (f) sodium oxalate
(c) ethyl benzoate (g) *m*-bromobenzoic acid
(d) diethyl phthalate (h) methyl formate

2. Name each of the following compounds:

(a) $CH_3(CH_2)_3CH(CH_3)CO_2H$ (e) $CH_3CH_2CH_2CONH_2$
(b) CCl_3CO_2H (f) CH_3COBr
(c) $CH_3CO_2^-K^+$ (g) $(CF_3CO)_2O$
(d) $CH_3CH_2CH_2CO_2CH_2CH_3$

3. Write equations for the reactions of *n*-butyric acid with the following reagents:

(a) CH_3CH_2OH/H^+ (c) NH_3/heat
(b) $SOCl_2$ (d) KOH

4. Write equations for the reactions of benzoic acid with the following reagents:

(a) $Ca(OH)_2$ (c) HNO_3/H_2SO_4
(b) Br_2/Fe (d) ethyl alcohol/H^+

SUGGESTED READING

Carter: The History of Barbituric Acid. Journal of Chemical Education, Vol. 28, p. 524, 1951.
On Sweetness and Sweeteners. Chemistry, Vol. 44, p. 21, 1971.
Sweet and Low Calorie. Chemistry, Vol. 47, No. 8, p. 24, 1974.

AMINES AND AMINE DERIVATIVES

CHAPTER 18

The *objectives* of this chapter are to enable the student to:

1. Distinguish primary, secondary, and tertiary amines.
2. Write the structures and names of the common alkyl and aromatic amines.
3. Describe the preparation of amines.
4. Recognize the more common reactions of amines.
5. Define a diazonium salt and explain its synthetic value in organic chemistry.
6. Recognize the important amines, amine derivatives, and their current uses.

    Amines are organic derivatives of ammonia in which one or more of the hydrogen atoms have been replaced by an alkyl or aryl group. The characteristic functional group present in amines is called the **amino group** and is written as —NH_2. Like alcohols, amines are classified as primary, secondary and tertiary amines. In the case of amines the classification is determined by the *number* of alkyl or aryl groups attached to the nitrogen atom. Some representative examples are illustrated below:

$$CH_3\ddot{N}H_2 \qquad CH_3\overset{\overset{\displaystyle H}{|}}{N}CH_2CH_3 \qquad (CH_3)_3N:$$

Methyl amine   Methylethyl amine   Trimethyl amine

    Amines can be named as amino derivatives of hydrocarbons according to the IUPAC system—as amino alkanes for example. This system, however, is little used. The common name nomenclature system is most widely used. It consists of naming the alkyl or aryl groups attached to the nitrogen atom, using the appropriate prefixes if two or more identical substituents are attached to the nitrogen, followed by the word "amine." The above examples illustrate this system of nomenclature (cf. Table 18–1).

    Simple aromatic amines are named as derivatives of the parent aromatic amine, **aniline.** The substituted anilines are named in the same manner as other substituted benzene derivatives.

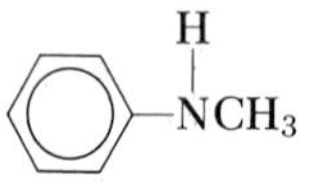

N-Methylaniline    *p*-Nitroaniline 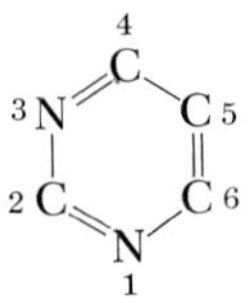   Pyrimidine nucleus

TABLE 18–1   PHYSICAL PROPERTIES OF AMINES

COMPOUND	STRUCTURE	M.P. °C	B.P. °C
Methylamine	CH_3NH_2	−93	−6
Ethylamine	$CH_3CH_2NH_2$	−81	+17
n-Propylamine	$CH_3CH_2CH_2NH_2$	−83	+48
Dimethylamine	$(CH_3)_2NH$	−92	+7
Diethylamine	$(CH_3CH_2)_2NH$	−50	+56
Trimethylamine	$(CH_3)_3N$	−117	+3
Triethylamine	$(CH_3CH_2)_3N$	−115	+90
Aniline	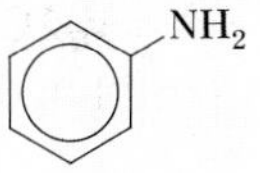	−6	+184
o-Toluidine	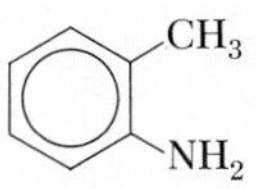	−15	+200
m-Toluidine	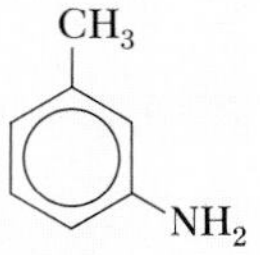	−30	+203
p-Toluidine	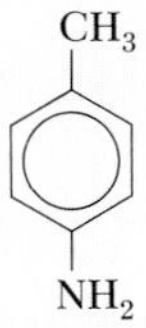	+44	+201
Nitrobenzene	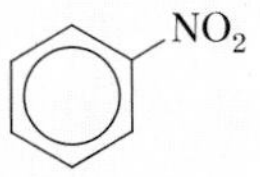	+6	+211

The nitrogen atom may also be part of a ring system, and both cyclic and aromatic types of amines are known (cf. p. 200).

PHYSICAL PROPERTIES OF AMINES

The lower molecular weight amines are soluble in water, and the solutions of the water-soluble amines are alkaline. The volatile amines have unpleasant odors that combine the odor of ammonia with that of decayed fish. The higher molecular weight ($>C_6$) amines are insoluble in water and soluble in organic solvents, and the unpleasant odor decreases with decreasing volatility. The properties of some typical amines are listed in Table 18–1.

Amines occur widely in living cells. For example, nucleic acids contain various pyrimidine and purine derivatives (Chapter 22). Also niacin and some coenzymes contain heterocyclic amines (pp. 327 and 328). The amino group of amino acids is also of importance in peptide and protein formation as discussed in the last chapter under amides, and is also important in protein metabolism (see p. 372). These important biochemical applications of amines will be discussed more thoroughly in the next section of this text.

Amines are basic like ammonia. Their base strength depends on the type of group (alkyl or aryl) attached to the nitrogen atom. Alkyl groups attached to the nitrogen atom increase the basicity of the amine relative to ammonia, and the alkyl amines are slightly stronger bases than ammonia. On the other hand, aromatic groups attached to the nitrogen atom decrease the basicity of the amine relative to ammonia. Aniline is a much weaker base than ammonia; diphenylamine is only weakly basic; and triphenylamine is essentially a neutral compound.

SALT FORMATION

Ammonia reacts with acids to form salts. The unshared pair of electrons on the nitrogen atom is used to form a new N—H bond, and the ammonia molecule is converted into the ammonium ion. The organic derivatives of ammonia, the amines, behave similarly. Aqueous mineral acids and carboxylic acids convert amines into salts. The amine salts are typical ionic saltlike compounds.

$$R\text{—}\ddot{N}H_2 \qquad R\overset{+}{N}H_3 \qquad R\ddot{N}H_2$$

$$R_2\ddot{N}H \xrightarrow{\text{HX}} R_2\overset{+}{N}H_2 \text{ X}^- \xrightarrow{\text{OH}^-} R_2\ddot{N}H + H_2O + X^-$$

$$R_3\ddot{N} \qquad R_3\overset{+}{N}H \qquad R_3\ddot{N}$$

They are nonvolatile solids, generally soluble in water and insoluble in nonpolar organic solvents.

PREPARATION OF AMINES

REDUCTION METHODS

The reduction of compounds containing the nitro ($-NO_2$) group affords a convenient path to primary amines, particularly the aromatic amines. The aliphatic nitro compounds are more difficult to prepare than the aromatic nitro compounds, and this method is not generally used for preparing aliphatic amines. The aromatic nitro compounds, however, can be easily prepared by nitration of the appropriate aromatic hydrocarbon. Reduction can be carried out catalytically (H_2 + Pt), but the more normal reducing agents used are a metal and a mineral acid, as shown below:

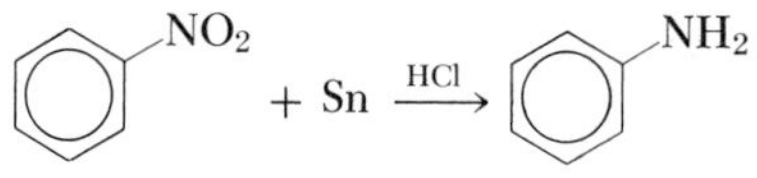

Nitrobenzene   Aniline

ALKYLATION OF ALKYL AMINES

The use of ammonia or amines in an anionic type displacement reaction with alkyl halides (see reactions of alkyl halides) provides a classical route to the aliphatic amines, as shown below.

$$R\text{—}X + \ddot{N}H_3 \rightarrow [R\text{—}\overset{+}{N}H_3]X^- \xrightarrow{\text{NaOH}} R\ddot{N}H_2 + Na^+X^- + H_2O$$
$$X = Cl, Br, I$$

Unfortunately, the alkylated product (RNH_2) also can react with the alkyl halide to produce some secondary amine, which in turn can give tertiary amine, which in turn can give a quaternary ammonium salt.

In practice, a mixture of all four products results. However, if a large excess of ammonia is used, the amount of the primary amine, RNH_2, formed is maximized.

REACTIONS OF AMINES

SALT FORMATION

As noted earlier in this chapter, amines, both aliphatic and aromatic, form salts with mineral acids. The free amines can be regenerated by treatment of the salt with a strong base.

$$(CH_3)_2\overset{..}{N}H + HBr \rightarrow (CH_3)_2\overset{+}{N}H_2Br^-$$

Dimethylamine hydrobromide

ALKYLATION

As noted in the preparation of amines, alkylation of amines occurs readily. Since the alkylation is difficult to control, this reaction is most useful in preparing quaternary ammonium salts.

$$(CH_3)_3\overset{..}{N} + CH_3I \rightarrow (CH_3)_4N^+I^-$$

ACYLATION

Carboxylic acid amides ($RCONH_2$) are prepared via the reaction of ammonia with an acid halide. Similar types of acylation reactions occur when primary and secondary amines react with acid halides. Since tertiary amines have no available N—H bond, they cannot be acylated. This method is superior as a laboratory preparation of amides compared to the pyrolysis method described earlier (see p. 240).

$$CH_3COCl + CH_3\overset{..}{N}H_2 \rightarrow CH_3-\overset{\overset{O}{\|}}{C}-\overset{H}{N}CH_3 + H^+Cl^-$$

Acetyl chloride   Methyl amine   N-Methyl acetamide

In addition to being acylated by carboxylic acid derivatives, ammonia and primary and secondary amines also react with sulfonic acid halides to form sulfonamides, as illustrated below for benzene sulfonyl chloride (the acid chloride of benzene sulfonic acid):

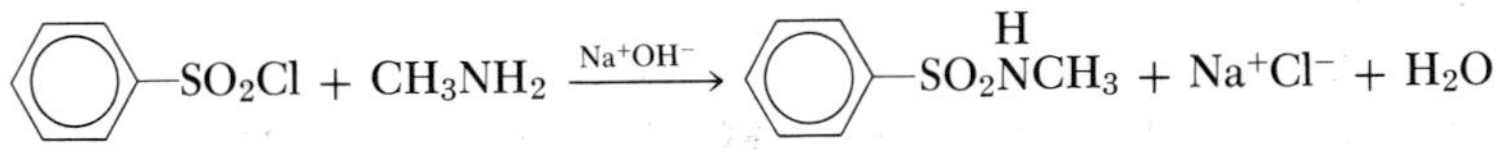

$$\langle\bigcirc\rangle\text{-}SO_2Cl + CH_3NH_2 \xrightarrow{Na^+OH^-} \langle\bigcirc\rangle\text{-}SO_2\overset{H}{N}CH_3 + Na^+Cl^- + H_2O$$

Benzene sulfonyl   Methyl amine   N-Methyl benzenesulfonamide
chloride

Sulfanilamide (p. 393) is the parent compound of the important class of chemotherapeutic agents known as the sulfa drugs. The majority of the drugs are prepared by reaction of a sulfanilamide derivative, as shown below:

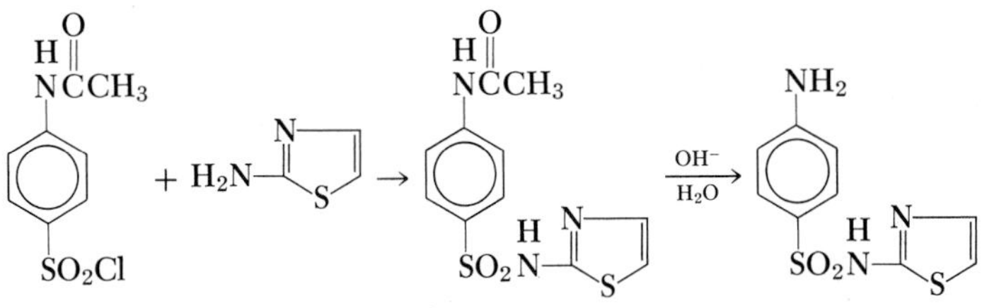

2-Aminothiazole   Sulfathiazole

The structure of the sulfa drug can be varied by varying the structure of the amine being condensed with the sulfonyl chloride (see p. 393).

AROMATIC SUBSTITUTION REACTIONS

The amino group ($-NH_2$) and amino derivatives ($-\overset{\overset{HO}{|}\overset{H}{||}}{N}CR$, $-\overset{\overset{H}{|}}{N}R$, and $-NR_2$) are *ortho-para* directing groups. In practice, the acylated derivative is usually employed, as this group, ($-\overset{\overset{HO}{|}\overset{||}{}}{N}CR$), is less of an activator than the amine group itself, and the substitution reaction is more easily controlled. The free amine can be regenerated by hydrolysis of the acylated compound. This type of reaction is illustrated below for the bromination of aniline:

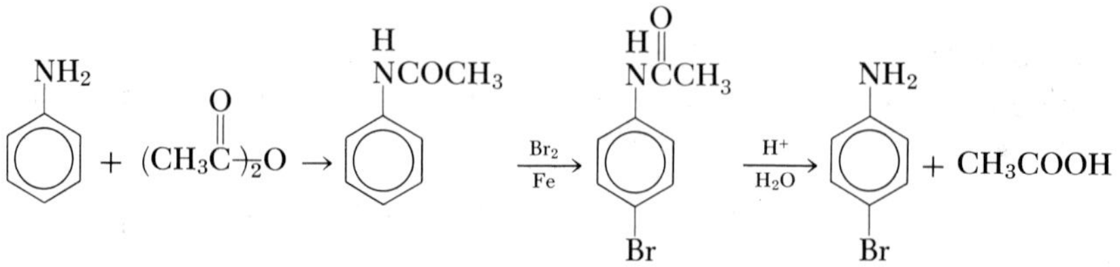

Aniline   Acetic anhydride   Acetanilide   Main product   *p*-Bromoaniline

Nitration, sulfonation, and Friedel-Crafts reactions proceed similarly.

REACTION WITH NITROUS ACID

Primary amines react with nitrous acid (HONO) to give a **diazonium** salt, and this type of reaction is known as **diazotization.** Since nitrous acid is not a stable acid, it is generated *in situ* from sodium nitrite and hydrochloric acid. Diazonium salts are generally not isolated, as they decompose easily and are explosive in the dry state. Aliphatic diazonium salts (R = alkyl) are usually not stable even at 0°C and are not as useful as the aromatic diazonium salts in synthesis. The aromatic diazonium compounds have a reasonable stability if kept at low temperatures (0° to 5°C), and undergo a variety of displacement and coupling reactions.

The main type of displacement reaction of aromatic diazonium salts is one which involves displacement of molecular nitrogen by an anion. Anions, such as CN^-, I^-, Br^-, Cl^-, H_2O, and ROH, are effective in this type of reaction. Displacement by these various types of anions is illustrated below for benzenediazonium chloride:

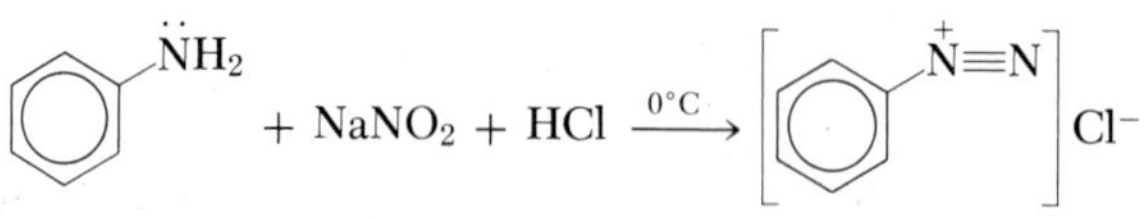

Aniline   Benzenediazonium chloride

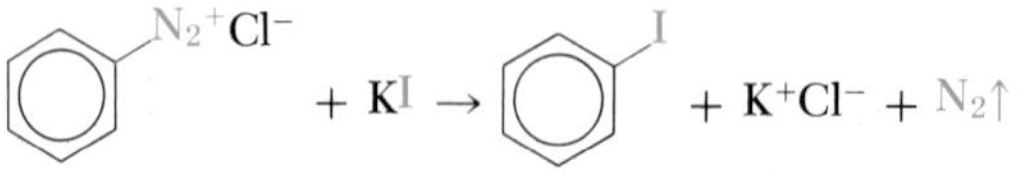

Iodobenzene

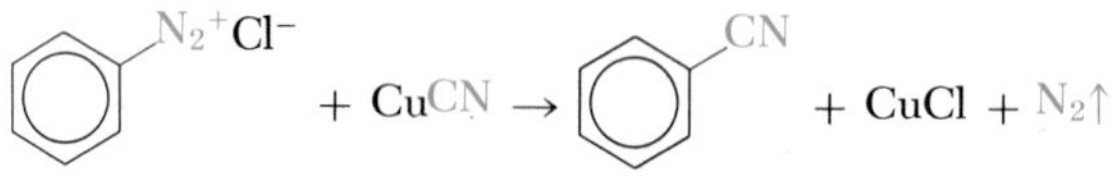

Benzonitrile

This type of displacement reaction is particularly useful for introducing cyano and iodo groups into an aromatic ring.

Another important reaction of diazonium salts is their **coupling** reaction in alkaline or neutral solution with reactive aromatic compounds such as phenols or aromatic amines. The coupling takes place between the diazonium salt and the *para* position of the aromatic compound, and the nitrogen is retained. The (—N=N—) linkage is known as the **azo** group, and azo compounds of this type are generally colored. Related compounds containing the azo group make up the important group of dyestuffs known as azo dyes.

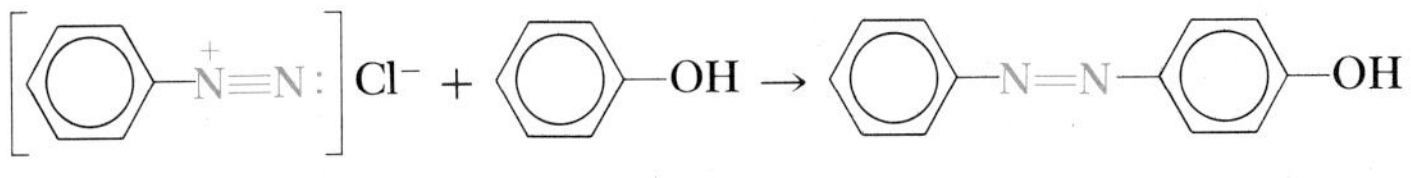

Benzenediazonium chloride   Phenol   *p*-Hydroxyazobenzene

IMPORTANT AMINES, AMINE DERIVATIVES AND THEIR USES

DIMETHYLAMINE

When allowed to react with nitrous acid, followed by reduction of the product by hydrogen, dimethyl hydrazine, $(CH_3)_2NNH_2$, is formed. Hydrazines of this type have been used as rocket propellants.

Di- and trimethyl amines are essential in the preparation of quaternary ammonium types of anion exchange resins. In general, amines are utilized in the preparation of dyes, drugs, herbicides, fungicides, soaps, disinfectants, insecticides, and photographic developers.

Among the aromatic amines, aniline is used to synthesize other important organic compounds used as dyes and dye intermediates, antioxidants, and drugs. Other derivatives of aromatic amines, such as *p*-aminosulfonic acid and *p*-toluidine, are used as dye intermediates.

Several derivatives of the aromatic amines have medicinal properties and are used as drugs. **Acetanilide** was used for many years as an antipyretic and analgesic drug. The toxicity of this compound in the body resulted in a search for similar compounds that were less toxic. **Phenacetin,** or **acetophenetidin,** which is closely related to acetanilide, was found to be less toxic and to possess the beneficial effects of acetanilide. Sulfanilamide and other sulfa drugs have been noted earlier for their therapeutic effects.

Another type of important organic compound that contains amino groups is **amino acids.** Amino acids may be considered as organic acids containing an amino group. Two simple amino acids are illustrated below.

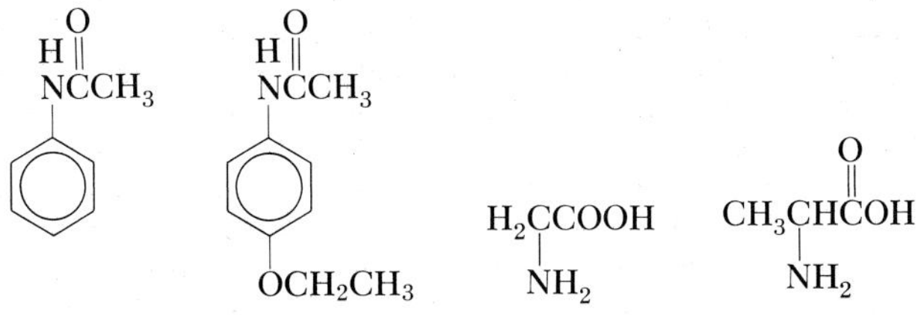

Acetanilide   Phenacetin   Glycine   Alanine

Amino acids are the fundamental units in the protein molecule and will be discussed later in detail under the chemistry of proteins.

FOOD ADDITIVES

A food additive can be defined as a substance or mixture of substances added intentionally by food manufacturers to prevent or retard spoilage, enhance flavor or nutritional quality, or improve the color or texture of the product. A careful reading of the labels of food products purchased at the supermarket will quickly make one aware of the multitude of additives consumed daily by any individual. It has been estimated that over 1500 different types of additives are available. In recent years the American public has questioned the utility, safety, and value of many of these chemicals which are added to our food, and the Food and Drug Administration (FDA) has established a list of chemicals which it feels to be Generally Recognized As Safe (GRAS list). But even most of these materials have not been thoroughly tested, and at the present time it is difficult to assess fully the dangers of any of these food additives.

Food additives are conveniently divided into classes—flavors, preservatives, colors, emulsifiers, and antioxidants—on the basis of the functions that they perform. The following discussion outlines some of the more common or controversial additives, although these illustrations are only an abbreviated list of the total number of additives used.

Flavors and flavor enhancers are the largest (1000) class of food additives. Many of these flavors are obtained from plants such as vanilla and peppermint and have been used for centuries. Yet little is known about the toxicological aspect of flavors. Today, the food chemist has prepared a variety of synthetic (imitation) flavor mixtures, such as imitation cherry, yet the FDA has no rigid standards for control or testing of these materials. Related to flavors are the flavor enhancers, of which monosodium glutamate (MSG) is the best known. Since MSG is a derivative of the amino acid glutamic acid, it was considered completely safe until recently when it was found to cause ailments such as headaches, chest pains, and shortness of breath in certain individuals. Since MSG is used in large amounts in Chinese food, and since some severe cases of these ailments have occurred in Chinese restaurants, this condition has become known as the "Chinese restaurant syndrome."

Preservatives are added to food to retard microbe growth and thus deter spoilage. For example, sodium propionate and calcium propionate are two common materials employed as preservatives in bread. They retard the growth of mold spores and prevent the bread from rapidly becoming stale. Similarly, benzoic acid and sodium benzoate are widely used in juices, margarine, and soft drinks as preservatives. A recent source of extensive controversy is the additive sodium nitrite, $NaNO_2$, which is added to cured meats such as hot dogs and ham to prevent the growth of the organism which causes botulism and to impart a nice red color to the meat. It has been suggested that the nitrite combines with certain secondary amines in the digestive tract to form nitrosamines, which have been proved to be cancer forming in animals. In fact, rats fed a diet containing nitrite and secondary amines developed tumors. Consequently, sodium nitrite is open to question as a safe additive, and since over 50 tons of this material are employed annually in cured meats, meat producers and consumer advocates are involved in a heated controversy over its continued use.

Color is an important factor in making food appetizing. For example, margarine is more acceptable when butter-yellow than if sold in its natural colorless state. Most of the colors used in the food industry are synthetic dyestuffs. Many similarly related synthetic dyes have been shown to be carcinogenic, yet no extensive testing of coloring agents has been completed by the FDA.

Emulsifiers are used to prevent or delay the separation of oil and water mixtures or to thicken or bind certain food products. For example, in the popular prepared salad dressings sold commercially, an emulsifier is added to help the oil globules remain dispersed in the mixture, and a uniform salad dressing is obtained. Similarly, in prewhipped toppings emulsifiers are used to strengthen the walls of the air bubbles in the topping and make the foam last longer.

Antioxidants are added to fatty foods to prevent rancidity. Some natural antioxidants, such as vitamin E, occur in certain plant oils such as olive oil. The most widely used synthetic antioxidants are butylated hydroxyanisole (BHA) and butylated hydroxytoluene (BHT), used in products such as cooking oils, potato chips, salted nuts, and margarine. Ascorbic acid (vitamin C) is another widely used antioxidant. It is used in preventing discoloration (darkening) in fruit juices, soft drinks, and frozen fruits.

At present, it is difficult to assess the potential hazards to man of these additives. They have become part of our everyday lives and in many cases have reduced the danger from food-carried microbial diseases and the waste caused by spoilage. Thus, they are not all bad. Concerned consumers will have to decide for themselves whether or not to limit their additive intake by reading product labels, limiting their use of convenience foods (which contain the highest percentage of additives), or not purchasing products that contain additives.

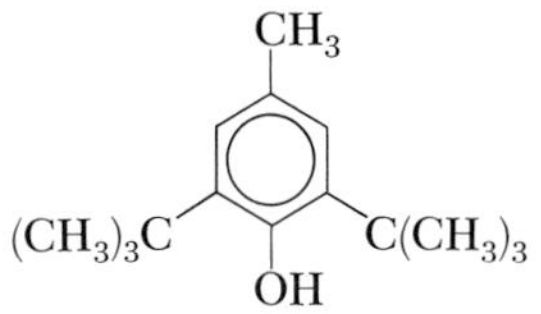

BHT

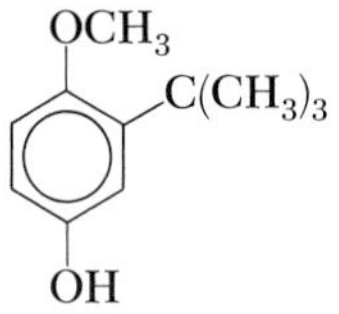

BHA

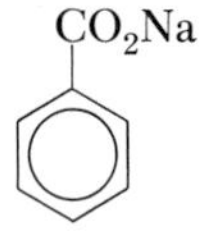

Sodium benzoate

$H_2N-CHCO_2H$
|
$CH_2CH_2CO_2Na$

MSG

$(CH_3CH_2CO_2)_2Ca$

Calcium propionate

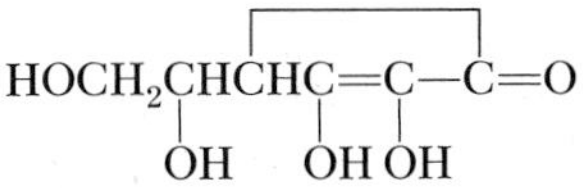

$HOCH_2CHCHC=C-C=O$
| | |
OH OH OH

Ascorbic acid

IMPORTANT TERMS AND CONCEPTS

amino group	coupling reaction	phenacetin	secondary amine
aniline	diazonium salt	primary amine	sulfanilamide
azo group	diazotization	quaternary ammonium salt	tertiary amine

QUESTIONS

1. Draw a structural formula for each of the following compounds:

   (a) diethylamine
   (b) tri-*n*-propyl amine
   (c) aniline
   (d) *m*-nitroaniline

   (e) isohexyl amine
   (f) *o*-toluidine
   (g) cyclobutyl amine
   (h) *p*-bromo-N,N-dimethylaniline

2. Give each of the following compounds an appropriate name:

(a)

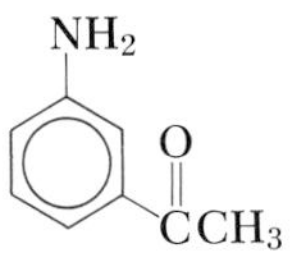

(c)

(e)

(b) $CH_3CH_2NCH_2CH_2CH_3$ (with H above N)

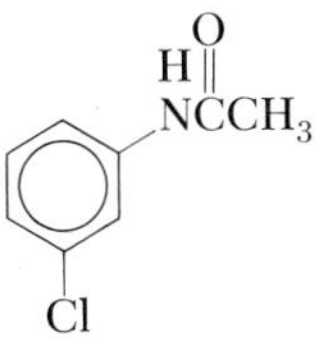

(d)

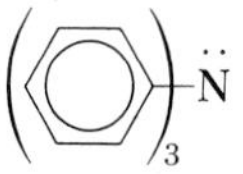

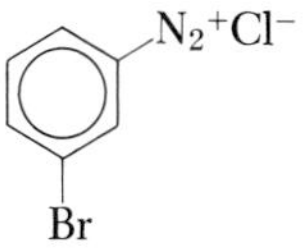

3. The fishy odor of a solution of dimethylamine in water is lost when an equimolar amount of hydrochloric acid is added. *Explain.*

4. Write equations for the reactions of *n*-butyl amine with each of the following reagents:

    (a) acetyl chloride   (c) ethylene oxide
    (b) HI

5. Write equations for the reactions of *p*-bromoaniline with the following reagents:

    (a) CH_3COCl (c) $NaNO_2$/HCl followed by N,N-dimethylaniline
    (b) $NaNO_2$/HCl   (d) benzene sulfonyl chloride/OH^-

SUGGESTED READING

Amundsen: Sulfanilamide and Related Chemotherapeutic Agents. Journal of Chemical Education, Vol. 19, p. 167, 1942.

Barron, Jarvik, and Sterling: Hallucinogenic Drugs. Scientific American, Vol. 210, No. 4, p. 29, 1964.

Kermode: Food Additives. Scientific American, Vol. 226, No. 3, p. 15, 1972.

Majtemyi: Food Additives—Food for Thought. Chemistry, Vol. 47, No. 5, p. 6, 1974.

Plummer and Yorkman: Antihypertensive and Diuretic Agents. Past, Present, and Future. Journal of Chemical Education, Vol. 37, p. 179, 1960.

Ray: Alkaloids. The World's Pain Killers. Journal of Chemical Education, Vol. 37, p. 451, 1960.

Sodium Nitrite-in-Meat Controversy. Research Reporter, Chemistry, Vol. 45, No. 11, p. 23, 1972.

POLYMERS AND POLYMERIZATION REACTIONS

CHAPTER 19

The *objectives* of this chapter are to enable the student to:

1. Define a polymer or macromolecule.
2. Define and recognize common monomer units.
3. Distinguish the difference between condensation and addition polymers.
4. Deduce that polymerization reactions represent the same reactions studied from a monofunctional viewpoint, but that the degree of reaction is different.
5. Recognize the common terms polyester, nylon, Bakelite, polypropylene, polystyrene, polyethylene, Teflon, and polyacrylate and to associate these terms with a repeating structural unit.
6. Recognize that a polymer may have different configurations and that these configurations influence the properties of the polymer.

A polymer (or macromolecule) is a large molecule of high molecular weight (from a few thousand to several million) that is composed of small components called monomers. Many macromolecules occur in nature, such as proteins, nucleic acids, polysaccharides, starch, and cellulose, and these polymers will be discussed in more detail in the biochemistry section of the text. The main concern of this chapter will be man-made (synthetic) polymers that have achieved widespread commercial use.

The monomers used in the preparation of polymers are usually fairly simple monofunctional or bifunctional compounds that have previously been encountered in this portion of the text. In most cases the reactions which yield polymers can be readily understood from the material which has been studied by students thus far in organic chemistry. The main difference between the organic reactions which have previously been studied and polymer-forming reactions is in the degree of reaction. Most of the reactions studied thus far have involved a reaction between two reactants to yield a relatively simple product. Polymerization reactions, however, are reactions which are functionally capable of proceeding indefinitely and which could in theory give a compound of infinite molecular weight. Consequently, in a polymerization reaction the initially formed intermediate or product is capable of further reaction and simple 1:1 products do not result. For example, in the free-radical polymerization of ethylene, shown on p. 256, the intermediate radical can add to a second molecule of ethylene to form a new radical, which in turn can add to a third molecule to form another radical, and this process can proceed until a reaction (called a *termination reaction*) occurs to interrupt this chain process.

$$R \cdot + CH_2{=}CH_2 \rightarrow RCH_2CH_2 \cdot \xrightarrow{CH_2=CH_2} RCH_2CH_2CH_2CH_2 \cdot$$

$$\text{Radical} \quad \text{Ethylene} \qquad\qquad\qquad\qquad \downarrow nCH_2{=}CH_2$$

$$R(CH_2CH_2)_{n+2}R \xleftarrow[\text{step}]{\underset{\text{termination}}{R\cdot}} R[CH_2CH_2]_{n+1}CH_2CH_2 \cdot$$

$$\text{Polyethylene}$$

Note that the ethylene unit is repeated numerous times and the molecular weight of the polymer depends upon the value of n. In most polymerizations, reaction conditions are adjusted to make n large so that high-molecular weight materials are obtained. Polymers which contain the same repeating unit, such as polyethylene, are called **homopolymers.** If two different monomers are used to form the polymer, a **copolymer** is obtained. For example, butadiene and styrene can be copolymerized to form a synthetic rubber used in automobile tires. Molecular weights of 25,000 to 500,000 can be attained by this type of copolymerization. If the two monomers are fed into the polymerization

$$CH_2{=}CH{-}CH{=}CH_2 + C_6H_5CH{=}CH_2 \xrightarrow{R\cdot} R{+}CH_2{-}\underset{\underset{H}{|}}{\overset{\overset{H}{|}}{C}}\overset{\overset{CH_2CHCH_2}{}}{\underset{\underset{C_6H_5}{|}}{}}{\Big]}_n R$$

$$\text{1,3-Butadiene} \qquad\qquad \text{Styrene} \qquad\qquad\qquad\qquad\qquad\qquad\qquad \text{SBR rubber}$$

reaction without any special precautions, a **random copolymer** is obtained in which no definite sequential pattern of monomer units is found. Special reactions, however, can be used to put large blocks of units of the same monomer back to back in the repeating chain. Copolymers of this type are called **block polymers** and have different physical properties from random copolymers.

Polymerization reactions can be conveniently divided into two types, namely, **condensation polymers** and **addition polymers.** In a condensation polymerization reaction two molecules are joined (condensed) together and a small molecule, such as water or an alcohol, is removed or eliminated in this reaction. In order for a condensation polymerization to form high molecular weight materials, the condensation reaction must occur over and over again. Consequently, the monomers employed in a condensation polymerization to form high molecular weight materials, the condensation reaction must occur responsible for building the polymer chain. Most of the chemical reactions involved in a condensation polymerization are well known reactions, such as esterification and amide formation, which have been encountered earlier in this treatment of organic chemistry.

An addition polymer has the same ratio of elements in the polymer repeating unit as was present in the monomer, since the monomeric units are merely added together to form the polymer, as shown earlier in the formation of polyethylene. Until recently, most addition polymers were generally formed from monomers which contained carbon-carbon double bonds. However, addition polymerizations involving other types of unsaturation, such as carbon-oxygen double bonds or carbon-nitrogen double bonds, are currently being investigated by polymer chemists and show promise of yielding useful polymers.

CONDENSATION POLYMERS

The type of bifunctional compounds which are condensed together to form a condensation polymer will determine the functional group linkage in the repeating unit of the polymer. Thus, if an acid and an alcohol are employed, a polyester will result. If an acid and an amine are condensed, a polyamide results, as shown below. Note that a molecule of water is eliminated in the condensation reaction.

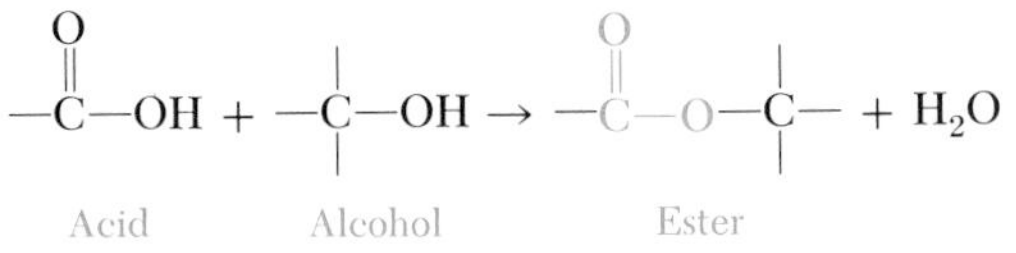

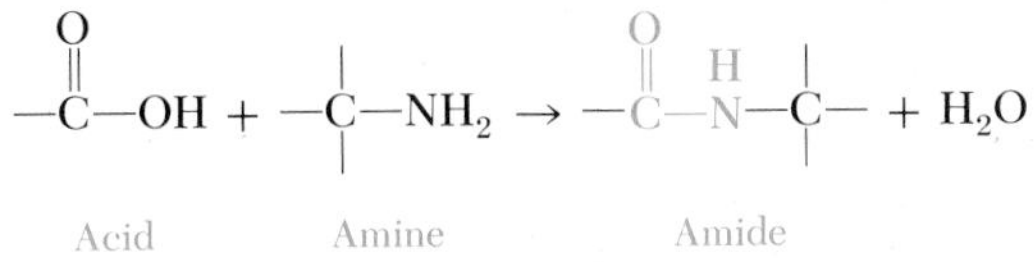

$$\underset{\text{Acid}}{-\overset{\overset{\displaystyle O}{\|}}{C}-OH} \; + \; \underset{\text{Alcohol}}{-\overset{\displaystyle |}{\underset{\displaystyle |}{C}}-OH} \; \rightarrow \; \underset{\text{Ester}}{-\overset{\overset{\displaystyle O}{\|}}{C}-O-\overset{\displaystyle |}{\underset{\displaystyle |}{C}}-} \; + \; H_2O$$

$$\underset{\text{Acid}}{-\overset{\overset{\displaystyle O}{\|}}{C}-OH} \; + \; \underset{\text{Amine}}{-\overset{\displaystyle |}{\underset{\displaystyle |}{C}}-NH_2} \; \rightarrow \; \underset{\text{Amide}}{-\overset{\overset{\displaystyle O}{\|}}{C}-\overset{\displaystyle H}{N}-\overset{\displaystyle |}{\underset{\displaystyle |}{C}}-} \; + \; H_2O$$

Probably the most familiar example of a polyester is the fiber **Dacron,** which is a polyester formed from terphthalic acid and ethylene glycol. Industrially, direct esterification between an acid and a glycol has been found to proceed poorly because of the difficulty in the removal of the water formed in the esterification reaction. Consequently, the transesterification reaction between methyl phthalate and ethylene glycol is employed, since the methyl alcohol formed is more easily removed. The polymerization is carried

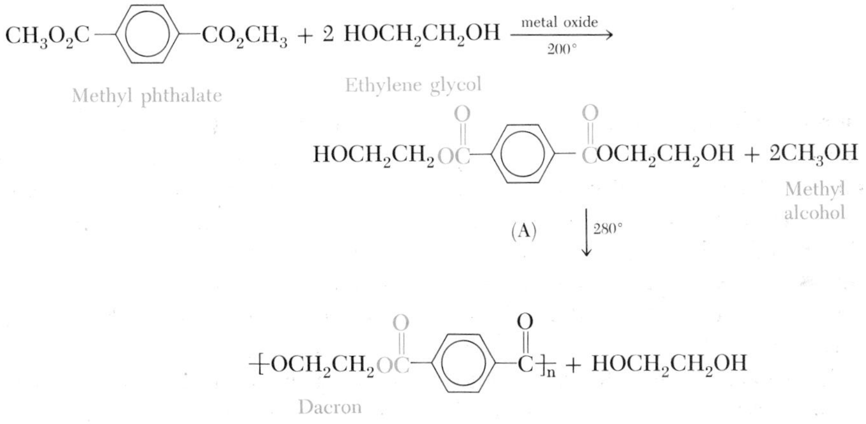

out in several steps. First, ester interchange is carried out at 200°C to give methyl alcohol (which is removed by distillation) and a new monomer (A). After removal of the methanol is completed, the temperature is raised to 280°C, and polymerization occurs to give Dacron and ethylene glycol which is also removed by distillation. Dacron fiber, also called Terylene or Teron, can be set in permanent creases and finds widespread commercial use in combination with wool in wash-and-wear fabrics. When cast into a film (Mylar, Cronar), a product of high tensile strength is obtained. Mylar is currently used extensively in making magnetic recording tapes.

The best known type of polyamides are the nylons, which were developed as a substitute for silk during World War II. The nylons are generally prepared by the copolymerization of a dibasic acid and a diamine. The different types of nylons are distinguished by the use of numbers. These numbers refer to the number of carbon atoms in the diamine and in the dibasic acid. For example, **Nylon 66** is prepared by a two-stage process from hexamethylenediamine (six carbons) and adipic acid (six carbons). When stoichiometric amounts of these two monomers are allowed to react, a 1:1 salt is formed. Further treatment of this salt at 270°C and 250 psi yields Nylon 66.

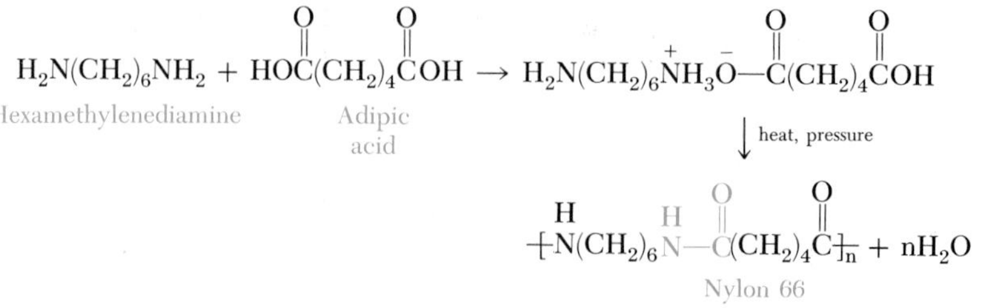

$$H_2N(CH_2)_6NH_2 + HOC(CH_2)_4COH \rightarrow H_2N(CH_2)_6\overset{+}{N}H_3\overset{-}{O}-\overset{O}{\overset{\|}{C}}(CH_2)_4COH$$

Hexamethylenediamine   Adipic
                          acid

$$\downarrow \text{ heat, pressure}$$

$$\left[\overset{H}{N}(CH_2)_6\overset{H}{N}-\overset{O}{\overset{\|}{C}}(CH_2)_4\overset{O}{\overset{\|}{C}}\right]_n + nH_2O$$

Nylon 66

This particular nylon has high strength because of its ability to form intermolecular hydrogen bonds. Consequently, it shows good resistance to breakage by stretching or abrasion and finds widespread use in carpet fabrics and machine parts, such as gear assemblies, which require no lubrication.

Other nylons, such as Nylon 610 from hexamethylenediamine and sebacic acid and Nylon 6 from ϵ-caprolactam, are also commercially useful.

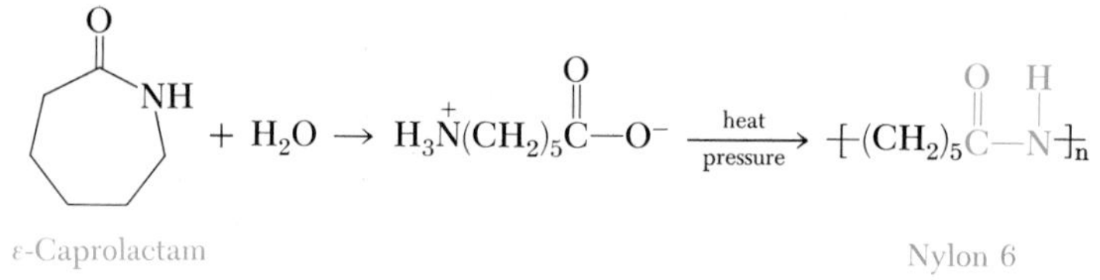

$$\text{(ring structure with NH and C=O)} + H_2O \rightarrow H_3\overset{+}{N}(CH_2)_5\overset{O}{\overset{\|}{C}}-O^- \xrightarrow[\text{pressure}]{\text{heat}} \left[(CH_2)_5\overset{O}{\overset{\|}{C}}-\overset{H}{N}\right]_n$$

ϵ-Caprolactam   Nylon 6

CROSS-LINKED POLYMERS

Simple polyesters and polyamides are long linear molecules without linkages interconnecting the individual strands of the polymer. If the individual strands can be connected so that a network of polymers results, this process is called *cross-linking* and the resultant polymers are called *cross-linked polymers*. For example, if a dibasic acid is copolymerized with a trihydroxylalcohol, such as glycerol, the third hydroxyl function is available for forming a connection between the individual strands of the polymers and cross-linking can occur. Polyester polymers with cross-links are called **alkyd resins.** For example, the alkyd resin **Glyptal** is commercially prepared from glycerol and phthalic anhydride. Generally, cross-linked polymers such as Glyptal are rigid, insoluble materials that do not flow nor soften when heated. When heated to their melting point, polymers of this type, called **thermosetting polymers,** undergo a permanent change and set to a

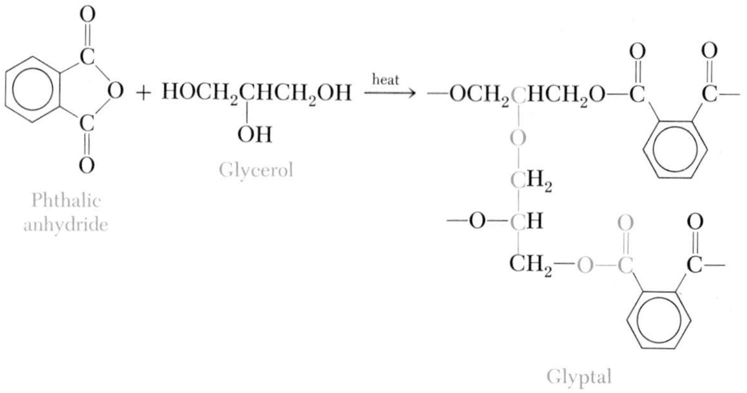

$$\text{(phthalic anhydride structure)} + HOCH_2CHCH_2OH \xrightarrow{\text{heat}} -OCH_2CHCH_2O-\overset{O}{\overset{\|}{C}}\ \overset{O}{\overset{\|}{C}}-$$

$$\underset{\text{OH}}{}$$

Phthalic   Glycerol
anhydride

$$\begin{array}{c} O \\ | \\ CH_2 \\ -O-CH \quad O \quad O \\ CH_2-O-\overset{O}{\overset{\|}{C}}\ \overset{O}{\overset{\|}{C}}- \end{array}$$

Glyptal

solid which cannot be remelted. In contrast to this behavior, **thermoplastic polymers** soften and flow when heated and can be remelted many times without change.

The resin **Bakelite** is a copolymer of phenol and formaldehyde. The resultant thermo-setting polymers find extensive use in heat and electrical resistant materials such as electric appliance handles and electrical switches.

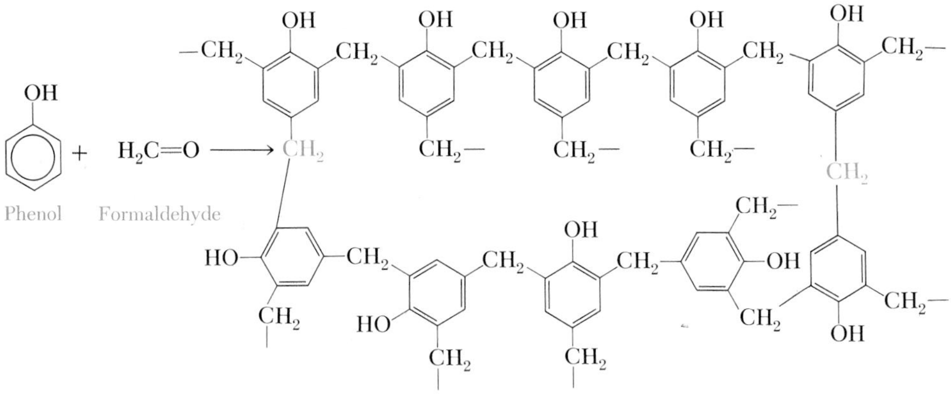

Bakelite

ADDITION POLYMERS

Addition polymers are formed by some sort of chain mechanism. These polymerizations may occur by anionic, cationic, or free-radical sequences, depending on the type of monomer employed.

Polypropylene is an addition polymer obtained by polymerization of propylene with a trialkylaluminum titanium tetrachloride (Ziegler-Natta) catalyst. This highly ordered polymer has a high melting point (165 to 170°C) and high tensile strength and is easily converted to molded objects and fibers.

$$CH_3CH=CH_2 \xrightarrow[TiCl_4]{R_3Al} \left[CH-CH_2\right]_n$$

Propylene   Polypropylene

Another important and well known addition polymer is polytetrafluoroethylene (**Teflon**). Since pure tetrafluoroethylene will polymerize explosively under free-radical conditions, special catalyst systems have been developed to control this polymerization. Teflon has excellent thermal stability over a wide temperature range, is chemically

$$nCF_2=CF_2 \xrightarrow{catalyst} \left[CF_2-CF_2\right]_n$$

Tetrafluoroethylene   Teflon

unreactive, and shows excellent low-friction properties. It is used in electrical insulators, in valve packings and gaskets in which chemical inertness is required, and in a wide variety of antistick applications, such as nonlubricated bearings and nonstick frying pans.

Another familiar and important addition polymer is *poly(vinyl chloride)* which is prepared by the polymerization of vinyl chloride. Poly(vinyl chloride), PVC, is used in sewer pipes to replace cast-iron pipes, in a variety of transparent items such as plastic raincoats, in phonograph records, and in insulation for electric wire.

$$nCH_2=CHCl \xrightarrow{catalyst} \left[CH_2CHCl\right]_n$$

Vinyl chloride   PVC

When vinyl chloride is copolymerized with vinylidine chloride, the polymers used in *Saran Wrap* are obtained. This copolymer has low moisture transmission and forms a tough film which is particularly useful in food packaging.

$$CH_2=CHCl + CH_2=CCl_2 \xrightarrow{\text{catalyst}} \left[CH_2CCl_2CH_2CHCl \right]_n$$

Vinyl chloride   Vinylidene   Saran wrap
chloride

Other important addition polymers are *polystyrene,* which is used in the form of a foam (Styrofoam) as a heat-insulating material and in many novelty items, *poly (methyl methacrylate),* which is used as a clear transparent plastic (Plexiglass, Lucite) in many

$$C_6H_5CH=CH_2 \xrightarrow{\text{catalyst}} \left[CHCH_2CHCH_2 \right]_n$$
$$\qquad\qquad\qquad\qquad\qquad C_6H_5 \quad C_6H_5$$

Styrene   Polystyrene

$$CH_2=C(CH_3)CO_2CH_3 \xrightarrow{\text{catalyst}} \left[CH_2C \right]_n$$

with CH$_3$ above and CO$_2$CH$_3$ below the repeating carbon.

Methyl methacrylate   Poly(methyl methacrylate)

automobile accessories, and *polyacrylonitrile,* which is the major constituent of acrylic fibers (Orlon, Acrilan).

$$CH_2=CHCN \xrightarrow{\text{catalyst}} \left[CH_2CH \right]_n$$
$$\qquad\qquad\qquad\qquad CN$$

Acrylonitrile   Polyacrylonitrile

CONFIGURATION OF POLYMERS

When a substituted vinyl monomer (CH$_2$=CHR) is polymerized, three possible configurations of the R-group (in relation to the backbone of the polymer) are possible as shown below.

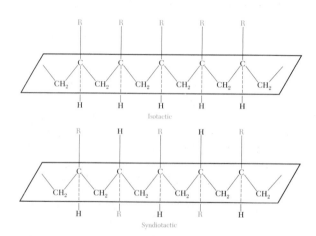

Isotactic

Syndiotactic

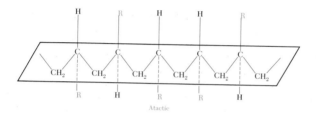

Atactic

When all the "R's" are above (or below) the plane of the backbone of the polymer chain, the polymer is called **isotactic.** When the R's alternate above and below the backbone, the polymer is called **syndiotactic.** Where there is a random arrangement of the R groups, the polymer is called **atactic.** The type of configuration of the polymer depends on the catalyst employed in the polymerization reaction. In general, isotactic and syndiotactic polymers are crystalline and exhibit good mechanical properties. In contrast, atactic polymers are generally noncrystalline and show weak mechanical properties. It wasn't until the discovery of the Ziegler-Natta catalyst system that polymer chemists were able to control the configuration of the resultant polymers and hence the properties of the polymers. With the Ziegler-Natta catalyst either isotactic or syndiotactic polymers can be produced depending on the exact type of catalyst use. For this important discovery, Ziegler and Natta were awarded the Nobel Prize.

TOPIC OF CURRENT INTEREST

PLASTICIZERS

Plasticizers are the chemicals that make plastic soft and flexible. Depending upon the type of plastic, plasticizers may comprise 30 to 50 per cent of the weight of the plastic. They are used in a variety of products, ranging from furniture, upholstery, food, packaging, soft drink and beer production, and construction products to medical blood storage bags. Over one billion pounds of these chemicals are used annually in the United States. Consequently, they appear in almost every phase of modern American life.

The most common type of plasticizer employed has been the phthalic acid esters (phthalates). Until recently, these plasticizing agents were not considered hazardous. However, evidence has been accumulated which shows that these compounds find

Phthalate ester

Di-(2-ethylhexyl) phthalate

their way into our environment. Indeed, of all the millions of chemical compounds developed by man, probably none is more widely dispersed throughout our environment than plasticizers. Recent studies have shown that these compounds appear in the water we drink, the food we eat, and the air we breathe. Though the amounts found in any given substance are too small to be lethal, their cumulative effect over a number of years may and should be a matter of public concern.

Plastic tubing is widely used in the food and beverage industries and in medicine for intravenous infusions. Passage of liquids through such tubing has been shown to contaminate the liquid with various amounts of the plasticizer. For example, kidney patients connected to a dialysis machine (a machine that cleanses the blood) by plastic tubing became nauseated, whereas the patients connected to similar machines without plastic tubing were not affected. The problem was finally traced to the plasticizers in the tubing, which were leaching out and finding their way into the patient's bloodstream. Similarly, it was found that human blood which was stored for three weeks in poly(vinylchloride) plastic bags containing di-(2-ethylhexyl) phthalate as the plasticizer contained 5 to 7 milligrams of the plasticizer, and that patients given transfusions of this blood showed significant quantities of the phthalate ester in the spleen, lung, liver, and abdominal fat. No phthalate esters of this kind were found in the blood of normal subjects who had not received the transfusions.

It should be emphasized that as yet phthalate esters have not been proved to be the cause of any actual disease. However, the widespread use of these compounds in such large amounts does cause some concern that potential risks do exist and will appear only later in this generation after millions of people have been exposed to them. It is obvious that much additional work will be required to determine the effects of plasticizers on man and to evaluate how these compounds aggravate common illnesses in our systems.

IMPORTANT TERMS AND CONCEPTS

addition polymer	homopolymer	polyamide	Styrofoam
atactic polymer	isotactic polymer	polyethylene	syndiotactic polymer
condensation polymer	macromolecule	polymer	thermoplastic polymer
copolymer	monomer	polymerization reaction	thermosetting polymer
cross-linking	Orlon	polypropylene	

QUESTIONS

1. How does an addition polymer differ from a condensation polymer?

2. What functional group(s) is contained in *each* of the following polymers?

   (a) Nylon 6   (d) Glyptal
   (b) Dacron   (e) Plexiglass
   (c) Orlon

3. Kel-F is an addition polymer of chlorotrifluoroethylene. Draw the repeating unit of this polymer.

4. Outline the preparation of Nylon 610.

5. What is the repeating unit in *each* of the following polymers?

   (a) Nylon 66   (f) Teflon
   (b) Dacron   (g) Orlon
   (c) Polystyrene   (h) Saran Wrap
   (d) Polyacrylonitrile (i) PVC
   (e) Lucite   (j) Polypropylene

SUGGESTED READING

Bruck: Thermally Stable Polymeric Materials. Journal of Chemical Education, Vol. 42, p. 18, 1965.

Fisher: New Horizons in Elastic Polymers. Journal of Chemical Education, Vol. 37, p. 369, 1960.

McGrew: Structure of Synthetic High Polymers. Journal of Chemical Education, Vol. 35, p. 178, 1958.

Price: The Effect of Structure on Chemical and Physical Properties of Polymers. Journal of Chemical Education, Vol. 42, p. 13, 1965.

BIOCHEMISTRY
OF THE CELL

CHAPTER 20

The *objectives* of this chapter are to enable the student to:

1. Explain how the electron microscope has affected the study of the biochemistry of the cell.
2. Describe the nature of the major subcellular components.
3. Recognize the similarities and differences between plant and animal cells.
4. State the functions of the major subcellular components.

The ultimate aim of biochemistry is a clear understanding of the chemical reactions that occur in the living cell and their relation to cellular function and structure. The use of radioisotope-labeled compounds and cytochemical techniques has assisted the biochemist in the location of the cellular sites of specific reactions, especially those involving enzymes. The *light microscope* has been invaluable in the study of staining reactions and the rough morphology of the cell. With the advent of the *electron microscope* the fine structure of the cell was revealed, and a whole new area of biochemical research was made available. A more complete understanding of the relationship of structure to function is now within the grasp of biochemists. In fact, Lehninger recently defined the living cell as a self-assembling, self-adjusting, self-perpetuating isothermal system of molecules which exchanges matter and energy with its environment. This system carries out many consecutive organic reactions that are promoted by enzymes produced by the cell. It operates on the principle of maximum economy of parts and processes, and its precise self-replication is ensured by a linear molecular code. The study of biochemistry concerns the many facets of Lehninger's definition, which should be referred to as each chapter of the section is studied.

Cells differ in size, appearance, and structure, depending on their function, but a typical animal cell has the features illustrated in Figure 20–1. A typical plant cell (Fig. 20–2) also includes the same structures. To understand the function of a plant, animal, or bacterial cell we must first study the subcellular components. Structures common to all cells include the cell membrane, nucleus, mitochondria, endoplasmic reticulum, ribosomes, and Golgi apparatus. Fortunately for cytological studies, the electron microscope can be focused on each subcellular component to reveal its structural details. A brief description of the biochemical processes occurring in each component may refer to compounds whose structures are not well known to the student, but which will be described in later chapters of this section.

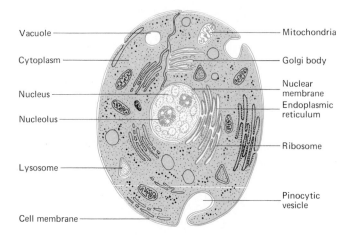

Vacuole
Cytoplasm
Nucleus
Nucleolus
Lysosome
Cell membrane

Mitochondria
Golgi body
Nuclear membrane
Endoplasmic reticulum
Ribosome
Pinocytic vesicle

Figure 20–1   A diagram of a typical cell based on an electron micrograph. (After the Living Cell, by Jean Brachet. Copyright © 1961 by Scientific American, Inc. All rights reserved.)

SUBCELLULAR COMPONENTS

CELL MEMBRANE

All the subcellular components of a cell are contained within a definite cell wall or membrane. This membrane plays a vital role in the passage of nutrient and waste material into and out of the cell. In addition to the cell membrane, plant cells have rigid walls that surround and protect the membrane. The cell walls consist of cellulose and other polysaccharides. The **cell membrane** is composed of lipids and protein arranged in such a fashion that water-soluble and lipid-soluble substances can pass through the membrane. The permeability of living membranes has never been adequately explained. Although many cells are bathed in a fluid rich in sodium and chloride ions and low in potassium ions, the cell contents are rich in potassium ions and low in sodium and chloride ions. Membranes must therefore be involved in a process of "active transport," or the movement of ions or molecules from a region of low concentration across the membrane into a region of higher concentration. This movement against a concentration gradient involves enzymes such as ATPase and energy in the form of adenosine triphosphate, or ATP. The different rates of absorption of monosaccharides and amino acids from the small intestine emphasize the importance of the membrane in selective permeability toward small ions and molecules.

CYTOPLASM

The **cytoplasm** is the general protoplasmic mass in which the definite subcellular components described above are embedded. At present all of the essential compounds

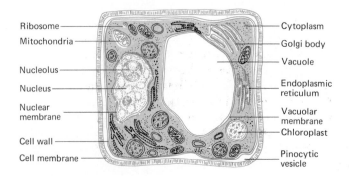

Ribosome
Mitochondria
Nucleolus
Nucleus
Nuclear membrane
Cell wall
Cell membrane

Cytoplasm
Golgi body
Vacuole
Endoplasmic reticulum
Vacuolar membrane
Chloroplast
Pinocytic vesicle

Figure 20–2   A diagram of a typical plant cell.

and macromolecules in the cell not associated with definite particles are thought to exist in the cytoplasm. Many soluble enzymes are found in the cytoplasm, particularly those associated with the conversion of glucose to pyruvic or lactic acids. Considerable research remains to be done on the components of the cell and the cytoplasm with respect to enzyme and coenzyme distribution and their role in various metabolic reactions.

NUCLEUS

The nucleus is roughly spherical in shape and is surrounded by a double layered membrane that is more porous than the cell membrane. In many cells the outer membrane is connected with the nuclear membrane by one or more channels through the cytoplasm. In addition there is usually a connection between the endoplasmic reticulum and the double-layered nuclear membrane. It has long been recognized that the **nucleus** serves as a site for the transmission and regulation of hereditary characteristics of the cell. This control is an essential feature of the **chromosomes** that are composed of **deoxyribose nucleic acid** (DNA), and basic protein (see p. 284). The nucleus also contains one or more small, dense, round bodies called nucleoli. These bodies contain DNA and **ribose nucleic acid** (RNA), and appear to be involved in the synthesis of RNA and proteins.

MITOCHONDRIA

These subcellular particles are shaped like an elongated oval 2 to 7 μ in length and 1 to 3 μ in diameter (Fig. 27–2). The walls of the mitochondria are double-layered membranes with projections that extend inward toward the center of the particle, increasing the surface of the membrane inside the mitochondrion. These projections inside the mitochondria are called **crista.** The liquid in the matrix of the particle contains protein, neutral fat, phospholipids, and nucleic acids. In contrast to the nucleus, the nucleic acids in the mitochondria are mostly RNA with only small amounts of DNA. The **mitochondria** have been called the *powerhouses of the cell,* since they are the site of major oxidative processes and oxidative phosphorylation which result in the formation of ATP.

CHLOROPLASTS

Plant cells contain highly pigmented particles 3 to 6 μ in diameter called **chloroplasts** (Fig. 20–2). These particles contain the green pigment chlorophyll and play a major role in photosynthesis. Inside the chloroplast membrane is a series of laminated membrane structures called **grana.** Chlorophyll and lipids are concentrated in the grana which are active in the photosynthetic process. The structure and function of chloroplasts parallel those of mitochondria.

ENDOPLASMIC RETICULUM AND RIBOSOMES

The **endoplasmic reticulum** is composed of a network of interconnected, thin, membranelike tubules or vesicles. In some areas of the cytoplasm the membranes are covered with dark round bodies about 0.015 μ in diameter which contain 80 per cent of the RNA in the cell and have therefore been termed ribosomes. These areas are known as rough endoplasmic reticulum contrasted to the smooth reticulum which does not have ribosomes adsorbed on its surface. The endoplasmic reticulum and accompanying ribosomes are also called microsomes or the microsomal region of a cell. The specific particles, the **ribosomes,** are the site for the synthesis of proteins within the cell.

THE GOLGI APPARATUS

This is also called the Golgi body or complex and consists of an orderly array of flattened sacs with smooth membranes associated with small vacuoles of varying size. Although it is similar to the smooth endoplasmic reticulum, it exhibits different staining properties and its membrane has a different composition. The **Golgi apparatus** is often connected to the cell membrane by a channel and serves as a way station in the transport of substances produced in other subcellular particles. In liver cells, for example, the Golgi apparatus is usually located close to the small bile canals and is involved in the transport and excretion of substances such as bilirubin glucuronide into the bile.

LYSOSOMES

These particles are spherical in shape with an average diameter of $0.4\ \mu$. They contain several soluble hydrolytic enzymes (hydrolases) that exhibit an optimum pH in the acid range. The lysosomal membrane is lipoprotein in nature and prevents the enzymes from escaping into the cellular cytoplasm. The membrane also prevents the substrates for the enzymes from entering the cell. When the cell is injured and the membrane is broken, the released enzymes cause cellular breakdown. In autolysis of tissue, whether normal (as involution of the thymus gland at puberty), pathological, or post-mortem, the lysosomal enzymes destroy cellular tissue. The processes of phagocytosis and pinocytosis involve the engulfment of foreign material into vesicles or vacuoles and the digestion of this material. These particles may be converted into lysosomes to assist in the hydrolysis of phagocytosed material.

VACUOLES AND VESICLES

These particles are roughly spherical in shape and vary in size from 0.1 to $0.7\ \mu$ in diameter. They are often found close to the Golgi apparatus and to channels involved in the entrance and excretion of material to and from the cell. Vacuoles may serve as temporary storage sacs, or as bodies involved in the removal of foreign material from the cell.

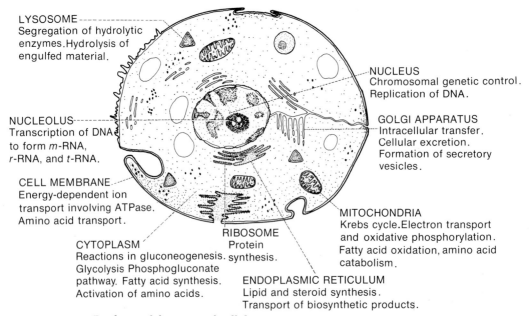

LYSOSOME
Segregation of hydrolytic enzymes. Hydrolysis of engulfed material.

NUCLEOLUS
Transcription of DNA to form m-RNA, r-RNA, and t-RNA.

CELL MEMBRANE
Energy-dependent ion transport involving ATPase. Amino acid transport.

CYTOPLASM
Reactions in gluconeogenesis. Glycolysis Phosphogluconate pathway. Fatty acid synthesis. Activation of amino acids.

RIBOSOME
Protein synthesis.

NUCLEUS
Chromosomal genetic control. Replication of DNA.

GOLGI APPARATUS
Intracellular transfer. Cellular excretion. Formation of secretory vesicles.

MITOCHONDRIA
Krebs cycle. Electron transport and oxidative phosphorylation. Fatty acid oxidation, amino acid catabolism.

ENDOPLASMIC RETICULUM
Lipid and steroid synthesis. Transport of biosynthetic products.

Figure 20-3   Biochemical functions of cellular components.

BIOCHEMICAL FUNCTION OF CELLULAR COMPONENTS

Many investigators have concentrated their research activities on the biochemical reactions that occur in a specific subcellular particle. It is obviously not possible at present to reconstruct the exact biochemical functions of the intact cell, but a combination of cytochemical techniques and research on reactions within the separated particles provide a greater understanding of the overall process. Some of the biochemical functions that have been associated with cellular components are outlined in Figure 20–3.

IMPORTANT TERMS AND CONCEPTS

chloroplasts   electron microscope   lysosomes   mitochondria
cytoplasm   Golgi apparatus   membrane   nucleus

QUESTIONS

1. What major instrumental development enabled the biochemist to study subcellular components?

2. What is the nature of the cell membrane? What evidence is there for selective permeability of the cell membrane?

3. The membrane surrounding the nucleus not only has pores but has a direct connection to the cytoplasm through the endoplasmic reticulum. Is this arrangement an advantage to the cell? Why?

4. Biochemists have intensively studied the mitochondria for several years. Why should they be so interested in this subcellular particle?

5. In an electron micrograph of the cell, how can rough endoplasmic reticulum be differentiated from smooth? What is one of the major functions of the rough form?

6. Discuss the nature and function of the Golgi apparatus.

7. What is the nature and function of the lysosomes?

SUGGESTED READING

Capaldi: A Dynamic Model of Cell Membranes. Scientific American, Vol. 230, No. 3, p. 26, 1974.
Everhart and Hayes: The Scanning Electron Microscope. Scientific American, Vol. 226, No. 1, p. 54, 1972.
Goodenough and Levine: The Genetic Activity of Mitochondria and Chloroplasts. Scientific American, Vol. 223, No. 5, p. 22, 1970.
Stent: Cellular Communication. Scientific American, Vol. 227, No. 3, p. 42, 1972.

PROTEINS

The *objectives* of this chapter are to enable the student to:

1. Differentiate between proteins, carbohydrates, and fats on the basis of their elementary composition.
2. Explain why naturally occurring amino acids are called alpha amino acids.
3. Recognize a sulfur-containing, an aromatic, and a heterocyclic amino acid.
4. Illustrate the formation and hydrolysis of a dipeptide.
5. Describe the separation of amino acids by ion exchange chromatography.
6. Describe the analysis of the insulin molecule by Sanger.
7. Distinguish between the primary, secondary, and tertiary structures of a protein.
8. Describe three methods of precipitating a protein from a solution.

Living cells are composed of approximately 70 per cent water and 30 per cent solid matter. The solid matter consists of organic compounds, half of which are proteins. It has been estimated that the average cell contains about 3000 different kinds of proteins which are fundamental constituents of all cells and tissues in the body and which are also required for the synthesis of body tissues, enzymes, certain hormones, and protein components of the blood.

Proteins are made by plant cells by a process starting with photosynthesis from carbon dioxide, water, nitrates, sulfates and phosphates. The complicated process of synthesis has not as yet been completely elucidated. Animals can synthesize only a limited amount of protein from inorganic sources and are mainly dependent on plants or other animals for their source of dietary protein (p. 369). Proteins are used in the body for growth of new tissue, for maintenance of existing tissue, and as a source of energy.

ELEMENTARY COMPOSITION

The five elements that are present in most naturally occurring proteins are **carbon, hydrogen, oxygen, nitrogen,** and **sulfur.** There is a wide variation in the amount of sulfur in proteins. Gelatin, for example, contains about 0.2 per cent, in contrast to 3.4 per cent in insulin.

Other elements, such as phosphorus, iodine, and iron, may be essential constituents of certain specialized proteins. Casein, the main protein of milk, contains phosphorus, an element of utmost importance in the diet of infants and children. Iodine is a basic constituent of the protein in the thyroid gland and is present in sponges and coral. Hemoglobin of the blood, which is necessary for the process of respiration, is an iron-containing protein. Most proteins show little variation in their elementary composition. The average content of the five main elements is as follows: carbon 53 per cent, hydrogen

7 per cent, oxygen 23 per cent, nitrogen 16 per cent, and sulfur 1 per cent. The relatively high content of nitrogen differentiates proteins from fats and carbohydrates.

MOLECULAR WEIGHT

Protein molecules are very large, as indicated by the approximate formula for oxyhemoglobin:

$$C_{2932}H_{4724}N_{828}S_8Fe_4O_{840}$$

The molecular weight of oxyhemoglobin would thus be about 68,000. The common protein egg albumin has a molecular weight of about 34,500. In general, protein molecules have weights that vary from 34,500 to 50,000,000. Their extremely large size can readily be appreciated when they are compared with the molecular weight of a fat such as tripalmitin, which is 807, or of glucose, which is 180.

AMINO ACIDS

In addition to their large size, protein molecules are also very complicated. Like any complex molecule, they may be broken down by hydrolysis into smaller molecules whose structure is more easily determined. Common reagents used for the hydrolysis of proteins are acids (HCl and H_2SO_4), bases (NaOH), and enzymes (proteases). The simple molecules that are formed by the complete hydrolysis of a protein are called **amino acids** (p. 240).

Before considering the properties and reactions of proteins, it may be well to study the individual amino acids. If a hydrogen is replaced by an amino group on the carbon atom that is next to the carboxyl group in acetic acid, CH_3COOH, the simple amino acid **glycine** will be formed.

$$\underset{\underset{NH_2}{|}}{CH_2COOH}$$

The carbon atom next to the carboxyl group is called the alpha (α) carbon, and since all the amino acids have an amino group attached to the alpha carbon atom, they are known as **alpha amino acids.**

In the following tabulation the amino acids are divided into groups according to their chemical structure. The common name for each amino acid is followed by the abbreviation used in sequence and structure models. The unit consisting of the carboxyl group, alpha carbon atom and alpha amino group common to all amino acids is emphasized.

Aliphatic

Glycine (Gly)

$$\underset{\underset{NH_2}{|}}{H—CHCOOH}$$

Alanine (Ala)

$$\underset{\underset{NH_2}{|}}{CH_3—CHCOOH}$$

Acidic

Aspartic acid (Asp)

$$\underset{\underset{NH_2}{|}}{HOOC—CH_2—CHCOOH}$$

Glutamic acid (Glu)

$$\underset{\underset{NH_2}{|}}{HOOC—CH_2—CH_2—CHCOOH}$$

Heterocyclic

Tryptophan (Trp)

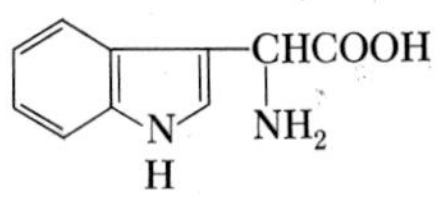

Aliphatic

Valine (Val)

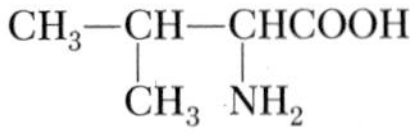

$$CH_3-CH-CHCOOH$$
$$\quad\quad CH_3 \quad NH_2$$

Leucine (Leu)

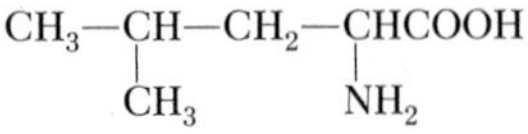

$$CH_3-CH-CH_2-CHCOOH$$
$$\quad\quad CH_3 \quad\quad\quad NH_2$$

Isoleucine (Ile)

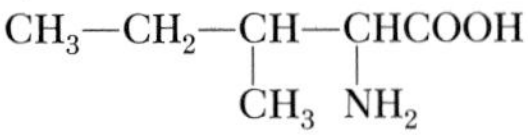

$$CH_3-CH_2-CH-CHCOOH$$
$$\quad\quad\quad\quad CH_3 \quad NH_2$$

Serine (Ser)

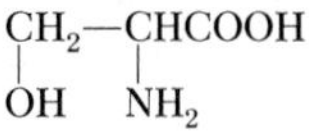

$$CH_2-CHCOOH$$
$$OH \quad NH_2$$

Threonine (Thr)

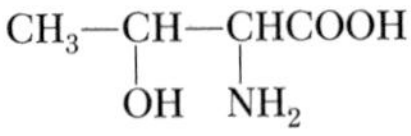

$$CH_3-CH-CHCOOH$$
$$\quad\quad OH \quad NH_2$$

Basic

Lysine (Lys)

$$CH_2-CH_2-CH_2-CH_2-CHCOOH$$
$$NH_2 \quad\quad\quad\quad\quad\quad\quad\quad NH_2$$

Arginine (Arg)

$$NH_2$$
$$C-NH-CH_2-CH_2-CH_2-CHCOOH$$
$$NH \quad\quad\quad\quad\quad\quad\quad\quad NH_2$$

Acidic

Asparagine (Asn)

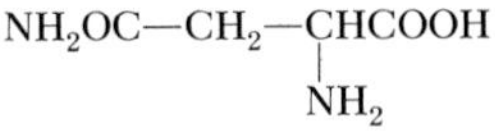

$$NH_2OC-CH_2-CHCOOH$$
$$\quad\quad\quad\quad\quad NH_2$$

Glutamine (Glu)

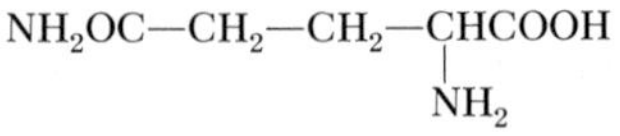

$$NH_2OC-CH_2-CH_2-CHCOOH$$
$$\quad\quad\quad\quad\quad\quad\quad NH_2$$

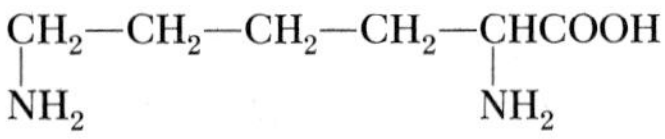

Cysteine (Cys)

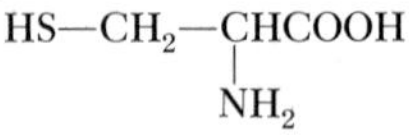

$$HS-CH_2-CHCOOH$$
$$\quad\quad\quad\quad NH_2$$

Cystine (Cys-Cys)

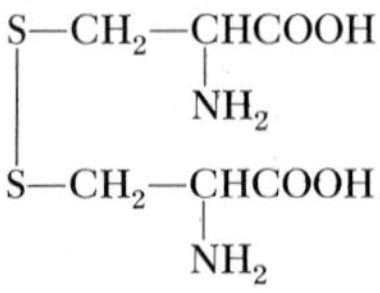

$$S-CH_2-CHCOOH$$
$$\quad\quad\quad\quad NH_2$$
$$S-CH_2-CHCOOH$$
$$\quad\quad\quad\quad NH_2$$

Methionine (Met)

$$CH_3-S-CH_2-CH_2-CHCOOH$$
$$\quad\quad\quad\quad\quad\quad\quad NH_2$$

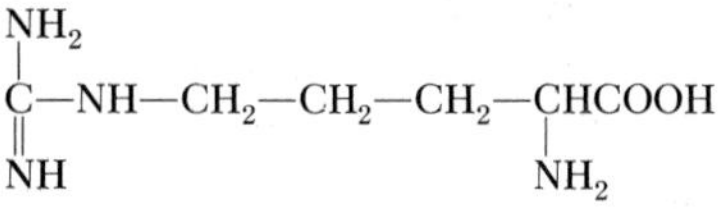

Aromatic

Phenylalanine (Phe)

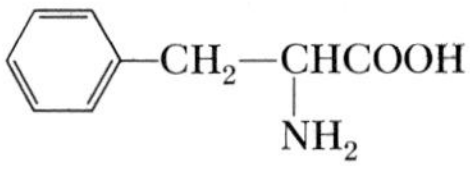

$$\text{⬡}-CH_2-CHCOOH$$
$$\quad\quad\quad\quad NH_2$$

Tyrosine (Tyr)

$$HO-\text{⬡}-CH_2-CHCOOH$$
$$\quad\quad\quad\quad\quad NH_2$$

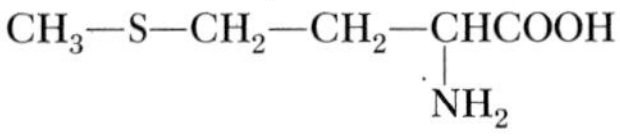

Heterocyclic

Histidine (His)

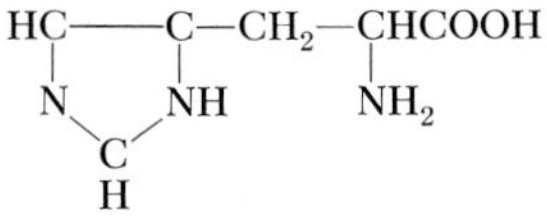

$$HC\quad\quad C-CH_2-CHCOOH$$
$$N\quad NH\quad\quad\quad NH_2$$
$$\quad C$$
$$\quad H$$

Proline (Pro)

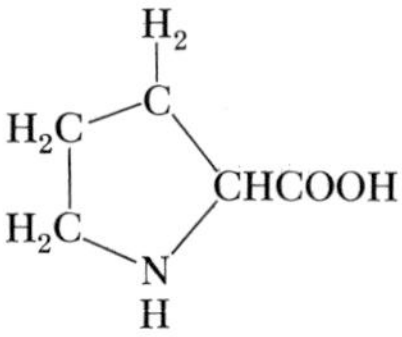

Hydroxyproline (Hypro)

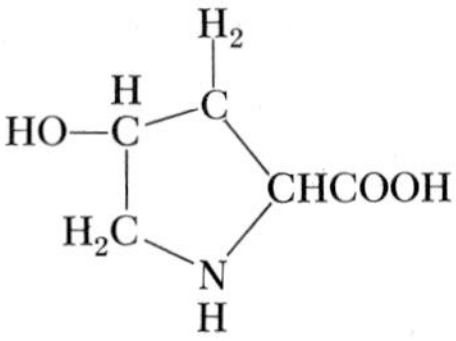

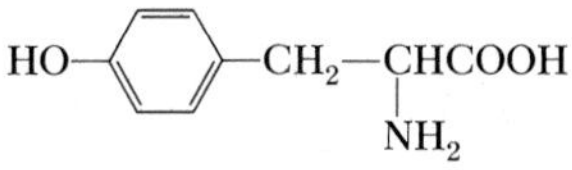

OPTICAL ACTIVITY OF AMINO ACIDS

All amino acids except glycine contain an asymmetric carbon atom in their formulas. For this reason they may exist in the D or L form. For an explanation of the D and L forms see page 303. Using alanine as an example of a simple amino acid, we may compare the D and L forms to those of glyceraldehyde and lactic acid:

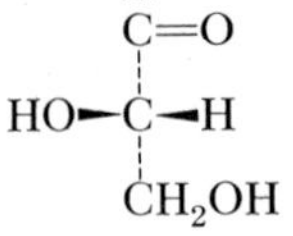

L-Glyceraldehyde 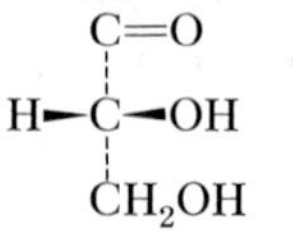   D-Glyceraldehyde 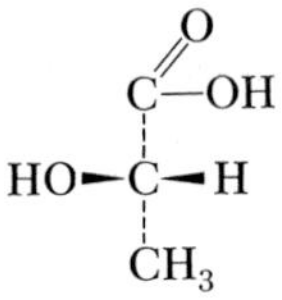   L-Lactic acid

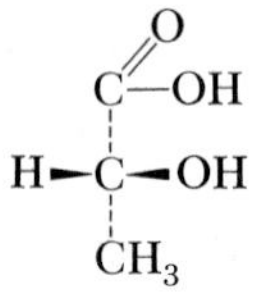

D-Lactic acid 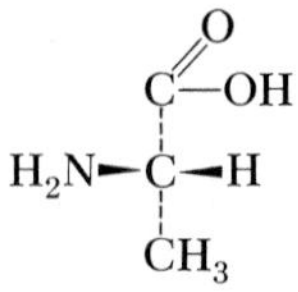   L-Alanine 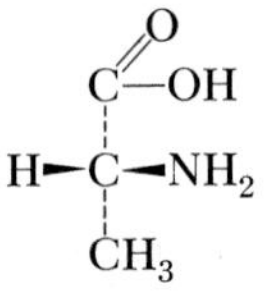   D-Alanine

Naturally occurring amino acids from plant and animal sources have the L configuration and would be designated L(+) or L(−), depending on their rotation of plane polarized light. For an explanation of the (+) and (−) terminology see page 301. D-Alanine and D-glutamic acid have been obtained from microorganisms, especially from their cell walls.

For an explanation of the (+) and (−) terminology see page 301.

AMPHOTERIC PROPERTIES OF AMINO ACIDS

Amino acids behave both as weak acids and as weak bases, since they contain at least one carboxyl and one amino group. Substances that ionize as both acids and bases in aqueous solution are called **amphoteric** (p. 142). An example is glycine, in which both the acidic and basic groups are ionized in solution to form dipolar ions or **zwitterions.**

$$CH_2COO^-$$
$$|$$
$$NH_3^+$$

The glycine molecule is electrically neutral, since it contains an equal number of positive and negative ions. The zwitterion form of glycine would therefore be isoelectric, and the pH at which the zwitterion does not migrate in an electric field is called the **isoelectric point.** Amphoteric compounds will react with either acids or bases to form salts. This is best illustrated by use of the zwitterion form of the amino acid.

$$CH_3{-}CH{-}COO^- + HCl \rightarrow CH_3{-}CH{-}COOH$$
$$\quad\quad | \quad\quad\quad\quad\quad\quad\quad\quad\quad | $$
$$\quad NH_3^+ \quad\quad\quad\quad\quad\quad\quad NH_3^+Cl^-$$

$$CH_3{-}CH{-}COO^- + NaOH \rightarrow CH_3{-}CH{-}COO^-\, Na^+ + H_2O$$
$$\quad\quad | \quad\quad\quad\quad\quad\quad\quad\quad\quad\quad | $$
$$\quad NH_3^+ \quad\quad\quad\quad\quad\quad\quad\quad NH_2$$

From these equations it can be seen that the addition of a H^+ to an isoelectric molecule results in an increased positive charge (NH_3^+), since the acid represses the ionization of the carboxyl group. Conversely, the addition of a base to an isoelectric molecule results in an increased negative charge (COO^-), since the base represses the ionization of the amino group. Since proteins are composed of amino acids, they are amphoteric substances with specific isoelectric points and are able to neutralize both acids and bases. This property of proteins is responsible for their buffering action in blood and other fluids.

REACTIONS OF AMINO ACIDS

The fact that amino acids can ionize as both weak acids and weak bases and contain amino groups and carboxyl groups suggests a very reactive molecule. Many of the common reactions of organic chemistry may be applied to amino acids.

Reaction with Nitrous Acid. This is the basis of the Van Slyke method for the determination of free primary amino groups (see p. 250).

$$R{-}CH{-}NH_2 + NaNO_2 + HCl \rightarrow [R{-}CH{-}\overset{+}{N}{\equiv}N]Cl^- \rightarrow R{-}CH{-}OH + N_2 + NaCl + H_2O$$
$$\quad | \quad\quad\quad\quad\quad\quad\quad\quad\quad\quad | \quad\quad\quad\quad\quad\quad\quad\quad | $$
$$\quad COOH \quad\quad\quad\quad\quad\quad\quad COOH \quad\quad\quad\quad\quad COOH$$

The nitrogen gas that is liberated in the reaction is collected and its volume measured. One half of this nitrogen comes from the amino acid and is used as a measure of the free amino nitrogen.

Reaction with 1-Fluoro-2,4-dinitrobenzene (FDNB). This compound, also called **Sanger's reagent,** reacts with the free amino group of an amino acid, as would an alkyl halide, to form a yellow colored dinitrophenylamino acid, DNP-amino acid.

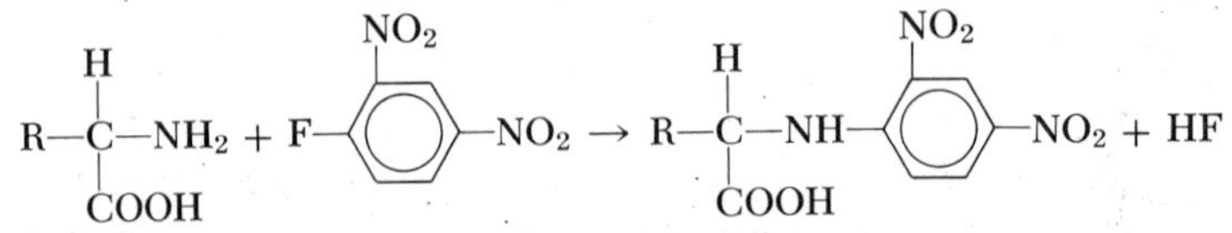

Dinitrophenylamino acid,
or DNP-amino acid

This reaction is very important in the determination of protein structure, since the reagent reacts with the free amino group of the terminal amino acid in a protein and thus identifies the end amino acid in the structure.

Reaction with Ninhydrin. Amino acids react with ninhydrin (triketohydrindene hydrate) to form CO_2, NH_3, and an aldehyde. The NH_3 that is formed in the reaction combines with a molecule of reduced and a molecule of oxidized ninhydrin and forms a blue-colored compound. This compound may be measured colorimetrically for the quantitative determination of amino acids.

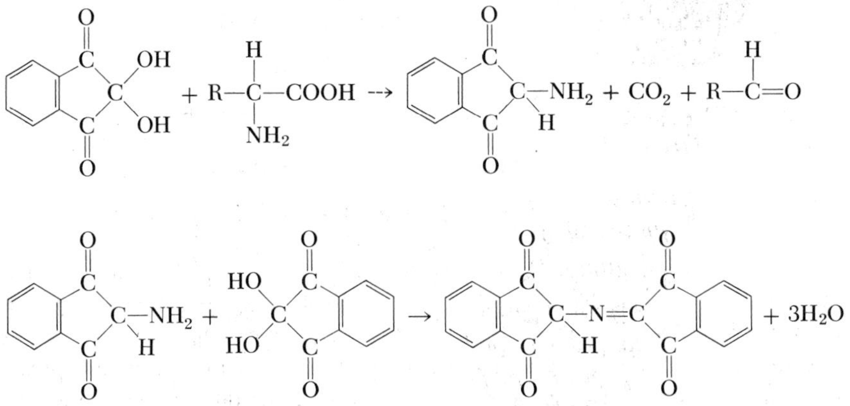

Blue-colored compound

Color Reactions of Specific Amino Acids. Certain amino acids, whether in the free form as in protein hydrolysates or combined in proteins, give specific color reactions that aid in their detection and determination. The **Millon test** depends on the formation of a red-colored mercury complex with tyrosine, whether free or in proteins, whereas tryptophan reacts with glyoxylic acid to produce a violet color in the **Hopkins-Cole test.** In the **Sakaguchi reaction,** the guanidino group in arginine forms a red color with α-naphthol and sodium hypochlorite, and cysteine and proteins that contain free sulfhydryl groups yield a red color with sodium nitroprusside. Both cystine and cysteine, free and in proteins, form a black precipitate of PbS in the **unoxidized sulfur test.**

POLYPEPTIDES

Amino acids are joined together in a polypeptide molecule by the peptide linkage. The **peptide linkage** is an amide linkage (p. 240) between the carboxyl group of one amino acid and the amino group of another amino acid, with the splitting out of a molecule of water. This type of linkage may be illustrated by the union of a molecule of alanine and a molecule of glycine:

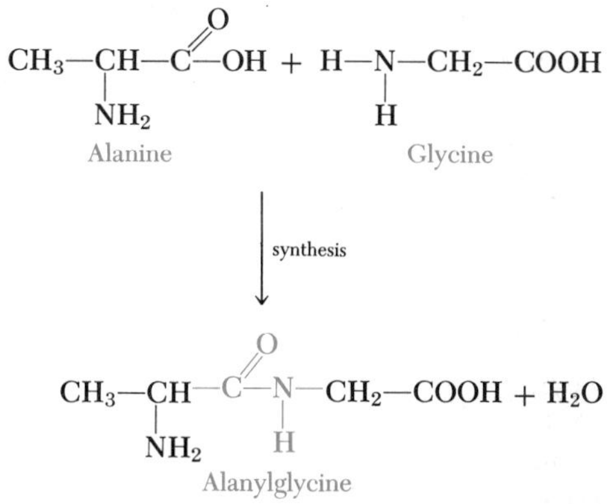

$$CH_3-CH-\overset{\overset{\textstyle O}{\|}}{C}-OH \;+\; H-\underset{\underset{\textstyle H}{|}}{N}-CH_2-COOH$$

Alanine   Glycine

↓ synthesis

$$CH_3-\underset{\underset{\textstyle NH_2}{|}}{CH}-\overset{\overset{\textstyle O}{\|}}{C}-\underset{\underset{\textstyle H}{|}}{N}-CH_2-COOH \;+\; H_2O$$

Alanylglycine

The compound alanylglycine, which results from this linkage, is called a **dipeptide.** The union of three amino acids would result in a tripeptide, and the combination of several amino acids by the peptide linkage would be called a **polypeptide.** Since each amino acid has lost a water molecule when it joins to two other amino acids in a polypeptide, the remaining compound is called an **amino acid residue.** A polypeptide chain illustrating the primary structure of a protein may be represented as follows:

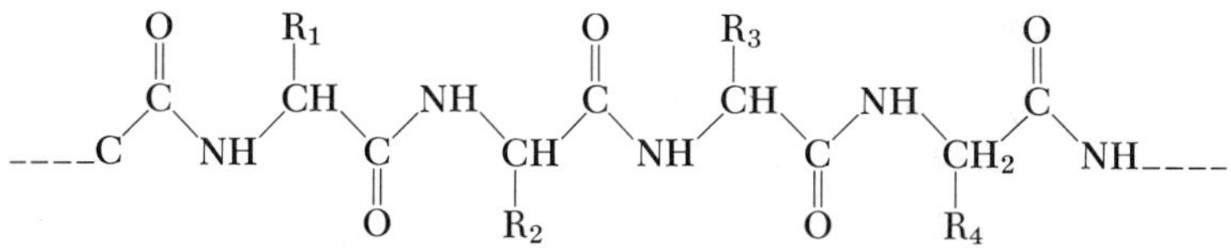

The R groups are the side chains of the specific amino acids in the polypeptide chain.

THE BIURET TEST

When a few drops of very dilute copper sulfate solution are added to a strongly alkaline solution of a peptide or protein, a violet color is produced. This is a general test for proteins and is given by peptides that contain two or more peptide linkages. **Biuret** is formed by heating urea and has a structure similar to the peptide structure of proteins:

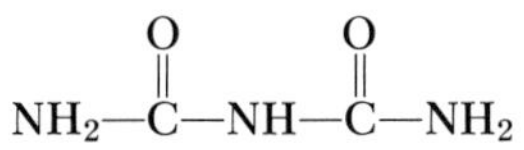

$$NH_2-\overset{\overset{\textstyle O}{\|}}{C}-NH-\overset{\overset{\textstyle O}{\|}}{C}-NH_2$$

Biuret

SEPARATION AND DETERMINATION OF AMINO ACIDS

Prior to 1950, it was not thought possible to unravel the amino acid sequence or to understand the structure of the complex protein molecules. Amino acid sequence determination was made possible by the development of the techniques of chromatography.

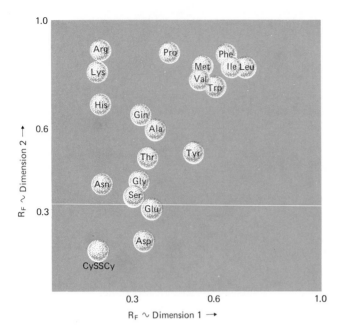

Figure 21-1   A schematic representation of a two-dimensional paper chromatogram. The solvent for dimension 1 is n-butanol-acetic acid-water (250 : 60 : 250 vol. per vol.) and for dimension 2, phenol-water-ammonia (120 : 20 : 0.3 per cent). Each solvent front moves with an R_F equal to 1 in each dimension. (After White et al.: Principles of Biochemistry. 4th ed. New York, McGraw-Hill, Inc., 1968, p. 114.)

PAPER CHROMATOGRAPHY

Paper chromatography is relatively simple and produces excellent separation, detection, and quantitation of the individual amino acids. The technique of ascending paper chromatography is often used. When a strip of filter paper is held vertically in a closed glass cylinder with its lower end dipped in a mixture of water and an organic solvent such as butyl alcohol, phenol, or collidine, the mixture of water and organic solvent moves up the paper. If a solution containing a mixture of amino acids is added as a small spot just above the solvent level, the individual amino acids will be affected by the water phase and the organic phase as they move up the paper. A solvent partition will occur, and each amino acid will be carried to a particular location on the paper. This location depends on many factors, including the pH, the temperature, the concentration of the solvents, and the time of chromatography. After drying the paper, it is sprayed with a solution of ninhydrin, which yields a blue to purple color with each amino acid. By comparison with known amino acids, separation and identification of the amino acids from a hydrolysate of a protein or polypeptide fragment may be achieved. This separation is often improved by a second chromatographic run during which different solvents are used and the dried paper from the first run is turned 90 degrees from the direction of the first migration. The results of a two-dimensional paper chromatography separation are shown in Figure 21-1.

ION EXCHANGE CHROMATOGRAPHY

The most successful separations and quantitative determinations of amino acids in mixtures are achieved with **ion exchange chromatography.** Ion exchange resins are insoluble synthetic resins containing acidic or basic groups, such as $-SO_3H$ or $-OH$. A sulfonated polystyrene resin may be used as a cation exchange resin by the addition of Na ions to produce $-SO_3Na$ groups on the surface of the resin. Basic amino acids react with a cation exchange resin as follows:

$$ResinSO_3^-Na^+ + NH_3^+R \rightarrow ResinSO_3^-NH_3^+R + Na^+Cl^-$$
$$Cl^-$$

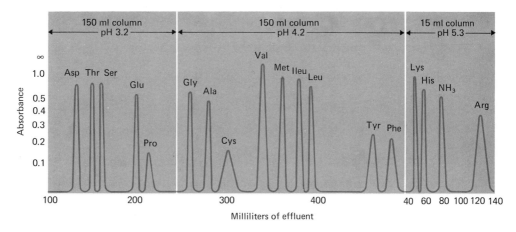

Figure 21–2   Chromatographic fractionation of a synthetic mixture of amino acids on columns of Amberlite IR-120. (After Moore et al.: Anal. Chem., *30:* 1186, 1958.)

The resin is placed in a column or long glass tube, and the mixture of amino acids, which are dissolved in a small volume of buffer, is placed on top of the column. The column is washed stepwise with buffer solutions of increasing pH values, and, as the amino acids pass down the column, the basic amino acids react with the —$SO_3^-Na^+$ groups, replacing Na^+, and are slowed in their passage. Glutamic and aspartic acids, which contain two carboxyl groups, are least affected by the column and come off in the first buffer. They are followed by the neutral amino acids and finally the basic amino acids. The **elution pattern** of representative amino acids from a cation exchange resin is illustrated in Figure 21–2.

ANALYSIS OF THE INSULIN MOLECULE BY SANGER

Sanger's efforts to establish the structure of insulin were made possible by the advances in the methods of protein chemistry, especially the technique of chromatography. In addition, he found that a dinitrophenyl (DNP) group could be attached to free amino groups to form a yellow compound, as was discussed earlier. This DNP group remained attached to the amino acid residue even after hydrolysis was used to split the peptides, therefore making it possible to identify the terminal residue. Employing the DNP method, he first established that each insulin molecule contained two amino acid residues with free amino groups. He concluded, therefore, that insulin consists of two chains. The two chains were held together by the disulfide —S—S— bonds of cystine residues, which could be broken by mild oxidation. The two intact chains were obtained, and it was proven that one contained 21 amino acids whereas the other contained 30. Each chain was hydrolyzed with acid into smaller pieces and the amino acids identified by chromatography.

The final difficult task was to determine the pairing of the half-cystine residues that joined the two chains. It was found that the shorter chain contained one disulfide linkage and the two chains were held together by two other disulfide bonds, as shown in Figure 21–3.

AMINO ACID SEQUENCE IN PEPTIDES AND PROTEINS

After Sanger laid the foundation for the attack on the amino acid sequence of a protein molecule, other workers studied peptides and proteins. Vincent du Vigneaud established the exact structure and sequence in two peptide hormones, **oxytocin** and **vasopressin,** that are elaborated by the posterior lobe of the pituitary gland. Oxytocin

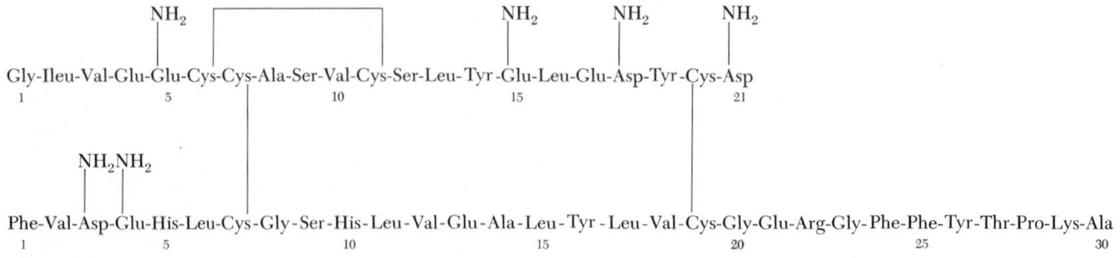

NH₂   NH₂   NH₂   NH₂

Gly-Ileu-Val-Glu-Glu-Cys-Cys-Ala-Ser-Val-Cys-Ser-Leu-Tyr-Glu-Leu-Glu-Asp-Tyr-Cys-Asp
1   5   10   15   21

NH₂ NH₂

Phe-Val-Asp-Glu-His-Leu-Cys-Gly-Ser-His-Leu-Val-Glu-Ala-Leu-Tyr-Leu-Val-Cys-Gly-Glu-Arg-Gly-Phe-Phe-Tyr-Thr-Pro-Lys-Ala
1   5   10   15   20   25   30

Figure 21-3   Amino acid sequence in beef insulin.

causes contraction of smooth muscles and is used in obstetrics to initiate labor. Vasopressin constricts vessels, raising blood pressure and affecting water and electrolyte balance. Each peptide contained eight amino acids, with the disulfide bridge of cystine across four of the amino acids.

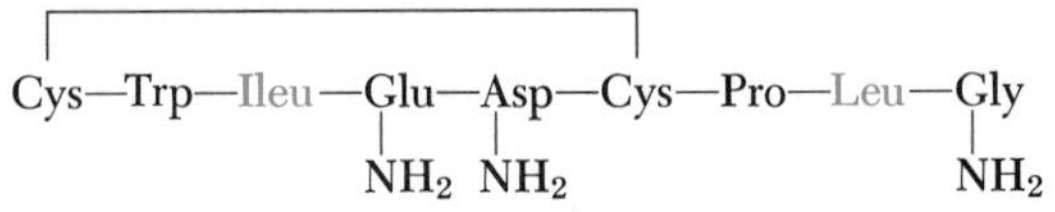

Cys—Trp—Ileu—Glu—Asp—Cys—Pro—Leu—Gly
NH₂   NH₂   NH₂

Oxytocin

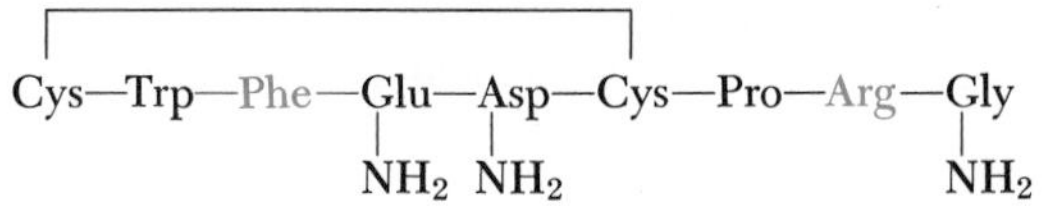

Cys—Trp—Phe—Glu—Asp—Cys—Pro—Arg—Gly
NH₂   NH₂   NH₂

Vasopressin

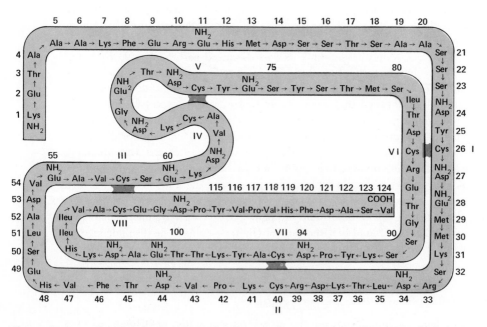

Figure 21-4   The complete amino acid sequence of enzyme ribonuclease. Standard three-letter abbreviations are used to indicate individual amino acid residues. (After Smyth et al.: J. Biol. Chem., 238: 227, 1963.)

The presence of two different amino acids in such a small peptide results in very different physiological activity. Du Vigneaud also succeeded in synthesizing these two molecules from amino acids and demonstrated the similar hormone activity of the synthetic peptides.

The amino acid sequence of the **adrenocorticotropic hormone, ACTH,** containing 39 amino acids, has been established. Larger protein molecules, such as the enzyme **ribonuclease** (Fig. 21–4), with 124 amino acids, the α and β polypeptide chains of **hemoglobin** (141 and 146 amino acid residues, respectively), and the **tobacco mosaic virus** protein with 158 amino acid residues, have also been characterized.

STRUCTURE OF PROTEINS

The chemical, physical, and biologic properties of specific proteins depend on the structure of the molecule as it exists in the native state. Proteins range in complexity from a simple polypeptide, such as vasopressin, with biologic activity, to a globular protein such as myoglobin, whose molecule involves cross linkages, helix formation, and folding and conformational forces.

PRIMARY STRUCTURE

The amino acid sequence determinations have established the exact structure of the polypeptide chain in simple proteins. The peptide linkage joining amino acids to produce a polypeptide is considered the **primary structure** of a protein (p. 273).

SECONDARY STRUCTURE

If only peptide bonds were involved in protein structure, the molecules would consist of long polypeptide chains coiled in random shapes. Most **native proteins,** however, are either fibrous or globular in nature, and consist of polypeptide chains joined together or held in definite folded shapes by hydrogen bonds. This influence of hydrogen bonding on the protein molecule is often called the **secondary structure** of the protein. Pauling studied the structure of the fibrous protein **α-keratin** and concluded that the polypeptide chains are regularly coiled to form a structure called the **α-helix** (Fig. 21–5A). The α-helix structure consists of a chain of amino acid units wound into a spiral which is held together by hydrogen bonds between a carbonyl group of one amino acid and the imino group of an amino acid residue further along the chain (Fig. 21–6). Each amino acid residue is 1.5 Å from the next amino acid residue, and the helix makes a complete turn for each 3.6 residues. The helix may be coiled in a right-handed or left-handed direction, but the right-handed helix is the most stable. The α-keratins of hair and wool consist of bundles or cables of 3 or 7 such α-helical coils twisted around each other (Fig. 21–5B).

In other proteins such as **fibroin,** the fibrous protein of silk, the polypeptide chains are in an extended zigzag configuration. These chains are arranged alongside each other to form a **pleated sheet** structure, in which the adjacent polypeptide chains run in opposite directions or are antiparallel to each other (Fig. 21–6). The adjacent chains in the pleated sheet are held together by hydrogen bonds.

TERTIARY STRUCTURE

The polypeptide chains of globular proteins are more extensively folded or coiled than those of fibrous proteins. This results from the activity of several types of bonds that hold the structure in a more complex and rigid shape. These bonds are responsible for the **tertiary structure** of proteins, and they exert stronger forces than hydrogen bonds

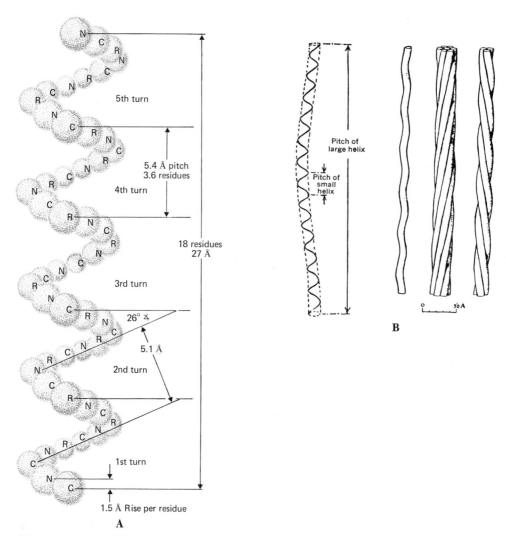

Figure 21-5 *A,* Representation of a polypeptide chain as an α-helical configuration. (After Pauling and Corey: Proc. Intern. Wool Textile Research Conf., *B,* 249, 1955.) *B,* Structure of compound α-helices, proposed to explain the structure of α-keratin. **Left,** The coiling of the axis of an α-helix into a long helix. **Right,** Diagrams of a compound α-helix, of a 7-strand α-cable, and a 3-strand α-rope. (From Pauling, in Edsall and Wyman: *Biophysical Chemistry,* Vol. I. Academic Press, 1958.)

in holding together polypeptide chains or folds of individual chains. A strong **covalent bond** is formed between two cysteine residues, resulting in the disulfide bond. **Salt linkages,** or **ionic bonds,** may be formed between the basic amino acid residues of lysine and arginine and the dicarboxylic amino acids such as aspartic and glutamic. Also, there are many examples of **hydrophobic bonding** that result from the close proximity of aromatic groups or of like aliphatic groups on amino acid residues. Examples of these bonds may be seen in Figure 21–7.

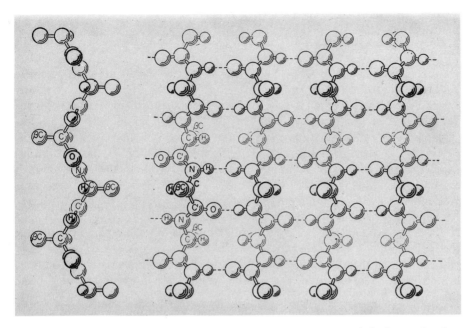

Figure 21-6   The antiparallel-chain pleated sheet structure with hydrogen bond arrangement. (From L. Pauling and R. B. Corey: Proc. Nat. Acad. Sci., 37:729, 1951.)

QUATERNARY STRUCTURE

This level of protein structure involves the polymerization, or degree of aggregation, of protein units. The hemoglobin molecule is a good example of subunit structure in proteins.

Hemoglobin is the respiratory protein of the red blood cell that has been described in detail by Perutz and his coworkers. This protein has a molecular weight of about 64,500,

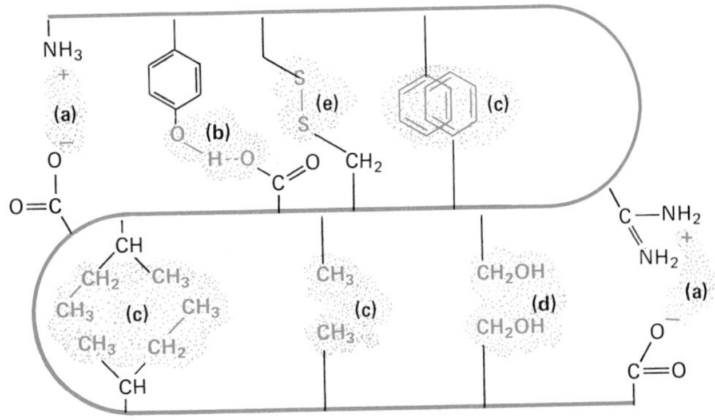

Figure 21-7   Some types of non-covalent bonds which stabilize protein structure: (*a*) Electrostatic interaction; (*b*) hydrogen bonding between tyrosine residues and carboxyl groups on side chains; (*c*) hydrophobic interaction of nonpolar side chains caused by the mutual repulsion of solvent; (*d*) dipole-dipole interaction; (*e*) disulfide linkage, a covalent bond. (After Anfinsen: The Molecular Basis of Evolution. New York, John Wiley and Sons, 1959, p. 102.)

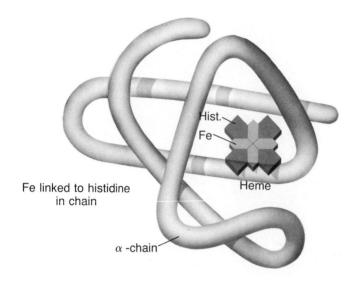

Fe linked to histidine
in chain

Hist.

Fe

Heme

α -chain

Figure 21–8   Artist's conception of
the α-chain of the hemoglobin molecule with its associated heme group,
illustrating the linkage of iron to histidine in the chain. (After Steiner, The
Chemical Foundations of Molecular
Biology. Princeton, N. J., D. Van
Nostrand, 1965.)

and is composed of four polypeptide chains and four heme molecules. Heme is a protoporphyrin derivative with one iron atom coordinated with each of the four pyrrole
nitrogen atoms.

Heme

The four polypeptide chains, two α and two β chains (p. 276), exist in the form of α-helices
which are folded and bent into three-dimensional structures. In the hemoglobin molecule,
the four chains are grouped in a tetrahedron-shaped structure with the four heme molecules embedded in hollows in the folded chains. The relationship of the iron atom, the
heme molecule, and one of the polypeptide chains may be seen in the model of one
fourth of a hemoglobin molecule (Fig. 21–8). The iron atom is linked to the imidazole
group of histidine at position 87 in the β chain.

CLASSIFICATION OF PROTEINS

Proteins are most often classified on the basis of their chemical composition or
solubility properties. Of the three main types, **simple proteins** are classified by solubility
properties, **conjugated proteins** by the nonprotein groups, and **derived proteins** by the
method of alteration.

Simple proteins such as **protamines** in the form of salmine and sturine from fish sperm,
histones in the form of the globin in hemoglobin, and **albumins** in the form of egg albumin
and serum albumin are all soluble in water, and protamines and histones contain a high

proportion of basic amino acids. The **globulins** as lactoglobulin in milk are insoluble in water but soluble in dilute salt solutions; the **glutelins** as glutenin in wheat are insoluble in water and dilute salt solutions, but are soluble in dilute acid or alkaline solutions; **prolamines** as zein in corn and gliadin in wheat are soluble in 70 to 80 per cent ethyl alcohol; whereas the **albuminoids** as keratin in hair, horn, and feathers are insoluble in all the solvents mentioned above and can be dissolved only by hydrolysis.

Conjugated proteins include **nucleoproteins,** which consist of a basic protein such as histones or protamines combined with nucleic acid. They are found in cell nuclei and mitochondria. **Phosphoproteins** as casein in milk and vitellin in egg yolk are proteins linked to phosphoric acid; **glycoproteins** are composed of a protein and a carbohydrate and occur in mucin in saliva and mucoids in tendon and cartilage, whereas **chromoproteins** such as hemoglobin and cytochromes consist of a protein combined with a colored compound. **Lipoproteins** are proteins combined with lipids such as fatty acids, fats, and lecithin, and are found in serum, brain, and nervous tissue.

Derived proteins are an indefinite type of protein produced, for example, by partial hydrolysis, denaturation, and heat, and are represented by proteoses, peptones, meta-proteins, and coagulated proteins.

DENATURATION OF PROTEINS

Denaturation of a protein molecule causes changes in the structure that result in marked alterations of the physical properties of the protein. When in solution, proteins are readily denatured by standing in acids or alkalies, shaking, heating, reducing agents, detergents, organic solvents, and exposure to x-rays and light. Some of the effects of **denaturation** are loss of biological activity, decreased solubility at the isoelectric point, increased susceptibility to hydrolysis by proteolytic enzymes, and increased reactivity of groups that had been masked by the folding of chains in the native protein. Examples of the last mentioned are the uncovered SH groups of cysteine and the OH groups of tyrosine.

The cleavage of several hydrogen bonds and of several possible disulfide bonds often results in the loss of biological activity by denaturation. In some proteins, the native configuration is so stable that denaturation changes are reversible. Hemoglobin, for example, can be reversibly denatured.

PRECIPITATION OF PROTEINS

One of the most important characteristics of proteins is the ease with which they are precipitated by certain reagents. Many of the normal functions in the body are essentially precipitation reactions; for example, the clotting of blood or the precipitation of casein by rennin during digestion. Since animal tissues are chiefly protein in nature, reagents that precipitate protein will have a marked toxic effect if introduced into the body. Bacteria, which are mainly protein, are effectively destroyed when treated with suitable precipitants. Many of the common poisons and disinfectants act in this way. The following paragraphs contain a brief summary of the most common methods of protein precipitation.

BY HEAT COAGULATION

When most protein solutions are heated, the protein becomes insoluble and precipitates, forming coagulated protein. Many protein foods coagulate when they are cooked. Tissue proteins and bacterial proteins are readily coagulated by heat. Routine examinations of urine specimens for protein are made by heating the urine in a test tube to coagulate any protein that might be present.

By Alcohol

Alcohol coagulates all proteins except the prolamines. A 70 per cent solution of alcohol is commonly used to sterilize the skin, since it effectively penetrates the bacteria. A 95 per cent solution of alcohol is not effective because it merely coagulates the surface of the bacteria and does not destroy them.

By Concentrated Inorganic Acids

Proteins are precipitated from their solutions by concentrated acids such as hydrochloric, sulfuric, and nitric acid. Casein, for example, is precipitated from milk as a curd when acted on by the hydrochloric acid of the gastric juice.

By Salts of Heavy Metals

Salts of heavy metals, such as mercuric chloride and silver nitrate, precipitate proteins. Since proteins behave as zwitterions, they will ionize as negative charges in neutral or alkaline solutions. The reaction with silver ions may be illustrated as follows:

$$R-\underset{\underset{NH_2}{|}}{CH}-COO^- + Ag^+ \rightarrow R-\underset{\underset{NH_2}{|}}{CH}-COOAg$$

Protein   Silver proteinate

These salts are used for their disinfecting action and are toxic when taken internally. A protein solution such as egg white or milk, when given as an antidote in cases of poisoning with heavy metals, combines with the metallic salts. The precipitate that is formed must be removed by the use of an emetic before the protein is digested and the heavy metal is set free to act on the tissue protein. A silver salt such as Argyrol is used in nose and throat infections, and silver nitrate is used to cauterize wounds and to prevent gonorrheal infection in the eyes of newborn babies by precipitating bacterial protein.

By Alkaloidal Reagents

Tannic, picric, and tungstic acids are common alkaloidal reagents that will precipitate proteins from solution. When in acid solution, the protein as a zwitterion ionizes as a positive charge. It will therefore react with picric acid (2,4,6-trinitrophenol) as shown:

$$R-\underset{\underset{NH_3^+}{|}}{CH}-COOH + \text{picric acid} \rightarrow R-\underset{\underset{NH_3-\text{picrate}}{|}}{CH}-COOH$$

Protein   Protein picrate

Tannic and picric acids are sometimes used in the treatment of burns. When a solution of either of these acids is sprayed on extensively burned areas, it precipitates the protein to form a protective coating; this excludes air from the burn and prevents the loss of water. In an emergency, strong tea may be used as a source of tannic acid for the treatment of severe burns. Many other therapeutic agents have been used in the treatment of burns, the most recent being penicillin.

IMPORTANT TERMS AND CONCEPTS

alpha amino acid	amino acid sequence	chromatography	heterocyclic amino acids
alpha helix	aromatic amino acids	hemoglobin	isoelectric point

peptide linkage   primary structure   tertiary structure   zwitterions
polypeptides   secondary structure

QUESTIONS

1. List the five main elements present in proteins with their average content. Using this list, how would you differentiate proteins from carbohydrates and fats?

2. What products are formed by complete hydrolysis of proteins? Which chemical agents would be used for complete hydrolysis?

3. Write the formula for an alpha amino acid containing three carbon atoms. How would you name this amino acid?

4. Write the formula and name of an amino acid that contains a heterocyclic ring.

5. Write the formula of an amino acid as a zwitterion and use the structure to explain the isoelectric point.

6. Illustrate with equations two reactions that involve the amino group of an amino acid.

7. Given the dipeptide alanylglycine as an unknown, explain the procedures you would employ to:
   a. Prove the identity of the two amino acids.
   b. Determine which amino acid in the dipeptide contained the free amino group.

8. Briefly explain the application of chromatography and Sanger's reagent to the analysis of the insulin molecule.

9. How does the primary structure differ from the tertiary structure of a protein? Explain.

10. Why is a protein solution used as an antidote in cases of poisoning with heavy metals? Explain.

11. Illustrate with equations the precipitation of proteins with tannic acid and with silver salts.

12. Why are preparations containing tannic or picric acids sometimes used in the treatment of burns?

SUGGESTED READING

Edelman: The Structure and Function of Antibodies. Scientific American, Vol. 223, No. 2, p. 34, 1970.
Fraser: Keratins. Scientific American, Vol. 221, No. 2, p. 86, 1969.
Isenberg and Grdinic: Cyclic Disulfides, Their Functions in Health and Disease. Journal of Chemical Education, Vol. 49, No. 6, p. 392, 1972.
McGuinness: Estimation of Protein Size, Weight, and Asymmetry by Gel Chromatography. Journal of Chemical Education, Vol. 50, No. 12, p. 826, 1973.
Research Reporter: Methionine and Origin of Life. Chemistry, Vol. 46, No. 2, p. 14, 1973.
Safrany: Nitrogen Fixation. Scientific American, Vol. 231, No. 4, p. 64, 1974.
Sharon: Glycoproteins. Scientific American, Vol. 230, No. 5, p. 78, 1974.
Vedvick and Coates: Hemoglobin: A Simple "Backbone" Type of Molecular Structure. Journal of Chemical Education, Vol. 48, p. 537, 1971.

NUCLEIC ACIDS

CHAPTER 22

The *objectives* of this chapter are to enable the student to:

1. Describe the structure of the products of complete hydrolysis of a nucleoprotein.
2. Illustrate hydrogen bonding and antiparallel chains in the DNA molecule.
3. Illustrate the tetranucleotide portion of one chain of DNA.
4. Distinguish between the different types of RNA molecules.

Nucleic acids were first isolated from cell nuclei; they make up about one fourth of the organic matter of living cells. They were thought to be fairly simple groups that were conjugated with proteins to form nucleoproteins. These proteins are characterized by their content of **basic amino acids** such as **arginine** and **lysine.** We know now that nucleic acids are polymers of large molecular weight with nucleotides as the repeating unit. Deoxyribonucleic acid (DNA) is present in the nucleus and ribonucleic acid (RNA) in the cytoplasm of all living cells. DNA and RNA are essentially responsible for the transmission of genetic information and the synthesis of protein by the cell, respectively. Progressive hydrolysis of a nucleoprotein would yield the protein and nucleic acid and its components as shown:

$$Nucleoprotein \rightarrow Nucleic\ acid + Protein$$
$$\downarrow$$
$$Nucleotides$$
$$\downarrow$$
$$Nucleosides + H_3PO_4$$
$$\downarrow$$
$$Purines\ and\ Pyrimidines + Pentose$$

THE PYRIMIDINE AND PURINE BASES

The heterocyclic rings that form the nucleus for both the pyrimidine and purine bases have already been described. The pyrimidine bases found in nucleic acids include cytosine, uracil, and thymine, which are represented as follows:

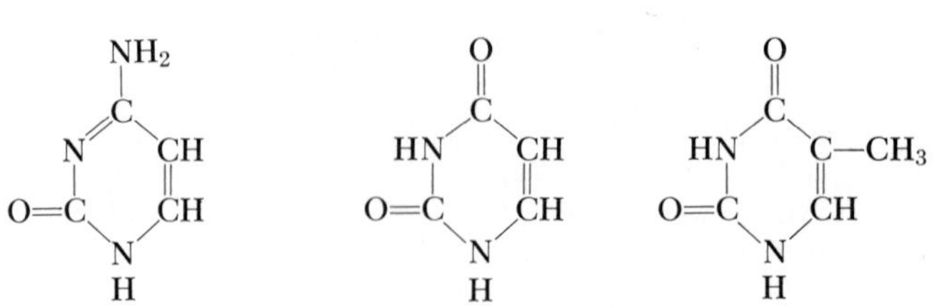

Cytosine	Uracil	Thymine
(2-oxy-4-amino pyrimidine)	(2,4-dioxy pyrimidine)	(5-methyl uracil)

Both DNA and RNA contain the purines adenine and guanine.

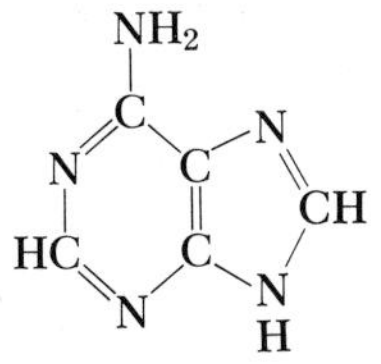

Adenine
(6-amino purine)

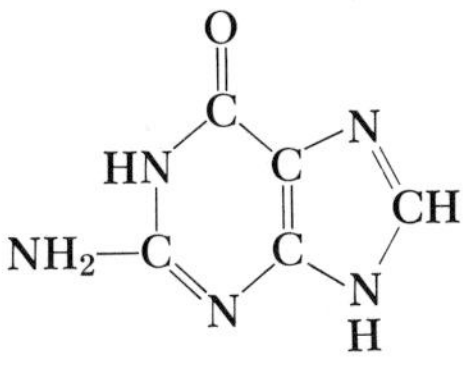

Guanine
(2-amino-6-oxy purine)

THE PENTOSE SUGARS

The pentose sugars that are essential components of the nucleic acids are simple sugars described on pages 304 and 305. On hydrolysis RNA yielded β-D-ribose, whereas DNA contained β-2-deoxy-D-ribose. Both pentose sugars occur in the furanose form.

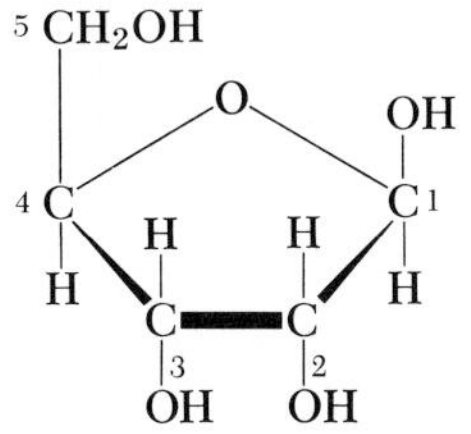

β-D-Ribose

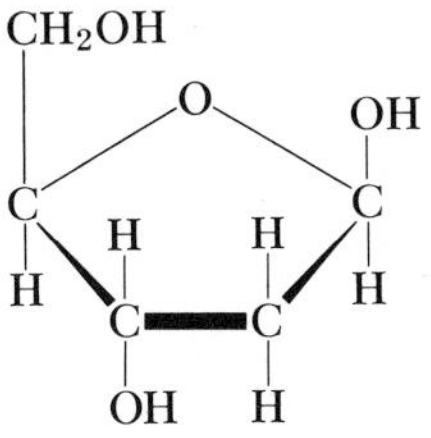

β-2-deoxy-D-Ribose

NUCLEOSIDES

When a purine or pyrimidine base is combined with β-D-ribose or β-2-deoxy-D-ribose, the resultant molecule is called a **nucleoside.** The linkage of the two components is from the nitrogen of the base (position 1 in pyrimidines, 9 in purines) to carbon 1' of the pentose sugars. Important examples of nucleosides are cytidine and adenosine:

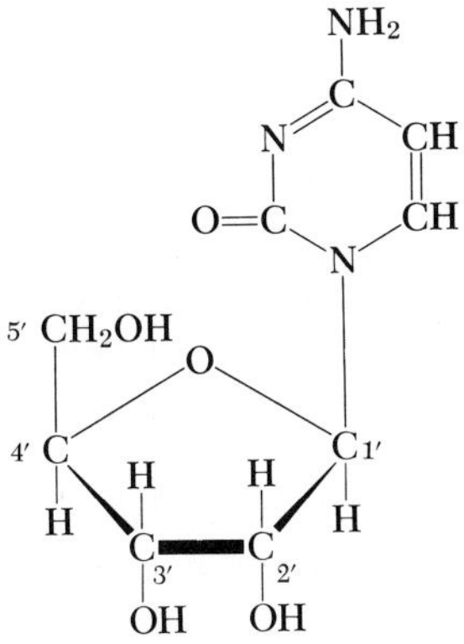

Cytidine

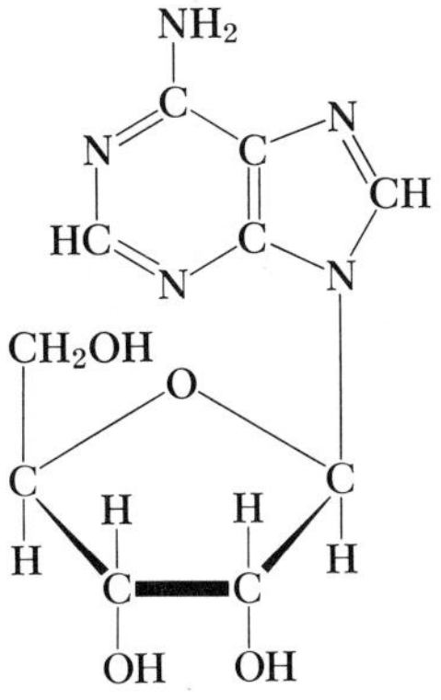

Adenosine

NUCLEOTIDES

When a phosphoric acid is attached to a hydroxyl group of the pentose sugar in the nucleoside by an ester linkage, the result is a **nucleotide.** The esterification may occur on the 2', 3', or 5' hydroxyl of ribose and the 3' or 5' hydroxyl of deoxyribose. In the nucleotides, the carbon atoms of ribose or deoxyribose are designated by prime numbers to distinguish them from the atoms in the purine or pyrimidine bases. These structures may be represented by adenylic acid and uridylic acid, two compounds of prime importance in muscle metabolism and carbohydrate metabolism:

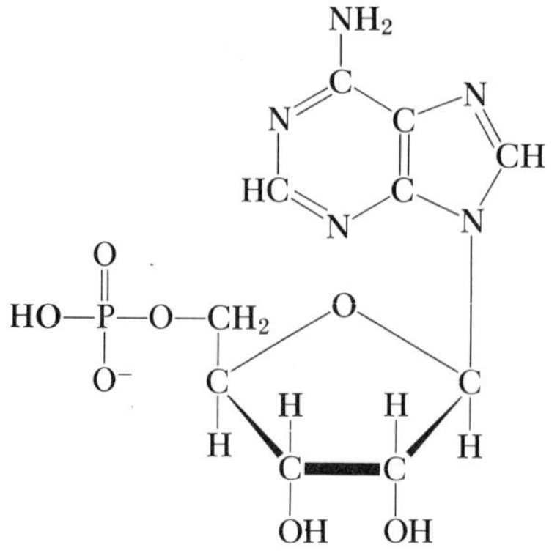

Adenylic acid
(adenosine-5′-monophosphate, AMP)

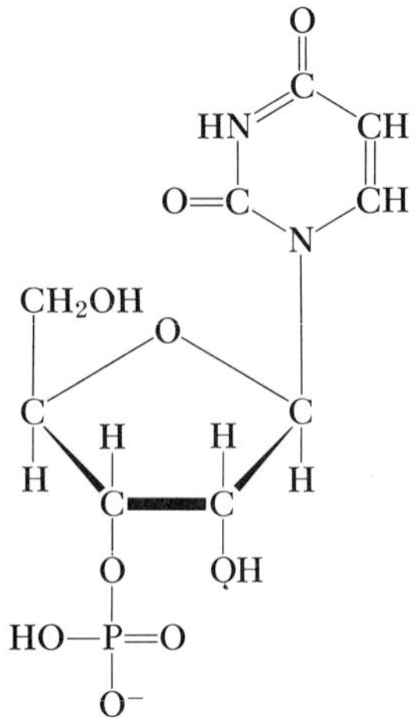

Uridylic acid
(uridine-3′-monophosphate, UMP)

The biochemist commonly uses abbreviations such as AMP and, for uridine monophosphate, UMP, to designate nucleotides. Adenylic acid may also exist as the diphosphate (adenosine diphosphate, ADP) or as the triphosphate (adenosine triphosphate, ATP) (p. 338). These two compounds are sources of high energy phosphate bonds and are involved in many metabolic reactions.

Nucleotides also combine with vitamins to form coenzymes, which will be discussed in Chapter 26.

NUCLEIC ACIDS

THE STRUCTURE OF DNA

The structure of DNA has been shown to consist of chains of nucleotides linked together with phosphate groups that connect carbon atom 3 of one sugar molecule to the number 5 carbon of the next sugar. A portion of the chain containing one each of the four nucleotides of DNA will be represented, although apparently the nucleotides can occur in random order in the chain. This structure is on page 287. The 3,5 linkage from one deoxyribose molecule to another is characteristic in DNA and is the predominant linkage in RNA. The molecular weights of DNA preparations from various sources exhibit a range from 6 to 16 million and illustrate the large size of the polymer. A series of hydrolytic experiments utilizing weak acids, bases, and enzymes has established a 1:1 ratio between cytosine and guanine nucleotides and a similar 1:1 ratio between adenine and thymine nucleotides.

By physical-chemical methods, such as x-ray diffraction, the DNA molecule has been shown to consist of two chains of nucleotides coiled in a double helical structure, forming a molecule that is very long by contrast to its diameter. Watson and Crick suggested that the pair of nucleotide chains run in opposite directions around the helix in such a way that an adenine of one chain is bonded to a thymine of the other by two hydrogen

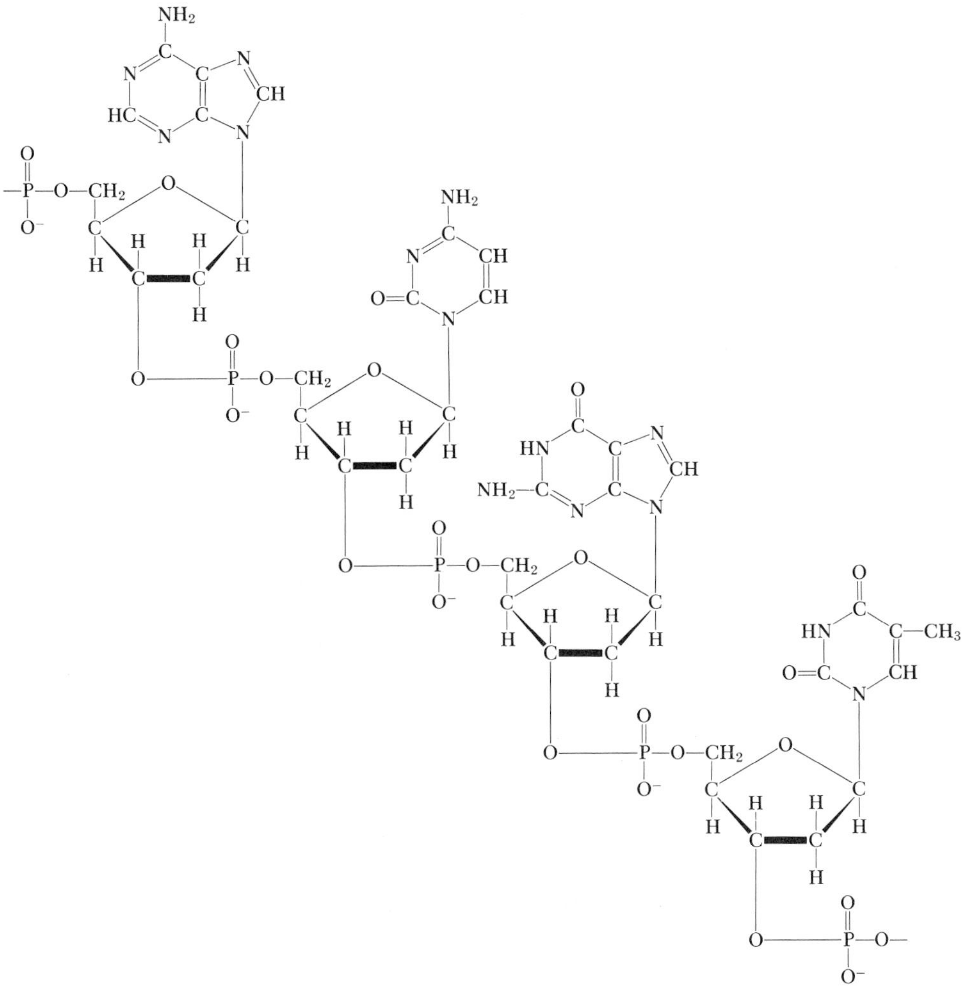

Tetranucleotide portion of one chain of DNA

bonds, and a guanine of one chain is joined by three hydrogen bonds to a cytosine of the other as shown in Figure 22–1. The chains consist of deoxyribose nucleotides joined together by phosphate groups with the bases projecting perpendicularly from the chain into the center of the helix. For every adenine or guanine projecting into the central portion, a thymine or cytosine molecule must project toward it from a second chain. Because of the base pairing, the two chains are not identical, and they do not run in the same direction with respect to the linkages between deoxyribose and the base. The chains in the DNA structure are therefore considered **antiparallel.** A portion of the helix emphasizing the hydrogen bonding and the antiparallel chain structure is shown in Figure 22–2.

The Structure of RNA

Information concerning the structure of RNA is not as extensive or as definite as that describing DNA. In general, RNA has been shown to have a structure similar to that of DNA, although the smaller molecules of viral RNA, and also messenger RNA, have single strands of nucleotides. The RNA molecules in the cytoplasm of the cell are of three major types. The largest molecules are **ribosomal RNA** (r-RNA), with molecular

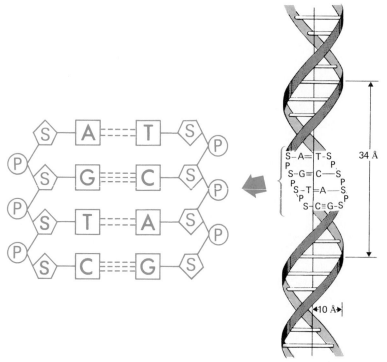

Figure 22–1   Double helix of DNA. Here P means phosphate diester, S means deoxyribose, A=T is the adenine-thymine pairing, and G≡C is the guanosine-cytosine pairing. (After *Outlines of Biochemistry*, 2nd ed., by E. E. Conn and P. K. Stumpf. Copyright © 1963. Reprinted by permission of John Wiley & Sons, Inc.)

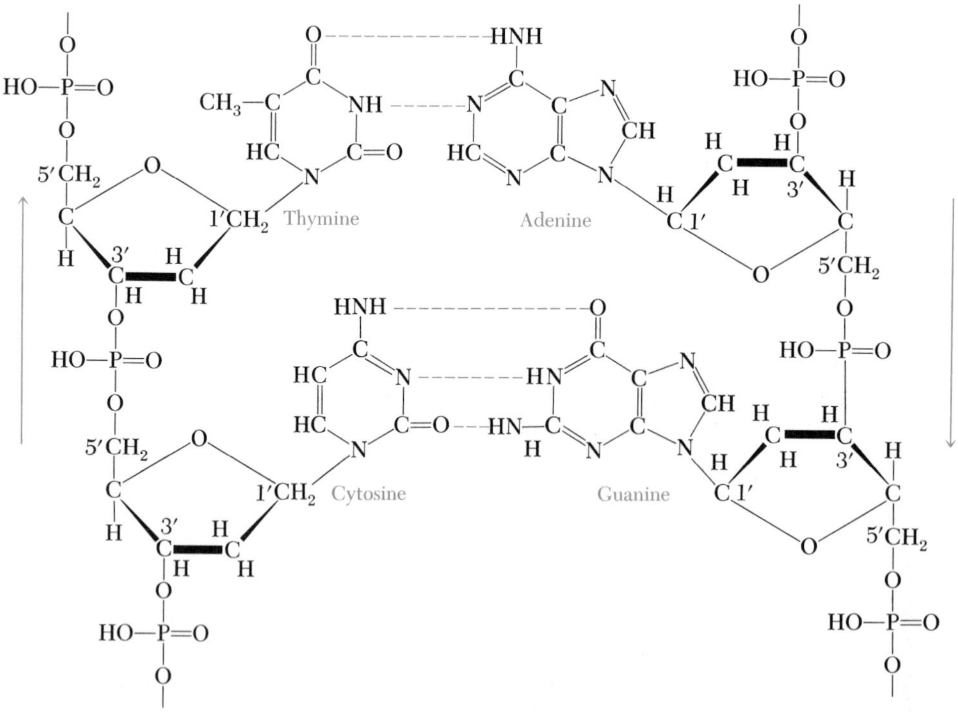

Figure 22–2   Hydrogen bonding with antiparallel chains. (Adapted from *Outlines of Biochemistry*, 2nd ed., by E. E. Conn and P. K. Stumpf. Copyright © 1963. Reprinted by permission of John Wiley & Sons, Inc.)

weights of a few million. They are associated with the structure of the ribosomes and serve as a template for protein synthesis within the cytoplasm. **Messenger RNA** (m-RNA) molecules are varied in size, with molecular weights from 300,000 to two million. The m-RNA molecules carry the genetic message from the DNA in the nucleus to the protein-synthesizing sites. The smallest RNA molecules are called **transfer RNA** (t-RNA) and have molecular weights from 25,000 to 40,000. Their function is to transport specific amino acids to their specific sites on the protein-synthesizing template. The basic composition of three transfer RNAs is known. They are composed of single strands of nucleotides bent into cloverleaf type structures to give the maximum number of hydrogen-bonded pairs. The closed loop of the chain at one end of the cloverleaf contains a sequence of three bases that serves as an anticodon for a specific amino acid (see Chapter 31). The role of these RNA molecules will be discussed in Chapter 31.

THE BIOLOGIC IMPORTANCE OF THE NUCLEIC ACIDS

Originally RNA was associated only with yeast and was thought to be restricted to plant sources. DNA from thymus tissue represented the nucleic acids of animal tissues. As methods for their determination have been developed, both RNA and DNA have been found in practically all types of cells. DNA appears to be restricted to the nucleus, most specifically to the chromosomes, whereas RNA occurs both in the cytoplasm and nucleus of a cell. The overall function of DNA and its relation to RNA and protein synthesis are outlined in Figure 22–3.

SYNTHESIS OF NUCLEIC ACIDS

From the composition of nucleic acids described on pages 286–287, it is apparent that the synthesis would involve the copolymerization of four ribonucleotides for RNA and four deoxyribonucleotides for DNA. The DNA molecule has been synthesized by Kornberg and his coworkers by treating a mixture of deoxyribonucleotides with an enzyme isolated from bacteria. In the presence of all four deoxyribonucleotides, a DNA primer, and Mg^{+2}, the enzyme **DNA polymerase** can synthesize DNA. Enzymes called **RNA polymerases** or **transcriptases** found in plants, animals, and bacteria catalyze the synthesis of RNA

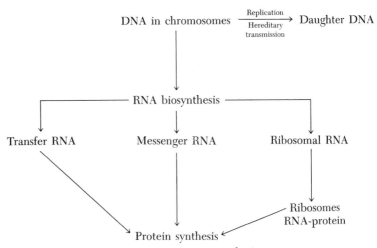

Figure 22–3   The relation of DNA to RNA in protein synthesis.

from ribonucleoside triphosphates, ATP, GTP, CTP, and UTP. The synthesis is DNA-dependent, and requires Mg^{+2}, in addition to all four ribonucleoside triphosphates. The DNA serves as a template, and the base sequence in the DNA is transcribed into a corresponding sequence in the RNA. These nucleic acid molecules are similar to those from natural sources and are used to study the properties, compositions, and reactions of these large molecules.

IMPORTANT TERMS AND CONCEPTS

adenine	guanine	nucleotide	pyrimidine
cytosine	hydrogen bonding	purine	RNA
DNA	nucleoside		

QUESTIONS

1. What type of protein is found in nucleoproteins? What products result from complete hydrolysis of nucleic acids?

2. Write the formula and name for a nucleoside containing β-2-deoxyribose.

3. How are the nucleotides linked together chemically when they form nucleic acids?

4. Prepare a sketch of the DNA molecule showing the double helix and the hydrogen bonding between the bases.

5. What is meant by the antiparallel chain structure of DNA?

6. How does the structure of RNA differ from that of DNA? What are the three major types of RNA in the cell?

SUGGESTED READING

Crick: The Genetic Code. Scientific American, Vol. 215, No. 4, p. 55, 1966.
Research Reporter: Portrait of a Gene. Chemistry, Vol. 42, No. 8, p. 20, 1969.
Research Reporter: Further Confirmation of Nucleic Acid Double Helix. Chemistry, Vol. 46, No. 7, p. 19, 1973.
Research Reporter: New DNA Structure Proposed. Chemistry, Vol. 47, No. 9, p. 22, 1974.
Sobell: How Actinomycin Binds to DNA. Scientific American, Vol. 231, No. 2, p. 82, 1974.
Temin: RNA-Directed DNA Synthesis. Scientific American, Vol. 226, No. 1, p. 24, 1972.
Watson: Double Helix. Atheneum, New York, 1968.
Yanofsky: Gene Structure and Protein Structure. Scientific American, Vol. 216, No. 5, p. 80, 1967.

ENZYMES

The *objectives* of this chapter are to enable the student to:

1. Define and describe the chemical nature of enzymes.
2. Recognize the relationship between the active site of an enzyme and its specificity of action.
3. Explain the value of the Michaelis constant, K_m, of an enzyme.
4. Describe the effect of pH, temperature, and end products on an enzyme reaction.
5. Recognize and illustrate the difference between a competitive and a noncompetitive inhibitor.

A rapidly developing field of biochemical research involves the study of enzyme-controlled reactions in the cells of animals, plants, and microorganisms. When classified according to biological function, enzymes make up the largest and most highly specialized class of proteins. Many of the thousands of chemical reactions that occur in living cells are catalyzed by enzymes, and it has been estimated that each cell contains in excess of 1000 separate enzymes. These enzymes are not randomly scattered in the cell as in a sack of enzymes, but are located in specific areas or subcellular components concerned with various types of metabolic activity. The enzymes of the Embden-Meyerhof pathway for glycolysis (see p. 354) are found in the cytoplasm, whereas the enzymes involved in the Krebs cycle (see p. 355) are located in the mitochondria.

THE CHEMICAL NATURE OF ENZYMES

Enzymes have always been considered as catalysts and are often compared to inorganic catalytic agents such as platinum and nickel that are often used in conjunction with high temperatures, high pressures, and favorable chemical conditions. Few of these conditions occur when an enzyme reacts in body tissue, at body temperature, and at the pH of body fluids. Enzymes were originally defined as catalysts, organic in nature, formed by living cells, but independent of the presence of the cells in their action. A more current definition would state that *enzymes are proteins, formed by a living cell, which catalyze a thermodynamically possible reaction by lowering the activation energy so the rate of reaction is compatible with the conditions in the cell.* An enzyme does not change the $\Delta G°$ (p. 337) or equilibrium constant of a reaction.

By now it is generally accepted that all enzymes are protein in nature. The purification, crystallization, and inactivation procedures exactly parallel those for pure proteins. For example, excessive heat, alcohol, salts of heavy metals, and concentrated inorganic acids will cause coagulation or precipitation of the protein material and thus inactivate an enzyme.

Originally enzymes were named according to their source or according to the method of separation when they were discovered. As the family of enzymes grew, they were named in a more orderly fashion by adding the ending **-ase** to the root of the name of the substrate. An enzyme's **substrate** is the compound or type of substance upon which it acts. For example, sucrase catalyzes the hydrolysis of sucrose, lipase is an enzyme that hydrolyzes lipids, and urease is the enzyme that splits urea. This system also was used to name types of enzymes such as proteases, oxidases, and hydrolases. The discovery of so many enzyme mechanisms in the past few years has resulted in a mass of complex substrates and enzyme nomenclature. The problem was assigned to a Commission on Enzymes of the International Union of Biochemistry. They were not in favor of eliminating all the names in common usage, but recommended the use of two names for an enzyme. One was the trivial name, either the one in common use or a simple name describing the activity of the enzyme. The other was constructed by the addition of the ending **-ase** to an accurate chemical name for the substrate.

CLASSIFICATION OF ENZYMES

As new enzymes are discovered and further enzyme mechanisms are elaborated the problem of classification becomes more complex. Perhaps the most satisfactory basis for classification depends on the type of reaction influenced by the enzyme. On this basis, the Commission on Enzymes of the International Union of Biochemistry proposed that enzymes be classified in six main divisions. Although new names are introduced for some of the groups of enzymes that catalyze specific chemical reactions, we will retain the familiar names of enzymes that are included as examples within the six groups.

1. **Oxidoreductases**—Enzymes in this group are involved in physiological oxidation processes in the body. They include the common oxidation-reduction enzymes, examples of which are the dehydrogenases, oxidases, peroxidases, and hydrases.
2. **Transferases**—This group of enzymes catalyzes the transfer of essential chemical groups from one compound to another. Important examples are transaminases, transmethylases, and transacylases.
3. **Hydrolases**—This large group of enzymes is involved in the hydrolysis of compounds by the introduction of water molecules. They include the digestive enzymes such as amylase, sucrase, lipase, and the proteases.
4. **Lyases**—These enzymes catalyze the removal of chemical groups from compounds nonhydrolytically. Decarboxylases that catalyze the removal of CO_2 from organic molecules and aldolase, an enzyme involved in glycolysis, are examples of lyases.
5. **Isomerases**—This group of enzymes catalyzes different types of isomerization. The names of the enzymes indicate their activity. For example, *cis-trans* isomerases, racemases, intramolecular transferases, and epimerases are included in this group.
6. **Ligases**—These enzymes catalyze the linking together of two molecules with the breaking of a pyrophosphate bond of ATP or a similar triphosphate. An example would be an amino acid, RNA ligase, which is involved in protein synthesis in the cell.

PROPERTIES OF ENZYMES

SPECIFICITY OF ACTION

Perhaps the major difference between the classical inorganic catalysts, such as platinum and nickel, and enzymes is the specificity of action of the latter. Platinum, for example, will act as a catalyst for several reactions. Enzymes may exhibit different types of specificities; for example, urease exhibits **absolute specificity** in action in that it catalyzes the splitting of a single compound, urea. Other enzymes exhibit **stereochemical specificity**; D-amino acid oxidase is specific for D-amino acids and will not affect the natural L-amino acids. Arginase catalyzes the hydrolysis of L-arginine to urea and ornithine, but will not act on the D-isomer.

Certain enzymes are **linkage specific;** for example, trypsin splits peptide bonds adjacent to lysine or arginine residues, whereas chymotrypsin splits bonds next to aromatic amino acids, such as tyrosine in a polypeptide chain. Enzymes that exhibit **group speci-ficity** are also valuable in sequence studies. Carboxypeptidase is specific for terminal amino acids containing a free carboxyl group, and aminopeptidase splits terminal amino acids with a free amino group off the end of a peptide chain. In general, the specificity of action accounts for the large numbers of enzymes found in cells and tissues, and for the fact that enzymes are involved in all the metabolic reactions that occur in the cell.

ENERGY OF ACTIVATION

Since all chemical reactions tend to proceed in the direction of forming products of a lower energy content than the reactants, the problem is how to initiate the reaction and make it proceed spontaneously. In the reaction

$$A + B \rightarrow \text{Activated state} \rightarrow C + D$$

energy is required to produce the activated state before the reaction can proceed to products $C + D$. The energy of activation, E_a, required to change $A + B$ to the activated state is affected by a catalyst (Fig. 7–1, p. 104). As an example of the effect of an enzyme on the energy of activation, to decompose hydrogen peroxide to water and oxygen without a catalyst requires 18.0 kcal/mole. An effective metallic catalyst such as platinum lowers the activation energy of the reaction to 12.0 kcal/mole, compared to the low activation energy required by the enzyme catalase of about 3.0 kcal/mole. In general, enzymes lower activation energies of reactions to the point where they can be readily carried out at body temperature under the conditions of living tissue.

ENZYME ACTIVITY

The activity of an enzyme is affected by many factors. Most important are the concentrations of the substrate and the enzyme, and the temperature and the pH of the reaction. In addition, the rate of enzyme reaction is affected by the nature of the end products, the presence of inhibitors, and light. The activity may be measured by following the chemical change that is catalyzed by the enzyme. The substrate is incubated with the enzyme under favorable conditions, and samples are withdrawn at short intervals for analysis of the end products or analysis of the decrease of substrate concentration. The enzyme lipase, for example, catalyzes the hydrolysis of fat molecules to fatty acids and glycerol. A simple method of measuring the activity of lipase would involve a determination of the rate of appearance of fatty acid molecules.

EFFECT OF SUBSTRATE

Michaelis and Menten in 1913 first expressed the concept of the **enzyme substrate complex** as a transition state in enzyme reactions.

$$E \quad + \quad S \quad \underset{k_{-1}}{\overset{k_1}{\rightleftharpoons}} \quad ES \quad \overset{k_2}{\longrightarrow} \quad E + P$$

Enzyme   Substrate   Enzyme substrate complex   Products

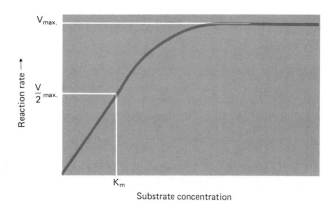

Reaction rate →

$V_{max.}$

$\dfrac{V}{2}$ max.

K_m

Substrate concentration

Figure 23–1   Effect of substrate concentration on the reaction rate when the enzyme concentration is held constant.

The formation of the ES complex permits the overall reaction to proceed at a lower energy of activation. A constant, K_m, known as the **Michaelis constant,** is related to the three rate constants, k_1, k_{-1}, and k_2. Under proper conditions of temperature and pH, the Michaelis constant, K_m, is approximately equal to the dissociation constant of the ES complex. Conversely, the reciprocal of K_m, or $1/K_m$, is a measure of the *affinity of an enzyme for its substrate.* The substrate concentration required to yield half the $v_{max.}$ can be readily determined from Figure 23–1, and is equal to K_m. A K_m of 0.001M would indicate that the active site of the enzyme is half saturated when its substrate is present at that concentration. The enzyme, therefore, has a high affinity for its substrate.

As seen from Figure 23–1, K_m by definition equals the substrate concentration [S] at one half the maximal rate, $v_{max.}$, where $v_{max.}$ equals the maximum rate at the saturation concentration of the substrate. The rate of the reaction, v, at a given substrate concentration can be expressed as:

$$v = \frac{v_{max.}[S]}{K_m + [S]}$$

This relationship between the rate of reaction, $v_{max.}$, K_m, and [S] is known as the Michaelis-Menten equation.

The curve shown in Figure 23–1 represents the ideal relationship between the substrate concentration [S] and the maximum velocity $v_{max.}$. Experimentally it is difficult to reproduce this curve for a particular enzyme to obtain an accurate measure of $v_{max.}$ or K_m. By inverting the Michaelis-Menten equation and expressing it as an equation for a straight line, the following relationship may be obtained:

$$\frac{1}{v} = \frac{K_m}{v_{max.}} \times \frac{1}{[S]} + \frac{1}{v_{max.}}$$

This equation may then be used to construct a double-reciprocal plot of $\dfrac{1}{v}$ versus $\dfrac{1}{[S]}$ to obtain a graphic evaluation of K_m and $v_{max.}$. This is called a Lineweaver-Burk plot, after the investigators who proposed its use (Fig. 23–2). Experimentally, it requires only a few points on the curve to determine K_m; therefore, the Lineweaver-Burk plot method is most often used for this purpose in the laboratory. In addition to the K_m representing a measure of the affinity of an enzyme for its substrate, it also is of practical value in the assay of enzymes. At a substrate concentration of 100 times the K_m value, the enzyme will exhibit a maximum rate of activity or $v_{max.}$. The K_m value, therefore, determines the amount of substrate to use in an enzyme assay.

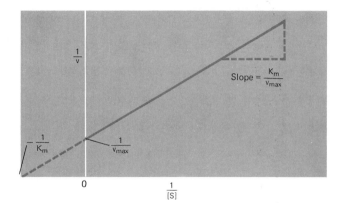

Figure 23–2   The Lineweaver-Burk plot of an enzyme reaction.

THE ACTIVE SITE

In view of the specificity of action of many enzymes, it is reasonable to assume that only a small portion of the enzyme protein is involved in its catalytic activity. The portion of the enzyme molecule to which the substrate binds is called the **active site.** This site often consists of specific amino acids or specific charges in a side chain. Trypsinogen, for example, is synthesized in the pancreas and is converted to the active enzyme trypsin by the enzyme enterokinase or even by trypsin itself. The primary structure of trypsinogen is a single polypeptide chain, and the conversion to trypsin is thought to involve the splitting off of a hexapeptide followed by a change in the conformation of the polypeptide to expose the catalytically active sites containing serine and histidine residues (Fig. 23–3).

EFFECT OF ENZYME

When a purified enzyme is used, the rate of reaction is proportional to the concentration of the enzyme over a fairly wide range (Fig. 23–4). The substrate concentration must be kept constant and remain in excess of that required to combine with the enzyme. This relationship may also be used to measure the amount of an enzyme in a tissue extract or a biologic fluid. At the proper substrate concentration ($100 \times K_m$), temperature, and pH, the measured rate of activity is proportional to the quantity of enzyme present.

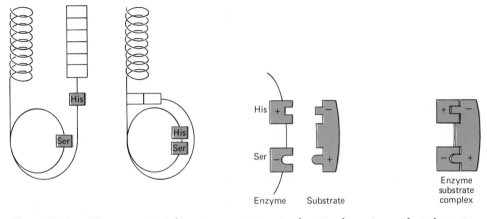

Figure 23–3   The conversion of trypsinogen to trypsin, showing the amino acids at the active site.

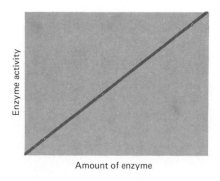

Amount of enzyme

Figure 23–4   The effect of increasing amounts of enzyme on the activity of the enzyme.

EFFECT OF pH

The hydrogen ion concentration, or pH, of the reaction mixture exerts a definite influence on the rate of enzyme activity. If a curve is plotted comparing changes in pH with the rate of enzyme activity, it takes the form of an inverted U or V (Fig. 23–5). The maximum rate occurs at the **optimum pH,** with a rapid decrease of activity on either side of this pH value. The optimum pH of an enzyme may be related to a certain electric charge on the surface or to optimum conditions for the binding of the enzyme to its substrate. Most enzymes exhibit an optimum pH value close to 7, although pepsin is most active at pH 1.6 and trypsin at pH 8.2. Pepsin has no activity in an alkaline solution, whereas trypsin is inactive in an acid solution.

EFFECT OF TEMPERATURE

The speed of most chemical reactions is increased two or three times for each 10°C rise in temperature. This is also true for reactions in which an enzyme is the catalyst, although the temperature range is fairly narrow. The activity range for most enzymes occurs between 10° and 50°C; the **optimum temperature** for enzymes in the body is around 37°C. The increased rate of activity observed at 50°C or above is short-lived, because the increased temperature first denatures and then coagulates the enzyme protein, thereby destroying its activity. The optimum temperature of an enzyme is therefore dependent on a balance between the rise in activity with increased temperature and the denaturation or inactivation by heat (Fig. 23–6). For any 10° rise in temperature the change in rate of enzyme activity is known as the Q_{10} value, or temperature coefficient. The Q_{10} value for most enzymes varies from 1.5 to 3.0.

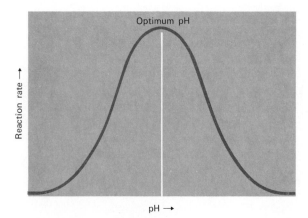

Optimum pH

Reaction rate →

pH →

Figure 23–5   The effect of pH on enzyme activity.

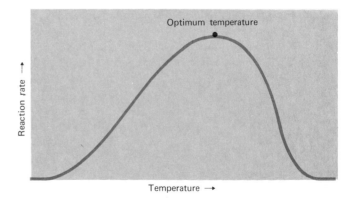

Figure 23-6   The effect of temperature on enzyme activity.

EFFECT OF END PRODUCTS

The end products of an enzyme reaction have a definite effect on the rate of activity of the enzyme. If they are allowed to increase in concentration without removal, they will slow the reaction. Some end products, when acid or alkaline in nature, may affect the pH of the mixture and thus decrease the rate of reaction. The effect of the end products on the activity of the enzyme is sometimes expressed as a chemical feedback system, with inhibition or decrease in rate called **negative feedback.** The activity of enzymes in the cell may be controlled to some extent by this chemical feedback system as in a sequence of cellular reactions in which the product D inhibits the enzyme reaction A → B.

$$A \rightarrow B \rightarrow C \rightarrow D$$

ACTIVATION OF ENZYMES

In the body many enzymes are secreted in an inactive form to prevent their action on the very glands and tissues that produce them. A proenzyme is the precursor of the active enzyme in the body. For example, pepsinogen is the proenzyme of pepsin and trypsinogen is the inactive form of trypsin. When pepsinogen is secreted into the stomach, it is converted into pepsin by hydrogen ions of the hydrochloric acid. The pepsin then activates more pepsinogen to form more of the active enzyme. Trypsinogen is secreted by the pancreas and is activated in the intestine by enterokinase (see p. 295 and Fig. 23–3). Also, during the process of purification an enzyme may become inactive. They may be activated by several agents: a change of pH, the addition of inorganic ions, or the addition of organic compounds.

The requirements for enzyme activity are further complicated by the fact that several enzymes require the presence of a metal ion for their activity. Enzymes have been characterized that require zinc, magnesium, iron, cobalt, and copper. Carbonic anhydrase, an enzyme that catalyzes the formation of carbonic acid from CO_2 and H_2O, requires zinc and is inactivated when this metal is removed.

ENZYME INHIBITORS

The activity of an enzyme may be inhibited by an increase in temperature, a change in pH, and the addition of a variety of protein precipitants. More specific inhibition can be achieved by the addition of an oxidizing agent to attack SH groups, or inhibitors such as iodoacetamide and p-chloromercuribenzoate that react with SH groups. Cyanide forms compounds with metals essential for enzyme action, whereas fluoride combines with magnesium and inhibits enzymes that require magnesium. Cyanide, for example, may remove a metal such as copper that is essential for the activity of the enzyme.

$$\text{Protein-Cu} + 2\,\text{CN}^- \rightleftharpoons \text{Cu(CN)}_2 + \quad \text{Protein}$$

Active enzyme   Inactive enzyme
(holoenzyme)   (apoenzyme)

Sodium azide and monoiodoacetate are also potent inhibitors. This type of compound usually combines with a group at the active site of the enzyme and cannot be displaced by additional substrate. These inhibitors are called **noncompetitive inhibitors,** since their degree of inhibition is not related to the concentration of the substrate.

Compounds that directly compete with the substrate for the active site on the enzyme surface in the formation of the enzyme-substrate complex are called **competitive inhibitors.** An example of competitive inhibition would be the action of sulfanilamide on the utilization of p-aminobenzoic acid in the body. The similarity of these two compounds may readily confuse the enzyme involved in the utilization of this B vitamin in the synthesis of tetrahydrofolic acid, the active coenzyme of folic acid.

p-Aminobenzoic acid   Sulfanilamide

The action of drugs in the body may depend on specific inhibitory effects on a particular enzyme in the tissues. The highly toxic nerve poison diisopropylfluorophosphate inhibits acetylcholine esterase, an enzyme essential for normal nerve function, by forming an enzyme inhibitor compound by attachment to a hydroxyl group on a serine residue in the enzyme.

Acetylcholine   Diisopropyl-   Enzyme inhibitor
esterase   fluorophosphate (DFP)   compound

Antibiotic drugs may act by inhibiting enzyme and coenzyme reactions in microorganisms. Penicillin, for example, adversely affects cell wall construction in bacteria. A similar mechanism may be involved in the action of insecticides and herbicides.

Inhibitors of enzyme action in the body are called **antienzymes.** The tapeworm is a classic example of a protein-rich organism that is not digested in the intestine of the host. Substances that inhibit the activity of pepsin and trypsin have been isolated from the tapeworm. A trypsin inhibitor has been found in the secretion of the pancreas and in milk made from fresh soya beans. This substance exhibits properties similar to those of enzymes, and its activity is destroyed by heat. It may be formed by the pancreas to control the production of trypsin.

IMPORTANT TERMS AND CONCEPTS

active site	enzyme	noncompetitive inhibitors	specificity
competitive inhibitors	Lineweaver-Burk plots	optimum pH	substrate
energy of activation	Michaelis constant	optimum temperature	

QUESTIONS

1. What is the nature of an enzyme and how may it be defined?

2. Briefly explain how enzymes are classified.

3. What is a substrate? Give an example of a substrate and enzyme using the trivial name for the enzyme.

4. What is meant by the energy of activation of a reaction? How is it affected by an enzyme?

5. What role does the formation of an enzyme-substrate complex play in the action of an enzyme? Explain.

6. What is meant by the active site of an enzyme? Explain.

7. Draw a graph representing the change in activity of an enzyme as the amount of its substrate is increased from zero to a maximum concentration. Explain the shape of the curve obtained.

8. What are the advantages of expressing an enzyme reaction as a Lineweaver-Burk plot?

9. How would you define the Michaelis constant, K_m, of an enzyme? Why is the K_m value important in enzyme reactions?

10. What is meant by (1) optimum pH? (2) optimum temperature of enzyme reactions?

11. Cyanide is a very potent poison. Explain how cyanide may exert its toxic properties.

12. Explain the difference between competitive and noncompetitive inhibitors, and give an example of each.

SUGGESTED READING

Koshland: Protein Shape and Biological Control. Scientific American, Vol. 229, No. 4, p. 52, 1973.

Miller and Cory: Activation Energies for a Base-Catalyzed and Enzyme-Catalyzed Reaction. Journal of Chemical Education, Vol. 48, p. 475, 1971.

Research Reporter: First Synthesis of an Enzyme, Ribonuclease. Chemistry, Vol. 42, No. 4, p. 21, 1969.

Stroud: A Family of Protein-Cutting Proteins. Scientific American, Vol. 231, No. 1, p. 74, 1974.

Thayer: Some Biological Aspects of Organometallic Chemistry. Journal of Chemical Education, Vol. 48, p. 807, 1971.

Wroblewski: Enzymes in Medical Diagnosis. Scientific American, Vol. 205, No. 2, p. 99, 1961.

CARBOHYDRATES

The *objectives* of this chapter are to enable the student to:

1. Distinguish between the $+$ and $-$ forms of an optically active compound.
2. Describe the structure of α-D($+$) glucose and its relation to α-D-glucopyranose.
3. Recognize the α and β isomers of the pyranose and furanose forms of sugars.
4. Describe the glycoside or acetal linkage between two monosaccharides.
5. Write the formula for sucrose and explain why it does not reduce Benedict's solution.
6. Illustrate the structure of a polysaccharide using several molecules of glucose.
7. Recognize the difference between the structures of cellulose and glycogen.

Carbohydrates are not as plentiful in the living cell as are proteins or nucleic acids. They make up about 10 per cent of the organic matter of the cell. There are probably less than 50 different kinds of carbohydrates in the cell, but they are involved in important physiological roles. They serve as a source of energy for the cell, as a storage form of chemical energy, as structural units in cell walls and membranes, and in cellular components responsible for function and growth.

As a class of compounds, they include simple sugars, starches, and celluloses. Simple sugars such as glucose, fructose, and sucrose are constituents of many fruits and vegetables. Starches are the storage form of carbohydrates in plants, and cellulose is the main supporting structure of trees and plants. About 75 per cent of the solid matter of plants consists of carbohydrates.

OPTICAL ACTIVITY

Stereoisomerism is a common phenomenon in organic chemistry. Examples of **structural** and **geometric isomers** were considered on p. 152 and 178 in Chapters 10 and 12. Optical isomerism is frequently encountered in organic compounds of biochemical interest and is essential in a study of the composition, properties, and reactions of carbohydrates. Many organic molecules, including the carbohydrates, exhibit the phenomenon of **optical activity.** Any optically active compound possesses the property of rotating a plane of polarized light. Ordinary light may be thought of as radiant energy propagated in the form of wave motion whose vibrations take place in all directions at right angles to the path of the beam of light. Certain minerals, such as Iceland spar and tourmaline, and Polaroid sheets or discs (properly oriented crystals embedded in a transparent plastic) allow only light vibrating in a single plane to pass through their crystals. When ordinary light is passed through a Nicol prism, which consists of two pieces of tourmaline cemented together, the resulting beam is traveling in one direction and in one plane, and is called **plane polarized** or just **polarized** light.

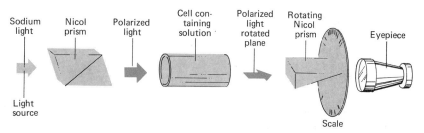

Figure 24–1   A diagrammatic sketch of the essential components of a polarimeter.

Carbohydrates in solution show the property of optical rotation, i.e., a beam of polarized light is rotated when it passes through the solution. The extent to which the beam is rotated, or the angle of rotation, is determined with an instrument called a **polarimeter.** A simple polarimeter is shown in Figure 24–1.

A substance whose solution rotates the plane of polarized light to the right is said to be **dextrorotatory;** one whose solution rotates the light to the left is called **levorotatory.** The rotation is designated $(+)$ for dextrorotatory or $(-)$ for levorotatory; for example, $(+)$lactic acid and $(-)$lactic acid.

To standardize experimental work with the polarimeter, the term **specific rotation** has been adopted. The specific rotation $[\alpha]_D^{20°}$ of a substance is the rotation in angular degrees produced by a column of solution 1 decimeter long whose concentration is 1 gram per ml. The terms $]_D^{20°}$ refer to the temperature of the solution, 20°C, and the source of plane polarized light, which is the monochromatic sodium light of wavelength 5890 to 5896 Å corresponding to the D line in the yellow part of the spectrum.

Van't Hoff and La Bel independently advanced the same theory to explain the fundamental reason for the optical activity of a compound. They postulated that the presence of an **asymmetric carbon atom** in a compound was responsible for the optical activity. An asymmetric carbon atom was defined as one which had four different groups attached to it. A simple compound such as lactic acid contains one asymmetric carbon atom, which is marked with an asterisk in the following illustration:

$$
\begin{array}{cc}
\text{COOH} & \text{COOH} \\
| & | \\
\text{H}-\text{C}^*-\text{OH} & \text{HO}-\text{C}^*-\text{H} \\
| & | \\
\text{CH}_3 & \text{CH}_3 \\
(+)\ \text{Lactic acid} & (-)\ \text{Lactic acid}
\end{array}
$$

The $(+)$ and $(-)$ forms of lactic acid are mirror images of each other, rotating the plane of polarized light an equal extent in opposite directions. These two compounds, which are identical in composition and differ only in spatial configuration, are known as **stereoisomers.** A molecular model, constructed as shown in Figure 24–2, in which the asymmetric carbon atom is the central sphere and is joined to four other spheres representing a hydrogen atom, a hydroxyl group, a methyl group, and a carboxyl group, may help to explain optical activity. This model emphasizes the asymmetric carbon atom and the fact that the $(+)$ and $(-)$ forms of lactic acid cannot be superimposed upon one another. One compound, however, resembles the other by being its **mirror image;** for if one model is held before a mirror, the image in the mirror corresponds to the other model. Such nonsuperimposable mirror image isomers as $(+)$ and $(-)$ lactic acid are an enantiomeric pair, and the $(+)$ and $(-)$ forms are **enantiomers.** In general, when a compound that exhibits optical activity is synthesized in the laboratory, a mixture of equal parts of the dextro and levo forms results. Such a mixture is called a **racemic mixture.**

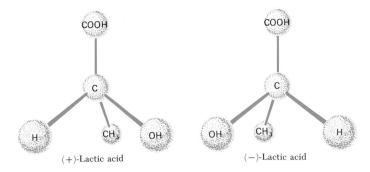

Figure 24-2   The spatial relationship of the groups attached to the asymmetric carbon atom of the (+) and (−) forms of lactic acid.

(+)-Lactic acid   (−)-Lactic acid

CLASSIFICATION

The classic definition of carbohydrates stated that they were compounds of C, H, and O in which the H and O were in the same proportion as in water. But a compound such as acetic acid, $C_2H_4O_2$, fits this definition and yet is not classed as a carbohydrate, whereas an important carbohydrate such as deoxyribose, $C_5H_{10}O_4$, a constituent of DNA (deoxyribonucleic acid) found in every cell, does not fit the definition. Carbohydrates are now defined as derivatives of polyhydroxyaldehydes or polyhydroxyketones. A sugar that contains an aldehyde group is called an **aldose,** and one that contains a ketone group is termed a **ketose.**

The simplest carbohydrates are known as **monosaccharides,** or simple sugars. Monosaccharides are derivatives of straight chain polyhydric alcohols and are classified according to the number of carbon atoms in the chain. A sugar with two carbon atoms is called a diose; with three, a triose; with four, a tetrose; with five, a pentose; and with six, a hexose. The ending -ose is characteristic of sugars. When two monosaccharides are linked together by splitting out a molecule of water, the resulting compound is called a **disaccharide.** The combination of three monosaccharides results in a **trisaccharide,** although the general term for carbohydrates composed of two to five monosaccharides is **oligosaccharide.** Polymers composed of several monosaccharides are called **polysaccharides.**

Carbohydrates which will be considered in this chapter may be classified as follows:

I. Monosaccharides
    Trioses—$C_3H_6O_3$
        Aldose—Glyceraldehyde
        Ketose—Dihydroxyacetone
    Pentoses—$C_5H_{10}O_5$
        Aldoses—Arabinose
                Xylose
                Ribose
    Hexoses—$C_6H_{12}O_6$
        Aldoses—Glucose
                Galactose
      Ketoses—Fructose
    Ascorbic acid
II. Disaccharides—$C_{12}H_{22}O_{12}$
      Sucrose (glucose + fructose)
      Maltose (glucose + glucose)
      Lactose (glucose + galactose)

III. Polysaccharides
    Hexosans
      Glucosans—Starch
                Glycogen
                Dextrin
                Cellulose
IV. Mucopolysaccharides
    Hyaluronic acid
    Chondroitin sulfate
    Heparin

TRIOSES

The trioses are important compounds in muscle metabolism, and are the basic sugars to which all monosaccharides are referred. The definition of a simple sugar may readily be illustrated by the use of the trioses. The polyhydric alcohol from which they are derived is glycerol. Oxidation on the end carbon atom produces the aldose sugar known as glyceraldehyde; oxidation on the center carbon yields the keto triose, dihydroxyacetone. It can be seen from the formula of glyceraldehyde that one asymmetric carbon atom

$$
\begin{array}{ccc}
& H & H \\
& | & | \\
H-C-OH & C=O & H-C-OH \\
| & | & | \\
H-C-OH \rightarrow & H-C-OH \quad and & C=O \\
| & | & | \\
H-C-OH & H-C-OH & H-C-OH \\
| & | & | \\
H & H & H
\end{array}
$$

Glycerol (polyhydric alcohol) Glyceraldehyde (aldose) Dihydroxyacetone (ketose)

is present. Therefore this sugar can exist in two forms, one of which rotates plane polarized light to the right, the other to the left. Modern terminology employs the D and L, written in small capital letters, for structural relationships, and a $(+)$ and $(-)$ for direction of rotation.

$$
\begin{array}{ccc}
H & H & H \\
| & | & | \\
C=O & C=O & C=O \\
| & | & | \\
H-C^{\circ}-OH & HO-C^{\circ}-H & H-C-OH \\
| & | & | \\
CH_2OH & CH_2OH & CH_2OH
\end{array}
$$

D $(+)$ Glyceraldehyde L $(-)$ Glyceraldehyde Perspective formula

Fischer projection formula

The isomeric forms of sugars are often represented as the **Fischer projection formula.** The asymmetric carbon atom ° of glyceraldehyde would represent the central sphere as the model in Figure 24–2, with the H and OH groups projecting in front of the plane of the paper and the aldehyde and primary alcohol group projecting behind. As ordinarily written, the horizontal bonds are understood to be in front of the plane and the vertical bonds behind the plane of the paper. The **perspective formula** emphasizes the position of the groups using dotted lines to connect those behind the plane and heavy wedges to represent groups in front of the plane. The Fischer projection formulas are always written with the aldehyde or ketone groups (the most highly oxidized groups) at the top of the structure; therefore, all monosaccharides with the hydroxyl group on the right of the carbon atom next to the bottom primary alcohol group are related to D-glyceraldehyde and are called D-sugars.

In like manner, if the hydroxyl group on the carbon atom next to the end primary alcohol group is on the left, it is related to L-glyceraldehyde and is an L-sugar. The direction of rotation of polarized light cannot be ascertained from the formula, but must be determined experimentally, and is designated $(+)$ for dextrorotatory and $(-)$ for levorotatory sugars.

Only two optical isomers of aldotriose exist, since it contains only one asymmetric carbon atom. The total number of isomers of a sugar is 2^n, where n is the number of different asymmetric carbon atoms. For example, pentoses contain three asymmetric carbon atoms and can form eight isomers; hexoses contain four asymmetric carbon atoms and can exist as 16 different isomers.

The pentoses are sugars whose molecules contain five carbon atoms; they occur in nature combined into polysaccharides, from which the monosaccharides may be obtained by hydrolysis with acids. Arabinose is obtained from gum arabic and the gum of the cherry tree, and xylose is obtained by hydrolysis of wood, corn cobs, or straw. Ribose and deoxyribose are constituents of the ribose nucleic acids, RNA, and deoxyribose nucleic acids, DNA, which are essential components of the cytoplasm and nuclei of cells.

HEXOSES

The hexoses are by far the most important monosaccharides from a nutritional and physiological standpoint. The bulk of the carbohydrates used as foods consists of hexoses free or combined in disaccharides and polysaccharides. Glucose, fructose, and galactose are the hexoses commonly occurring in foods. Glucose, also called **dextrose,** is the normal sugar of the blood and tissue fluids and is utilized by the cells as a source of energy. Fructose often occurs free in fruits and is the sweetest sugar of all the monosaccharides. Galactose is a constituent of **milk sugar** and is found in brain and nervous tissue. All these monosaccharides are D-sugars, since the hydroxyl group on the carbon next to the primary alcohol is on the right. Fructose is a ketose, and the others are aldoses.

Although glucose and galactose are represented as simple aldehyde structures, this form does not explain all the reactions they undergo. Both of these aldoses, for example, do not give a characteristic test for aldehyde. Also, when a glucose solution is allowed to stand, a change in its specific rotation may be observed.

$$
\begin{array}{ccc}
\begin{array}{c}
\text{H} \\
| \\
\text{C}{=}\text{O} \\
| \\
\text{H}{-}\text{C}{-}\text{OH} \\
| \\
\text{HO}{-}\text{C}{-}\text{H} \\
| \\
\text{H}{-}\text{C}{-}\text{OH} \\
| \\
\text{H}{-}\text{C}{-}\text{OH} \\
| \\
\text{CH}_2\text{OH}
\end{array}
&
\begin{array}{c}
\text{CH}_2\text{OH} \\
| \\
\text{C}{=}\text{O} \\
| \\
\text{HO}{-}\text{C}{-}\text{H} \\
| \\
\text{H}{-}\text{C}{-}\text{OH} \\
| \\
\text{H}{-}\text{C}{-}\text{OH} \\
| \\
\text{CH}_2\text{OH}
\end{array}
&
\begin{array}{c}
\text{H} \\
| \\
\text{C}{=}\text{O} \\
| \\
\text{H}{-}\text{C}{-}\text{OH} \\
| \\
\text{HO}{-}\text{C}{-}\text{H} \\
| \\
\text{HO}{-}\text{C}{-}\text{H} \\
| \\
\text{H}{-}\text{C}{-}\text{OH} \\
| \\
\text{CH}_2\text{OH}
\end{array}
\\
\text{D-Glucose} & \text{D-Fructose} & \text{D-Galactose} \\
\text{(aldose)} & \text{(ketose)} & \text{(aldose)}
\end{array}
$$

Freshly prepared aqueous solutions of crystalline glucose often yield a specific rotation as high as +113 degrees, whereas glucose crystallized from pyridine exhibits a specific rotation as low as +19 degrees. On standing, both of these solutions change their rotation until an equilibrium value of +52.5 degrees is reached. This change in rotation is called **mutarotation.** Since the specific rotation of an organic compound is related to its structure, as are its melting point, boiling point, and other properties, it may be suspected that glucose exists in two different isomeric forms. This has been shown to be true and is explained by the existence of an **intramolecular bridge structure** involving carbon atoms 1 and 5. It can be seen from these formulas that the free aldehyde group no longer exists and a new asymmetric carbon is produced. To indicate the position of the hydroxyl group on the first carbon and to distinguish between the two new isomers, the α-isomer has the OH on the right and the β-isomer on the left as shown. When the α- or β-isomer is dissolved in water, an equilibrium mixture of 37 per cent α and 63 per cent β, with a specific rotation of 52.5 degrees, is formed. This intramolecular bridge structure and the phenomenon of mutarotation are common to all aldohexoses, and since the structure contains an additional asymmetric carbon atom, the number of possible isomers is doubled.

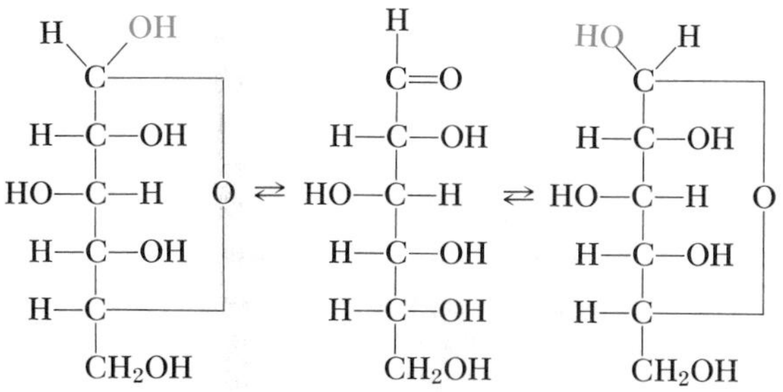

α-D-Glucose
$+113°$

D-Glucose,
straight chain
form

β-D-Glucose
$+19°$

These oxygen ring structures suggested to Haworth that the sugars may be represented as derivatives of the heterocyclic rings pyran and furan (pp. 199 and 200). The relation between the straight chain structure of glucose and Haworth's **glucopyranose** may best be understood by writing glucose in a chain as shown in A. The chain is folded and a rotation of groups occurs around carbon-5 to bring the primary alcohol group (carbon-6) into the proper spatial relation to the other groups (B). The hemiacetal is then formed between the aldehyde on carbon-1 and the OH group on carbon-5 to form α-D-glucopyranose, shown in C.

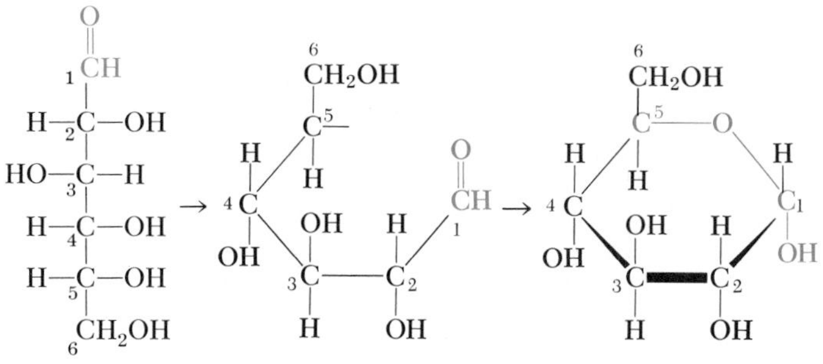

Glucose

A

B

α-D-Glucopyranose

C

The heavy lines represent the base of a space model in which the five carbons and one oxygen are in the same plane perpendicular to the plane of the paper. The thick bonds of the ring extend toward the reader, whereas the thin bonds of the ring are behind the plane of the paper. Groups which are ordinarily written on the right of the oxide ring structure appear below the plane of the pyranose ring, and those to the left of the carbon chain appear above the plane. The glucopyranose structure, C, has the OH on carbon-1 below the plane of the ring and would therefore be an α-isomer.

Monosaccharides such as the pentoses and fructose, whose oxide rings enclose four carbon atoms, are written as derivatives of furan as shown:

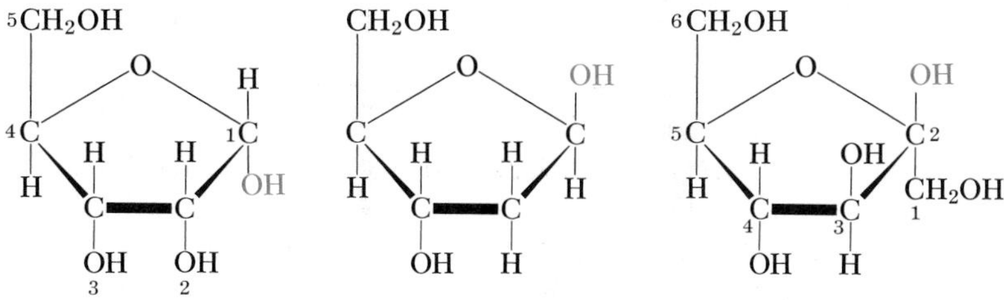

α-D-Ribose

β-D-Deoxyribose

β-D-Fructofuranose

REACTIONS OF CARBOHYDRATES

DEHYDRATION

When aldohexoses or aldopentoses are heated with strong acids, they are dehydrated to form furfural derivatives. Pentoses yield furfural (p. 199) whereas aldohexoses form hydroxymethyl furfural.

Furfural   Hydroxymethyl furfural

The furfural derivatives formed in this reaction combine with α-naphthol to give a purple color. This color is the basis of the Molisch test, a general test for carbohydrate. Furfural reacts with orcinol to yield a green color which constitutes Bial's test for pentoses.

ACETAL OR GLYCOSIDE FORMATION

When monosaccharides are treated with an alcohol in a strong acid solution, they form **glycosides.** The hemiacetal structure reacts with alcohols or an alcoholic hydroxyl group to form an acetal or glycoside (see p. 232).

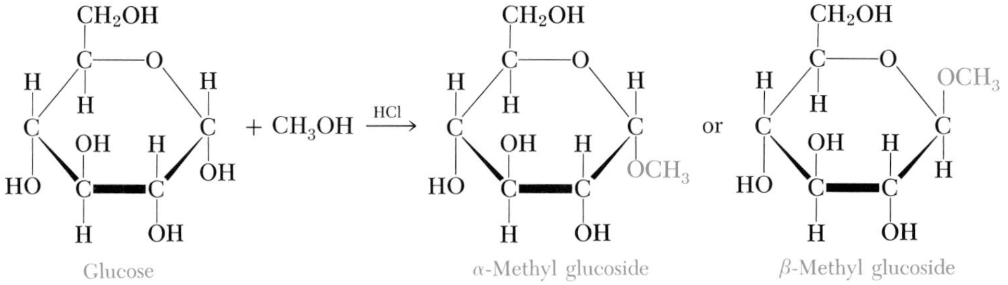

Glucose α-Methyl glucoside β-Methyl glucoside

The position of the methyl group below or above the plane of the ring indicates α- or β-methyl glucoside in that order. This is a very important reaction, since many of the disaccharides and polysaccharides are glycosides in which one of the alcoholic hydroxyl groups in the second monosaccharide reacts with the hydroxyl on carbon-1 of the first monosaccharide.

OXIDATION

In Chapter 16 one of the important reactions of carbonyl compounds was the oxidation to carboxylic acids. Sugars that contain *free or potential aldehyde or ketone groups* in the hemiacetal type of structure are oxidized in alkaline solution by Cu^{+2} and Ag^+. The reaction of aldehydes in Benedict's, Fehling's, or the silver mirror test has already been described (p. 230). Sugars that undergo oxidation in these reactions are called **reducing sugars.**

All the reducing sugars that are capable of reducing Cu^{+2} to Cu^+ are oxidized in the reaction described in the preceding section. If the aldehyde group of glucose is oxidized to a carboxyl group by a weak oxidizing agent, gluconic acid is formed. Oxidation of the primary alcohol group, either by chemical agents or enzymes, produces glucuronic acid. Glucuronic acid combines with drugs and toxic compounds in the body, and the conjugated glucuronides are excreted in the urine.

FERMENTATION

The enzyme mixture called **zymase** present in common bread yeast will act on some of the hexose sugars to produce alcohol and carbon dioxide. The fermentation of glucose may be represented as follows:

$$C_6H_{12}O_6 \xrightarrow{\text{zymase}} 2C_2H_5OH + 2CO_2$$

Glucose   Ethyl alcohol

The common hexoses (with the exception of galactose) ferment readily, but pentoses are not fermented by yeast. Disaccharides must first be converted into their monosaccharide constituents by other enzymes present in yeast before they are susceptible to fermentation by zymase.

There are many other types of fermentation of carbohydrates besides the common alcoholic fermentation. When milk sours, the lactose of milk is converted into lactic acid by a fermentation process. Citric acid, acetic acid, butyric acid, and oxalic acid may all be produced by special fermentation processes.

ESTER FORMATION

Esters formed between the hydrogen atom of a hydroxyl group of phosphoric acid and the hydroxyl group of a monosaccharide are common (p. 207), and several of these phosphorylated sugars are encountered in carbohydrate metabolism (Chapter 29).

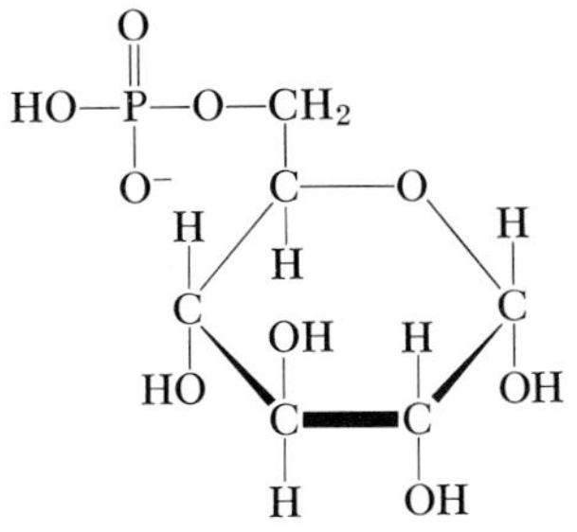

α-D-Glucose-6-phosphate 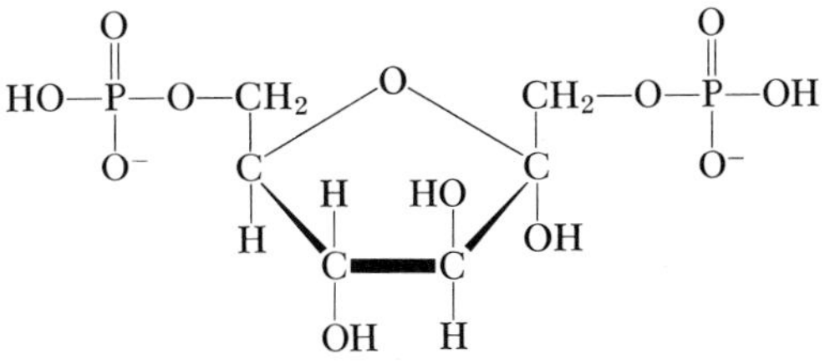   α-D-Fructose-1, 6-diphosphate

DISACCHARIDES

A disaccharide is composed of two monosaccharides whose combination involves the splitting out of a molecule of water. The acetal linkage is always made from the aldehyde group of one of the sugars to a hydroxyl or ketone group of the second. In order to reduce Benedict's solution, disaccharides must have a potential aldehyde or ketone group that is not involved in the acetal linkage between the two sugars.

SUCROSE

Sucrose is commonly called **cane sugar** and is the ordinary sugar that is used for sweetening purposes in the home. It is found in many plants such as sugar beets, sorghum cane, the sap of the sugar maple, and sugar cane. Commercially it is prepared from sugar cane and sugar beets.

Sucrose is composed of a molecule of glucose joined to a molecule of fructose in such a way that the linkage involves the reducing groups of both sugars (carbon-1 of glucose and carbon-2 of fructose). It is the only common mono- or disaccharide that will

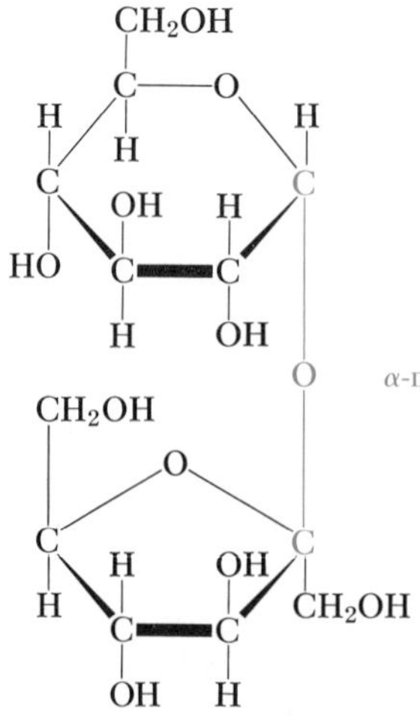

α-D-Glucopyranose-β-D-fructofuranoside

Sucrose

not reduce Benedict's solution. When sucrose is hydrolyzed, either by the enzyme sucrase or by an acid, a molecule of glucose and a molecule of fructose are formed.

LACTOSE

The disaccharide present in milk is lactose, or **milk sugar.** It is synthesized in the mammary glands of animals from the glucose in the blood. Commercially, it is obtained from milk whey and is used in infant foods and special diets. Lactose, when hydrolyzed by the enzyme lactase or by an acid, forms a molecule of glucose and a molecule of galactose. Lactose will reduce Benedict's solution, but is not fermented by yeast. From its reducing properties, it is obvious that the linkage between its constituent monosaccharides does not involve both potential aldehyde groups (carbon-1 of galactose is connected to carbon-4 of glucose).

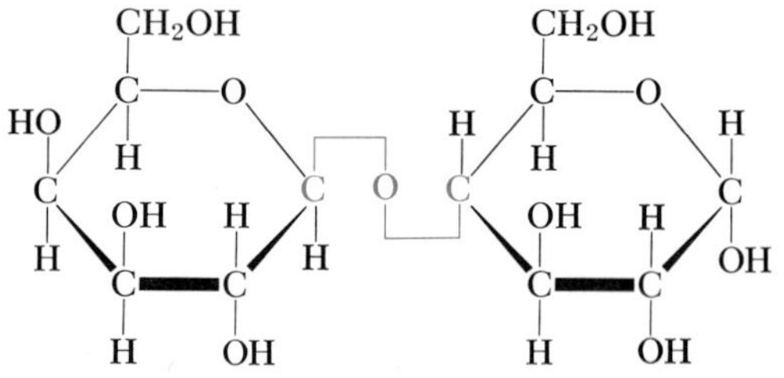

β-D-Galactopyranosyl-α-D-glucopyranose

Lactose

MALTOSE

Maltose is present in germinating grains. Since it is obtained as a product of the hydrolysis of starch by enzymes present in malt, it is often called **malt sugar.** It is also formed in the animal body by the action of enzymes on starch in the process of digestion. Commercially, it is made by the partial hydrolysis of starch by acid in the manufacture of corn syrup. Maltose reduces Benedict's solution and is fermented by yeast. On hydrolysis it forms two molecules of glucose.

Cellobiose is similar to maltose since it is composed of two glucose molecules joined in a carbon-1 to carbon-4 linkage. In contrast to maltose, the glucoses in cellobiose are joined in a β-1,4 linkage.

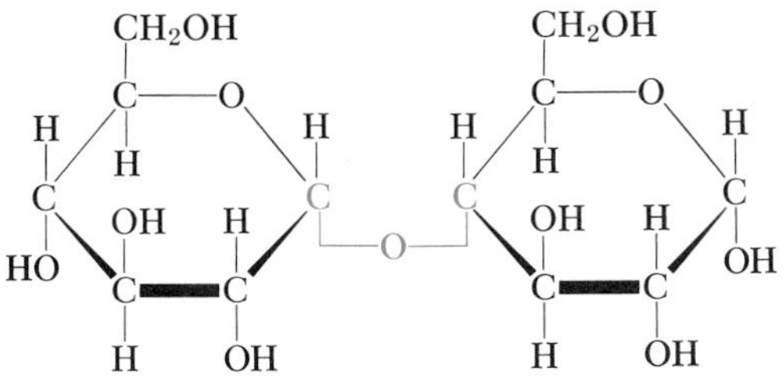

$$\text{CH}_2\text{OH} \qquad\qquad \text{CH}_2\text{OH}$$

α-D-Glucopyranosyl-α-D-glucopyranose

Maltose

POLYSACCHARIDES

The polysaccharides are complex carbohydrates that are made up of many monosaccharide molecules and therefore possess a high molecular weight. They differ from the simple sugars in many ways. They do not have a sweet taste, do not reduce Benedict's solution, are usually insoluble in water, and when dissolved by chemical means form colloidal solutions because of their large molecules.

There are polysaccharides formed from pentoses or from hexoses, and there are also mixed polysaccharides. Of these, the most important are composed of the hexose glucose and are called **hexosans,** or more specifically, **glucosans.** As in a disaccharide, whenever two molecules of a hexose combine, a molecule of water is split out. For this reason, a hexose polysaccharide may be represented by the formula $(C_6H_{10}O_5)_x$. The x represents the number of hexose molecules in the individual polysaccharide. Because of the complexity of the molecules, the number of glucose units in any one polysaccharide is still an estimate. In addition, the molecular weight values obtained for polysaccharides from different plant and animal sources show considerable variation. For this reason the molecular weights of the polysaccharides described in the following section are only approximations.

STARCH

From a nutritional standpoint, starch is the most important polysaccharide. It is made up of glucose units and is the storage form of carbohydrates in plants. It consists of two types of polysaccharides: **amylose,** composed of a chain of glucose molecules connected

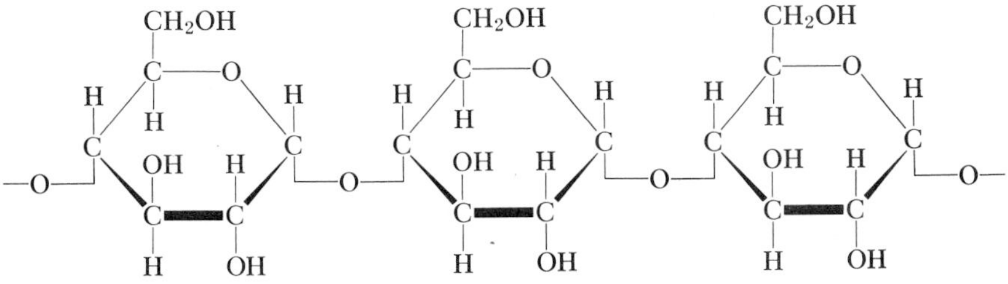

Amylose

by α-1,4 linkages, and **amylopectin,** which is a branched chain or polymer of glucose with both α-1,4 and α-1,6 linkages. The repeating structure of glucose molecules in amylose is usually represented as glucopyranose units as shown in the accompanying diagram. Amylose has a molecular weight of about 50,000, compared to about 300,000 for amylopectin. The branching of the glucose chain in amylopectin occurs about every 24 to 30 glucose molecules.

Starch will not reduce Benedict's solution and is not fermented by yeast. When starch is hydrolyzed by enzymes or by an acid, it is split into a series of intermediate compounds possessing small numbers of glucose units. The product of complete hydrolysis is the free glucose molecule. A characteristic reaction of starch is the formation of a blue compound with iodine. This test is often used to follow the hydrolysis of starch, since the color changes from blue through red to colorless with decreasing molecular weight:

$$\text{starch} \rightarrow \text{amylodextrin} \rightarrow \text{erythrodextrin} \rightarrow \text{achroodextrin} \rightarrow \text{maltose} \rightarrow \text{glucose}$$

| blue | blue | red | colorless | colorless with iodine |

DEXTRINS

Dextrins are found in germinating grains, but are usually obtained by the partial hydrolysis of starch. Those formed from amylose have straight chains, whereas those derived from amylopectin exhibit branched chains of glucose molecules. The larger branched chain molecules give a red color with iodine and are the erythrodextrins. They are soluble in water and have a slightly sweet taste. Large quantities of dextrins are used in the manufacture of adhesives because they form sticky solutions when wet. An example of their use is the mucilage on the back of postage stamps.

GLYCOGEN

Glycogen is the storage form of carbohydrate in the animal body and is often called animal starch. It is found in liver and muscle tissue. It is soluble in water, does not reduce Benedict's solution, and gives a red-purple color with iodine. The glycogen molecule is similar to the amylopectin molecule in that it has branched chains of glucose with α-1,4 and α-1,6 linkages that occur about every 12 to 18 glucose molecules. The branched

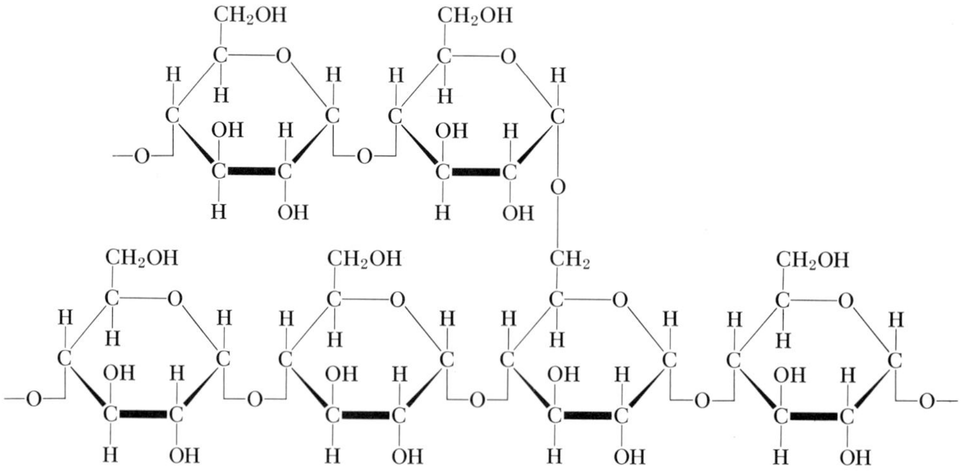

Branched chain of glucose molecules in amylopectin and glycogen

chain structure common to both glycogen and amylopectin is shown above and represented in Figure 24–3.

The molecular weight of glycogen is greater than that of amylopectin, and often exceeds 5,000,000. When glycogen is hydrolyzed in the animal body, it forms glucose to help maintain the normal sugar content of the blood.

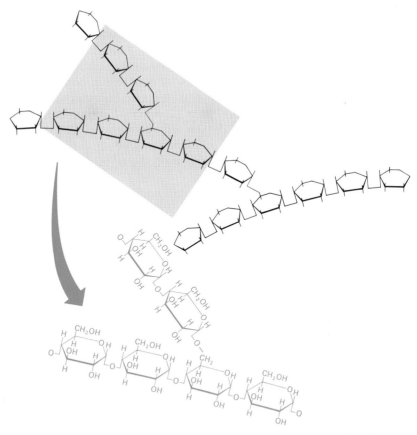

Figure 24–3 Representation of the branched chain structure of amylopectin and glycogen. The shaded portion is illustrated in detail in the lower half of the figure (arrow).

CELLULOSE

Cellulose is a polysaccharide that occurs in the framework, or supporting structure, of plants. It is composed of a straight chain polymer of glucose molecules similar in structure to that pictured for amylose. The major difference concerns the linkage of the glucose molecules. In amylose the linkage is α-1,4, as in maltose, whereas in cellulose the linkage is β-1,4, which occurs in **cellobiose.** The maltose type of structure is hydrolyzed by enzymes and serves as a source of dietary carbohydrate. In contrast, the cellobiose

Cellulose

structure is insoluble in water, will not reduce Benedict's solution and is not attacked by enzymes present in the human digestive tract. The molecular weight of cellulose has been estimated to be between 200,000 and 2,000,000. The structure of cellulose is represented in the accompanying diagram.

The chemical treatment of cellulose has resulted in several important commercial products. For example, cotton, which is almost pure cellulose, takes on a silk like luster and increases in strength when treated under tension with a concentrated solution of sodium hydroxide. This process, called mercerization, produces mercerized cotton, which is used in large quantities in the manufacture of cotton cloth.

Other important products related to cellulose are polymers (see Chapter 19). When treated with nitric acid, cotton is converted into cellulose nitrates, which are esters of commercial importance. **Rayon** is produced by treating cellulose with sodium hydroxide and carbon disulfide. The solution that results from this treatment is forced through fine holes into dilute sulfuric acid to make the rayon fibers. **Cellophane** is made by a process similar to that used for rayon.

In addition to the esters of cellulose, certain ethers have become important. Methylcellulose, ethylcellulose, and carboxymethylcellulose are examples of these ethers. Methylcelluloses are used as sizing and finish for textiles; ethylcellulose has properties that make it a desirable material in the manufacture of plastics, coatings, and films; and carboxymethylcellulose is used as a sizing agent for textiles and as a builder in the manufacture of synthetic detergents.

MIXED POLYSACCHARIDES

Heparin is a mixed polysaccharide that possesses anticoagulant properties. It prevents the clotting of blood by inhibiting the conversion of prothrombin to thrombin. Thrombin acts as a catalyst in converting plasma fibrinogen into the fibrin clot. A structural polysaccharide found in higher animals is the mucopolysaccharide hyaluronic acid. It is an essential component of the ground substance, or intercellular cement, of connective tissue. Chondroitin sulfates are structural polysaccharides also found in the ground substance and cartilage of mammals.

IMPORTANT TERMS AND CONCEPTS

aldose	fructofuranose	ketose	optical activity
asymmetric carbon atom	glucopyranose	monosaccharides	polysaccharides
disaccharides	hemiacetal	mutarotation	reducing sugar

QUESTIONS

1. Write the formula for the aldose, glyceraldehyde. Would this compound exhibit optical activity in solution? Explain.

2. Write the formula for lactic acid. Would this compound be classified as a carbohydrate? Explain.

3. Write the formulas for D-fructose and L-fructose. Star the asymmetric carbon atoms in each molecule.

4. Explain fully what is meant by (1) D (+) glucose, and (2) L (−) fructose.

5. What is (1) an aldose, (2) a hexose, (3) a pentose, (4) a ketose, and (5) a disaccharide?

6. How does the phenomenon of mutarotation complicate the representation of the formulas for carbohydrates? Explain with an example.

7. How are the α- and β-isomers of the pyranose and furanose ring forms of the sugars indicated in the structures?

8. Write an equation to illustrate the formation of an acetal or a glucoside between β-ᴅ-galacto-pyranose and the OH on carbon-4 of α-ᴅ-glucopyranose.

9. When a reducing sugar reacts with Benedict's solution, what other products are formed besides Cu_2O? Why is the formation of Cu_2O important in the test?

10. Write the formula for dihydroxyacetone. Use the formula to illustrate the formation of dihydroxyacetone-3-phosphate.

11. Write the formula for sucrose and use it to explain why sucrose will not reduce Benedict's solution.

12. Write a partial polysaccharide structure that illustrates both α-1,4 and α-1,6 linkages between the monosaccharides.

13. How would you account for the great difference in properties between starch and cellulose?

14. Compare starch, dextrins, glycogen, and cellulose as to size of molecule, chemical composition, color with iodine, and importance.

SUGGESTED READING

Elias: The Natural Origin of Optically Active Compounds. Journal of Chemical Education, Vol. 49, No. 7, p. 448, 1972.

McCord and Getchell: Cotton. Chemical & Engineering News, November 14, 1960, p. 106.

Mowery: Criteria for Optical Activity in Organic Molecules. Journal of Chemical Education, Vol. 46, p. 269, 1969.

Pauling: Vitamin C and the Common Cold. San Francisco, W. H. Freeman and Company, 1970.

Slocum, Sugarman, and Tucker: The Two Faces of ᴅ and ʟ Nomenclature. Journal of Chemical Education, Vol. 48, p. 597, 1971.

Vennos: Construction and Uses of an Inexpensive Polarimeter. Journal of Chemical Education, Vol. 46, p. 459, 1969.

LIPIDS

CHAPTER 25 ████████████████████████████████

The *objectives* of this chapter are to enable the student to:

1. Illustrate the formation of a triglyceride from glycerol and three molecules of a fatty acid.
2. Write an equation illustrating the process of saponification.
3. Describe the separation of lipids by thin-layer and gas-liquid chromatography.
4. Distinguish between phospholipids and glycolipids.
5. Recognize the sterol nucleus in the steroid hormones.
6. Distinguish between the structures of estrone, testosterone, and cortisone.

The **lipids** serve two major physiological roles in the living cell—as structural components of cell membranes, and as a storage form of energy-rich fuel. As with the carbohydrates, there are fewer than 50 different kinds of lipids in the cell, and they compose slightly more than 5 per cent of the total organic matter. The cells of the brain and nervous tissue are especially rich in lipids.

Lipids are characterized by the presence of fatty acids or their derivatives and by their solubility in fat solvents such as acetone, alcohol, ether, and chloroform. Chemically, lipids are composed of five main elements: carbon, hydrogen, oxygen, and, occasionally nitrogen and phosphorus. At present there is no generally accepted method of classification of lipids. Some schemes divide them into simple lipids, compound lipids, and steroids, but a more practical classification may be that followed in this chapter:

> Fats—esters of fatty acids with glycerol.
> Phospholipids—compounds that contain phosphorus, fatty acids, glycerol, and a nitrogenous compound.
> Sphingolipids—compounds that contain a fatty acid, phosphoric acid, choline, and an amino alcohol, sphingosine.
> Glycolipids—composed of a carbohydrate, a fatty acid, and an amino alcohol.
> Steroids—high molecular weight cyclic alcohols.
> Fat soluble vitamins—vitamins A, D, E, and K.
> Waxes—esters of fatty acids with alcohols other than glycerol.

FATTY ACIDS

Since all fats are esters of fatty acids and glycerol, it may be well to consider the composition and properties of these substances before discussing lipids in general. The fatty acids that occur in nature almost always have an even number of carbon atoms in their molecules. They are usually straight chain organic acids that may be saturated or unsaturated. Some of the important fatty acids that occur in natural fats are listed in Table 25–1.

TABLE 25–1 SOME IMPORTANT FATTY ACIDS OCCURRING IN NATURAL FATS

NAME	FORMULA	CARBON ATOMS	POSITION OF DOUBLE BONDS	OCCURRENCE
Saturated				
Butyric	C_3H_7COOH	4		Butter fat
Caproic	$C_5H_{11}COOH$	6		Butter fat
Caprylic	$C_7H_{15}COOH$	8		Coconut oil
Capric	$C_9H_{19}COOH$	10		Palm kernel oil
Lauric	$C_{11}H_{23}COOH$	12		Coconut oil
Myristic	$C_{13}H_{27}COOH$	14		Nutmeg oil
Palmitic	$C_{15}H_{31}COOH$	16		Animal and vegetable fats
Stearic	$C_{17}H_{35}COOH$	18		Animal and vegetable fats
Arachidic	$C_{19}H_{39}COOH$	20		Peanut oil
Unsaturated				
Palmitoleic (1 =)°	$C_{15}H_{29}COOH$	16	Δ9†	Butter fat
Oleic (1 =)	$C_{17}H_{33}COOH$	18	Δ9	Olive oil
Linoleic (2 =)	$C_{17}H_{31}COOH$	18	Δ9, 12	Linseed oil
Linolenic (3 =)	$C_{17}H_{29}COOH$	18	Δ9, 12, 15	Linseed oil
Arachidonic (4 =)	$C_{19}H_{31}COOH$	20	Δ5, 8, 11, 14	Lecithin

° Number of double bonds.

† Δ9 indicates a double bond between carbon 9 and 10, Δ12 between carbon 12 and 13, and so forth.

In the series of saturated fatty acids, those up to and including capric acid are liquid at room temperature. The most important saturated fatty acids are **palmitic** and **stearic acids.** They are components of the majority of the common animal and vegetable fats.

Unsaturated fatty acids are characteristic constituents of oils. **Oleic acid,** which contains one double bond, is the most common unsaturated fatty acid (p. 182).

The **prostaglandins** are cyclic fatty acids found in seminal plasma and other tissues. They are derived from polyunsaturated fatty acids such as arachidonic. A typical example of the prostaglandins is compound PGE_2.

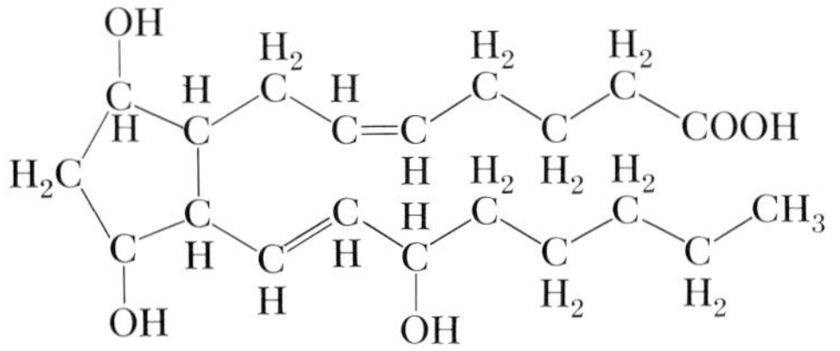

PGE$_2$ (Prostaglandin)

These compounds are involved in the control of lipid metabolism (Chapter 30) and are thought to depress the action of cyclic-3′,-5′-AMP (p. 352). Their control of cyclic AMP levels may play an important role in protein, RNA, and even DNA synthesis.

FATS

From a chemical standpoint fats are esters of fatty acids and glycerol (p. 210). This combination of 3 molecules of fatty acid with 1 molecule of glycerol may be illustrated as shown in the accompanying diagram.

Tristearin is called a **simple glyceride** because all the fatty acids in the fat molecule are the same. Other examples of simple glycerides would be tripalmitin and triolein. In most naturally occurring fats, different fatty acids are found in the same molecule. These are called **mixed glycerides** and may contain both saturated and unsaturated fatty acids. The glycerides are classed as **neutral lipids** since their molecules are not charged.

Both fats and oils are esters of fatty acids and glycerol. In general, fats are solid at room temperature and are characterized by a relatively high content of saturated fatty acids. Oils are liquids that contain a high concentration of unsaturated fatty acids. A fat that contains short chain saturated fatty acids may also exist as a liquid at room temperature.

Most of the common animal fats are glycerides that contain saturated and unsaturated fatty acids. Since the saturated fatty acids predominate, these fats are solid at room

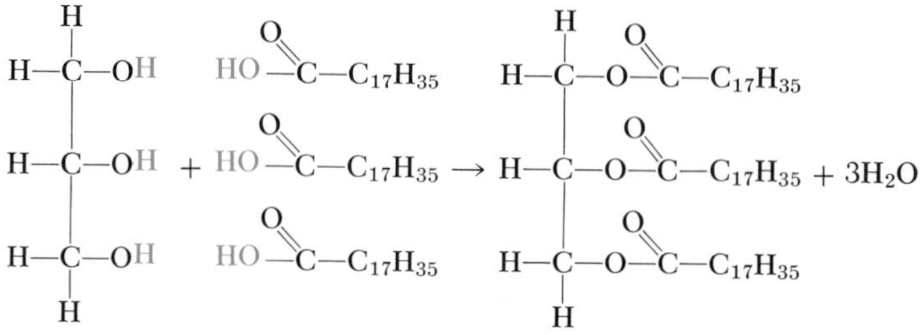

Glycerol   3 molecules of stearic acid   Tristearin, a fat

temperature. Vegetable oils such as olive oil, corn oil, cottonseed oil, and linseed oil are characterized by their high content of oleic, linoleic, and linolenic acids (see Topic of Current Interest on polyunsaturates, p. 182). Coconut oil, like butter fat, contains a relatively large percentage of short chain fatty acids.

REACTIONS OF FATS

Glycerol Portion. When glycerol or a liquid containing glycerol is heated with a dehydrating agent, **acrolein** is formed. Acrolein has a very pungent odor and is sometimes formed by the decomposition of glycerol in the fat of frying meats.

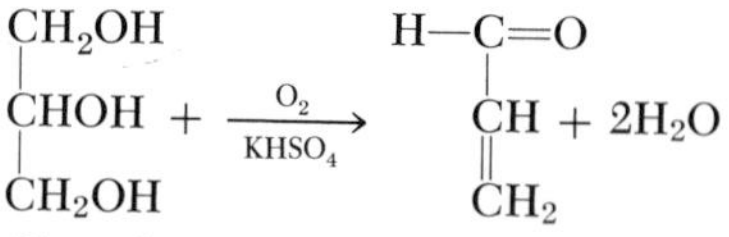

Glycerol   Acrolein

The formation of acrolein is often used as a test for fats, since all fats yield glycerol when they are heated.

Rancidity. Many fats develop an unpleasant odor and taste when they are allowed to stand in contact with air at room temperature. The two common types of rancidity are **hydrolytic** and **oxidative.** Hydrolytic changes in fats are the result of the action of enzymes or microorganisms producing free fatty acids. If these acids are of the short chain variety, as is butyric acid, the fats develop a rancid odor and taste. This type of rancidity is common in butter.

The most common type of rancidity is the oxidative type. The unsaturated fatty acids in fats undergo oxidation at the double bonds. The combination with oxygen results in the formation of peroxides, volatile aldehydes, ketones, and acids. Heat, light, moisture, and air are factors that accelerate oxidative rancidity. The prevention of rancidity of lard and vegetable shortenings that are used in the manufacture of crackers, pretzels, pastries, and similar food products has long been an important problem. Modern packaging has helped considerably in this connection, although a more important contribution has been the development of "**antioxidants**" such as vitamin E (p. 335), which effectively inhibits the autoxidation of unsaturated fatty acids. The majority of the vegetable shortenings on the market as well as certain brands of lard are protected from rancidity by the addition of antioxidants.

Hydrogenation. It has already been stated that the main difference between oils and fats is the number of unsaturated fatty acids in the molecule. Vegetable oils may be converted into solid fats by the addition of hydrogen to the double bonds of the unsaturated fatty acids. This process has already been illustrated and discussed on page 182.

Hydrolysis. Fats may be hydrolyzed to form free fatty acids and glycerol by the action of acid, alkali, superheated steam, or the enzyme lipase. In hydrolysis of a fat, the 3 water molecules (that were split out when the 3 fatty acid molecules combined with 1 glycerol molecule in an ester linkage to make the fat molecule) are replaced with the resultant splitting of the fat into glycerol and fatty acids. Commercially, fats are a cheap source of glycerol for use in the manufacture of high explosives and pharmaceuticals.

SAPONIFICATION

Hydrolysis by an alkali is called **saponification,** and produces glycerol and salts of the fatty acids that are called soaps. In the laboratory, fats are usually saponified by an alcoholic solution of an alkali. The fats are more soluble in hot alcohol and the reaction is therefore more rapid. **Soaps** may be defined as metallic salts of fatty acids. The saponification of a fat may be represented as follows:

$$
\begin{array}{ccc}
\begin{array}{l}
CH_2\!-\!O\!-\!\overset{\displaystyle O}{\overset{\|}{C}}\!-\!C_{17}H_{35} \\[6pt]
CH\!-\!O\!-\!\overset{\displaystyle O}{\overset{\|}{C}}\!-\!C_{17}H_{35} \\[6pt]
CH_2\!-\!O\!-\!\overset{\displaystyle O}{\overset{\|}{C}}\!-\!C_{17}H_{35}
\end{array}
& + \ 3NaOH \ \rightarrow &
\begin{array}{l}
CH_2OH \\[6pt]
CHOH \\[6pt]
CH_2OH
\end{array}
+ \ 3C_{17}H_{35}\overset{\displaystyle O}{\overset{\diagup\!\diagdown}{C}}\!-\!ONa
\end{array}
$$

Tristearin   Glycerol   Sodium stearate (soap)

Sodium salts of fatty acids are known as **hard soaps,** whereas potassium salts form **soft soaps.** The ordinary cake soaps used in the home are sodium soaps. White laundry soap in the form of bars, soap chips, or powdered soap contains sodium silicate and a water-softening agent such as sodium carbonate or sodium phosphate. Tincture of green soap, commonly used in hospitals, is a solution of potassium soap in alcohol. When sodium soaps are added to hard water, the calcium and magnesium salts present replace sodium to form insoluble calcium and magnesium soaps. The familiar soap curd formed in hard water is due to these **insoluble soaps.**

Detergents. These compounds are a mixture of the sodium salts of the sulfuric acid esters of lauryl and cetyl alcohols. They may be used in hard water because they do not form insoluble compounds with calcium and magnesium.

Extensive research on new detergents and emulsifying agents has resulted in the development of several hundred products possessing almost any desired property. The chemical reaction involved in the formation of detergents and the alteration in structure that results in biodegradable detergents is outlined on page 196. At present over four billion pounds of detergents are sold each year with synthetic detergents, called **syndets,** outselling soap by more than four to one.

Soaps and detergents are excellent emulsifying agents which exhibit the property of converting a mixture of oil and water into a permanent emulsion. The **cleansing power** of soaps and detergents is related to their action as emulsifying agents and to their ability to lower surface tension. By emulsifying the grease or oily material that holds the dirt on the skin or clothing, one can rinse off the particles of grease and dirt with water. The ability of soaps and detergents to break or stabilize oil and water emulsion has been given the name **"detergency."**

ANALYSIS OF LIPIDS

Thin-layer chromatography and gas-liquid chromatography are presently the methods of choice for the analysis of lipids.

Thin-layer chromatography is carried out on a thin, uniform layer of silica gel spread on a glass plate and activated by heating in an oven (100° to 250°C). Samples of lipid material in the proper solvent are spotted along one edge of the plate with micropipettes. After evaporation of the solvent, the plates are placed vertically in a covered glass tank which contains a layer of suitable solvent on the bottom. Within a few minutes the lipids are separated by the solvent rising through the thin layer carrying the spots to different locations on the silica gel by a combination of adsorption on the gel and varying distribution in the solvent system. The plates are removed, dried, and sprayed with various

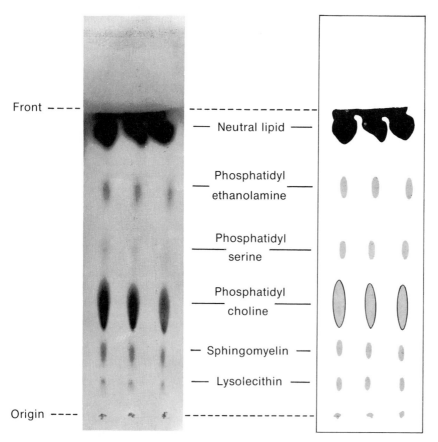

Front

—— Neutral lipid ——

Phosphatidyl
ethanolamine

Phosphatidyl
serine

Phosphatidyl
choline

—— Sphingomyelin ——

—— Lysolecithin ——

Origin

Figure 25-1   Separation of phospholipids by thin-layer chromatography.

detection agents to visualize the lipid components. An example of the thin-layer chro-matographic separation of several of the phospholipids is shown in Figure 25–1. This technique is very sensitive and can be made quantitative by removing the spots and measuring the concentration of the component by gas-liquid chromatography.

Gas-liquid chromatography is another powerful tool of the lipid chemist. Any substance that is volatile or can be made into a volatile derivative, for example fatty acids being converted into their methyl ester, can be separated and analyzed by this technique. The volatile substance is injected into a long column which contains a nonvolatile liquid on a finely divided inert solid. The column is heated and the volatile material is carried through the tube by an inert gas such as helium. Separation depends on the difference in vapor pressure and the partition coefficients of the components in the nonvolatile liquid. As the fractionated components reach the end of the column, they pass over a detection device that is extremely sensitive to differences in organic material carried by the gas, and it records the changes in the gas flow as peaks on a recorder chart. By the use of helium gas alone as the control, and known lipid components as standards to determine the position and area under the peaks, quantitative analysis of lipids can be achieved.

PHOSPHOLIPIDS

The phospholipids are found in all animal and vegetable cells. They are composed of glycerol, fatty acids, phosphoric acid, and a nitrogen-containing compound. More spe-cifically, they are esters of **phosphatidic acid** with choline, ethanolamine, serine, or inositol (hexahydroxycyclohexane).

PHOSPHATIDYL CHOLINE, OR LECITHINS

The lecithins are esters of phosphatidic acid and choline.

L-α-Phosphatidic acid   α-Lecithin

The formula is written with the fatty acid on the left side of the central, or β-carbon, to indicate optical activity and an asymmetric carbon atom. Naturally occurring phospha-tides have the L form and may contain at least five different fatty acids; however, the β-carbon usually is attached to an unsaturated fatty acid. In addition, the formula indicates that lecithin exists in the dissociated state since phosphoric acid is a fairly strong acid and choline is a strong base. **Choline** is a quaternary ammonium compound whose basicity in aqueous solution is similar to that of KOH.

The lecithins are constituents of brain, nervous tissue, and egg yolk. From a physio-logical standpoint they are important in the transportation of fats from one tissue to another and are essential components of the protoplasm of all body cells. In industry lecithin is obtained from soybeans and finds wide application as an emulsifying agent.

If the oleic acid on the central carbon atom of lecithin is removed by hydrolysis, the resulting compound is called **lysolecithin.** Disintegration of the red blood cells, or hemoly-sis, is caused by intravenous injection of lysolecithin. The venom of snakes such as the cobra contains an enzyme capable of converting lecithins into lysolecithins, which accounts for the fatal effects of the bite of these snakes. A few insects and spiders produce toxic effects by the same mechanism.

PHOSPHATIDYL ETHANOLAMINE, OR CEPHALINS

The cephalins are found in brain tissue and are essentially mixtures of phosphatidyl ethanolamine and phosphatidyl serine.

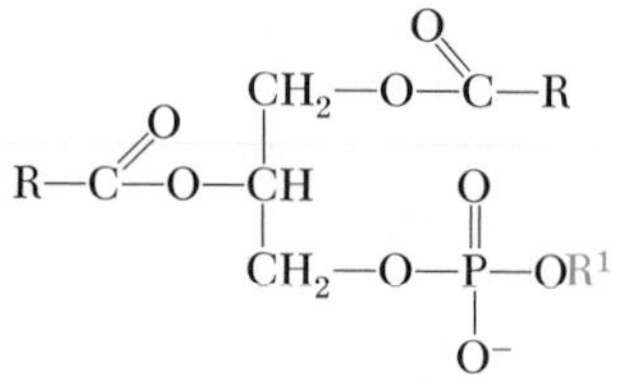

$$CH_2-O-C-R$$

$$R-C-O-CH$$

$$CH_2-O-P-OR^1$$

Basic structure

For Phosphatidyl serine,

R^1 is $-CH_2CHNH_2$

$\quad\quad\quad\quad COOH$

For Phosphatidyl ethanolamine,

R^1 is $-CH_2CH_2NH_2$

The cephalins are involved in the blood-clotting process and are therefore essential constituents of the body.

Other types of phospholipids related to the lecithins and cephalins are the phosphatidyl inositols and the plasmologens, which are derivatives of phosphatidyl ethanolamine. These compounds are not completely characterized at present but are found in brain, heart, and liver tissue.

SPHINGOLIPIDS AND GLYCOLIPIDS

The common structure in both sphingolipids and glycolipids is **ceramide,** which is composed of sphingosine and a fatty acid (R). Sphingomyelin is an example of a sphingolipid and consists of ceramide joined to a phosphate and lecithin.

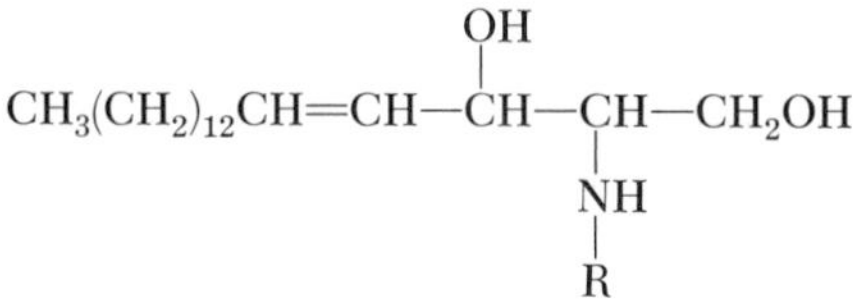

$$CH_3(CH_2)_{12}CH=CH-CH-CH-CH_2OH \quad\quad Ceramide-O-P-O-CH_2CH_2N(CH_3)_3$$

Ceramide

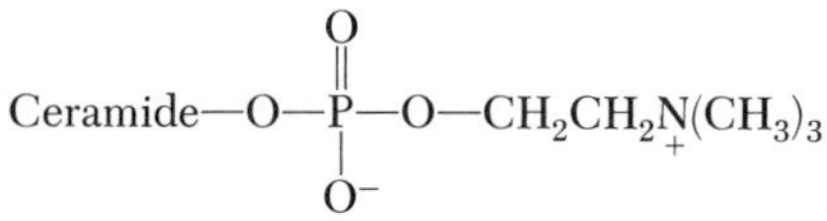

Sphingomyelin

The sphingomyelins are found in large amounts in brain and nervous tissue and are essential constituents of the protoplasm of cells. The glycolipids are compound lipids that contain a carbohydrate and are often called **cerebrosides** because they are found in brain and nervous tissue. They are composed of a carbohydrate and a ceramide. There are four different cerebrosides, each containing a different fatty acid. **Kerasin** contains lignoceric acid ($C_{23}H_{47}COOH$), **phrenosin** a hydroxy lignoceric acid, **nervon** an unsaturated lignoceric acid with one double bond called nervonic acid, and **oxynervon,** a hydroxy derivative of nervonic acid. The carbohydrate in these lipids is usually galactose, although glucose is sometimes present. The structure of a typical cerebroside may be represented as shown on the next page.

STEROIDS

The steroids are derivatives of cyclic alcohols of high molecular weight that occur in all living cells. The lipid material from tissue that is not saponifiable by alkaline hydrolysis contains the steroids. The parent hydrocarbon compound for all the steroids

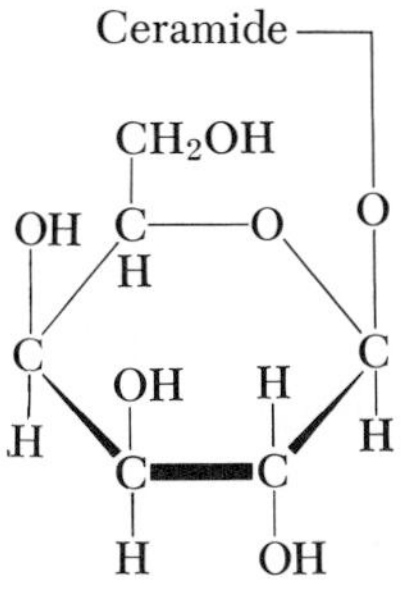

Cerebroside

is the cyclopentanophenanthrene nucleus, also called the sterol nucleus. This structure is an integral part of the cholesterol molecule which may be used to illustrate the lettering system for the rings and the number system for the carbon atoms.

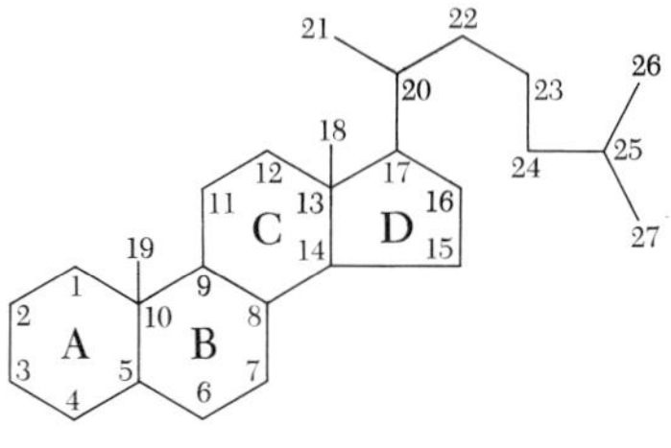

Cholesterol structure designation

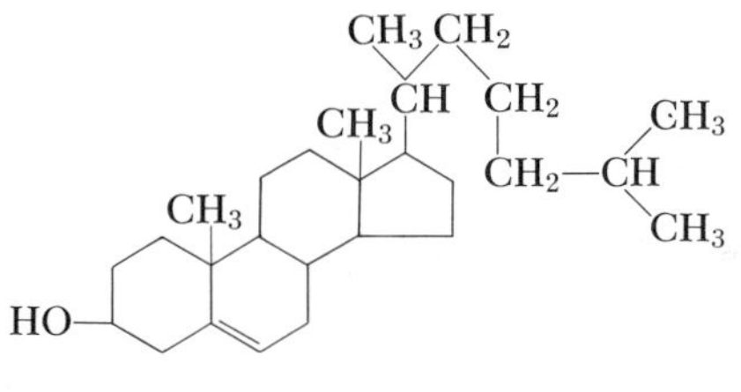

Cholesterol

The most common sterol is **cholesterol,** which is found in brain and nervous tissue and in gallstones. Cholesterol reacts with acetic anhydride and sulfuric acid in a dry chloroform solution to yield a green color. This is called the **Liebermann-Burchard reaction** and is the basis for both qualitative detection and quantitative methods for cholesterol. Cholesterol is the precursor of bile acids, sex hormones, hormones of the adrenal cortex, and vitamin D, which will be discussed in the following sections.

BILE SALTS

The bile salts are natural emulsifying agents found in the bile, a digestive fluid formed by the liver. Cholesterol and bile pigments are also important constituents of the bile. Bile is stored in the gall bladder and released at intervals to assist in the digestion and absorption of fats. **Cholic acid** and **deoxycholic acid** are the major bile acids that are combined with glycine or taurine by an amide linkage to form bile salts such as glyco-cholate or taurocholate.

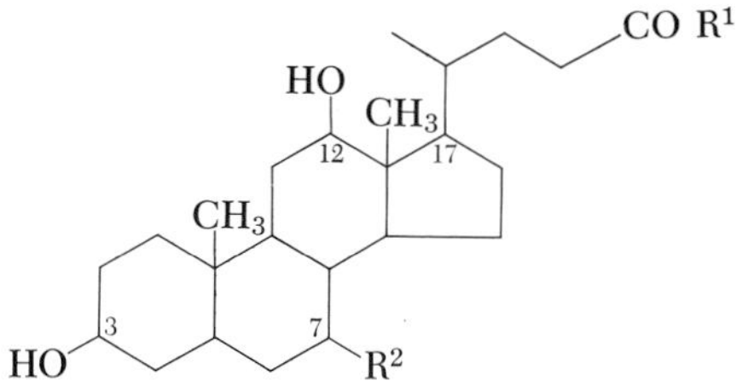

Basic structure

Cholic acid: R^1 is OH and R^2 is OH
Deoxycholic acid: R^1 is OH and R^2 is H
Glycocholic acid: R^1 is NH_2CH_2COOH (Glycine)
Taurocholic acid: R^1 is $NH_2CH_2CH_2SO_3H$ (Taurine)

Chenodeoxycholic acid with hydroxyl groups at positions 3 and 7 in rings A and B, and **lithocholic acid** with a single hydroxyl group at position 3, are also found in human bile.

HORMONES OF THE ADRENAL CORTEX

The cortex of the adrenal gland is an endocrine gland which produces a group of hormones with important physiological functions. If the gland exhibits decreased function, as in Addison's disease, electrolyte and water balance are abnormal, carbohydrate and protein metabolisms are adversely affected, and the patient is more sensitive to cold and stress. Typical steroid hormones of the gland are represented as follows:

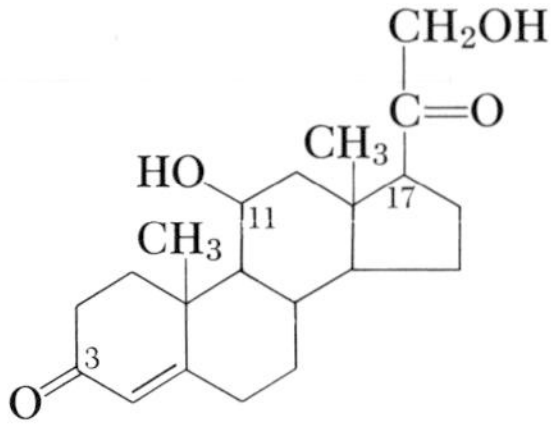

Corticosterone

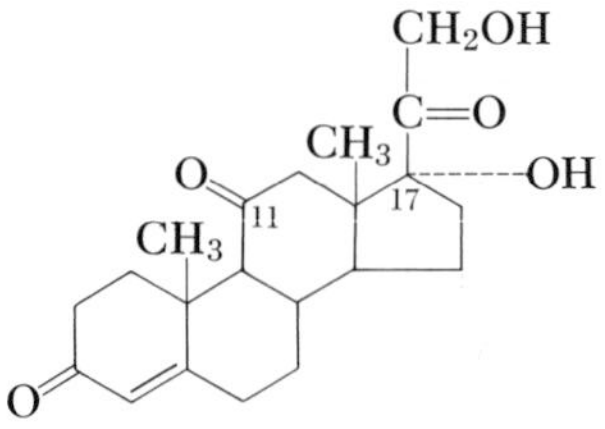

11-Dehydro-17-hydroxycorticosterone
(compound E, cortisone)

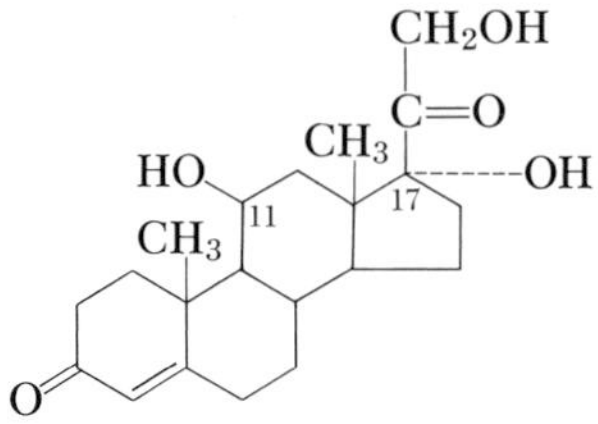

17-Hydroxycorticosterone
(hydrocortisone, cortisol)

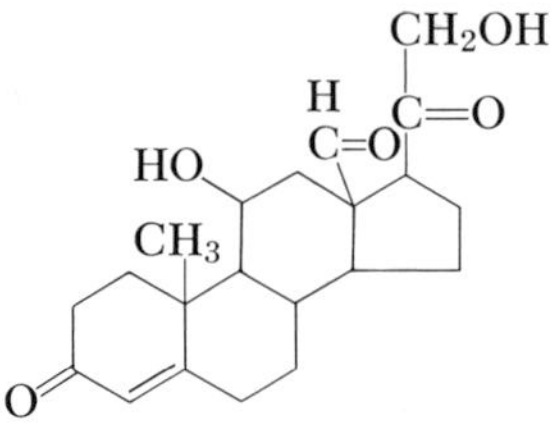

Aldosterone
(aldehyde form)

Corticosterone was the original name of the first adrenal cortical hormone, which accounts for the naming of other hormones as derivatives of this compound. Three major types of adrenal cortical hormones illustrate the relation of structure to physiological activity.

1. Compounds containing an oxygen on the C-11 position (C—OH, or C=O) exhibit greatest activity in carbohydrate and protein metabolism. Examples are **corticosterone, cortisone,** and **cortisol.**

2. Hormones without an oxygen on the C-11 position have their greatest effect on electrolyte and water metabolism. Examples are **11-deoxycorticosterone** and **11-deoxy-cortisol.**

3. **Aldosterone** is the only compound without a methyl group at C-18. It is replaced by an aldehyde group that can exist in the aldehyde form or in the hemiacetal form. Aldosterone has a very potent effect on electrolytes and is called a **mineralocorticoid.** In higher doses it also acts on carbohydrate and protein metabolism.

Corticosterone, cortisol, and aldosterone are the major hormones found in the blood, with **cortisol** exerting the greatest effect on carbohydrate and protein metabolism, and aldosterone on the body fluid electrolytes.

FEMALE SEX HORMONES

The female sex hormones are steroid in structure and are formed in the ovaries, which are glands lying on the sides of the pelvic cavity. Follicles and corpus lutea in different stages of development are located in the cortex of the ovary and form hormones that regulate the **estrus** or **menstrual cycle** and function in pregnancy. Follicular hormones are also responsible for the development of the secondary sexual characteristics that occur at puberty.

HORMONES OF THE FOLLICLE

The liquid within the follicle contains at least two hormones, known as **estrone** and **estradiol.** Estrone (theelin) was the first hormone to be isolated from the follicular liquid, but estradiol (dihydrotheelin) is more potent than estrone and may be the principal hormone.

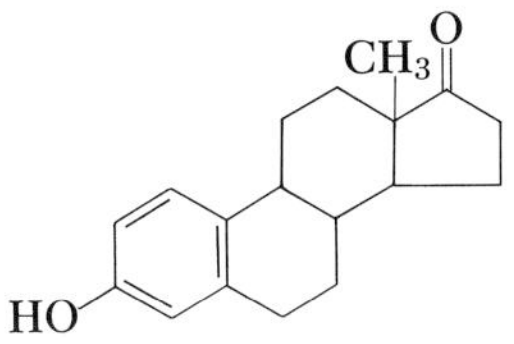

Estrone 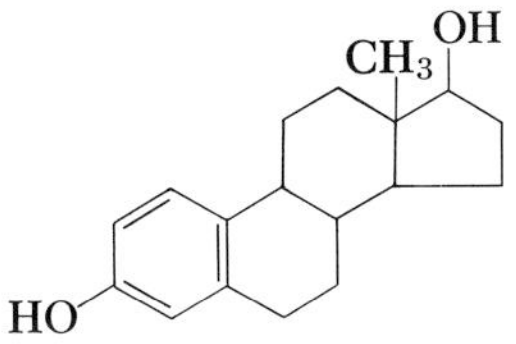   Estradiol

These two compounds are excreted in the urine in increased amounts during pregnancy.

THE HORMONE OF THE CORPUS LUTEUM

The hormone produced by the corpus luteum is called **progesterone.** In the body progesterone is converted into **pregnanediol** by reduction before it is excreted in the urine.

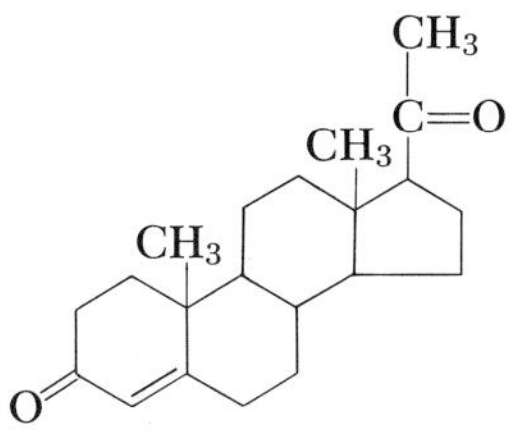

Progesterone 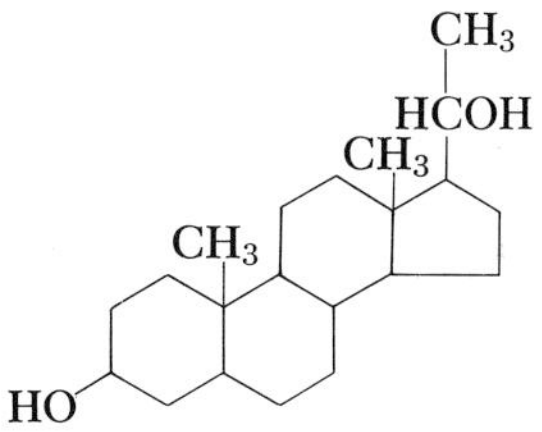   Pregnanediol

The main function of progesterone is the preparation of the uterine endometrium for implantation of the fertilized ovum. If pregnancy occurs, this hormone is responsible for the retention of the embryo in the uterus and for the development of the mammary glands prior to lactation. In the normal menstrual cycle the administration of progesterone inhibits ovulation, a property used in the development of "the pill" (p. 394).

MALE SEX HORMONES

The male sex hormones are produced by the testes, which are two oval glands located in the scrotum of the male. Small glands in the testes form spermatozoa, which are capable of fertilizing a mature ovum. Between the cells that manufacture spermatozoa are the **interstitial cells,** which produce a hormone called **testosterone.** This hormone is probably converted into other compounds such as **androsterone** before being excreted in the urine. **Dehydroandrosterone** has also been isolated from male urine but is much less active than the other two hormones. The male sex hormones, or **androgens,** have structures similar to the estrogens.

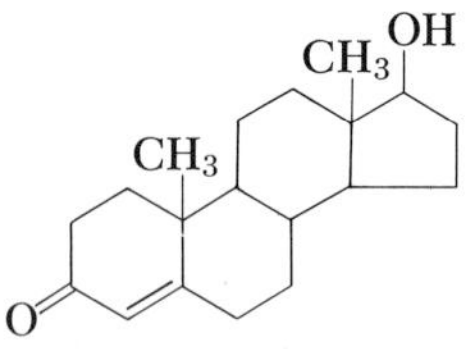

Testosterone 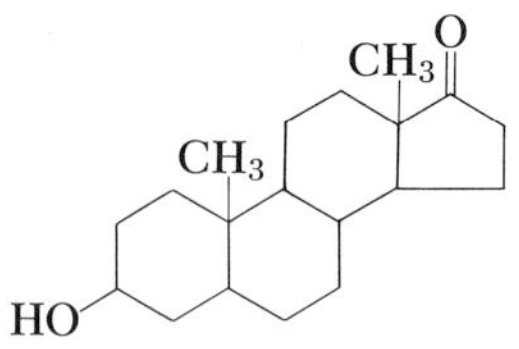   Androsterone

The main function of the androgens in men is the development of masculine sexual characteristics, such as deepening of the voice, the growth of a beard, and distribution of body hair at puberty. They also control the function of the glands of reproduction (seminal vesicles, prostate, and Cowper's gland).

WAXES

Waxes are simple lipids that are esters of fatty acids and high molecular weight alcohols. Common, naturally occurring waxes are **beeswax, lanolin, spermaceti,** and **carnauba wax.** Lanolin, from wool, is the most important wax from a medical standpoint, since it is widely used as a base for many ointments, salves, and creams. Spermaceti, obtained from the sperm whale, is used in cosmetics, some pharmaceutical products, and in the manufacture of candles. Carnauba wax is obtained from the carnauba palm and is widely used in floor waxes and in automobile and furniture polishes.

IMPORTANT TERMS AND CONCEPTS

cephalin	detergents	rancidity	testosterone
cerebroside	estrone	saponification	triglyceride
cholesterol	lecithin	steroids	unsaturated fatty acid

QUESTIONS

1. Name and write the formulas for the three most commonly occurring fatty acids.

2. Are most commonly occurring fats composed of simple glycerides or mixed glycerides? Explain.

3. Illustrate the formation of an ester linkage between glycerol and three molecules of butyric acid. How would you name the product?

4. Write the formula for a triglyceride that would exist as a solid at room temperature and for one that would exist as an oil.

5. Explain how you would test for the presence of glycerol in the laboratory and why this test can be used as a general test for fats.

6. What type of rancidity occurs in common shortenings? How is this prevented? Explain.

7. Write an equation illustrating the process of saponification. Name all compounds in the equation.

8. Discuss the advantages and disadvantages of the new detergents and emulsifying agents.

9. How do insoluble soaps, soft soaps, and hard soaps differ from each other?

10. What analytical procedure would you employ to separate a mixture of phospholipids and neutral lipids into its components? Explain.

11. Write the formula for a typical phosphatidyl ethanolamine. What function does this compound serve in the body?

12. Explain the relation between the structure of the adrenal cortical steroids and their physiological function.

13. What is the major structural difference between aldosterone and the other steroid hormones of the adrenal cortex?

14. Write the structure for estrone and indicate how it differs from the male sex hormones.

SUGGESTED READING

Beyler: Some Recent Advances in the Field of Steroids. Journal of Chemical Education, Vol. 37, p. 497, 1960.

Gaucher: An Introduction to Chromatography. Journal of Chemical Education, Vol. 46, p. 729, 1969.

Karasek, DeDecker, and Tiernay: Qualitative and Quantitative Gas Chromatography for the Undergraduate. Journal of Chemical Education, Vol. 51, No. 12, p. 816, 1974.

Kushner and Hoffman: Synthetic Detergents. Scientific American, Vol. 185, No. 4, p. 26, 1951.

Magliulo: Prostaglandins. Journal of Chemical Education, Vol. 50, No. 9, p. 602, 1973.

Mancott and Tietjen: Polyunsaturation in Food Products. Chemistry, Vol. 47, No. 10, p. 29, 1974.

Ruchelman: Gas Chromatography: Medical Diagnostic Aid. Chemistry, Vol. 43, No. 11, p. 14, 1970.

VITAMINS AND COENZYMES

CHAPTER 26

The *objectives* of this chapter are to enable the student to:

1. Define and describe the properties of vitamins.
2. Explain the relationship between water-soluble vitamins and coenzymes.
3. Explain the difference between a coenzyme, a prosthetic group, and an apoenzyme.
4. Describe the chemical nature of a specific coenzyme and its function in the body.
5. Describe the relation between vitamin A and Δ^{11} *cis*-retinal.
6. Describe the formation of vitamin D_2 from ergosterol.

In order to support normal growth, function, and reproduction in the living cell, the diet must supply not only carbohydrates, lipids, proteins, water, and inorganic salts, but also trace amounts of organic molecules called **vitamins.** The lack of one or more vitamins in the diet will result in a deficiency disease in man or animals; however, overdosing with vitamin preparations can result in a condition of hypervitaminosis. Excessive ingestion of vitamin A produces nausea, vomiting, skin irritations, and mental disturbances, while overdoses of vitamin D lead to increased blood calcium, depression of brain function, and kidney damage.

COENZYMES

In early studies of the enzymes of yeast it was observed that dialysis of a solution of yeast inactivated the enzymes. When the dialyzed material was added to the enzymes, they again exhibited activity. The cofactor in the dialysate was called a **coenzyme.** Since that time several coenzymes have been discovered, and they have been found to consist of small organic molecules. If the organic molecule, or nonprotein portion, is readily separated from the enzyme, it is called a coenzyme. If it is firmly attached to the protein portion of the enzyme, it is called a **prosthetic group.** Most enzymes may therefore be considered as conjugated proteins composed of an inactive protein molecule called the apoenzyme combined with the prosthetic group or coenzyme. The complete, conjugated, active molecule is called a **holoenzyme.**

WATER-SOLUBLE VITAMINS

Vitamins or derivatives of vitamins serve in intermediary metabolism as coenzymes or prosthetic groups in enzymatic reactions involving oxidation, reduction, and decarboxylation. The water-soluble B vitamins contain several vitamins that exhibit the properties of coenzymes.

VITAMIN B$_1$

Vitamin B$_1$, or thiamine, contains a pyrimidine ring and another heterocyclic ring containing sulfur, called thiazole.

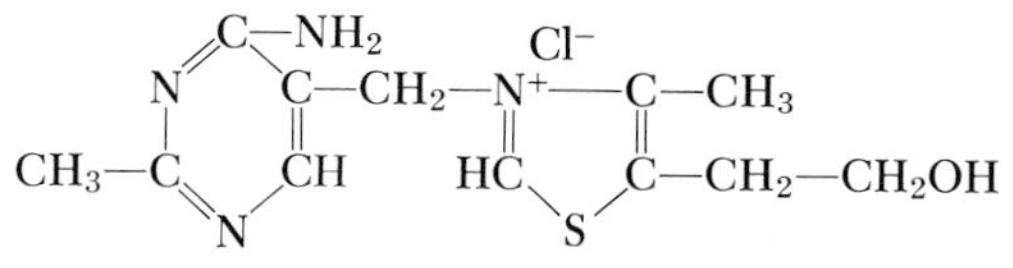

Thiamine chloride

A deficiency of the vitamin in the diet results in a disease called **polyneuritis** in animals and **beriberi** in man. The peripheral nerves of the body are involved, with muscle cramps, numbness of the extremities, pain along the nerves, and eventually atrophy of muscles, edema, and circulatory disturbances occurring in the body. Thiamine occurs free in cereal grains, but occurs as the coenzyme, **thiamine pyrophosphate,** in yeast and meat.

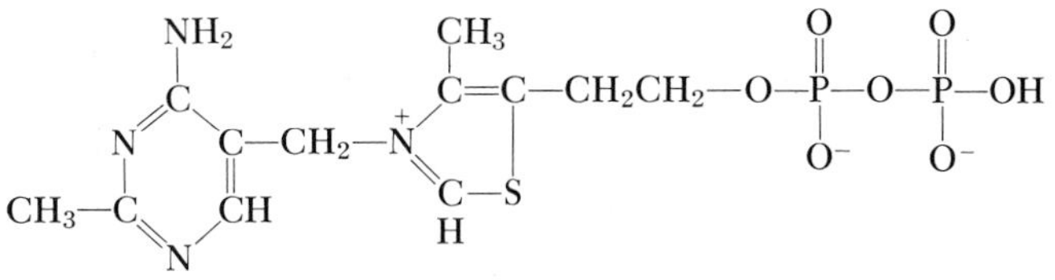

Thiamine pyrophosphate (cocarboxylase) (TPP)

Cocarboxylase functions in the oxidative decarboxylation of pyruvic acid to form acetaldehyde and carbon dioxide. The thiazole ring is the active site of this function. If this reaction does not occur at a normal rate, pyruvic acid may accumulate in the blood and tissues and give rise to the neuritis that is common in thiamine deficiency. **Thiamine pyrophosphate, TPP,** also serves as a coenzyme for enzymes such as α-keto acid oxidase, phosphoketolase, and transketolase.

RIBOFLAVIN

Riboflavin, or vitamin B$_2$, is composed of a pentose alcohol, ribitol, and a pigment, flavin.

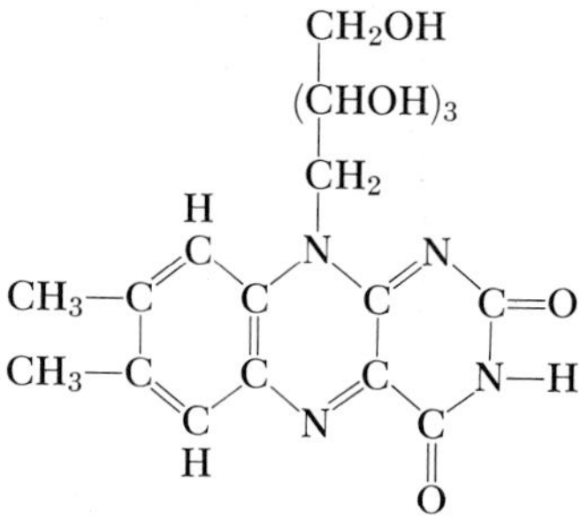

Riboflavin

A deficiency of vitamin B$_2$ in the diet of animals causes lack of growth, hair loss, and cataracts. Lack of the vitamin in man affects vision and causes inflammation of the cornea and sores and cracks in the corners of the mouth.

Foods rich in riboflavin are yeast, liver, eggs, and leafy vegetables. Milled cereal products lose both their thiamine and vitamin B$_2$, and at present there is a trend toward the fortification of white flour with these two vitamins.

The vitamin functions as a coenzyme; in fact, it occurs in foods as a component of two flavin coenzymes, FMN and FAD. The structure of **flavin mononucleotide (FMN)** and the manner in which it is incorporated in **flavin adenine dinucleotide (FAD)** are represented as follows:

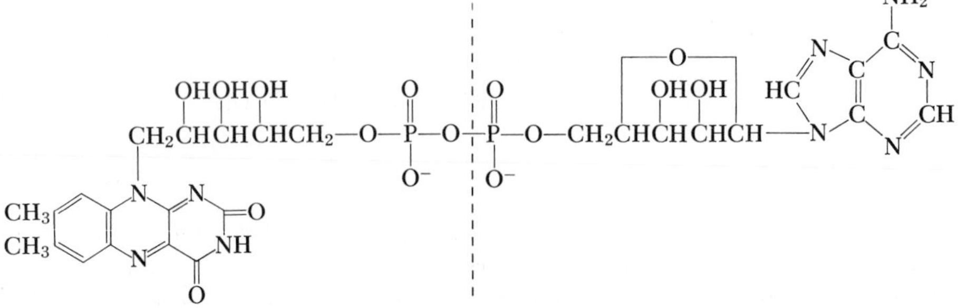

Flavin mononucleotide (FMN)

Flavin adenine dinucleotide (FAD)

Both FMN and FAD serve as coenzymes for a group of enzymes which catalyze oxidation-reduction reactions, such as glutathione reductase, succinic dehydrogenase, and D-amino acid oxidase. The flavin portion of the molecule is the active site for the oxidation-reduction reactions.

NICOTINIC ACID AND NICOTINAMIDE

These two compounds have comparatively simple structures, and as vitamins are called **niacin.**

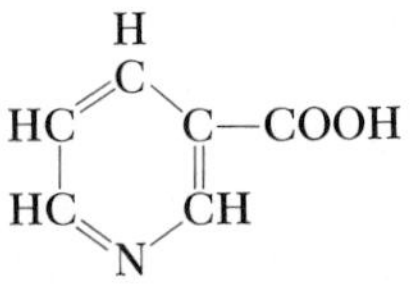

Nicotinic acid

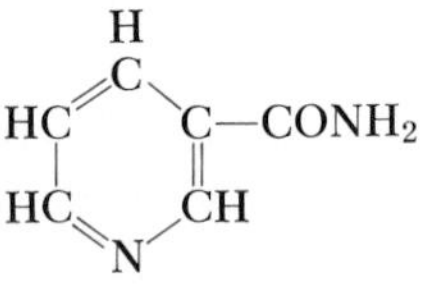

Nicotinamide

A deficiency of niacin results in **pellagra,** a disease characterized by skin lesions on parts of the body that are exposed to sunlight. A sore and swollen tongue, loss of appetite, diarrhea, and nervous and mental disorders are typical symptoms of the disease. Liver, lean meat, and yeast are good sources of niacin.

Niacin is an essential component of two important coenzymes, **nicotinamide adenine dinucleotide, NAD,** and **nicotinamide adenine dinucleotide phosphate, NADP.**

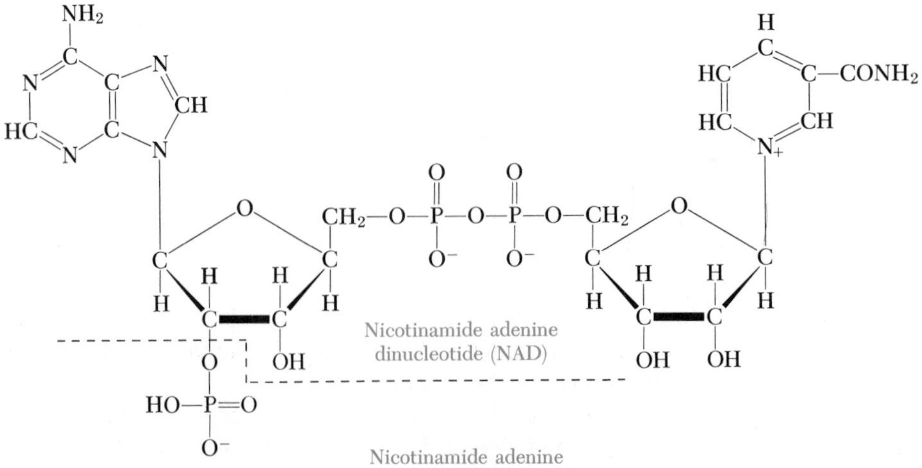

Nicotinamide adenine dinucleotide (NAD)

Nicotinamide adenine dinucleotide phosphate (NADP)

The nicotinamide portion of NAD and NADP is involved in the mechanism of the oxidation-reduction reactions with which these coenzymes are involved.

Both NAD and NADP are coenzymes for dehydrogenases, which catalyze oxidation-reduction reactions. Lactic dehydrogenase, for example, catalyzes the oxidation of lactic acid to pyruvic acid with NAD serving as a coenzyme and being reduced to NADH in the reaction. Alcohol dehydrogenase and glucose-6-phosphate dehydrogenase also require NAD as coenzyme. Many enzymes of clinical diagnostic significance, such as lactic dehydrogenase, may be determined quantitatively in body fluids by the change in form of the coenzyme that occurs in the reaction.

$$\text{NAD} \quad \rightleftarrows \quad \text{NADH}$$

Zero absorption of   **Strongly absorbs**
light at 340 nm   **light at 340 nm**

PYRIDOXINE

The original name for this vitamin was vitamin B_6, which is a general name for **pyridoxine** and two closely related compounds, **pyridoxal** and **pyridoxamine.** These compounds, like nicotinic acid, are pyridine derivatives.

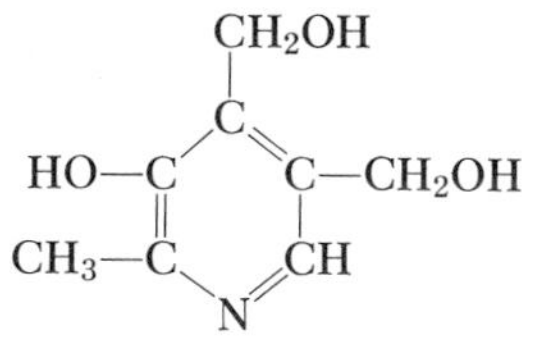

Pyridoxine 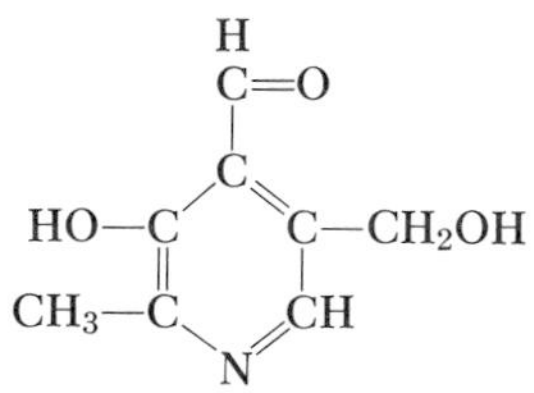   Pyridoxal 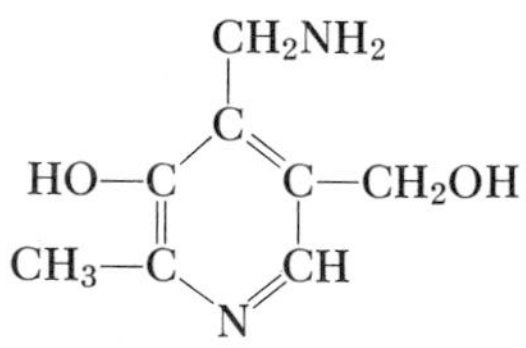   Pyridoxamine

Vitamin B_6 is widely distributed in nature, with yeast, eggs, liver, cereals, legumes, and milk serving as good sources. The phosphate derivatives of pyridoxal and pyridoxamine occur in vitamin B_6 sources and serve as the coenzyme forms of the vitamin.

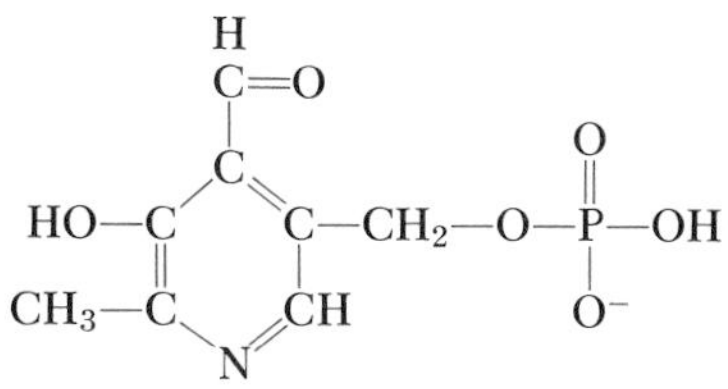

Pyridoxal phosphate 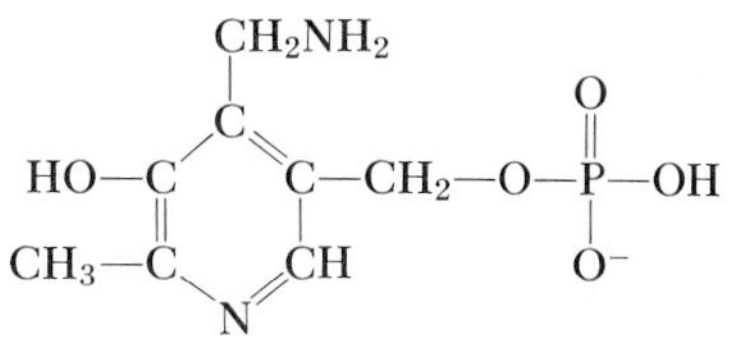   Pyridoxamine phosphate

Pyridoxal phosphate is the major coenzyme for several enzymes involved in amino acid metabolism. Processes such as transamination, decarboxylation, and racemization of amino acids require pyridoxal phosphate as a cofactor. The active sites in pyridoxal phosphate are the aldehyde group and the adjacent hydroxyl group.

PANTOTHENIC ACID

Pantothenic acid is an amide of dihydroxydimethylbutyric acid and alanine.

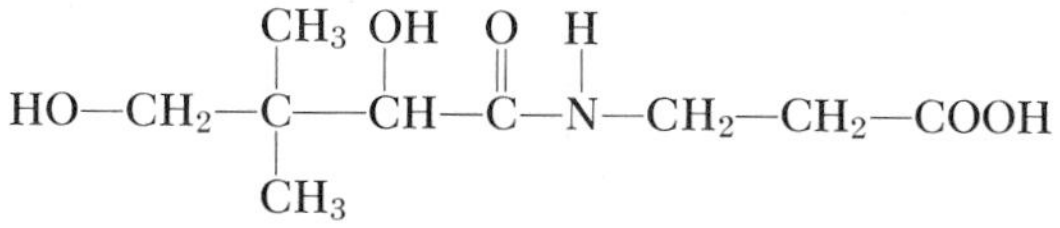

Pantothenic acid

Many animals show deficiency symptoms on diets lacking pantothenic acid; for example, the rat fails to grow, and exhibits a dermatitis, graying of hair, and adrenal cortical failure. In recent dietary research on pantothenic acid deficiency in man, such symptoms as emotional instability, gastrointestinal tract discomfort, and a burning sensation in the hands and feet have been observed.

Pantothenic acid is so widespread in nature that it was named from the Greek word **pantos,** meaning everywhere. Yeast, eggs, liver, kidney, and milk are good sources. The coenzyme form of this vitamin is known as **coenzyme A.**

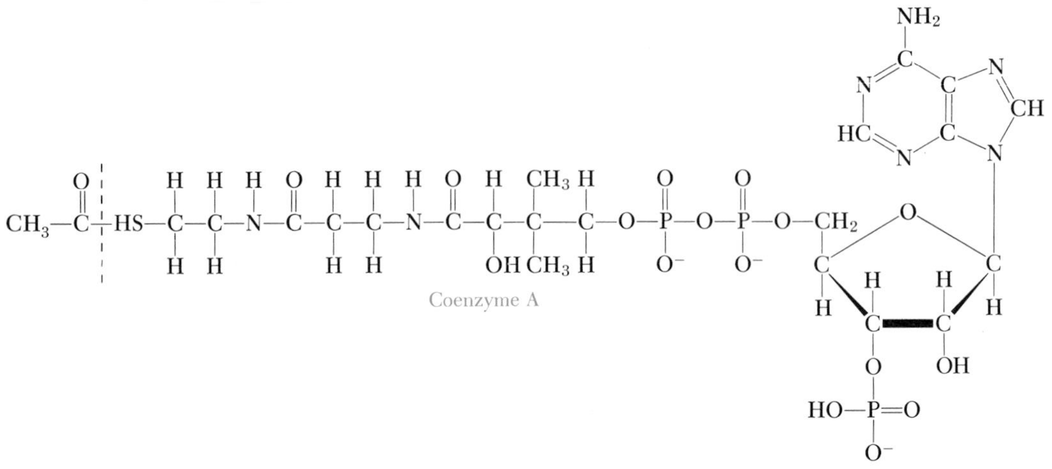

Acetyl coenzyme A

The functional group of the coenzyme is the —SH group, resulting in the abbreviated form, CoASH. In biological systems it functions mainly as **acetyl CoA,** and is involved in acetylation reactions, synthesis of steroids, and the metabolic reactions that will be discussed in subsequent chapters. The formation of acetyl CoA involves a reaction of the functional —SH group with acetic acid. Acetyl CoA is also important as a source of acetate for the Krebs cycle.

FOLIC ACID

Folic acid is a complex molecule consisting of three major parts: a yellow pigment called a pteridine, *p*-aminobenzoic acid, and glutamic acid. Its composition led to the name **pteroylglutamic acid;** its structure may be represented as follows:

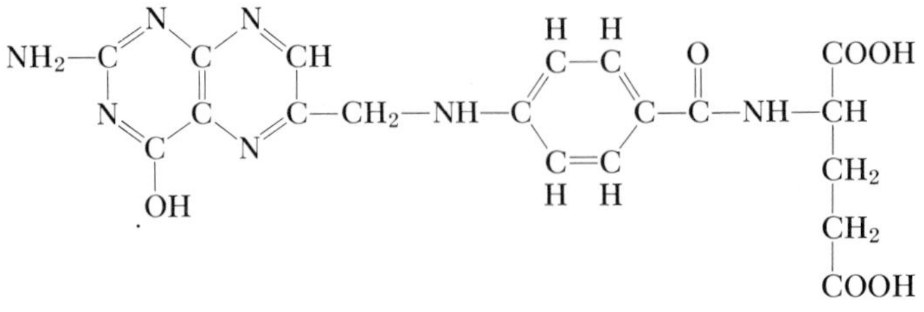

Folic acid

A lack of this vitamin in the diets of young chickens and monkeys causes anemia and other blood disorders. Recently favorable clinical results have been reported in man following the use of folic acid in macrocytic anemias, which are characterized by the presence of giant red corpuscles in the blood.

Folic acid occurs in many plant and animal tissues, especially in the foliage of plants, from which it was named. Yeast, soybeans, wheat, liver, kidney, and eggs are good sources of this vitamin.

VITAMIN B$_{12}$

Vitamin B$_{12}$ has a complex chemical structure that is centered about an atom of cobalt bound to the four nitrogen atoms of a tetrapyrrole, to a nucleotide, and to a cyanide group. It is called **cyanocobalamin,** and is represented as follows:

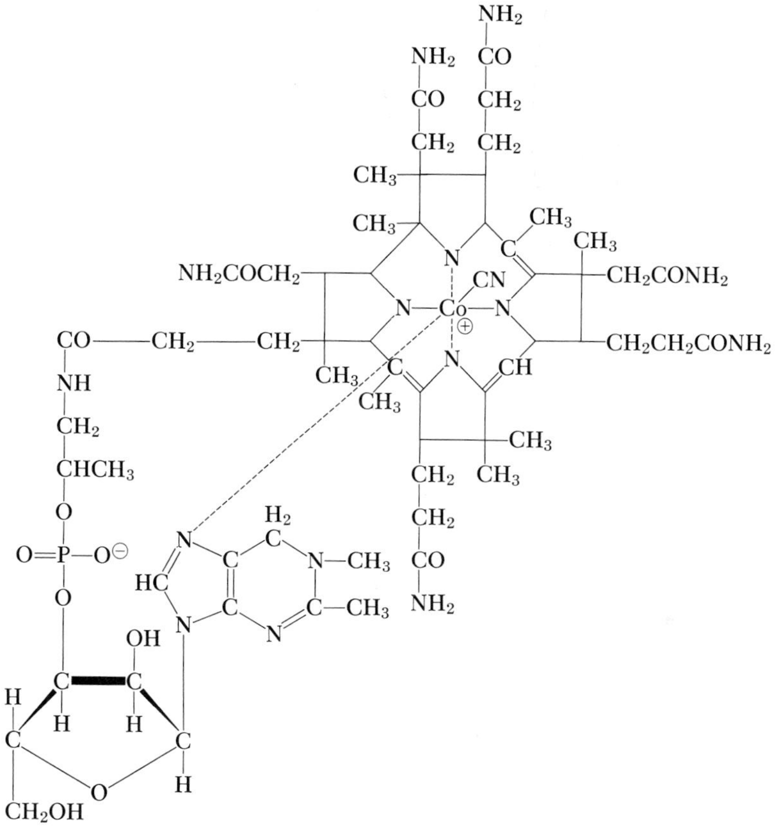

Vitamin B$_{12}$ (cyanocobalamin)

Vitamin B$_{12}$, like folic acid, is useful in the treatment of human and animal anemias, particularly **pernicious anemia.** In addition to increasing the hemoglobin and the red cell count, vitamin B$_{12}$ administration also produces a remission of the neurological symptoms of anemia.

The best source of vitamin B$_{12}$ is liver. Other sources include milk, beef extract, and culture media of microorganisms. The coenzyme form of the vitamin occurs in nature and is known as **coenzyme B$_{12}$.** It is an unstable compound in which the CN or OH group attached to the cobalt atom in vitamin B$_{12}$ is replaced by the nucleoside adenosine, as shown on page 332.

The coenzyme is readily converted into either cyano- or hydroxycobalamin in the presence of cyanide or light. Coenzyme B$_{12}$ functions in several important reactions in metabolism, including the isomerization of dicarboxylic acids, the conversion of glycols and glycerol to aldehydes, the biosynthesis of methyl groups, and the synthesis of nucleosides.

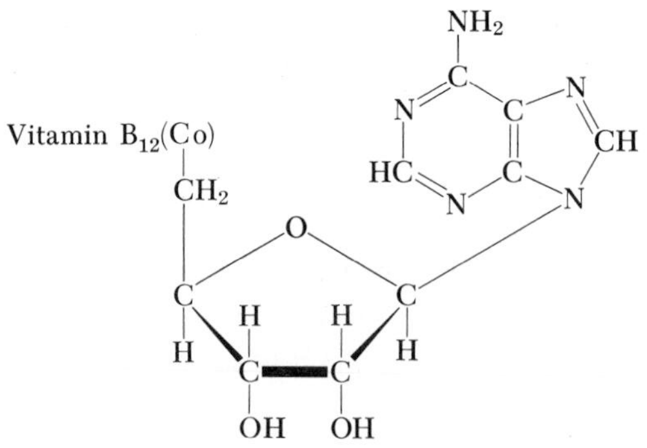

Vitamin $B_{12}(Co)$

Coenzyme B_{12}

ASCORBIC ACID (VITAMIN C)

Ascorbic acid is an enediol of a hexose sugar acid. The reduced or enediol form is readily oxidized to form **dehydroascorbic acid.** Both forms are biologically active.

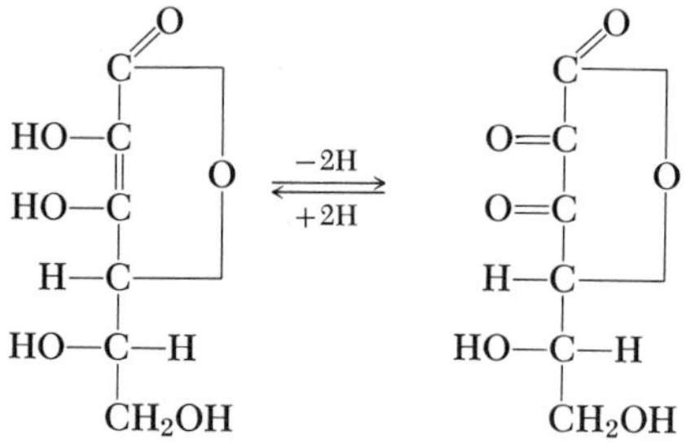

L-Ascorbic acid   Dehydroascorbic acid

A deficiency of ascorbic acid in the diet results in the disease known as **scurvy.** As early as 1720, citrus fruits were used as a cure for scurvy. The fact that all British ships were later required to carry stores of lime juice to prevent scurvy on long voyages led to the use of the term "limey" for sailors. Early symptoms of scurvy are loss of weight, anemia, and fatigue. As the disease progresses, the gums become swollen and bleed readily, and the teeth loosen. The bones become brittle, and hemorrhages develop under the skin and in the mucous membranes. Extreme scurvy is not commonly seen today, although many cases of subacute, or latent, scurvy are recognized. Symptoms such as sore receding gums, sores in the mouth, tendency to fatigue, lack of resistance to infections, defective teeth, and pains in the joints are indicative of **subacute scurvy.**

The richest sources of ascorbic acid are citrus fruits, tomatoes, and green leafy vegetables. A large percentage of the ascorbic acid in foods is destroyed or lost in cooking. Prolonged boiling and the addition of sodium bicarbonate to maintain the green color of vegetables can destroy 70 to 90 per cent of the vitamin C content.

Ascorbic acid may function in oxidation or reduction processes in the body, since it is a powerful reducing agent. The adrenal cortex contains appreciable amounts of ascorbic acid, which may function in the synthesis of steroid hormones in the adrenal gland.

FAT-SOLUBLE VITAMINS

It was noted in experiments prior to 1920 that certain animal fats such as cod liver oil were capable of promoting growth in young rats which were fed a purified fat-free diet. The vitamins found in these animal fats were first collectively called vitamin A, but now include vitamins A, D, E, and K.

VITAMIN A

Vitamin A is closely related to the carotenoid pigments, alpha, beta, and gamma carotene, and cryptoxanthin, which are polyunsaturated hydrocarbons. The carotene pigment, beta carotene, has an all-*trans* structure and is an active precursor of the vitamin.

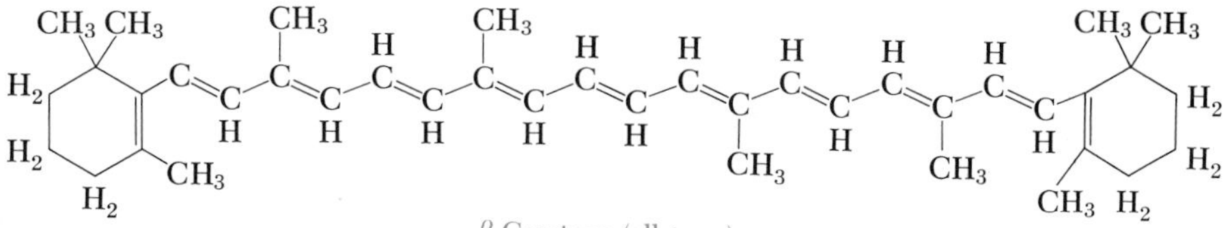

β-Carotene (all-*trans*)

Vitamin A represents half the beta carotene molecule with the ends oxidized to primary alcohol groups.

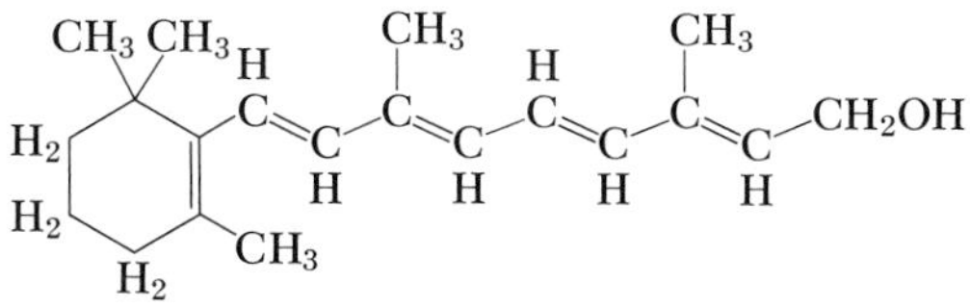

Vitamin A (all-*trans*)

The carotene pigments and cryptoxanthine can be converted into vitamin A in the animal body.

A diet deficient in vitamin A will not support growth, and the deficiency produces keratinization, a hardening and drying of the mucous membranes. One of the first symptoms of the lack of the vitamin is **night blindness.** Later the eyes develop a disease called **xerophthalmia,** characterized by inflamed eyes and eyelids. The eyes ultimately become infected, and when this infection involves the cornea and lens, sight is permanently lost. A continued deficiency of vitamin A results in extensive infection in the respiratory, digestive, and urinary tracts. Vitamin A deficiency also causes sterility, since it affects the lining of the genital tract.

Fish liver oils are potent sources of vitamin A. Eggs, liver, milk and dairy products, green vegetables, and tomatoes are good food sources of the vitamin. The body has the ability to store vitamin A in the liver when it is present in the food in excess of the body requirements. Infants obtain a store of the vitamin in the first milk (colostrum) of the mother, which is ten to one hundred times as rich in vitamin A as ordinary milk.

THE BIOCHEMISTRY OF VISION

The visual process in the eyes of man and animals involves two types of photoreceptors: rods for vision in dim light, and cones for vision in bright light. Little is known of the mechanism involving the visual process in the cones. We owe to George Wald's extensive research on vision our more complete knowledge of the visual mechanism in the rods.

Vitamin A is involved in the visual process in the formation of the visual pigment **rhodopsin,** which is a complex composed of retinal (formerly called retinine) and opsin (a protein). **Retinal** has been identified as vitamin A aldehyde; it may exist in the *cis* or all-*trans* form. All-*trans*-retinal is vitamin A with the primary alcohol group oxidized to an aldehyde. The structure of Δ^{11} *cis*-retinal is shown as follows:

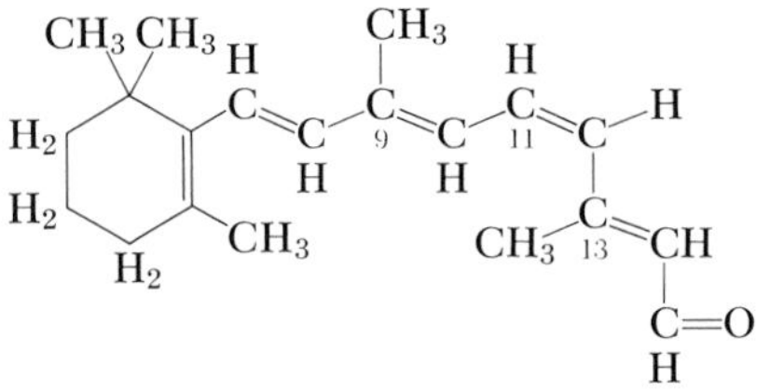

Δ^{11} cis-retinal

The relation between rhodopsin, retinal, and vitamin A and the visual cycle is shown in the following scheme:

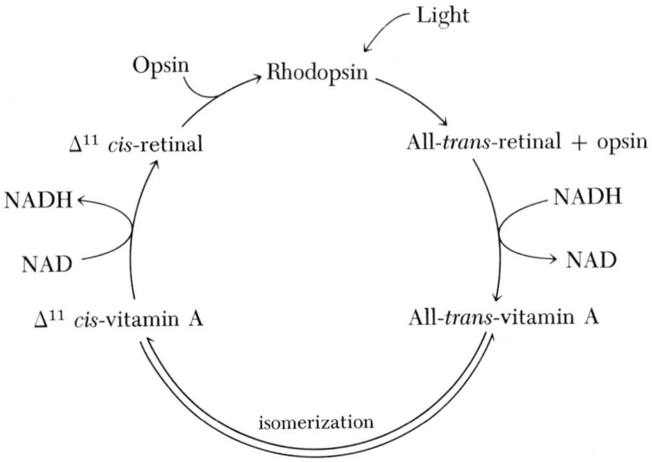

When light strikes rhodopsin, isomerization of Δ^{11} cis-retinal to the all-*trans*-retinal occurs, and the complex splits into the protein opsin and all-*trans*-retinal. The latter compound is reduced to all-*trans*-vitamin A. In the regeneration of the visual pigment rhodopsin, the all-*trans*-vitamin A is first isomerized to Δ^{11} cis-vitamin A, then oxidized to Δ^{11} cis-retinal. There is a loss of vitamin A during the regeneration of rhodopsin after exposure to light. Since this supply of vitamin A must come from the blood, a normal rate of rhodopsin synthesis is therefore dependent on the vitamin A concentration in the blood. In the vitamin-conscious culture of the United States, most of our citizens have a more than adequate dietary intake of vitamin A. Dietary deficiencies that exist in underdeveloped countries result in many cases of xerophthalmia.

VITAMIN D

Several compounds with vitamin D activity exist, although only two of them commonly occur in foods. These two compounds are produced by the irradiation of ergosterol and 7-dehydrocholesterol with ultraviolet light. Ergosterol is a sterol that occurs in yeast and molds, whereas 7-dehydrocholesterol is found in the skin of animals. Irradiated ergosterol is called **calciferol**, or vitamin D_2; irradiated 7-dehydrocholesterol is called vitamin D_3. The structures of the two forms of vitamin D are very similar.

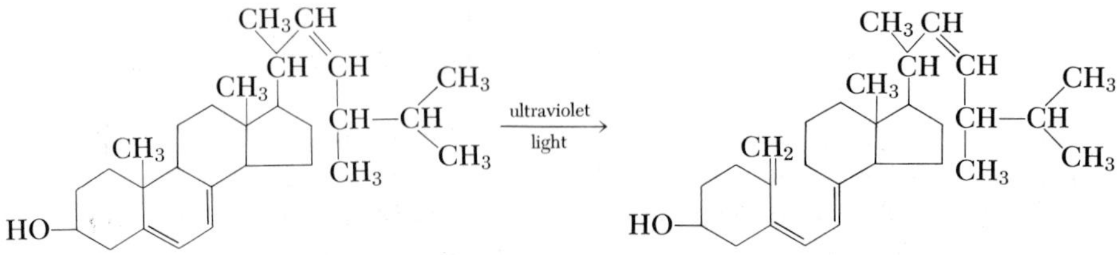

Ergosterol

Vitamin D_2 (calciferol)

The main function of vitamin D in the body is to increase the utilization of calcium and phosphorus in the normal formation of bones and teeth. The exact mode of action of the vitamin is not known, although it increases calcium and phosphorus absorption from the intestine, stimulates the activity of the enzyme phosphatase, and is essential for normal growth.

The lack of vitamin D in the diet of infants and children results in a disease called **rickets,** characterized by abnormal formation of the bones and poor tooth development. Rickets does not occur in adults after bone formation is complete, although the condition of **osteomalacia** may occur in women after several pregnancies. In osteomalacia, the bones soften and abnormalities of the bony structure may occur.

The fish liver oils are the most potent sources of vitamin D. The ultraviolet rays in sunlight form vitamin D by irradiation of 7-dehydrocholesterol in the skin. Children who play outdoors in the summer materially increase the vitamin D content of their bodies. The vitamin D content of some foods, particularly milk, is increased by the addition of small amounts of irradiated ergosterol.

VITAMIN E

Vitamin E is chemically related to a group of compounds called **tocopherols.** Alpha-, beta-, and gamma-tocopherol have vitamin E activity, but alpha-tocopherol is the most potent.

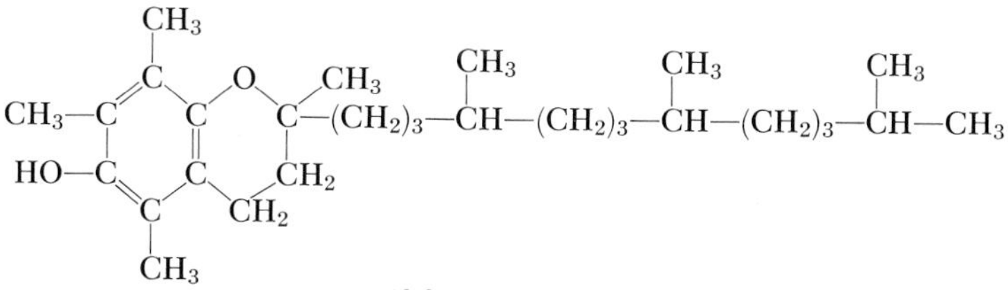

Alpha-tocopherol

The other two tocopherols differ only in the number and position of the CH_3 groups on the aromatic ring. Beta- is a 1,4-di-CH_3, and gamma- a 1,2-di-CH_3 derivative. Vitamin E is stable to heat but is destroyed by oxidizing agents and ultraviolet light.

The richest source of vitamin E is wheat germ oil. Corn oil, cottonseed oil, egg yolk, meat, and green leafy vegetables are good sources of this vitamin. The tocopherols are excellent antioxidants and prevent the oxidation of several substances in the body, including unsaturated fatty acids and vitamin A. As an antioxidant, vitamin E may protect mitochondrial systems in the cell from irreversible oxidation by lipid peroxides. It may also protect lung tissue from damage by oxidants present in smog-contaminated atmospheres.

VITAMIN K

Vitamin K is a derivative of 1,4-naphthoquinone, as is illustrated in the formula:

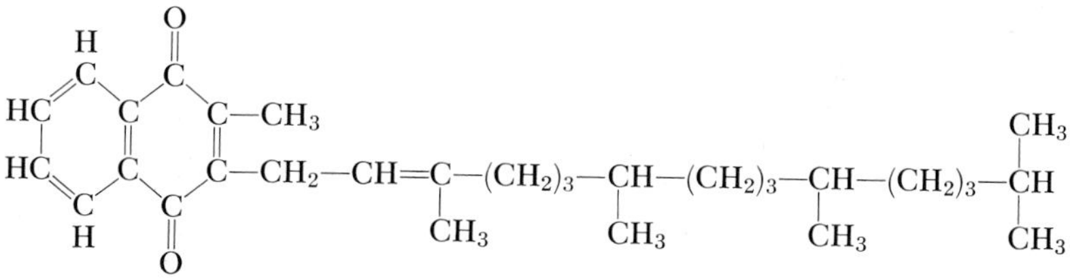

Vitamin K

The 2-methyl-1,4-naphthoquinones and naphthohydroquinones possess vitamin activity. Vitamin K is stable to heat, but is destroyed by alkalies, acids, oxidizing agents, and light.

A diet lacking in vitamin K will cause an increase in the clotting time of blood. This condition produces hemorrhages under the skin and in the muscle tissue. The abnormality in the clotting mechanism is due to a reduction in the formation of **prothrombin,** one of the factors in the normal process.

Richest sources of vitamin K_1 are alfalfa, spinach, cabbage, and kale. The vitamin K_2 is present in bacteria. The bacteria present in putrefying fish meal are capable of synthesizing vitamin K_2 and are potent sources.

IMPORTANT TERMS AND CONCEPTS

calciferol
Δ^{11} *cis*-retinal
coenzyme A
cyanocobalamin

flavin adenine dinucleotide
nicotinamide adenine dinucleotide
pyridoxal phosphate

tetrahydrofolic acid
thiamine pyrophosphate
tocopherol

QUESTIONS

1. Describe the symptoms of a common deficiency disease caused by the lack of a vitamin.

2. What is the difference between a coenzyme and a prosthetic group? Why is an apoenzyme inactive?

3. What type of vitamin is often found as a part of the coenzyme molecule? Explain.

4. Name the important coenzymes that are involved in oxidation-reduction reactions in the body.

5. Write the structure for acetyl coenzyme A, pointing out its component parts.

6. Why are adequate amounts of vitamin C in the diet important in nutrition?

7. What is the relation between Δ^{11} *cis*-retinal, vitamin A, and vision?

8. Why are babies and young children often given doses of cod liver oil?

9. Discuss the function of tocopherols in the body and in food products.

SUGGESTED READING

Dowling: Night Blindness. Scientific American, Vol. 215, No. 4, p. 78, 1966.
Fulkrod: Vitamin C and the Diet of a Student. Journal of Chemical Education, Vol. 49, No. 11, p. 738, 1972.
Hubbard and Kropf: Molecular Isomers in Vision. Scientific American, Vol. 216, No. 6, p. 64, 1967.
Johnson and Williams: Action of Sight upon the Visual Pigment Rhodopsin. Journal of Chemical Education, Vol. 47, p. 736, 1970.
Mellinkoff: Chemical Intervention. Scientific American, Vol. 229, No. 3, p. 102, 1973.
Young: Visual Cells. Scientific American, Vol. 233, No. 4, p. 80, 1970.

BIOCHEMICAL ENERGY

CHAPTER 27

The *objectives* of this chapter are to enable the student to:

1. Calculate the $\Delta G°$ for a coupled reaction that utilizes ATP as the driving force.
2. Recognize ATP as a high energy compound and its relation to ADP and AMP.
3. Explain the process of oxidative phosphorylation and its site of action in the mitochondria.
4. Recognize the difference in moles of ATP formed by NADH and FADH in the electron transport system.

To understand more fully the reactions catalyzed by enzymes in the metabolic processes of the living cell, we must consider the energy relationships that are involved. The energy released from one reaction within a cell may be used almost simultaneously in another reaction that is essential in cellular economy. Energy produced in the cell may also be used as heat—for mechanical work as in muscular contraction or as an electric impulse in nerve transmission. Many of the reactions in metabolism produce chemical energy which is stored in **high-energy compounds** such as ATP. These high-energy compounds are used to drive essential reactions in the metabolic cycles of carbohydrate, lipid, and protein metabolism.

COUPLED REACTIONS

The free energy, ΔG, of a reaction is a measure of the driving force of the reaction and the tendency for the reaction to occur spontaneously. For a chemical reaction to proceed spontaneously to the right, the free energy must decrease; that is, the ΔG must be negative. Such a reaction is called exergonic $(-\Delta G)$, in contrast to an endergonic reaction $(+\Delta G)$, which does not proceed spontaneously. Free-energy changes are usually determined under a set of standard conditions and are known as the *standard free-energy change, $\Delta G°$*, of the reaction. There are many chemical reactions in a cell that have a $+\Delta G$ that will not proceed without assistance of additional energy. These endergonic reactions may be coupled to a highly exergonic reaction, with a $-\Delta G$, so that energy is delivered to the endergonic reaction. In such coupled reactions the algebraic sum of the free-energy changes in the two reactions must be negative in sign (a net decline in free energy) for the coupled reaction to occur. The energy in ATP (adenosine triphosphate), as represented by a $-\Delta G°$ of 8.0 kcal when ATP is converted to ADP (adenosine diphosphate), is often used to drive endergonic reactions. In the formation of glucose-6-phosphate from glucose and

337

phosphate, about 3.0 kcal are required. If this reaction is coupled with the ATP → ADP reaction in the presence of the enzyme hexokinase, the following $\Delta G°$ results:

$$\text{Glucose} + \text{ATP} \xrightarrow{\text{hexokinase}} \text{Glucose-6-PO}_4 + \text{ADP}$$

$$\Delta G° = -8.0 + 3.0 = -5.0 \text{ kcal (approximately)}$$

Many **coupled reactions** in the cell involve the formation of a common intermediate with the assistance of ATP. The formation of sucrose from glucose and fructose, for example, has a $\Delta G°$ of $+5.5$ kcal and requires the conversion of ATP to ADP to drive the reaction to completion. The coupled reaction and formation of a common intermediate, glucose-1-PO$_4$, can be represented as follows:

$$\text{Glucose} + \text{ATP} \rightarrow \text{Glucose-1-PO}_4 + \text{ADP}$$
$$\text{Glucose-1-PO}_4 + \text{Fructose} \rightarrow \text{Sucrose} + \text{P}_i$$
$$\Delta G° = +5.5 - 8.0 = -2.5 \text{ kcal (approximate)}$$

In the first reaction, the terminal PO$_4$ group of ATP was transferred to glucose and with it some of the energy of the ATP. In the second reaction, the energy-enriched glucose-1-PO$_4$ reacts with fructose to form sucrose.

HIGH-ENERGY COMPOUNDS

Early investigations on the nature of muscular contraction revealed that the presence of the high-energy compound creatine phosphate has a driving force in muscle reactions. Studies on the oxidation of glucose and especially the metabolic cycles of carbohydrate oxidation emphasized the role of **adenosine triphosphate, ATP,** and this energy-rich compound has become the key in linking endergonic processes to those that are exergonic.

High-energy compounds are often complex phosphate esters that yield large amounts of free energy on hydrolysis. A more detailed consideration of the energy released by the stepwise hydrolysis of ATP will illustrate the high-energy concept.

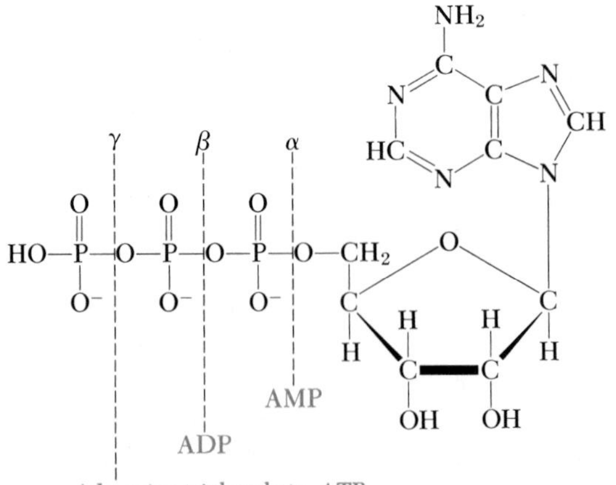

Adenosine triphosphate, ATP

$$\text{ATP} \xrightarrow{\text{hydrolysis}} \text{ADP} + \text{H}_3\text{PO}_4 \qquad \Delta G° = -8.0 \text{ kcal}$$

$$\text{ADP} \xrightarrow{\text{hydrolysis}} \text{AMP} + \text{H}_3\text{PO}_4 \qquad \Delta G° = -6.5 \text{ kcal}$$

$$\text{AMP} \xrightarrow{\text{hydrolysis}} \text{Adenosine} + \text{H}_3\text{PO}_4 \qquad \Delta G° = -2.2 \text{ kcal}$$

Several explanations have been proposed for the release of energy on the hydrolysis of high-energy compounds. These include the fact that these compounds are unstable in acid and alkaline solutions and are readily hydrolyzed. Also, the hydrolysis products, inorganic phosphate, ADP, and AMP, have many more resonance possibilities than the parent ATP. A major reason for the release of energy involves the *type of bond structure* in these compounds. The β and γ bonds in ATP are anhydride linkages that involve a large amount of repulsion energy between the phosphates, which is released on hydrolysis.

Other phosphorus-containing, high-energy compounds include:

Acetyl phosphate
$\Delta G^\circ = -10.0$ kcal

Phosphoenolpyruvic acid
$\Delta G^\circ = -12.0$ kcal

Creatine phosphate
$\Delta G^\circ = -10.0$ kcal

Acetyl coenzyme A
$\Delta G^\circ = -8.2$ kcal

The top two compounds in the previous illustrations have anhydride linkages between a phosphate and either a carbonyl or acid enol group. Creatine phosphate, the high-energy compound in muscle, has a direct linkage between phosphate and nitrogen, whereas the acyl mercaptide linkage in acetyl coenzyme A is also characteristic of an energy-rich compound. In every instance the high-energy compound is readily hydrolyzed to products that undergo spontaneous reactions. These reactions result in forms that are thermodynamically more stable.

Simple phosphate esters, such as AMP, glucose-6-phosphate, and 3-phosphoglyceric acid, are not considered as high-energy compounds, and yield less energy on hydrolysis. The distinction between high-energy compounds and low-energy phosphate esters is arbitrary, but a dividing line of 5.0 kcal per mole is usually set by biochemists.

In the discussion of metabolism that follows, there will be many examples of the use of high-energy compounds in the storage of energy, the transmission of energy, and the coupling of energy obtained from foodstuffs to the utilization of that energy for cellular reactions.

THE FORMATION OF ATP

Since adenosine triphosphate has been marked as a key compound in the storage of chemical energy and in the coupling of exergonic reactions to endergonic reactions in the cell, it is a major driving force in the metabolic reactions in the tissue. Although ATP can be formed by light energy in the process of photosynthesis, which will be discussed in Chapter 29, the present discussion will consider its formation in the cytoplasm in the absence of oxygen (substrate level phosphorylation) and in the **mitochondria** by the process of oxidative phosphorylation.

In the Embden-Meyerhof anaerobic pathway (p. 354), glucose is phosphorylated and is eventually broken down into 3-carbon phosphorylated derivatives. Two of these derivatives, 1,3-diphosphoglyceric acid and phosphoenolpyruvic acid, react with ADP with the assistance of enzymes to form 3-phosphoglyceric acid and pyruvic acid, respectively, plus a molecule of energy-rich ATP in each reaction. These reactions can take place in the absence of O_2 and in the cytoplasm, and are termed **substrate level phosphorylations.**

OXIDATIVE PHOSPHORYLATION ATP

One of the major aerobic or oxidative schemes of carbohydrate metabolism (Krebs cycle, p. 355) involves the reaction of intermediate compounds with the production of several moles of ATP. An **electron transport system** in the mitochondria of the cell actively transports electrons from a reduced metabolite to oxygen with the assistance of enzymes and coenzymes, as shown in Figure 27–1. P_i is inorganic phosphate. NAD is nicotinamide adenine dinucleotide, FMN is flavin mononucleotide, and FAD is flavin adenine dinucleotide; these three coenzymes were discussed in Chapter 26. Other intermediate compounds in Figure 27–1 between the reduced metabolite and oxygen, besides NAD and FAD, are coenzyme Q and the cytochromes. The overall reactions involve first

$$NAD + Metabolite \cdot H_2 \rightarrow Metabolite + NADH$$

then

$$NADH + 3\,ADP + 3\,P_i + \tfrac{1}{2}O_2 \rightarrow NAD + 3\,ATP + H_2O$$

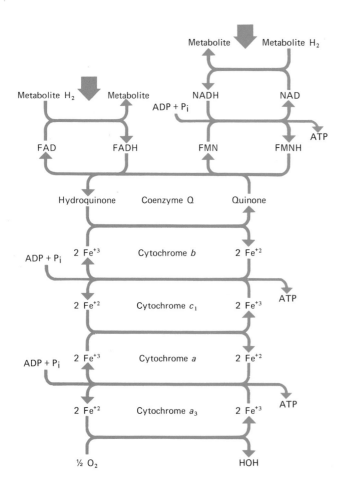

Figure 27–1   Oxidative phosphorylation and the electron transport system.

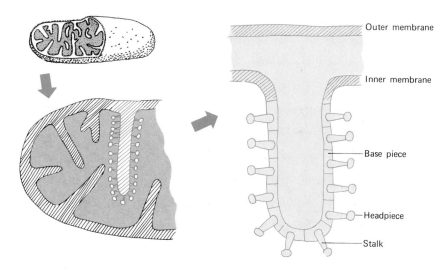

Figure 27–2   Proposed structure of the inner membrane of mitochondria.

As may be seen from Figure 27–1, ATP is generated at three sites on the electron transport chain with the following electron transfers: from NADH to coenzyme Q through flavoprotein FMN, from coenzyme Q to cytochrome c through cytochrome b, and from cytochrome c to O_2 through cytochrome a_3. Three moles of ATP may therefore be generated by the passage of electrons from a mole of substrate through NAD to molecular oxygen, but only two moles may be generated if electrons are transferred directly from the substrate to coenzyme Q through FAD because this transfer bypasses the first phosphorylation site.

The enzymes of electron transport are located on the inner membrane of the mitochondria. This membrane is thought to consist of repeating units, each composed of a headpiece which projects into the matrix, joined by a stalk to a basepiece (Fig. 27–2). The basic electron transport chain is located in the basepieces. Each basepiece is considered a complex containing a portion of the electron transport enzymes. Four such complexes plus NADH, coenzyme Q, and cytochrome c constitute a complete electron transport system. The energy generated by electron transport in the basepiece complex is transmitted through the stalk protein to the headpiece, where it is converted into the high-energy bond of ATP.

NAD, **nicotinamide adenine dinucleotide** (structure shown in Chapter 26), is a dinucleotide composed of AMP linked to nicotinamide-ribose-phosphate. The nicotinamide portion of the molecule is involved in the oxidation and reduction reactions in oxidative phosphorylation.

FAD, **flavin adenine dinucleotide** (structure shown in Chapter 26), is a dinucleotide composed of a flavin-ribose-phosphate linked to AMP. The reduction of FAD in the electron transport system involves the flavin portion of the molecule.

Coenzyme Q is a lipid soluble quinone, sometimes called ubiquinone-10 for the ten isoprene units found in the side chain (the number may vary from 0 to 10). This coenzyme is readily reduced to the hydroquinone form during the transport of hydrogen.

The **cytochromes** are oxidation-reduction pigments that consist of iron-porphyrin complexes known as **heme,** which is also an integral part of hemoglobin, the respiratory pigment of the red blood cells. The heme in cytochrome c, for example, is attached to a protein molecule by coordination with two basic amino acid residues, and by thioether linkages formed by the addition of a sulfhydryl group from each of two molecules of cysteine in the protein molecule. Cytochrome c is an electron carrier in the oxidative phosphorylation cycle, in which the iron atom of heme is changed from Fe^{+++} to Fe^{++} as shown:

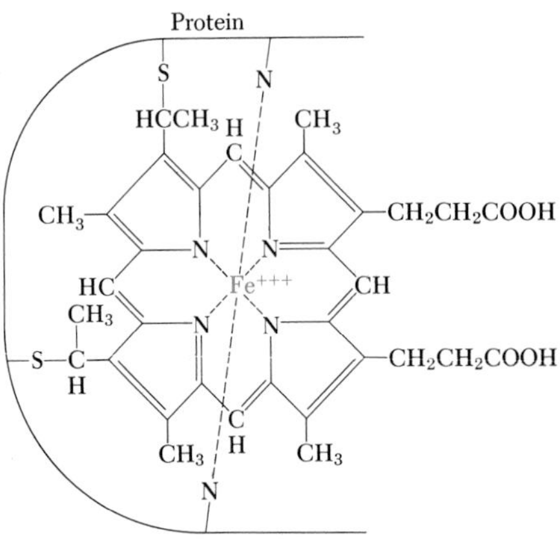

Cytochrome c
(oxidized)

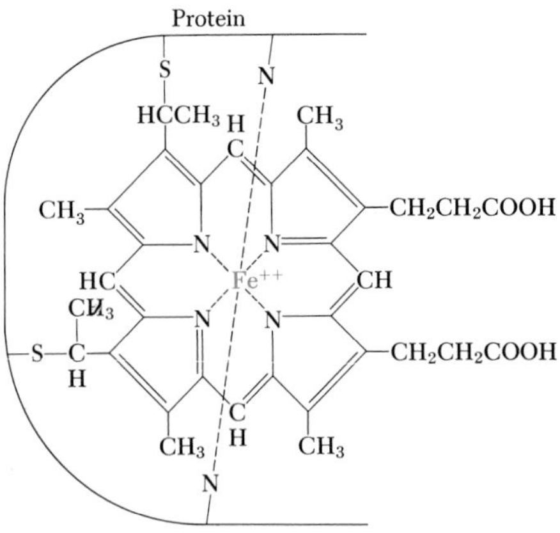

Cytochrome c
(reduced)

Biochemical energy in the form of ATP is an essential driving force in many metabolic reactions in the cells and tissues. As has been described, several complex reactions are involved in the synthesis of this vital compound, and it should be emphasized that three moles of ATP are formed when the electrons from NADH are transported through the system to oxygen. Also two moles of ATP are formed when electrons from FADH are transported to oxygen. These relationships will assist in the understanding of the energy balance in the metabolic cycles.

IMPORTANT TERMS AND CONCEPTS

ATP	electron transport system	NADH
coenzyme Q	FADH	oxidative phosphorylation
coupled reactions	free-energy change	
cytochromes	high-energy compounds	

QUESTIONS

1. What are the different forms of energy that may be produced in the living cell?

2. The formation of an ester from an acid and an alcohol resulted in a $\Delta G°$ of $+2.0$ kcal. ATP formed an intermediate with the acid to drive the reaction to completion. Represent the reaction and calculate the new $\Delta G°$.

3. Briefly explain why phosphoenolpyruvic acid is a high-energy compound.

4. Why is the oxidative phosphorylation mechanism also called the electron transport system? Explain.

5. Explain why only two moles of ATP are formed in the electron transport system when electrons from FADH are transported to oxygen.

6. Trace the process of the electron transport system from the basepiece to the headpiece on the inner membrane of the mitochondria.

SUGGESTED READING

Alberty: Thermodynamics of the Hydrolysis of Adenosine Triphosphate. Journal of Chemical Education, Vol. 46, p. 713, 1969.

Changeux: The Control of Biochemical Reactions. Scientific American, Vol. 212, No. 4, p. 36, 1965.

Dickerson: The Structure and History of an Ancient Protein. Scientific American, Vol. 226, No. 4, p. 58, 1972.

Kirschbaum: Biological Oxidations and Energy Conservation. Journal of Chemical Education, Vol. 45, p. 28, 1968.

INTRODUCTION
TO METABOLISM

CHAPTER 28 ═══════════════════════════════════

The *objectives* of this chapter are to enable the student to:

1. Describe the processes of digestion and absorption of carbohydrates.
2. Describe the processes of digestion and absorption of dietary fat.
3. Describe the processes of digestion and absorption of dietary protein.
4. Define the processes of intermediary metabolism in the cell.
5. Distinguish between the processes of catabolism and anabolism in the outline of intermediary metabolism.

A simple definition of **metabolism** would be the consideration of all the enzymatic reactions occurring in the living cell. Since we are obviously not aware of all the reactions that occur in the cell, we may begin by breaking metabolism into several component parts. Two major divisions are **anabolism,** or the biosynthetic processes, and **catabolism,** or the biodegradative processes that result in metabolites needed by the cell and the chemical energy used in cellular reactions. Prior to the metabolic reactions in the cell, the major foodstuffs of the diet must undergo the processes of digestion and absorption.

SALIVARY DIGESTION

Food taken into the mouth is broken into smaller pieces by chewing and is mixed with saliva, the first of the digestive fluids. Saliva contains **mucin,** a glycoprotein that makes the saliva slippery, and **ptyalin,** an enzyme that catalyzes the hydrolysis of starch to maltose. Since this enzyme has little time to act on starches in the mouth, its main activity takes place in the stomach before it is inactivated by the acid gastric contents. There are no enzymes in the saliva that act on dietary fats or proteins.

GASTRIC DIGESTION

When food is swallowed it passes through the esophagus into the stomach. During the process of digestion the food is mixed with gastric juice, which is secreted by small tubular glands located in the walls of the stomach. Gastric juice is a pale yellow, strongly acid solution containing the enzymes **pepsin** and **rennin.** Pepsin initiates the hydrolysis of large protein molecules into smaller, more soluble molecules of **proteoses** and **peptones,** whereas rennin converts casein of milk into a soluble protein. The optimum pH of pepsin is 1.5 to 2; thus, it is ideally suited for the digestion of protein in normal stomach contents, the pH of which is 1.6 to 1.8. The mixing action of the stomach musculature and the process of digestion produce a liquid mixture called chyme, which passes through the pyloric opening into the intestine.

INTESTINAL DIGESTION

The acid chyme is neutralized by the alkalinity of the three digestive fluids, **pancreatic juice, intestinal juice,** and **bile,** in the first part of the small intestine, the duodenum. When fat enters the duodenum, the gastrointestinal tract hormone **cholecystokinin** is secreted and is carried by the blood to the gallbladder, where it stimulates that organ to empty its bile into the small intestine. Bile acids and bile salts are good detergents and emulsify fats for digestion by **pancreatic lipase.** Another hormone that is secreted when the chyme enters the duodenum is **secretin.** This hormone enters the circulation and stimulates the pancreas to release pancreatic juice into the intestine. There are enzymes in the pancreatic juice that are capable of digesting proteins, fats, and carbohydrates. The pancreatic proteases are **trypsin, chymotrypsin,** and **carboxypolypeptidase,** whereas the pancreatic lipase is called **steapsin.** The enzyme **amylopsin** in pancreatic juice is an amylase similar to ptyalin in the saliva. Pancreatic lipase is activated by bile salts and splits fats into fatty acids, glycerol, soaps, mono- and diglycerides. Native proteins and the proteoses and peptones that result from the action of pepsin are gradually split into amino acids by the proteases of the pancreatic juice and **aminopolypeptidase** and **dipeptidase** in the intestinal juice. The intestinal juice also contains three disaccharide-splitting enzymes, **sucrase, lactase,** and **maltase.** Cane sugar is the main source of dietary sucrose; milk contains lactose; and maltose comes from the partial digestion of starch by ptyalin and amylopsin. Sucrase, lactase, and maltase split these disaccharides into their constituent monosaccharides, thus completing the digestion of carbohydrates.

ABSORPTION

The monosaccharides glucose, fructose, and galactose are absorbed directly into the bloodstream through the capillary blood vessels of the **villi.** The villi are fingerlike projections on the inner surface of the small intestine that greatly increase the effective absorbing surface. There are approximately five million villi in the human small intestine, and each villus is richly supplied with both lymph and blood vessels. Considerable evidence exists to indicate that the intestinal mucosa possesses the property of **selective absorption,** which is not possessed by a nonliving membrane. Enzyme mechanisms requiring energy are involved in the selective absorption of all molecules that pass through the intestinal mucosa. The rate of absorption is not determined by the size of the molecule but by the specific mechanism.

In the absorption process, as the end products of fat digestion pass through the villi of the intestinal mucosa they are reconverted into triglycerides, which then enter the lymph circulation. Bile salts are essential in absorption, both because of their effect on the solubility of fatty acids and because of their direct involvement in the absorption process.

Amino acids are absorbed through the intestinal mucosa directly into the bloodstream by an active process that requires energy and enzymes. After absorption, the amino acids are carried by the portal circulation to the liver and subsequently to all the tissues of the body.

INTERMEDIARY METABOLISM

Intermediary metabolism is concerned with the molecules presented to the cells and tissues following the process of absorption and involves a multitude of enzyme-catalyzed reactions in different parts of the cell. The specific details of carbohydrate, lipid, and protein metabolism will be presented in subsequent chapters; however, an overview of metabolism in the body as outlined in Figure 28–1 will assist in understanding the details.

Following absorption into the cell, the end-products of digestion are further converted into simpler molecules by the process of catabolism. Hexoses, pentoses, and glycerol are converted into the 3-carbon phosphorylated sugar glyceraldehyde-3-phosphate and then into the 2-carbon acetyl group of acetyl coenzyme A. The fatty acids and amino acids from digestion and absorption are also broken down in the process of catabolism to **acetyl coenzyme A.** This common end-product of catabolism, acetyl CoA, is then fed into the Krebs cycle, where the acetyl group is eventually catabolized into CO_2 with the production of chemical energy in the form of ATP. From the above discussion and Figure 28–1 it may be seen that the various pathways of catabolism flow toward a final common pathway, the Krebs cycle.

In Figure 28–1 it is apparent that the flow of anabolism is opposite that of catabolism. In the Krebs cycle α keto acids are formed, which may be aminated to form amino acids used in the synthesis of protein molecules. Acetyl groups from acetyl CoA may be assembled into fatty acids and eventually into lipids. Also, the acetyl groups from acetyl CoA may be converted back to pyruvate and proceed up the chain of reactions to the polysaccharide glycogen.

The regulation of intermediary metabolism in the cells depends on regulatory or allosteric enzymes, regulatory hormones, and the control of the concentration of a specific enzyme in the cell. **Allosteric enzymes** are usually located at the beginning of a series

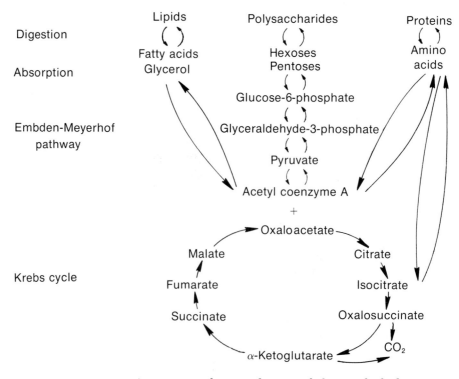

Figure 28–1 An overview of intermediary metabolism in the body.

of metabolic enzyme-catalyzed reactions and catalyze the rate-limiting step of the series. **Regulatory hormones** produced by endocrine glands are chemical messengers that stimulate or inhibit specific metabolic processes in other tissues or cells. The concentration of a specific enzyme in a cell at any instant results from a balance between the rates of its synthesis and degradation. The metabolism of a compound may be initiated by the synthesis of a specific enzyme responding to the presence of that compound as its substrate.

Catabolism of the major foodstuffs is always accompanied by conservation of some of the chemical energy in the form of ATP. This compound serves as a ready source of chemical energy to initiate catabolic and anabolic reactions in the cell.

IMPORTANT TERMS AND CONCEPTS

absorption   anabolism   digestion   Krebs cycle
acetyl coenzyme A   catabolism   Embden-Meyerhof pathway   metabolism

QUESTIONS

1. Outline the process of digestion and absorption of carbohydrates in the gastrointestinal tract.

2. Briefly describe the digestion and absorption of dietary fat.

3. Outline the process of protein digestion and absorption.

4. Define and compare the terms metabolism, catabolism, and anabolism.

5. Outline the general pathways of metabolism of the three major foodstuffs in the body.

6. What factors are involved in the regulation of intermediary metabolism in the cell? Explain.

SUGGESTED READING

Clark and Marcker: How Proteins Start. Scientific American, Vol. 218, No. 1, p. 36, 1968.
Grünewald: The Evolution of Proteins. Chemistry, Vol. 41, No. 1, p. 11, 1968.
Horecker: Pathways of Carbohydrate Metabolism and Their Physiological Significance. Journal of Chemical Education, Vol. 42, p. 244, 1965.
Kretchmer: Lactose and Lactase. Scientific American, Vol. 227, No. 4, p. 70, 1972.
Research Reporter: Why the Stomach Does Not Digest Itself. Chemistry, Vol. 46, No. 5, p. 20, 1973.

CARBOHYDRATE METABOLISM

CHAPTER 29

The *objectives* of this chapter are to enable the student to:

1. Recognize the factors involved in the control of the normal blood sugar level.
2. Discuss the role of hormones in the control of the blood sugar level.
3. Describe the process of glycogenesis.
4. Recognize the essential reactions in the Embden-Meyerhof pathway.
5. Outline the essential reactions in the Krebs cycle.
6. Account for the total number of moles of ATP formed in the Embden-Meyerhof and Krebs pathways.
7. Recognize the relationship between the phosphogluconate and the Embden-Meyerhof pathways.
8. Recognize the relationship between intermediates of the Embden-Meyerhof and phosphogluconate pathways and the process of photosynthesis.

After the digestion of the larger dietary carbohydrate molecules to monosaccharides and the absorption of the smaller molecules, they are presented to the living cell mainly as glucose and pentoses. There is a constant circulation of appreciable amounts of glucose in the blood stream. The maintenance of the normal blood sugar level depends on many metabolic factors. The pathways of reactions that affect carbohydrate metabolism in the cell are in a sense general pathways involved in the metabolism of all foodstuffs and should be considered essential chains of reactions in the cell.

Since one of the major functions of catabolism is to produce chemical energy that may be used in driving cellular reactions to produce metabolites that are needed in the chain of metabolic reactions, the role of pathways in producing ATP will be considered.

THE BLOOD SUGAR

After the monosaccharides are absorbed into the blood stream, they are carried by the portal circulation to the liver. Fructose and galactose are phosphorylated by liver enzymes and are either converted into glucose or follow similar metabolic pathways. The metabolism of carbohydrates, therefore, is essentially the metabolism of glucose.

The concentration of glucose in the general circulation is normally 70 to 90 mg per 100 ml of blood. This is known as the **normal fasting level** of blood sugar. After a meal containing carbohydrates, the glucose content of the blood increases, causing a temporary condition of **hyperglycemia.** In cases of severe exercise or prolonged starvation, the blood sugar value may fall below the normal fasting level, resulting in the state of **hypoglycemia.**

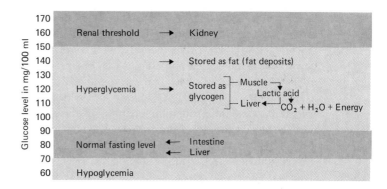

Figure 29–1 Factors involved in the regulation of the glucose level of the blood.

The hyperglycemic levels following a meal are returned to normal by three major processes: storage of glucose as glycogen or as fat; oxidation to produce energy; or excretion by the kidneys.

The operation of the processes that counteract hyperglycemia after a meal is illustrated in Figure 29–1. During active absorption of carbohydrates from the intestine the blood sugar level rises, causing a temporary hyperglycemia. In an effort to bring the glucose concentration back to normal, the liver may remove glucose from the blood stream, converting it into glycogen for storage. The muscles will also take glucose from the circulation to convert it to glycogen or to oxidize it to produce energy. If the blood sugar level continues to rise, the glucose may be converted into fat and stored in the fat depots. These four processes usually control the hyperglycemia; but if large amounts of carbohydrates are eaten and the blood sugar level exceeds an average of 160 mg of glucose per 100 ml, the excess is excreted by the kidneys. The blood sugar level at which the kidney starts excreting glucose is known as the **renal threshold** and has a value from 150 to 170 mg per 100 ml.

In addition to the above factors, there are more specific reactions of the liver and the hormones to bring about regulation of the level of the glucose in the blood. The liver, for example, functions both in the removal of sugar from the blood and in the addition of sugar to the blood. During periods of hyperglycemia the liver stops pouring sugar into the blood stream and starts to store it as liver glycogen. During fasting the liver supplies glucose to the blood by breaking down its glycogen and by forming glucose from other food material such as amino acids or glycerol. The liver is assisted in this control process by several hormones.

HORMONES AND THE BLOOD SUGAR LEVEL

The properties and action of enzymes have already been discussed. In metabolism there are many related chemical reactions under the influence of enzymes. Another important group of regulating agents is the **hormones.** Enzyme action is more specific than hormone action, and the factors involved in the action of a hormone appear to be related to the action of other hormones. In a normal individual, major cellular processes depend on an endocrine or **hormone balance,** and a disturbance in this balance results in metabolic abnormalities. In the regulation of a body process, a hormone probably has control over several specific enzyme-catalyzed reactions.

INSULIN

Although it was demonstrated as early as 1889 that removal of the pancreas of an animal would result in diabetes mellitus, it was not until 1922 that Banting and Best developed a method for obtaining active extracts of the pancreas. Within a short time insulin became available in sufficient quantities for the treatment of diabetes. It was first crystallized in 1926. More recently, as a result of the brilliant work of Sanger and his coworkers, a molecule consisting of two chains of amino acids with a molecular weight of 6000 has been described. The native molecule is thought to consist of four chains with a molecular weight of 12,000. Since it is a protein, it is not effective when taken by mouth, because the proteolytic enzymes of the gastrointestinal tract cause its hydrolysis and destroy its activity. Insulin is usually injected subcutaneously when administered to a diabetic.

Insulin lowers the blood sugar level by increasing the conversion of glucose into liver and muscle glycogen, by regulating the proper oxidation of glucose by the tissues, and by preventing the breakdown of liver glycogen to yield glucose. In muscle and adipose tissue, insulin acts by increasing the rate of transport of glucose across membranes into the cells. Also, there is considerable evidence that in liver tissue insulin acts by controlling the phosphorylation of glucose to form glucose-6-phosphate, which is the first step in the formation of glycogen. In the absence of an adequate supply of insulin the transformation of extracellular glucose to intracellular glucose-6-phosphate is retarded.

Diabetes Mellitus. If the pancreas fails to produce sufficient insulin, the condition of **diabetes mellitus** results. The failure of the storage mechanisms in the absence of insulin causes a marked increase in the blood sugar level. Glucose is ordinarily excreted in the urine because the renal threshold is exceeded. The impairment of carbohydrate oxidation causes the formation of an excess of ketone bodies. Some of these ketone bodies are acid in nature, and the severe acidosis that results from the lack of insulin causes **diabetic coma,** which is sometimes fatal to a diabetic patient. When the correct dosage of insulin is injected, carbohydrate metabolism is properly regulated and the above symptoms do not appear.

Glucagon

Glucagon is a hormone that is produced by the α-cells of the pancreas. It is a polypeptide of known amino acid sequence with a molecular weight of about 3500. Glucagon causes a rise in the blood sugar level by increasing the activity of the enzyme liver phosphorylase, which is involved in the conversion of liver glycogen to free glucose. The activation of phosphorylase depends on the presence of the compound cyclic-3′,5′-adenosine monophosphate (AMP), whose formation is increased by the action of glucagon.

Epinephrine

This hormone is produced by the central portion, or medulla, of the adrenal glands. Epinephrine is antagonistic to the action of insulin in that it causes glycogenolysis in the liver with the liberation of glucose. It stimulates an enzyme to produce cyclic-3′,5′-AMP from ATP and is also involved in the activation of phosphorylase. In addition to hyperglycemia, it also increases blood lactic acid by converting muscle glycogen to lactic acid. Continued secretion of epinephrine occurs under the influence of strong emotions such as fear or anger. This mechanism is often used as an emergency function to provide instant glucose for muscular work. The hyperglycemia that results often exceeds the renal threshold, and glucose is excreted in the urine.

Adrenal Cortical Hormones

Hormones such as **cortisone** and **cortisol** are produced by the outer layer or cortex of the adrenal gland. These hormones, especially those with an oxygen on position 11, have an effect on carbohydrate metabolism. In general they stimulate the production of glucose in the liver by increasing gluconeogenesis from amino acids. The cortical hormones are therefore antagonistic to insulin.

ANTERIOR PITUITARY HORMONES

Of the many hormones secreted by the anterior lobe of the pituitary gland, the growth hormone, ACTH and the diabetogenic hormone affect the blood sugar level. The **growth hormone** causes the liver to increase its formation of glucose, but at the same time it stimulates the formation of insulin by the pancreas. Its action is complex and not completely understood. **ACTH,** the adrenocorticotropic hormone, stimulates the function of the hormones of the adrenal cortex and their action on the blood sugar level. The **diabetogenic hormone,** when injected into an animal, causes permanent diabetes and exhaustion of the islet tissue of the pancreas.

Although the overall control of the blood sugar level depends on the action of the liver and a balanced action of several hormones, it is readily apparent that insulin plays a major role in the normal process and is an important factor in the control of diabetes mellitus.

GLYCOGEN

As may be recalled from Chapter 24, glycogen is a polysaccharide with a branched structure composed of linear chains of glucose units joined by α-1,4 linkages and with α-1,6 linkages at the branch points. During absorption of the carbohydrates, the excess glucose is stored as glycogen in the liver. Normally this organ contains about 100 g of glycogen, but it may store as much as 400 g. The glycogen in the liver is readily converted into glucose and serves as a reservoir from which glucose may be drawn if the blood sugar level falls below normal. The formation of glycogen from glucose is called **glycogenesis,** whereas the conversion of glycogen to glucose is known as **glycogenolysis.** The muscles also store glucose as glycogen, but muscle glycogen is not as readily converted into glucose as is liver glycogen.

GLYCOGENESIS

The process of glycogenesis is not just a simple conversion of glucose to glycogen. As we have learned previously, insulin is involved in the action of glucokinase in the phosphorylation of glucose to glucose-6-phosphate. The glucose-6-phosphate is then converted to glucose-1-phosphate with the aid of the enzyme phosphoglucomutase. The glucose-1-phosphate then reacts with uridine triphosphate (UTP) to form an active nucleotide, uridine diphosphate glucose (UDPG). In the presence of a branching enzyme and the enzyme UDPG-glycogen-transglucosylase, the activated glucose molecules of UDPG are joined in glucosidic linkages to form glycogen. These reactions may be represented as shown on page 352.

GLYCOGENOLYSIS

The process of glycogenolysis liberates glucose into the blood stream to maintain the blood sugar level during fasting and to supply energy for muscular contraction. In the liver the reaction is initiated by the action of the enzyme phosphorylase, which splits the 1,4 glucosidic linkages in glycogen. The enzyme phosphorylase exists in two forms: an active form, **phosphorylase a,** and an inactive form, **phosphorylase b.** Phosphorylase

$$\text{Glucose} \xrightarrow[\text{glucokinase}]{\text{(insulin)}} \text{Glucose-6-phosphate} \xrightarrow{\text{phosphoglucomutase}} \text{Glucose-1-phosphate}$$

$$\text{ATP} \quad \text{ADP}$$

Glucose-1-phosphate + Uridine triphosphate (UTP)

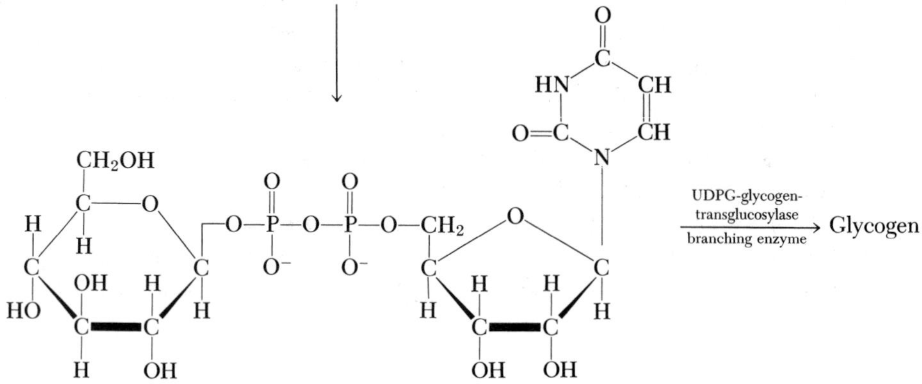

Uridine diphosphate glucose (UDPG)

$$\xrightarrow[\text{branching enzyme}]{\substack{\text{UDPG-glycogen-} \\ \text{transglucosylase}}} \text{Glycogen}$$

b is converted to the active form of the enzyme by ATP in the presence of Mg^{+2} and phosphorylase b kinase, as shown:

$$2\ \text{Phosphorylase b} + 4\ \text{ATP} \xrightarrow[\text{cyclic-3',5'-AMP}]{\text{kinase, } Mg^{+2}} \text{Phosphorylase a} + 4\ \text{ADP}$$

The phosphorylase b kinase is activated by cyclic-3′,5′-AMP, a derivative of adenylic acid.

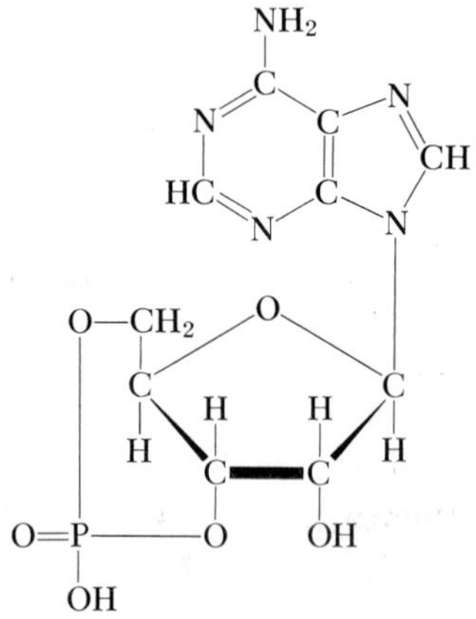

Cyclic-3′,5′-AMP

Cyclic-3′,5′-AMP is formed from ATP by the action of an enzyme called *cyclase enzyme*. This enzyme is activated by epinephrine and glucagon which therefore indirectly activate the phosphorylase responsible for the initiation of glycogenolysis. Other enzymes assist the breakdown to glucose-1-phosphate, which is subjected to the reversed action of phosphoglucomutase to yield glucose-6-phosphate. A specific enzyme in the liver, glucose-6-phosphatase, acts on glucose-6-phosphate to produce glucose. This enzyme is not present in muscle; therefore, muscle glycogen cannot serve as a source of blood glucose. These reactions may be illustrated as follows:

$$\text{Glycogen} \xrightarrow[\text{debranching enzyme}]{\text{phosphorylase}} \text{Glucose-1-phosphate}$$

$$\text{Glucose-1-phosphate} \xrightarrow{\text{phosphoglucomutase}} \text{Glucose-6-phosphate}$$

$$\text{Glucose-6-phosphate} \xrightarrow[\text{in liver}]{\text{glucose-6-phosphatase}} \text{Glucose}$$

OXIDATION OF CARBOHYDRATES

Glucose is ultimately oxidized in the body to form CO_2 and H_2O with the liberation of energy. **Glucose-6-phosphate** is a principal compound in the metabolism of glucose. As discussed earlier, it may be formed by the phosphorylation of glucose under the control of insulin. Once it is formed, it may be converted to glycogen or to free glucose, or it may be metabolized by several mechanisms or pathways. The two major pathways of glucose-6-phosphate metabolism are the **anaerobic,** or **Embden-Meyerhof, pathway** followed by the **aerobic,** or **Krebs, cycle.**

Glycolysis

The ready availability of muscle preparations and their use in the development of physiology led to an early study of the biochemical changes associated with muscular contraction. It was observed that when a muscle contracts in an anaerobic medium, glycogen disappears and pyruvic and lactic acids are formed. In the presence of oxygen, or under aerobic conditions, the glycogen is re-formed, and the pyruvic and lactic acids disappear. Further studies demonstrated that one fifth of the lactic acid formed during glycolysis is oxidized to CO_2 and water, whereas the remaining four fifths is converted to glycogen.

Substances other than the carbohydrates in food and the lactic acid from muscular contraction may be converted into glycogen. These glycogenic compounds are formed by the process of **gluconeogenesis,** which is the conversion of noncarbohydrate precursors into glucose. Examples of these precursors are the glycogenic amino acids, the glycerol portion of fat, and any of the metabolic breakdown products of glucose, such as pyruvic acid, which may form glucose by reversible reactions in metabolism. The reactions discussed in the above section can be summarized in the **lactic acid cycle.**

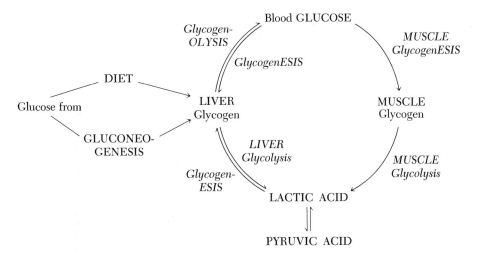

The Lactic Acid Cycle

ANAEROBIC, OR EMBDEN-MEYERHOF, PATHWAY OF GLYCOLYSIS

The chemical reactions in metabolic pathways like the Embden-Meyerhof pathway are detailed and complex, and may lead to confusion at first examination. An understanding of this and other metabolic pathways may be expedited by consideration of a preliminary outline of the essential reactions. Glucose from any source is converted to glucose-6-phosphate, which in turn is converted into fructose-6-phosphate, which is further phosphorylated to fructose-1,6-diphosphate. This set of reactions has converted the hexose glucose into a hexose diphosphate. The pathway diverges at this point with the splitting of fructose diphosphate into two triose monophosphates. One of these, glyceraldehyde-3-phosphate, is first transformed into 1,3-diphosphoglyceric acid, then successively into 3-phospho- and 2-phosphoglyceric acid. The last compound then forms phosphoenolpyruvic acid which is converted into pyruvic acid. The pyruvic acid ($CH_3COCOOH$) is a key compound which may be reduced to lactic acid (p. 241) or further oxidized in the Krebs cycle.

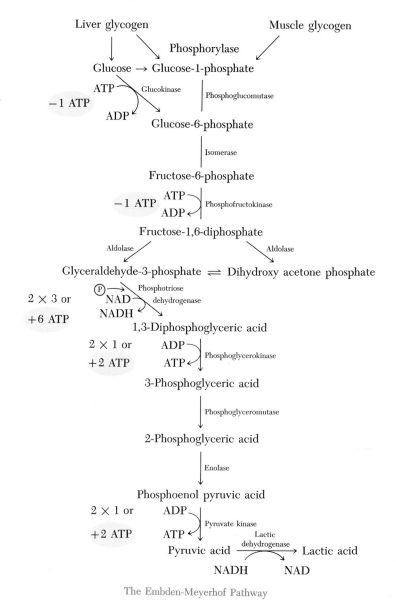

The Embden-Meyerhof Pathway

The names of the chemical compounds and enzymes involved in the Embden-Meyerhof pathway are shown on page 354.

The requirement for and liberation of ATP in this anaerobic pathway is emphasized by the shaded areas. One mole of ATP is required for the phosphorylation of glucose and one more for the conversion of fructose-6-phosphate to fructose-1,6-diphosphate. The reaction of 1,3-diphosphoglyceric acid to form 3-phosphoglyceric acid liberates 1 mole of ATP per triose molecule or 2 moles of ATP per glucose molecule. The conversion of phosphoenolpyruvic acid to pyruvic acid also yields 2 moles of ATP per glucose molecule. In addition, the reaction of glyceraldehyde-3-phosphate to form 1,3-diphosphoglyceric acid liberates 1 mole of NADH per triose molecule or 2 moles per glucose molecule. In the oxidative phosphorylation process (p. 340), 3 moles of ATP are liberated for each mole of NADH. For each mole of glucose broken down in the Embden-Meyerhof pathway, therefore, *2 moles of ATP are consumed and 10 moles are liberated, with a net gain of 8 moles of ATP.*

AEROBIC, OR KREBS, CYCLE

The pyruvic acid formed in the Embden-Meyerhof pathway and the lactic acid from the lactic acid cycle or from the reduction of pyruvic acid are eventually oxidized with the formation of CO_2 and energy in the form of ATP. Pyruvic acid from any source forms acetyl Co A, which transfers the acetyl group to oxaloacetic acid to make the tricarboxylic acid, citric acid. Citric acid is successively transformed into *cis*-aconitic acid, then to

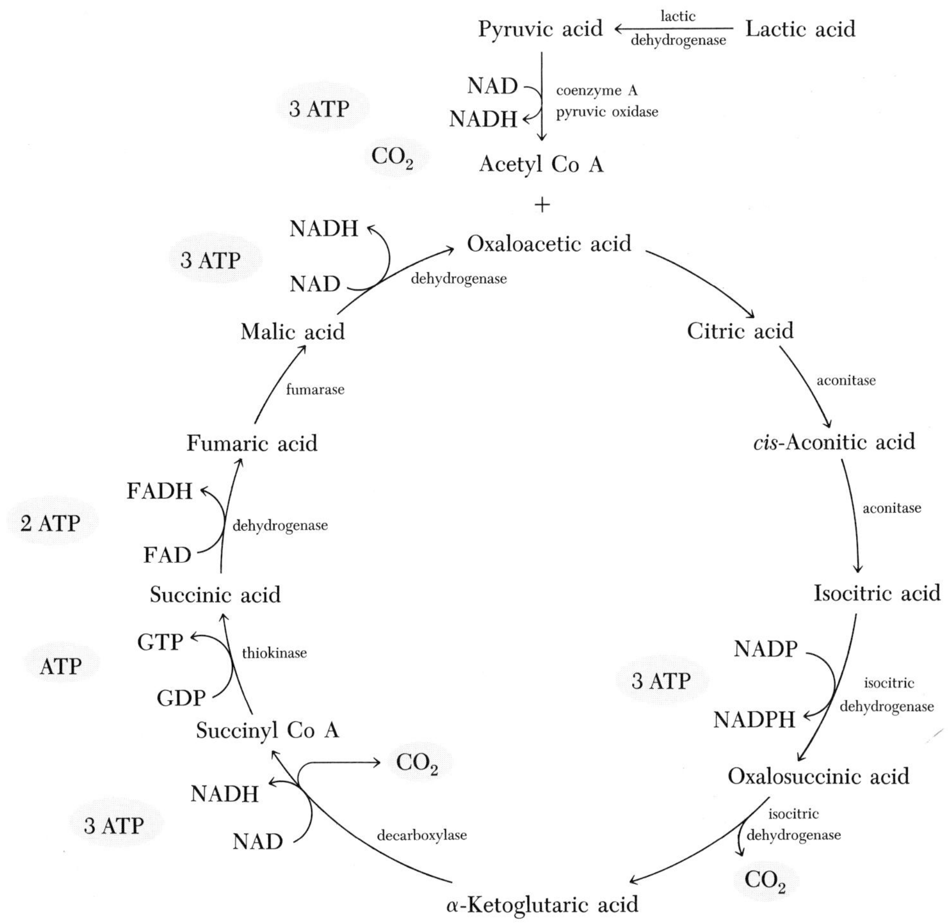

The Krebs Cycle

isocitric and to oxalosuccinic, all tricarboxylic acids. The oxalosuccinic acid then loses CO_2 to form α-ketoglutaric acid, which is converted to succinyl Co A and then to succinic acid and a series of dicarboxylic acids, including fumaric acid and malic acid, back to oxaloacetic acid, and the cycle is completed. The names of the compounds, enzymes, and coenzymes of the Krebs cycle are shown in the diagram on page 355.

The overall reaction for the conversion of pyruvic acid to carbon dioxide and water may be written as:

$$CH_3COCOOH + \tfrac{5}{2}O_2 + 15ADP + 15P_i \rightarrow 3CO_2 + 2H_2O + 15ATP$$

The moles of ATP formed and CO_2 liberated in one turn of the Krebs cycle are shown in shaded areas. In the oxidative phosphorylation process (p. 340), 2 moles of ATP are liberated for each mole of FADH.

When one molecule of glucose is completely oxidized it liberates 686.0 kilocalories. Each molecule of glucose subjected to the Embden-Meyerhof pathway liberates 8 moles of ATP (6 moles from the NADH formed in the conversion of glyceraldehyde-3-phosphate to 1,3-diphosphoglyceric acid, and 2 moles net yield of ATP formed directly). Since each mole of glucose forms 2 moles of pyruvic acid, the Krebs cycle will yield 2 × 15 or 30 moles of ATP per molecule of glucose. *A total of 38 moles of ATP are therefore formed by the oxidation of a molecule of glucose in the Embden-Meyerhof and Krebs cycles.* Since each mole of ATP will yield approximately 8.0 kcal on hydrolysis, the 38 moles are equivalent to 304.0 kcal. This series of reactions is therefore capable of storing about 44 per cent of the available calories in the form of the high energy compound ATP, to be used in muscular work and for other energy requirements.

ALTERNATE PATHWAYS OF CARBOHYDRATE OXIDATION

Pathways other than the Embden-Meyerhof and Krebs cycles are involved in the oxidation of carbohydrates. The most generally accepted alternate pathway is the **phosphogluconate pathway** which is also called the **hexose monophosphate shunt** or the

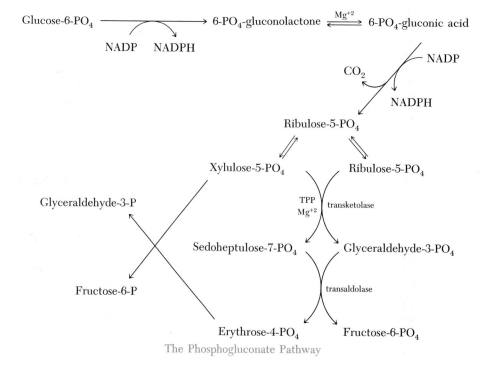

The Phosphogluconate Pathway

pentose phosphate pathway. The key metabolic compound, glucose-6-phosphate, is oxidized in this alternate pathway as outlined below (see also diagram on opposite page):

Glucose-6-phosphate from any source is oxidized to 6-phosphogluconolactone and then to 6-phosphogluconic acid. The acid is converted to a pentose, ribulose-5-phosphate, by the loss of CO_2, and the pentose is transformed into two other pentoses, xylulose-5-phosphate and ribose-5-phosphate. The two latter compounds are converted into a seven-carbon, sedoheptulose-7-phosphate, and the three-carbon glyceraldehyde-3-phosphate. These two compounds are then transformed into a four-carbon, erythrose-4-phosphate, and the hexose fructose-6-phosphate. The final reaction involves xylulose-5-phosphate and erythrose-4-phosphate, forming glyceraldehyde-3-phosphate and more fructose-6-phosphate.

PHOTOSYNTHESIS

Carbohydrates are formed in the cells of plants from carbon dioxide and water. In the presence of sunlight and chlorophyll, the green pigment of leaves, these two compounds react to form pentoses, trioses, fructose, and more complex sugars. **Chlorophyll** is a protoporphyrin derivative containing magnesium that is located in the chloroplasts of green leaves.

Originally the reaction between carbon dioxide and water to form carbohydrates was represented as follows:

$$CO_2 + H_2O \xrightarrow[\text{chlorophyll}]{\text{sunlight}} C_6H_{12}O_6 + 6\ O_2$$
$$\text{Simple sugar}$$

This process by which plants convert the energy of sunlight to form food material is called **photosynthesis.** Although photosynthesis is represented as a simple chemical reaction, it is more complex and includes several intermediates of the phosphogluconate and Embden-Meyerhof pathways.

The use of isotopes and radioactive tracers has greatly assisted the research workers in this field. When a green leaf is grown in an atmosphere of $^{14}CO_2$, the radioactive carbon appears very rapidly in a three-carbon atom and a five-carbon atom intermediate, and later in glucose and starch.

The reaction involving the conversion of light energy into chemical energy is called the **light reaction.** This transformation of energy occurs during photosynthetic phosphorylation and takes place in the chloroplasts of plants. The essential reaction that occurs in **photophosphorylation** can be represented as follows:

$$ADP + P_i \xrightarrow{\text{light energy}} ATP$$

The photochemical process is initiated by the absorption of light by chlorophyll, which produces an excited-state molecule in which several electrons are raised from their normal energy level to a higher level in the double bond structure of chlorophyll. These excited electrons flow from chlorophyll to an iron-containing protein, **ferredoxin,** and bring about the reduction of NADP to form NADPH, which is used in the CO_2 fixation reactions of photosynthesis. Some of the excited electrons flow from ferredoxin through flavin pigments to a quinone structure called **plastoquinone,** then to **cytochrome pigments,** and then back to chlorophyll and their normal energy level. During this cycle some of the energy is given up by coupling in the reaction of ADP with P_i to form ATP.

The incorporation of carbon into carbohydrates has been called the **dark reaction,** since it is not dependent on light energy. This fixation of carbon dioxide during photosynthesis takes place in a cycle of reactions. A pentose phosphate, **ribulose-1,5-diphosphate,** is the key compound that combines with carbon dioxide, with the formation of two molecules of 3-phosphoglyceric acid. The phosphoglyceric acid is then reduced to 3-phosphoglyceraldehyde by a dehydrogenase enzyme with NADP as a coenzyme. This triose phosphate is condensed to fructose phosphate and finally to glucose by the action of **aldolase.** The reactions occurring in this cycle that convert carbon dioxide to glucose are similar to those in the phosphogluconate oxidative pathway (see p. 356).

IMPORTANT TERMS AND CONCEPTS

ATP formation	epinephrine	glycolysis	phosphogluconate pathway
blood sugar level	glucose-6-phosphate	insulin	phosphorylase
cyclic-3′,5′-AMP	glycogenesis	Krebs cycle	photosynthesis
Embden-Meyerhof pathway	glycogenolysis	lactic acid cycle	

QUESTIONS

1. Discuss the factors involved in counteracting the normal hyperglycemia that occurs after a meal.

2. Discuss the role of insulin in the control of the normal blood sugar level.

3. List the hormones other than insulin involved in the control of the blood sugar level. Discuss the function of one of these hormones.

4. Describe the process of glycogenesis.

5. What is cyclic-3′,5′-AMP? What role does it play in glycogenesis?

6. How do the reactions of the lactic acid cycle explain the fate of the lactic acid formed by the process of glycolysis? Explain.

7. Outline the essential reactions in the Embden-Meyerhof pathway of glycolysis.

8. How many moles of ATP per glucose molecule are required to run the Embden-Meyerhof pathway, and how many moles are produced?

9. Outline the essential reactions in the aerobic Krebs cycle.

10. One turn of the Krebs cycle will yield how many moles of ATP? Why is this number so large compared to the Embden-Meyerhof pathway?

11. Outline the essential reactions in the hexose monophosphate shunt or pentose phosphate pathway.

12. What are the major products of the hexose monophosphate shunt? How are they used in other pathways and cycles?

13. What is photosynthesis? Explain the difference between the so-called "light reaction" and the "dark reaction."

SUGGESTED READING

Govindjee: The Absorption of Light in Photosynthesis. Scientific American, Vol. 231, No. 6, p. 68, 1974.

Horecker: Pathways of Carbohydrate Metabolism and Their Physiological Significance. Journal of Chemical Education, Vol. 42, p. 244, 1965.

Margaria: The Sources of Muscular Energy. Scientific American, Vol. 226, No. 3, p. 84, 1972.

Paston: Cyclic AMP. Scientific American, Vol. 227, No. 2, p. 97, 1972.

Smith and York: Stereochemistry of the Citric Acid Cycle. Journal of Chemical Education, Vol. 47, p. 588, 1970.

LIPID METABOLISM

CHAPTER 30

The *objectives* of this chapter are to enable the student to:

1. Recognize the normal blood lipids and their role in metabolism and storage of fat.
2. Describe the reactions involved in the oxidation of fatty acids.
3. Account for the total number of moles of ATP formed in the oxidation of a fatty acid.
4. Outline the reactions involved in the synthesis of fatty acids.
5. Describe the reactions that take place when ketone bodies are formed in the liver.
6. Discuss the synthesis of cholesterol and its relationship to atherosclerosis.

The bulk of the dietary lipids are fats which produce fatty acids and glycerol in hydrolysis. The glycerol molecules follow carbohydrate metabolic pathways, whereas other pathways in the cell are required for the catabolism of fatty acids. Fats are energy-rich molecules, and the caloric value of fats is more than twice that of carbohydrates or proteins. Chemical energy production is therefore an important feature of the catabolic processes of fat metabolism. Other lipids that are involved in both catabolic and anabolic reactions in the cell are the phospholipids, glycolipids, and sterols. These lipids also function as components of cell membranes, nerve tissue, and the membranes of subcellular particles such as microsomes and mitochondria. Sterols like cholesterol can be readily synthesized by the cells and are currently a topic of considerable controversy, since there may be a relation between dietary cholesterol, blood cholesterol levels, and atherosclerosis (see Topic of Current Interest, p. 182).

BLOOD LIPIDS

The blood lipids to a certain extent parallel the behavior of the blood sugar. Their concentration in the blood increases after a meal and the level is returned to normal by processes of storage, oxidation, and excretion.

The lipids of the blood are constantly changing in concentration as lipids are added by absorption from the intestine, by synthesis, and by removal from the fat depots; they are removed by storage in the fat depots, oxidation to produce energy, synthesis to produce tissue components, and excretion into the intestine. The **normal fasting level** of blood lipids is usually measured in the plasma. Average values for young adults are: total lipids 510, phospholipids 200, triglycerides 150, and total cholesterol 160 mg/100 ml. The triglycerides, phospholipids, and cholesterol in the plasma are combined with protein as lipoprotein complexes. These **lipoproteins** are bound to the α- and β-globulin fractions of the plasma proteins and are transported in this form. A small amount of **nonesterified**

fatty acids (**NEFA**) is always present in the blood and is bound to the albumin fraction of the plasma for transportation. These free fatty acids are thought to be the most active form of the lipids involved in metabolism. Their concentration is affected by the mobilization of fat from fat depots and by the action of several hormones.

FAT STORAGE

Fats may be removed from the blood stream by storage in the various fat depots. When fat is stored under the skin, it is usually called **adipose tissue.** However, considerable quantities of fat may be stored around such organs as the kidneys, heart, lungs, and spleen. This type of depot fat acts as a support for these organs and helps to protect them from injury. Recent studies employing the electron microscope reveal two major types of storage fat. One type is composed almost entirely of fat globules and has the characteristics of a storage depot. The second type contains many cells and a more extensive blood circulation, and is metabolically active, converting glycogen to fat and releasing fatty acids to other tissues as energy sources. An increase in this type of storage fat in overweight individuals places increased stress on the heart and circulatory system.

Obesity. Obesity is the condition in which excessive amounts of fat are stored in the fat depots. In a small percentage of cases, obesity is due to a disorder of certain endocrine glands, but as a general rule it results from eating more food than the body requires. Most of the food consumed by an adult is used to produce energy, and food in excess of that necessary to fulfill the energy requirements of the body is stored as fat. Thin people generally are more active than fat people and are able to eat larger amounts of food without putting on weight.

Many people apparently eat all they want and yet maintain a fairly constant weight over long periods of time. This weight control may be due to the appetite, which is abnormally increased in people who are gaining weight and decreased in those who are losing weight. Recent investigations, however, have cast doubt on the simple explanation of overeating as the only factor responsible for obesity. Apparently there are some individuals who can maintain obesity or increase their weight on a low calorie diet. If they attempt to lose weight by decreasing the intake of food, their rate of metabolism decreases and they require fewer calories to maintain their activity. This combination of *endocrine balance, rate of metabolism,* and *difference in requirement for calories* is as yet not completely understood.

THE SYNTHESIS OF TISSUE LIPIDS

Lipids such as **phospholipids, glycolipids,** and **sterols** are essential constituents of cells, protoplasm, and tissues in various parts of the body. They are also involved in specialized functions, i.e., blood clotting mechanisms and in transportation of lipids in the blood. The adipose tissue that is stored around organs of the body does not contain the same proportion of saturated or unsaturated fatty acids as the food fat and therefore must also be synthesized. The most important organ in the body concerned with lipid synthesis is the liver. It is able to synthesize phospholipids and cholesterol and to modify all blood fats by lengthening or shortening, and saturating or unsaturating, the fatty acid chains.

Lecithin is used in transporting fats to the various tissues and may be involved in the oxidation of fats. Another essential phospholipid is **cephalin,** which is a vital factor in the clotting of blood. Special fats and oils in the body such as milk fat, various sterols, the natural oil of the scalp, and the wax of the ear are examples of lipids synthesized from the fats of food.

OXIDATION OF FATTY ACIDS

Fatty acids that arise from the breakdown of any lipid, but especially from fats, are oxidized completely to form CO_2, water, and energy. The glycerol portion of fats is phosphorylated in the liver to form glycerophosphate, which is then oxidized to dihydroxyacetone phosphate. Both of these products can enter the Embden-Meyerhof pathway of carbohydrate metabolism.

The oxidation of fatty acids occurs in a series of reactions that require several enzymes and cofactors, with the production of acetyl coenzyme A. The acetyl CoA molecules then enter the Krebs cycle to form CO_2, H_2O, and energy. Early research by Knoop in 1904 established the fact that fatty acids were oxidized on the beta-carbon atom with the subsequent splitting off of two carbon fragments. In his **theory of beta-oxidation** he stated that acetic acid was split off in each stage of the process that reduced an 18-carbon fatty acid to a two-carbon acid.

In the past few years the detailed reactions, with their enzymes and cofactors, have been worked out, and Knoop's theory has been confirmed. Instead of acetic acid, the key compound in the reactions is acetyl CoA. Five reactions are involved in the conversion of a long-chain fatty acid into a CoA derivative with two less carbon atoms and a molecule of acetyl CoA. These reactions are outlined in the scheme shown below.

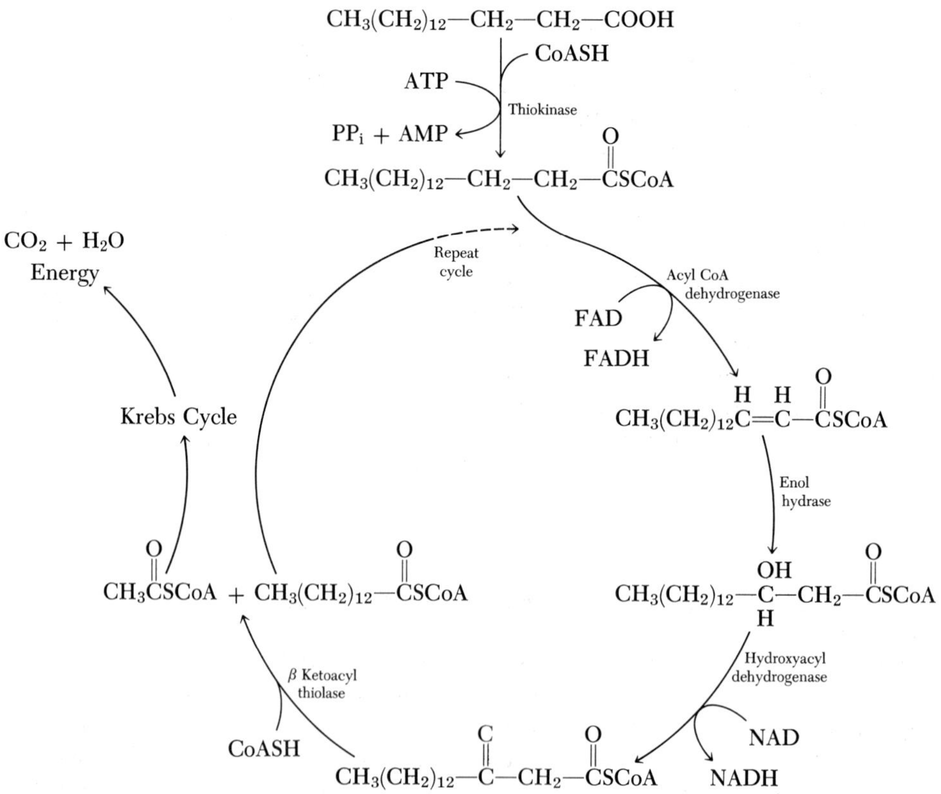

The first reaction initiates the series and involves the activation of a fatty acid molecule by conversion into a coenzyme A derivative. A dehydrogenase enzyme, with FAD as the coenzyme, desaturates the fatty acid; then a hydration is catalyzed by an enol hydrase. The hydroxyl group on the β-carbon atom is oxidized by a dehydrogenase with NAD as a coenzyme. The oxidized derivative plus coenzyme A is split into a fatty acid molecule with 2 less carbons, and acetyl CoA is formed. The acetyl CoA enters the Krebs cycle to form CO_2 and H_2O, plus energy. The new fatty acid coenzyme A derivative does not have to be activated, but directly re-enters the cycle and again loses an acetyl CoA molecule. *Palmitic acid would require seven turns of the cycle to form 8 acetyl CoA moles.*

During the oxidation of palmitic acid, 7 FADH and 7 NADH moles would be formed. When these compounds enter the electron transport chain, they would form ATP as shown:

$$
\begin{aligned}
7 \text{ FADH} &\rightarrow & 14 \text{ ATP} \\
7 \text{ NADH} &\rightarrow & \underline{21 \text{ ATP}} \\
& & 35 \text{ ATP} \\
& & -\ \underline{1 \text{ ATP}} \text{ used in first reaction} \\
& & 34 \text{ Net ATP}
\end{aligned}
$$

In the seven turns of the cycle, 8 acetyl CoA moles are formed. As may be recalled from the oxidation of this compound in the Krebs cycle (p. 355), each mole of acetyl CoA will give rise to 12 moles of ATP. The acetyl CoA formed from the oxidation of palmitic acid will therefore account for the formation of $8 \times 12 = 96$ moles of ATP.

The sum, then, is $34 + 96 = 130$ ATP for the complete oxidation of palmitic acid in the above scheme. The total combustion of palmitic acid yields 2338.0 kcal, compared to 130 ATP's times 8.0 kcal or 1040.0 kcal, which represents 48 per cent of the total. This represents a very efficient conservation of energy in the form of ATP molecules when palmitic acid is completely oxidized by the tissues. Also, a contributing factor to this efficient utilization is the fact that all the enzymes utilized in the β-oxidation scheme, the Krebs cycle, oxidative phosphorylation, and electron transport are found in the **mitochondria** of the cell.

SYNTHESIS OF FATTY ACIDS

The β-oxidation pathway in the mitochondria can be reversed to form fatty acid molecules, but this accounts for only a small percentage of the fatty acids synthesized in the tissues. The **cytoplasm** of the cell is the major site, and acetyl coenzyme A is the starting material, for the synthesis. Acetyl coenzyme A is carboxylated to form **malonyl coenzyme A** under the influence of acetyl CoA carboxylase in the presence of ATP and the vitamin biotin. An **enzyme-biotin complex** adds CO_2 with the help of ATP. Acetyl CoA then reacts with this complex to form malonyl CoA, as follows:

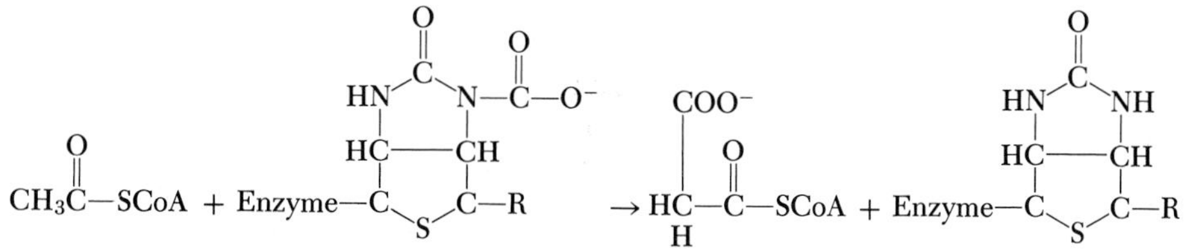

Acetyl CoA   Biotin-enzyme-CO_2 Malonyl CoA
               complex

Malonyl CoA and acetyl CoA then form complexes with a multienzyme system called *fatty acid synthetase*, which includes an acyl carrier protein (ACP) that binds acyl intermediates during the formation of long-chain fatty acids. These two complexes then condense to form acetoacetyl-S-ACP, which is reduced to β-hydroxybutyryl-S-ACP with the assistance of NADPH, followed by the loss of a molecule of water to form an

α,β-unsaturated-S-ACP. The unsaturated compound is reduced to butyryl-S-ACP, which combines with another molecule of malonyl-S-ACP to continue elongation of the chain.

SYNTHESIS OF TRIGLYCERIDES

Triglycerides are synthesized in the tissues from glycerol and fatty acids in activated forms. The active form of glycerol is **glycerophosphate,** which reacts with two fatty acid CoA derivatives to form a **diglyceride,** which then reacts with another mole of fatty acid CoA to form a **triglyceride.** An outline of the process can be illustrated by the following scheme:

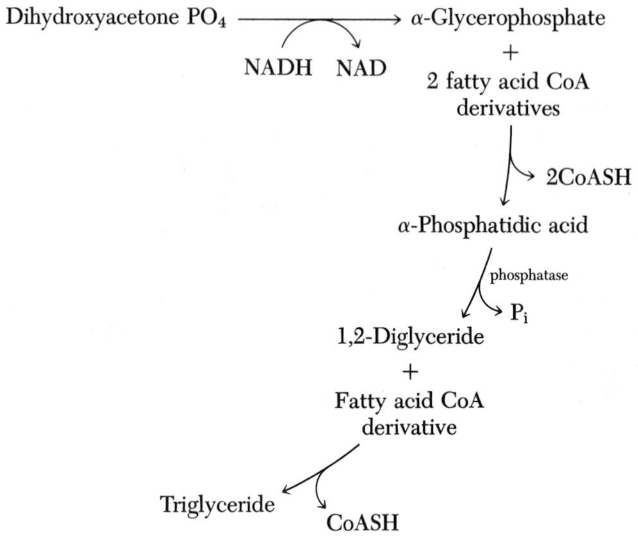

$$\text{Dihydroxyacetone PO}_4 \xrightarrow[\text{NADH} \quad \text{NAD}]{} \alpha\text{-Glycerophosphate}$$

+

2 fatty acid CoA
derivatives

↘ 2CoASH

α-Phosphatidic acid

phosphatase ↘ P_i

1,2-Diglyceride

+

Fatty acid CoA
derivative

Triglyceride ↙ CoASH

FORMATION OF KETONE BODIES

The ketone, or acetone, bodies consist of **acetoacetic acid, β-hydroxybutyric acid,** and **acetone.** In a normal individual they are present in the blood in small amounts, averaging about 0.5 mg per 100 ml. Also, about 100 mg of ketone bodies is excreted per day in the urine. This low concentration in the blood and the small amount excreted in the urine are insignificant. But large amounts are present in the blood and urine during starvation and in the condition of diabetus mellitus. In general, any condition that results in a restriction of carbohydrate metabolism, with a subsequent increase in fat metabolism to supply the energy requirements of the body, will produce an increased formation of ketone bodies. This condition is called **ketosis.**

The precursor of the ketone bodies is acetoacetic acid which is formed in the liver from acetoacetyl CoA, a normal intermediate in the beta-oxidation of fatty acids. It may also be formed by the condensation of two molecules of Acetyl CoA. Both methods of formation can be represented in the normal reversible reaction as follows:

$$2\ \underset{\text{Acetyl CoA}}{CH_3\overset{O}{\overset{\|}{C}}SCoA} \underset{}{\overset{\text{thiolase}}{\rightleftharpoons}} \underset{\text{Acetoacetyl CoA}}{CH_3\overset{O}{\overset{\|}{C}}CH_2\overset{O}{\overset{\|}{C}}SCoA} + CoASH$$

The liver contains a deacylase enzyme which readily converts acetoacetyl CoA to the free acid.

The other ketone bodies are formed from acetoacetic acid; acetone by decarboxylation and β-hydroxybutyric acid by the action of a specific enzyme, as shown in the accompanying diagram.

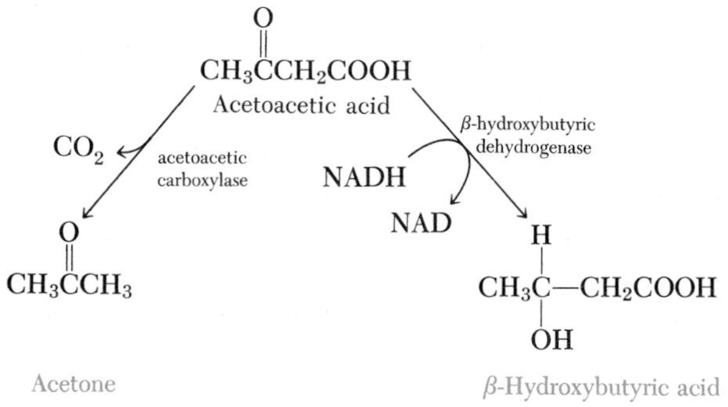

PHOSPHOLIPID METABOLISM

Knowledge concerning the metabolism of the phospholipids is incomplete, although they are known to serve many important functions in the body. Because their molecules are more strongly dissociated than any of the other lipids, they tend to be more soluble in water, to lower surface tension at oil-water interfaces, and to be involved in the electron transport system in the tissues. They would have a tendency to concentrate at cell membranes, and are probably involved in the transport mechanisms for carrying fatty acids and lipids across the intestinal barrier and from the liver and fat depots to other body tissues. Further evidence for their function in transporting lipids is found in their presence in the lipoproteins of the plasma. Phospholipids are essential components of the blood clotting mechanism, and sphingomyelin is one of the principal components of the myelin sheath of nerves.

Dietary phospholipids are probably broken into their constituents by enzymes in the gastrointestinal tract. The synthesis of most of the phospholipids has been established in recent years by the use of isotopes to tag precursors and intermediate compounds. For example, the synthesis of lecithin involves the reaction of cytidine diphosphate choline and a 1,2-diglyceride. The phosphatidyl ethanolamines, or cephalins, are synthesized by similar reactions, and the sphingomyelins are synthesized by the reaction of N-acylsphingosine with **cytidine diphosphate choline.** The synthesis of lecithin may be outlined as illustrated utilizing the 1,2-diglyceride as formed in triglyceride synthesis.

$$\text{1,2-Diglyceride} + \text{CDP-choline} \xrightarrow[\text{enzyme}]{\text{transferase}} \text{Lecithin} + \text{Cytidine monophosphate}$$

STEROL METABOLISM

The metabolism of sterols is mainly concerned with cholesterol and its derivatives. The **synthesis of cholesterol** and its relation to the other steroids of the body has been the subject of considerable research. Using either stable or radioactive isotopes, it has been shown that cholesterol can be synthesized from two-carbon compounds such as acetyl CoA. It can also be synthesized from acetoacetyl CoA and other intermediates. Although the synthesis of cholesterol occurs in many tissues in the body, the liver is the main site of cholesterol formation.

Cholesterol is a key compound in the synthesis of essential steroids such as bile acids, sex hormones, adrenal cortical hormones, and vitamin D. Not only is cholesterol converted to bile acids by the liver, but it is also excreted as such in the bile. As mentioned previously, cholesterol in the bile can give rise to gall stones by accumulating on insoluble objects or particles. The concentration of cholesterol in the blood is apparently dependent on the dietary intake of sterols and neutral fats, and the synthesis of cholesterol by the liver. The normal level in the blood gradually increases with age and ranges from 150 to 200 mg/100 ml. Blood cholesterol levels are often determined in patients to assess their cholesterol status. Many methods have been devised for this determination, and many of them are modifications of the Liebermann-Burchard reaction described in Chapter 25. If the cholesterol level in the blood is maintained at an abnormally high concentration, such as 200 to 300 mg/100 ml, deposition of cholesterol plaques may occur in the aorta and lesser arteries. This condition, known as **atherosclerosis** or **arteriosclerosis,** is seen in older persons and often results in circulatory or heart failure. Considerable research effort is being directed at this problem in an attempt to reduce the cholesterol level in the blood of these patients and thus alleviate the symptoms of the disease (see Topic of Current Interest, p. 182).

CORRELATION OF CARBOHYDRATE AND FAT METABOLISM

From a nutritional standpoint it has long been apparent that carbohydrate can be converted into fat in the body. When glucose tagged with ^{14}C was fed to animals, the fatty acids of liver and other tissue fat were found to be labeled with ^{14}C. The conversion of fat to carbohydrate has long been open to question. The glycerol portion of fat is closely related to the three-carbon intermediates of carbohydrate metabolism, but it has been more difficult to demonstrate a direct relation between fatty acids and glucose. Since the role of acetyl CoA has been established in both carbohydrate and fat metabolism, it is apparent that the acetyl CoA from fatty acid oxidation can enter the Krebs cycle in the same fashion as this compound formed from pyruvic acid. The correlation between carbohydrate, fat, and protein metabolism in the body is represented in the scheme shown on page 377.

IMPORTANT TERMS AND CONCEPTS

acetyl CoA	ATP formation	cholesterol synthesis	fatty acid synthesis
adipose tissue	blood lipids	fatty acid oxidation	ketone bodies

QUESTIONS

1. List the important lipids in a normal individual's blood and their approximate concentration.

2. Discuss some of the advantages and disadvantages of a generous supply of adipose tissue.

3. Outline the essential reactions in the scheme for oxidation of fatty acids.

4. How do you account for the large number of moles of ATP formed in the oxidation of fatty acids?

5. Outline the reactions involved in the synthesis of fatty acids.

6. Describe the process of synthesis of triglycerides.

7. Discuss the reactions that may take place in the liver for the formation of acetoacetic acid.

8. What is ketosis? Explain the cause of this condition in the body.

9. Why is atherosclerosis receiving so much attention in our society?

10. Name four important compounds in the body that are synthesized from cholesterol.

SUGGESTED READING

Brady: Hereditary Fat-Metabolism Diseases. Scientific American, Vol. 229, No. 2, p. 88, 1973.
Gibson: The Biosynthesis of Fatty Acids. Journal of Chemical Education, Vol. 42, p. 236, 1965.
Green: The Synthesis of Fat. Scientific American, Vol. 202, No. 2, p. 46, 1960.
Magliulo: Prostaglandins. Journal of Chemical Education, Vol. 50, No. 9, p. 602, 1973.
Spain: Atherosclerosis. Scientific American, Vol. 215, No. 2, p. 48, 1966.

PROTEIN
METABOLISM

CHAPTER 31

The *objectives* of this chapter are to enable the student to:

1. Recognize the relationship between the amino acid pool and the dynamic state of tissue proteins.
2. Describe the activation of amino acids and their combination with t-RNA prior to protein synthesis.
3. Illustrate the processes of transcription and translation in protein synthesis.
4. Explain the relationship between codons and anticodons in protein synthesis.
5. Recognize the difference between deamination and transamination.
6. Outline the essential reactions in the urea cycle.
7. Recognize the reactions involved in purine and pyrimidine metabolism.
8. Describe the correlation between carbohydrate, lipid, and protein metabolism.

The metabolism of proteins in the living cell is concerned with a large variety of complex molecules. Tissue proteins of various species of animals, plants, and micro-organisms all have specific structures and compositions. Protein enzymes and hormones, plasma proteins, and the protein of hemoglobin and of various nucleoproteins represent other types of proteins. The synthesis of new proteins for growth and development, or anabolism, involves the building of different amino acids into the proper sequences and spatial arrangements to produce specific protein molecules. The process of catabolism of proteins to produce energy involves many general metabolic reactions and many that are specific for the metabolism of each of the twenty different amino acids.

AMINO ACID POOL

In contrast to carbohydrate and fat metabolism, there are no storage depots for proteins or amino acids. The increased concentration of amino acids that occurs from the process of absorption, synthesis, or catabolism represents a temporary pool of amino acids which may be used for metabolic purposes. This pool of amino acids is available to all tissues and may be synthesized into new tissue proteins, blood proteins, hormones, enzymes, or nonprotein nitrogenous substances such as creatine and glutathione. The relationship that exists between the **amino acid pool** and protein metabolism in general may be represented as follows:

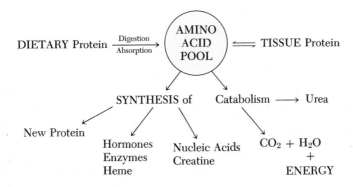

THE DYNAMIC STATE OF BODY PROTEIN

Until the late 1930's it was believed that the body proteins of the adult human were stable molecules and that the majority of the amino acids from the diet were catabolized to produce energy. A small proportion was thought to be used for maintenance and repair of the existing tissue proteins. When isotopes became available, Schoenheimer and his associates demonstrated that tissue proteins exist in a *dynamic state of equilibrium.* When the nitrogen of an amino acid was labeled with ^{15}N and incorporated in the diet of an animal, about 50 per cent of the ^{15}N was found in the tissues of the animal, and a greater percentage was found in the nitrogen of amino acids other than that specifically fed. This indicated that the amino acids of tissue proteins were constantly changing places with those in the amino acid pool, and that the body proteins were extremely labile molecules.

More recent research using isotopically labeled amino acids has shown that tissue proteins vary considerably in their rate of turnover of amino acid molecules. The **turnover rate** represents the amount of protein synthesized or degraded per unit time, and the **turnover time** is usually expressed as the half-life of a protein in the tissues. Liver and plasma proteins have turnover times (a half-life) of 2 to 10 days, in contrast to 180 days for muscle protein and 1000 days for some collagen proteins.

TOPIC OF CURRENT INTEREST

ESSENTIAL AMINO ACIDS AND PROTEIN SYNTHESIS

The synthesis of protein is always occurring in the body, especially in those tissues with a rapid turnover rate. A growing child or animal is continually building new tissue and therefore makes the greatest demand on the amino acid pool. The individual amino acids required for protein synthesis are apparently sorted out by the body and used to construct specific protein molecules. Although the tissues, particularly the liver, are able to synthesize some amino acids, others must be present in the diet to assure a complete and proper assortment for synthetic purposes.

The amino acids that cannot be synthesized by the body and must therefore be supplied by dietary protein are called **essential amino acids.** If an essential amino acid is lacking in the diet, the body is unable to synthesize tissue protein. If this condition occurs for any length of time, a negative nitrogen balance will exist, and there will be weight loss, lowered serum protein level, and marked edema. Extensive feeding experiments on laboratory rats have established the following amino acids as essential for growth:

Histidine	Isoleucine
Methionine	Leucine

Arginine	Lysine
Tryptophan	Valine
Threonine	Phenylalanine

From the studies of Rose on the amino acid requirements of man, it was proposed that all of the above ten except histidine and possibly arginine are essential to maintain nitrogen balance. An individual is in **nitrogen balance** when the nitrogen excreted equals the nitrogen intake in a given period of time. A growing child or a patient recovering from a prolonged illness is in **positive nitrogen balance.** Starvation, a wasting disease, or a diet lacking sufficient amounts of essential amino acids can result in a **negative nitrogen balance.**

Many common dietary proteins are deficient in one or more of these essential amino acids. Gelatin, for example, lacks tryptophan and is therefore an **incomplete protein.** If gelatin were the sole source of protein in the diet, a growing child could eat large quantities of this protein every day without building new body tissues. Zein and gliadin, the prolamines of corn and wheat respectively, are deficient in lysine, and zein is also low in tryptophan. The supplementing of these proteins with small amounts of the deficient amino acid (lysine or tryptophan) will result in adequate growth. Although an incomplete protein will not support growth when it is the only protein in the diet, we seldom confine ourselves to the consumption of a single protein. In an ordinary mixed diet the essential amino acids are best supplied by protein of animal origin, such as meat, eggs, milk, cheese, and fish.

In other parts of the world, particularly in Africa, Asia, and South America, the diet is often deficient in high quality protein. Subnormal growth in children and prevalence of disease and early death are common in these populations. Small children develop **kwashiorkor** when deprived of adequate protein, and their bloated bellies and wrinkled skin are often seen in photographs pleading for food for hungry nations. Many of these children die before they can be adequately nourished. Agricultural research aimed at the production of high-yield wheat, high-lysine corn, and edible protein from fish meal and seaweed is being conducted to increase the world supply and availability of protein food.

MECHANISM OF PROTEIN SYNTHESIS

A very active field of research at the present time involves the mechanism of protein synthesis, the sequence of amino acids in the protein being synthesized, and the nature of the genetic code responsible for this sequence. Protein synthesis is initiated by the activation of amino acids. This process occurs by the combination of the amino acid with ATP and an enzyme specific for the amino acid, with the splitting off of two molecules of phosphoric acid.

$$R-CH-COOH \xrightarrow[\text{ATP} \quad \text{PP}_i]{\text{aminoacyl synthetase}} Enz\text{-}Adenine\text{-}Ribose-O-\overset{\overset{O}{\|}}{P}-O-\overset{\overset{O}{\|}}{C}-CH-R$$

R—CH—COOH
|
NH₂

Amino acid (AA)

Amino acyl AMP enzyme complex
(E-AMP-AA)

The second step of the process involves the transfer of the activated amino acid to a **transfer RNA molecule.** The t-RNA molecules are small (mol. wt. 30,000) nucleic acid molecules with a terminal grouping of cytidylic-cytidylic-adenylic acid represented as s-RNA-C-C-A. The transfer of the activated amino acid is under the influence of the same enzyme that produced the activation and CTP, or cytidine triphosphate.

$$E\text{-}AMP\text{-}AA + t\text{-}RNA \longrightarrow t\text{-}RNA\text{-}AA + AMP + E$$

(activated amino acid)   transfer RNA   transfer RNA- amino acid complex

The third step involves the transfer of t-RNA-bound amino acids to the ribosomes of the cellular cytoplasm. Ribosomes are nucleoprotein particles composed of approximately 60 per cent basic proteins and 40 per cent ribosomal RNA (r-RNA). There are two subunits, a 30 S and a 50 S ribosome, that are joined to make a complete 70 S ribosome. In the cytoplasm ribosomes are joined by strands of messenger RNA to form polysomes, which consist of 4 to 6 ribosomes. Both the t-RNA and m-RNA molecules are bound by the ribosomes to achieve a synthesis of polypeptide from the activated amino acids carried by the t-RNA.

The active template that controls protein synthesis on the ribosomes is called messenger RNA (m-RNA). It is synthesized by RNA polymerase under the direction of DNA in the nucleus, and it carries information to the ribosomes to direct the sequence of alignment of t-RNA-bound amino acids. This first process of overall information transfer from DNA to m-RNA is called **transcription** (transcribing the message). The specific programming of the amino acids on the polysomes or m-RNA molecules to synthesize a protein containing a definite sequence of amino acids is called **translation** (translating the code).

The attachment of the activated amino acids bound to t-RNA to a specific area or site on the m-RNA is governed by the genetic code. There is a specific site on m-RNA consisting of three consecutive bases that binds a particular amino acid; this site is called a **codon.** The t-RNA for that particular amino acid has a complementary triplet of bases called an **anticodon** that binds the t-RNA to the site on the m-RNA on the ribosome. Since there are four different nucleotide residues or bases in m-RNA, and a sequence of three bases is involved in coding for each amino acid, a total of 4^3 or 64 different combinations of the three bases would be available for coding. The genetic code is *nonoverlapping* in that it requires the action of a specific group of three bases on the m-RNA chain, and it is also said to be *degenerate* in that more than one codon may be employed by m-RNA to insert a specific amino acid into the peptide chain. The codons currently proposed for specific amino acids are shown in Table 31–1. Note that there is a specific codon for the initiation of a protein chain and three chain termination codons.

TABLE 31–1   CODONS FOR SPECIFIC AMINO ACIDS

AMINO ACIDS	CODONS FOR m-RNA
Alanine	GCA, GCC, GCG, GCU
Arginine	AGA, AGG, CGA, CGG, CGC, CGU
Asparagine	AAC, AAU
Aspartic acid	GAC, GAU
Cysteine	UGC, UGU
Glutamic acid	GAA, GAG
Glutamine	CAG, CAA
Glycine	GGA, GGC, GGG, GGU
Histidine	CAC, CAU
Isoleucine	AUA, AUC, AUU
Leucine	CUA, CUC, CUG, CUU, UUA, UUG
Lysine	AAA, AAG
Methionine	AUG—Chain initiation
Phenylalanine	UUU, UUC
Proline	CCA, CCC, CCG, CCU
Serine	AGC, AGU, UCA, UCG, UCC, UCU
Threonine	ACA, ACG, ACC, ACU
Tryptophan	UGG
Tyrosine	UAC, UAU
Valine	GUA, GUG, GUC, GUU
Chain termination	UAA, UAG, UGA

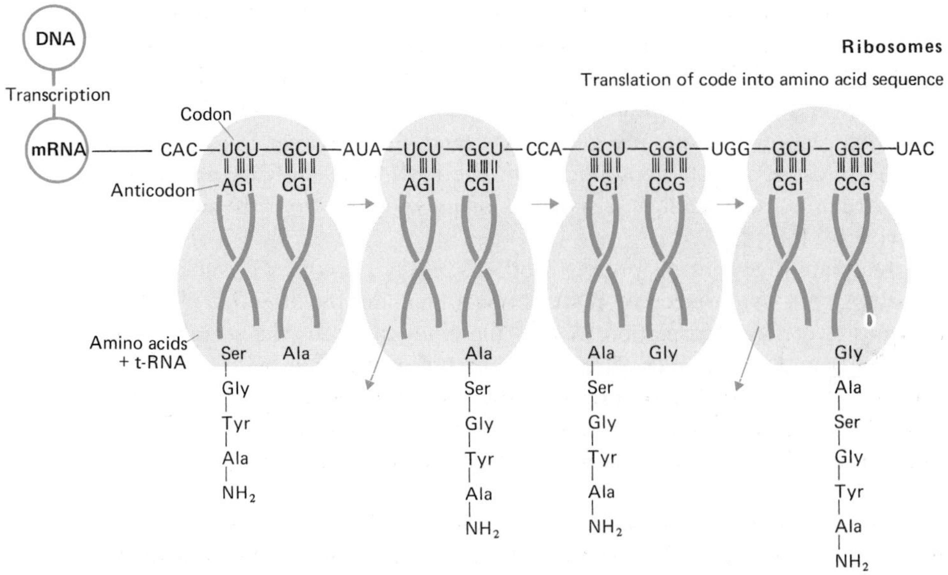

Figure 31–1   The figure demonstrates the two binding sites on the ribosome and the movement of the ribosomes on m-RNA (to the right) or movement of m-RNA (to the left). As the chain grows by the addition of one amino acid, a t-RNA molecule moves off the ribosome and a new single t-RNA amino acid moves onto the ribosome next to the polypeptide carrying the t-RNA.

Each 70 S ribosome has two binding sites. One binds the t-RNA with a growing polypeptide chain, the other binds a t-RNA with a single activated amino acid. A sequence in the synthesis of a polypeptide chain is shown in Figure 31–1.

The scheme outlined in Figure 31–1 is capable of synthesizing polypeptide molecules with a specific amino acid sequence governed by the m-RNA that was coded by the DNA in the nucleus of the cell. The initiation and termination of the synthesis is governed by specific codons (Table 31–1), and the number and types of proteins synthesized is controlled by the supply of specific m-RNA molecules. It has been estimated that it requires about two minutes for a ribosome to synthesize a small protein molecule, and since there may be 4 to 6 ribosomes in a polysome or even larger numbers of ribosomes involved in the synthesis of a high-molecular weight protein, the time of synthesis may be less than 30 seconds.

METABOLIC REACTIONS OF AMINO ACIDS

The amino acids in the metabolic pool that are not immediately used for synthesis can undergo several metabolic reactions. They may follow the path of catabolism through deamination, urea formation, and energy production, or they may assist in the synthesis of new amino acids by the process of reamination and transamination. **Deamination, reamination, transamination,** and **urea formation** are processes common to all amino acids and are therefore very important to protein metabolism.

DEAMINATION

A general reaction of catabolism is the splitting off of the amino group of an amino acid, with the formation of ammonia and a keto acid. This process is called **oxidative deamination** and is catalyzed by enzymes found in liver and kidney tissue called **amino**

acid oxidases. These enzymes are generally flavoprotein enzymes containing either flavin adenine dinucleotide, FAD, or flavin mononucleotide, FMN. The enzyme dehydrogenates the amino acid to form an imino acid, which is hydrolyzed to a keto acid and ammonia.

$$
\underset{\substack{\text{Amino acid}}}{\text{R}-\underset{\underset{\text{NH}_2}{|}}{\text{CH}}-\text{COOH}} \xrightarrow[\text{FAD}\quad\text{FADH}]{\substack{\text{amino acid}\\\text{oxidase}}} \underset{\substack{\text{Imino acid}}}{\text{R}-\underset{\underset{\text{NH}}{\|}}{\text{C}}-\text{COOH}} \xrightarrow[\text{H}_2\text{O}]{\text{hydrolysis}} \underset{\substack{\text{Keto acid}}}{\text{R}-\underset{\underset{\text{O}}{\|}}{\text{C}}-\text{COOH} + \text{NH}_3}
$$

The fate of the keto acid depends on the amino acid from which it is derived. In general the catabolism of each amino acid must be studied separately. Glycine, for example, is the simplest amino acid, yet it can be transformed metabolically to formate, acetate, ethanolamine, serine, aspartic acid, fatty acids, ribose, purines, pyrimidines, and protoporphyrin. This amino acid may therefore play a role in carbohydrate, lipid, protein, nucleic acid, and hemoglobin metabolism, and it admirably illustrates the interrelationships that exist among the different types of metabolism in the body. Other amino acids undergo complex metabolic reactions that are beyond the scope of this book.

TRANSAMINATION

The process of deamination results in the formation of many keto acids that are capable of accepting an amino group to form a new amino acid. That this process of **reamination** occurs was readily apparent from the isotope-labeling experiments of Schoenheimer. He observed a ready exchange of amino groups of dietary amino acids and tissue amino acids. A major mechanism for the conversion of keto acids to amino acids in the body is known as **transamination.** Glutamic acid, for example, could react with oxaloacetic acid in the presence of a transaminase to form a new keto acid, α-ketoglutaric, and the new amino acid, aspartic acid.

$$
\begin{array}{cccc}
\text{COOH} & \text{COOH} & & \text{COOH} & \text{COOH} \\
| & | & & | & | \\
\text{CH}_2 & \text{C}=\text{O} & & \text{CH}_2 & \text{CHNH}_2 \\
| & | & \xrightleftharpoons[\text{pyridoxal phosphate}]{\text{transaminase}} & | & | \\
\text{CH}_2 & + \text{CH}_2 & & \text{CH}_2 & + \text{CH}_2 \\
| & | & & | & | \\
\text{CHNH}_2 & \text{COOH} & & \text{C}=\text{O} & \text{COOH} \\
| & & & | & \\
\text{COOH} & & & \text{COOH} & \\
\text{Glutamic} & \text{Oxaloacetic} & & \alpha\text{-Ketoglutaric} & \text{Aspartic} \\
\text{acid} & \text{acid} & & \text{acid} & \text{acid}
\end{array}
$$

The coenzyme **pyridoxal phosphate** is required in the reaction and **pyridoxamine phosphate** is formed. The enzyme that catalyzes the reaction in the serum is called SGOT, serum glutamic-oxaloacetic transaminase, and a sharp rise in its concentration in the serum is indicative of myocardial infarction, a heart condition involving the cardiac muscle. There are many specific transaminases that serve as catalysts in the transfer of amino groups from amino acids to a variety of keto acids. Transamination reactions serve as important links joining carbohydrate, fat, and protein metabolism. A keto acid from any source can be used for the synthesis of an amino acid to be incorporated in tissue protein. For example, α-ketoglutaric acid is a constituent in the Krebs cycle and serves as a direct link from protein to carbohydrate and fat metabolism.

FORMATION OF UREA

The ammonia, carbon dioxide, and water that result from the deamination and oxidation of the amino acids are combined to form urea. Urea formation takes place in the liver by a fairly complicated series of reactions, first described by Krebs and his coworkers. The ammonia and carbon dioxide combine with the amino acid ornithine to form citrulline which combines with another molecule of ammonia to form arginine, which is then hydrolyzed by means of the enzyme arginase, present in the liver, to form urea and ornithine. The ornithine may then enter the beginning of the cycle.

In recent years the detailed mechanism of the cycle has been worked out. Apparently ornithine does not react directly with CO_2 and NH_3 to form citrulline, but reacts with a compound called **carbamyl phosphate.** This compound is synthesized from ATP, CO_2, and NH_3 in the presence of the specific enzyme carbamyl phosphate synthetase and the cofactors N-acetylglutamate and Mg^{+2}.

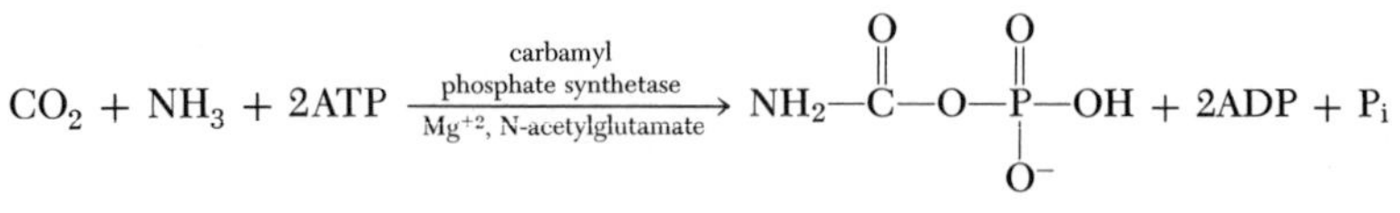

$$CO_2 + NH_3 + 2ATP \xrightarrow[\text{Mg}^{+2},\text{ N-acetylglutamate}]{\text{carbamyl phosphate synthetase}} NH_2-\overset{\overset{O}{\|}}{C}-O-\overset{\overset{O}{\|}}{\underset{\underset{O^-}{|}}{P}}-OH + 2ADP + P_i$$

Carbamyl phosphate

The formation of arginine is also not a simple reaction of citrulline and NH_3 but involves a combination with aspartic acid to form argininosuccinic acid, which then splits into arginine and fumaric acid. The currently accepted **urea cycle** can be represented as follows:

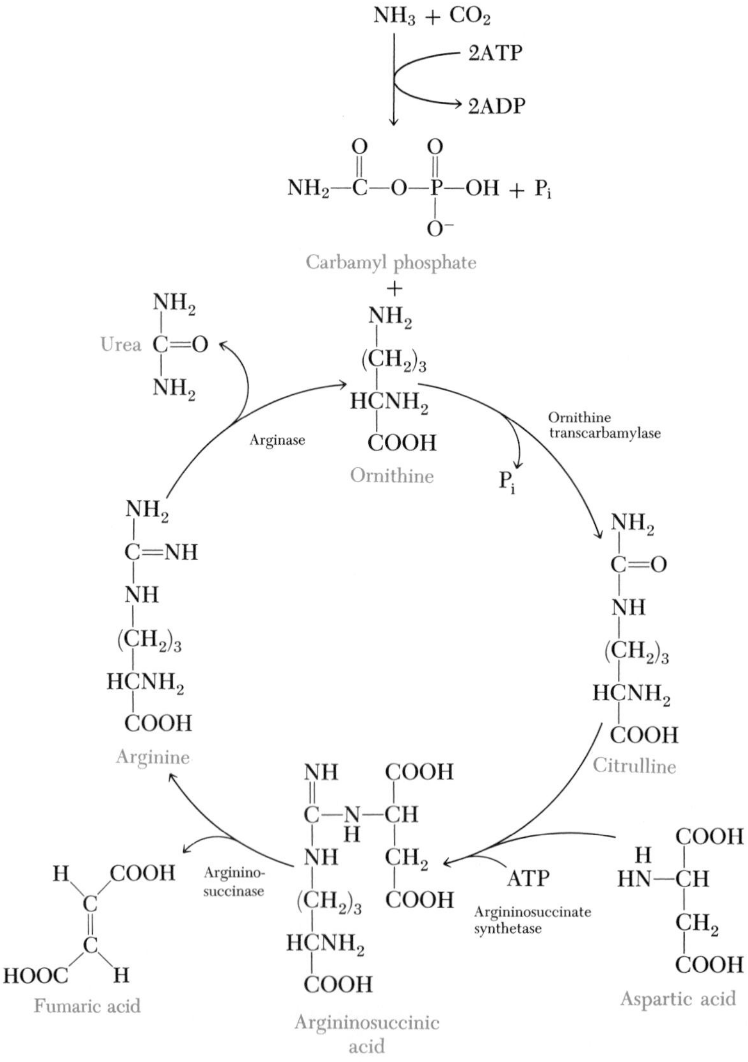

As urea is formed in the liver it is removed by the blood stream, carried to the kidneys, and excreted in the urine. Urea is the main end product of protein catabolism and accounts for 80 to 90 per cent of the nitrogen that is excreted in the urine.

NUCLEOPROTEIN METABOLISM

The important nucleic acids DNA and RNA are essential constituents of the cell nucleus, the chromosomes, and viruses and are involved in the synthesis of protein. During the process of digestion the protein is split from the nucleic acids and is broken down to amino acids. The nucleic acids are first attacked by ribonuclease and deoxyribonuclease to form nucleotides that are further hydrolyzed by nucleotidases to form phosphates and nucleosides. The nucleosides are absorbed through the intestinal mucosa and split by nucleosidases of the tissues into D-ribose, deoxyribose, purines, and pyrimidines. In metabolism the amino acids and sugar follow the ordinary process of protein and carbohydrate utilization. The phosphoric acid is used to form other phosphorus compounds in the body or may be excreted in the urine as phosphates.

PURINE METABOLISM

The synthesis of purines has been elaborated in recent years. This is a very complex process involving several steps, specific enzymes, and cofactors. Inosinic acid in the form of a mononucleotide is formed first and serves as an intermediate in the synthesis of adenylic acid and guanylic acid.

The purines that are formed by the hydrolysis of nucleosides in the tissues undergo catabolic changes, forming uric acid, which is excreted in the urine. The nucleosides **adenosine, inosine, guanosine,** and **xanthosine** are split into ribose plus adenine, hypoxanthine, guanine, and xanthine, respectively. These purines are not completely broken down to NH_3, CO_2, H_2O, and energy, but are progressively oxidized with the assistance of specific enzymes.

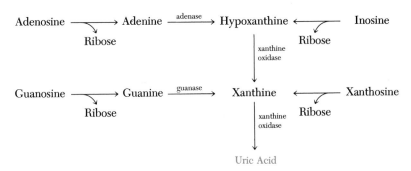

In most mammals other than man and apes the uric acid is converted into **allantoin,** a more soluble substance.

PYRIMIDINE METABOLISM

Pyrimidine synthesis starts with ammonia and carbon dioxide reacting with ATP to form **carbamyl phosphate** (see p. 374). This compound combines with aspartic acid to form carbamyl aspartic acid, which is converted to dihydroorotic acid, which is reduced to orotic acid. Orotic acid then reacts with 5-phosphoribosyl-1-pyrophosphate to form the nucleotide orotidine-5-phosphate. Decarboxylation of this nucleotide produces the primary pyrimidine, uridine-5-phosphate, or uridylic acid. The mononucleotide of uridylic acid apparently serves as the starting material for the synthesis of other pyrimidine nucleotides.

The pyrimidines that result from nucleoside hydrolysis in the tissues can be broken into small molecules in catabolism. Cytosine loses ammonia to form uracil, which is reduced to dihydrouracil utilizing the coenzyme NADPH. The ring is then opened to form a ureidopropionic acid, which is further split to β-alanine plus ammonia and carbon dioxide. These end products are eventually converted to urea for excretion.

CREATINE AND CREATININE

Creatine and creatinine are two nitrogen-containing compounds that are usually associated with protein metabolism in the body. **Creatine** is widely distributed in all tissues but is especially abundant in muscle tissue, where it is combined with phosphoric acid as **phosphocreatine,** or **creatine phosphate.** In the contraction of muscles, phosphocreatine apparently plays an important role as a reservoir of high-energy phosphate bonds readily convertible to ATP. The energy for the initial stages of muscular contraction probably

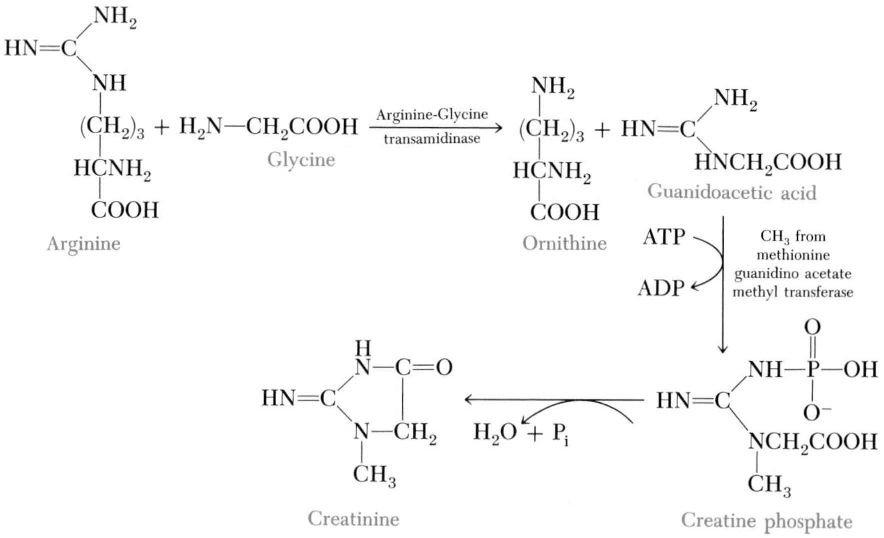

comes from the hydrolysis of this compound to form creatine and phosphoric acid. These two substances are later combined during the recovery period of the muscle. Creatine is synthesized from the amino acids glycine, arginine, and methionine.

Creatinine is also present in the tissues but is found in much larger amounts in the urine. It is formed from either creatine phosphate or creatine and is an end product of creatine metabolism in muscle tissue.

CORRELATION OF CARBOHYDRATE, LIPID, AND PROTEIN METABOLISM

The correlation between carbohydrate and lipid metabolism has already been discussed. Since the catabolism of amino acids results in keto acids such as pyruvic acid, it can readily be seen that these products could enter the metabolic scheme of the carbohydrates. The over all correlation of the three major types of metabolism is represented in the diagram on the next page.

CARBOHYDRATE METABOLISM

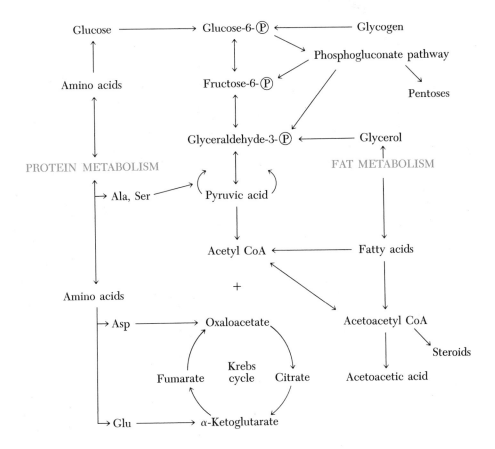

IMPORTANT TERMS AND CONCEPTS

amino acid pool	deamination	t-RNA	translation
anticodon	essential amino acids	transamination	urea cycle
codon	ribosomes	transcription	uric acid
creatinine	m-RNA		

QUESTIONS

1. Explain the concept of the amino acid pool.

2. What is meant by the dynamic state of tissue proteins in the body?

3. What is an essential amino acid? What would happen to a growing child that was deprived of adequate amounts of these amino acids? Why?

4. Describe the process of activation of amino acids and their combination with t-RNA prior to the synthesis of protein.

5. Illustrate the process of translation in protein synthesis.

6. What roles do ribosomes and m-RNA play in protein synthesis? Explain.

7. In protein synthesis, explain the relationship between codons and anticodons.

8. What are the processes of deamination and transamination? Illustrate one process with equations.

9. Outline the essential reactions in the urea cycle.

10. What products would result from the complete hydrolysis of RNA? Briefly describe the metabolic fate of each of the products.

11. Briefly outline the process for the synthesis of creatine.

12. Explain how the process of transamination can serve as a common link between carbohydrate, fat, and protein metabolism.

SUGGESTED READING

Clark and Marcker: How Proteins Start. Scientific American, Vol. 218, No. 1, p. 36, 1968.
Grünewald: The Evolution of Proteins. Chemistry, Vol. 41, No. 1, p. 11, 1968.
Harpstead: High-Lysine Corn. Scientific American, Vol. 225, No. 2, p. 34, 1971.
Howe: Amino Acids in Nutrition. Chemical and Engineering News, p. 74, July 23, 1962.
Roth: Ribonucleic Acid and Protein Synthesis. Journal of Chemical Education, Vol. 38, p. 217, 1961.
Scrimshaw and Behar: Protein Malnutrition in Young Children. Science, Vol. 133, p. 2039, 1961.
Yanotsky: Gene Structure and Protein Structure. Scientific American, Vol. 216, No. 5, p. 80, 1967.

THE BIOCHEMISTRY OF GENETICS

CHAPTER 32

The *objectives* of this chapter are to enable the student to:

1. Explain why single cells such as bacteria are used in the study of genetics.
2. Explain the relation between DNA of the cell and genetics.
3. Describe the function of genes and their relation to the chromosomes of the cell.
4. Illustrate the process of replication of DNA.
5. Describe and illustrate the regulation of enzyme synthesis in the cell.
6. Recognize that inborn errors of metabolism are genetic diseases.
7. Outline the inborn errors of metabolism related to phenylalanine metabolism.
8. Discuss the possibility of genetic engineering for the reversal of cancer and repair of genetic disease.

In previous chapters the structure of the living cell, its subcellular particles, its chemical composition, and many of its chemical reactions have been considered. An essential role of the cell that has not been discussed is its role in the transmission of genetic information. The composition and structure of the RNA and DNA molecules of the cell were covered in Chapter 22 and the relation between DNA, chromosomes, and genes was suggested. In this chapter we will explore briefly the biochemistry of heredity and the genetic information transmitted by the DNA and RNA molecules.

CELLS AND HEREDITY

Most of the biochemical studies in genetics have been carried out with single cells such as bacteria, which are capable of producing many generations in a short period of time. Bacteria are simple systems that contain relatively little genetic information compared to the cells of man and animals. The number of genes in a bacterial cell would be very low, and there should be a small amount of DNA as compared to that in cells of higher organisms. These facts make bacteria ideal systems for research in genetics. Extensive biochemical research investigations have established the genetic makeup of these cells, and proposals for the regulation of the transcription of genetic information by the cells have been advanced. This type of research has developed into a new area of science called **molecular biology**, which is concerned with explaining genetics on a molecular level. It should be realized that knowledge about DNA and about RNA and

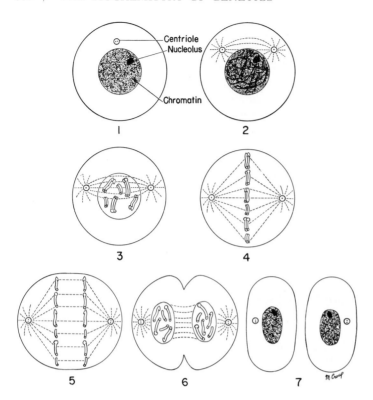

Figure 32–1   Mitosis in a cell of a hypothetical animal with a diploid number of six (haploid number = 3); one pair of chromosomes short, one pair long and hooked, and one pair long and knobbed. 1, Resting stage. 2, Early prophase: centriole divided and chromosomes appearing. 3, Later prophase: centrioles at poles, chromosomes shortened and visibly doubled. 4, Metaphase: chromosomes arranged on equator of spindle. 5, Anaphase: chromosomes migrating toward poles. 6, Telophase: nuclear membranes formed; chromosomes elongating; cytoplasmic divisions beginning. 7, Daughter cells: resting phase. (From Mazur, A. and Harrow, B.: *Textbook of Biochemistry.* 10th edition. Philadelphia: W. B. Saunders Company, 1971, p. 477.)

DNA replication occurring in single cells may not be directly applied to human genetics, but it greatly increases our understanding of the genetic processes in the body.

The nucleus of the cell, which contains most of the DNA (Chapter 20), is of prime importance in genetics. The **chromosomes,** which contain the genes, are thought to be large DNA molecules. A developing embryo contains **germ cells,** which are **haploid** and possess only one set of chromosomes, and **somatic cells,** which are **diploid** and possess two sets of chromosomes. The somatic cells give rise to the tissues and organs characteristic of the species, while the germ cells, at sexual maturity, are responsible for the development of eggs or sperm and thus for the passing of the genetic characteristics of the species. As the cells divide during the process of reproductive replication, the chromosomes and genes divide into exact duplicates of themselves which are finally deposited in the nucleus of each daughter cell, as shown in Figure 32–1. The chromosomes and genes of the germ cells and somatic cells, by this process of replication, transmit the genetic information from the parents to the offspring.

DNA AND GENETICS

The relation of DNA to genetics was first recognized in the 1940s. In 1943 Avery observed that the DNA from a virulent strain of bacteria transformed a nonvirulent strain of the same microorganism into a virulent strain. It is significant that the haploid cells contain only half the amount of DNA found in diploid somatic cells, and that in a given species the amount of DNA per diploid cell is fairly constant from one type of cell to another. Also, it has been shown that the purine and pyrimidine base composition of DNA varies from one species to another. The DNA in the cells or tissues of a single species has the same base composition and does not change with nutritional status, age, or a change in environment. As discussed in Chapter 22, the number of guanine bases is equal to the number of cytosine residues, and the number of adenine residues is equal to the number of thymine residues in all DNA's. In addition, the sum of the purine bases equals the sum of the pyrimidine bases, and the G-C and A-T pairs with their accompanying hydrogen bonds (Fig. 22–2) not only fit best in the double helix structure (Fig. 22–1), but also exhibit the strongest hydrogen bonding and stability.

CHROMOSOMES AND GENES

In experiments with bacteria and bacterial viruses, it has been demonstrated that single chromosomes of DNA-containing bacteria and viruses consist of single very large double helical DNA molecules. For example, the chromosome of a commonly studied bacteria, *E. coli*, consists of a single DNA molecule with a molecular weight of about 2.8 billion, about 4.2 million base pairs, and a contour length of over 1200 μ. Cells of higher organisms contain several chromosomes, each of which consists of one or more large DNA molecules. The DNA in a single human cell has been calculated to contain 5.5 billion base pairs and to have a total length of about 2 meters.

Genes are segments of a chromosome that codes for a single polypeptide chain of a protein or enzyme. Since the sequence of three nucleotides in DNA—a codon (p. 371)—is required to code for a single amino acid residue, the size of the gene which determines a specific protein may be calculated by multiplying the number of amino acid residues in the protein by 3. In proteins or enzymes that have two or more different polypeptide chains, each coded by a different gene, more than one gene may be required to code for these proteins. The exact location of a gene in the chromosome of a bacterial cell that codes for a specific protein can be established by complex genetic mapping methods that are beyond the scope of this book. Fortunately, the genes that code for the individual enzymes of a multienzyme system are often located adjacent to each other in the chromosomes and are transcribed and translated as a group. The single chromosome of bacteria or bacterial viruses may contain from 5 to as high as 4000 different genes.

REPLICATION OF DNA

The mechanism by which genetic information can be accurately replicated was proposed by Watson and Crick, based on their double helical structure of DNA. Since the two strands of DNA are structurally complementary to each other and therefore contain complementary information in their base sequences, the **replication** of DNA during cell division was suggested to occur by separation of the two strands, each becoming a template to specify the base sequence of a new complementary strand. The enzyme **DNA polymerase,** first extracted from bacteria (*E. coli*), catalyzes the replication. An illustration of this process with the formation of two daughter double helical DNA's, each containing one strand from the parent DNA, is shown in Figure 32–2.

In reduplication of cell nuclei, which is necessary in cell division, the double helix may unravel, and each of the original chains may serve as a template for the synthesis of another chain. It has been shown experimentally that by adding labeled purine and pyrimidine intermediates to a synthesizing system one chain of newly synthesized DNA contains labeled intermediates from the system, whereas the original chain is unlabeled. Since the base-pairing pattern of DNA is followed, the newly synthesized chain will possess the exact nucleotide sequence of the original parent chain. The result is the synthesis of two pairs of DNA chains (Fig. 32–2) in which each pair is identical in nucleotide sequence and genetic coding information to the original pair of parent chains. In the laboratory it has been demonstrated that pure DNA preparations from a particular

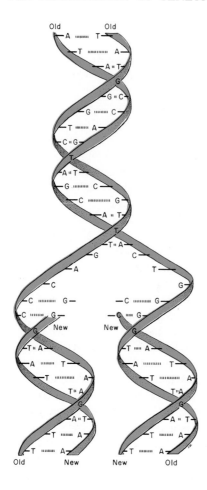

Figure 32–2 Replication of DNA as suggested by Watson and Crick. The complementary strands are separated and each forms the template for synthesis of a complementary daughter strand. (From Villee, C.: *Biology.* 5th ed. Philadelphia: W. B. Saunders Company.)

species of bacteria or bacteriophage, when added to another species of bacteria, will serve as a template to direct the recipient cells to develop the characteristics of the donor bacteria. It is quite possible that the multiplication of viruses within cells may occur by the same process. For example, type 1 poliomyelitis virus, which has been crystallized, contains an RNA that serves as a template to infect cells with this type of virus. This knowledge suggests that RNA functions mainly in the cytoplasm of a cell as a template for the synthesis of specific cellular proteins. A close relation exists between DNA of the nucleus and RNA of the cytoplasm, since one chain of DNA and one chain of RNA could twist around each other to form a double helix and thus influence the RNA template.

TRANSCRIPTION OF DNA

The replication of DNA in the nucleus of cells to preserve and pass on its genetic information to the DNA of daughter cells is obviously of prime importance in genetics. The next step, that of directing protein-synthesizing machinery to produce specific molecules such as enzymes and hormones required in metabolic reactions in the body, is called **transcription** (p. 371). In earlier studies of protein synthesis in intact cells, Jacob and Monod observed an increased rate of synthesis of cytoplasmic RNA. They proposed that this species of RNA served as a messenger carrying genetic information from the DNA of the chromosomes to the surface of the ribosomes. In addition, they suggested that this **messenger RNA** is formed enzymatically so that it has a base sequence complementary to that of one strand of RNA. The m-RNA was believed to contain the complete message necessary for specifying one or more polypeptide chains and for binding the ribosomes to serve as a functioning template for protein synthesis (p. 372).

The discovery of DNA polymerase that was required for DNA replication stimulated a search for **RNA polymerase,** the catalyst for DNA transcription. This enzyme was also obtained from extracts of *E. coli.* Messenger RNA synthesis was found to require RNA polymerase, ribonucleoside triphosphates and double-stranded DNA molecules. Only one strand of the DNA is transcribed as it imparts its genetic message to m-RNA.

REGULATION OF PROTEIN SYNTHESIS

After the DNA of the chromosomes imparts its encoding message to the m-RNA and this molecule directs the synthesis of specific protein molecules on the ribosomes, what process directs the amount and type of proteins produced? It seems obvious that the living cell must possess a mechanism for regulating the relative amounts of the different protein molecules that are synthesized. Enzymes, for example, that are required in the main metabolic cycles in the cells are necessarily synthesized in greater quantities than those needed for the synthesis of coenzymes. There must also be a mechanism for turning on or off the synthesis of proteins as they are needed for cellular functions.

Most of our information about regulation comes from studies on the control of synthesis of enzyme proteins. Jacob and Monod postulated that the chromosomes carry three types of genes, **structural genes, operator genes,** and **regulator genes.** The structural gene DNA directs the synthesis of protein molecules as described previously. The operator gene controls the action of adjacent structural genes in the synthesis of specific m-RNA molecules. The structural genes and their operator gene are designated an **operon.** The operator gene itself is controlled by a regulator gene which directs the synthesis of protein molecules called **repressors.** When the repressor combines with its operator gene, the structural genes cannot function in protein synthesis, and the operon is said to be repressed. If a situation in the cell inactivates the repressor and permits the operon to function, the operon is said to be derepressed (Fig. 32–3). As suggested in Chapter 23, the

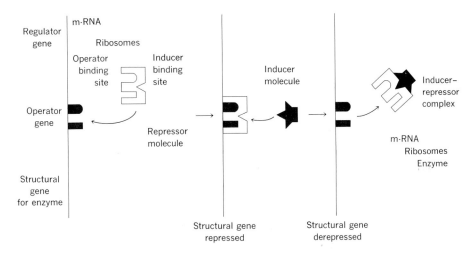

Figure 32–3   Proposed mechanism for regulation of enzyme synthesis.

end products of enzyme reactions, which are often metabolites, can affect the activity of enzyme systems. They may carry out this function by control of the repressor and the enzyme synthesis mechanism described previously.

THE CENTRAL DOGMA OF GENETICS

Since about 1955, a major objective in biochemistry has been to document and clarify the details of what is called the **central dogma of genetics:** that DNA is the hereditary material; that its information is encoded in the sequences of its subunits, the genes; and that this information is transcribed onto RNA and then translated into protein. Recently, in studies of the bacterial genes of *E. coli*, electron micrographs were obtained that show both an inactive chromosome segment and an active chromosome segment. In the active segment the DNA is seen being transcribed onto messenger RNA and the RNA being translated into protein. The electron micrographs obtained in these studies bear a striking resemblance to diagrams of transcription and translation that have been proposed by recognized research workers in this area.

In another recent study of how cells proceed through division in the cell cycle, an electron micrograph clearly depicting chromosome replication was obtained. The DNA double helix does not unwind as neatly as shown in Figure 32–2 but appears as tangled strands with segments of replicating DNA molecules joining the original DNA strand at several junctures in the micrograph.

Multiplication of viruses within cells can occur by a similar process in which the viral genes consist of DNA that transmits information to the RNA of the cell and then into the cell proteins. Several common viruses which cause poliomyelitis, the common cold, and influenza are called **RNA viruses,** since the RNA replicates directly into new copies of RNA and translates information directly to proteins of the cell without DNA involvement in their replication. Recently evidence for a reverse flow of genetic information from RNA to DNA has been obtained in experiments using the Rous sarcoma virus. It has been suggested that in normal cells there are regions of DNA that serve as templates for the synthesis of RNA, and that this RNA serves in turn as a template for the synthesis of DNA which then becomes integrated with the cellular DNA. These experiments may be valuable in an explanation of the origin of cancer in humans. Since it was shown that cancer-causing RNA viruses can produce a DNA transcript of the viral RNA, it is possible that the viral RNA may transmit genetic information to the genes that will eventually surface as spontaneous cancer.

From these and other studies a science of **genetic engineering,** which would have as its first objective the reversal of cancer growth and the repair of inborn deficiencies in genetic diseases, may result. Just as viral RNA may transmit cancer in man, as mentioned earlier, scientists may be able to introduce chemotherapeutic RNA molecules into the body tissues for repair or curative purposes.

TOPIC OF CURRENT INTEREST

INBORN ERRORS OF METABOLISM

As early as 1906, Garrod, an English physician, described several abnormal patterns of metabolism. He called these abnormalities **inborn errors of metabolism,** since he recognized that the conditions were inherited. To the present time, approximately one hundred of these genetic diseases have been reported. Many of them are extremely rare. The diseases due to a metabolic block in amino acid catabolism are of special interest since they occur somewhat more frequently. A common clinical manifestation of such conditions is mental deficiency. Three of these inborn errors, **phenylketonuria, alkaptonuria,** and **tyrosinemia,** will be considered in more detail as they are concerned with the metabolism of phenylalanine, an essential amino acid. Each involves a deficiency of a different single, specific enzyme whose synthesis is genetically controlled.

PHENYLKETONURIA

Phenylketonuria, or PKU, was first recognized by Fölling in 1934 when he detected large amounts of **phenylpyruvic acid** in the urine of several mentally retarded patients. All individuals who have so far been found to excrete phenylpyruvic acid in their urine daily have shown some degree of mental deficiency; about 1 per cent of the patients in institutions for the mentally retarded excrete phenylpyruvic acid in their urine. Since the condition was recognized as a genetic disease, attention has been focused on parents as carriers of the defect and on detection and treatment of newborn infants with the disease.

The normal metabolism of phenylalanine involves its transformation into tyrosine with the aid of **phenylalanine hydroxylase.** If sufficient quantities of this enzyme are not synthesized, as in phenylketonuria, the concentration of phenylalanine in the blood, spinal fluid, and urine increases. In the tissues, the phenylalanine is converted into phenylpyruvic, phenyllactic, and phenylacetic acids as shown in the following scheme. These metabolites of phenylalanine are excreted in the urine in large amounts in phenylketonuria.

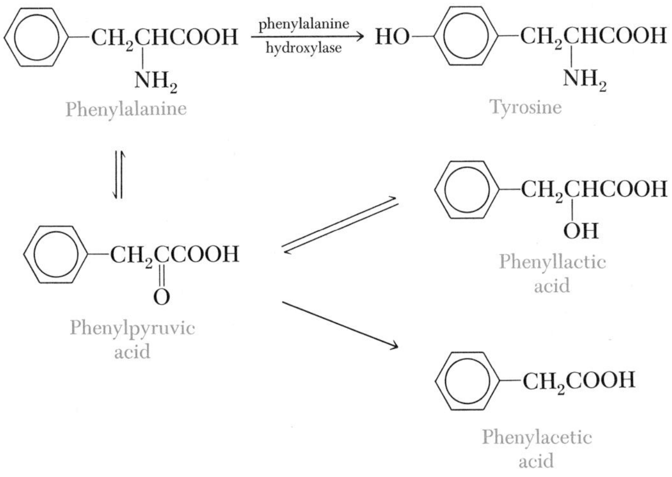

As the disease itself is rather well understood today, the practical problem consists of identifying infants with the disease and initiating adequate treatment. Simple tests for the detection of phenylpyruvic acid in the urine or, preferably, tests for the increased concentrations of phenylalanine in the blood are available and are required in several states. The most successful treatment consists of restricting the amount of phenylalanine in the diet of the PKU infant, providing only the minimum quantity essential for normal growth and development. Fortunately, this regime allows the child to develop normally; the diet may apparently be relaxed somewhat after the age of six without serious effects on the child.

ALKAPTONURIA

Alkaptonuria is a rare genetic abnormality of tyrosine metabolism that is readily recognized by the characteristic changes in color which occur in the urine. When freshly passed the color is normal, but on standing it begins to darken. Alkalinity speeds up the change, and the urine passes through shades of brown to a final black color. Diapers wet with urine become darkly stained and as a result the condition is frequently recognized in early infancy. In adults, alkaptonuria may first be detected in a life insurance examination or routine medical check. The urine has strong reducing properties, which gives rise to a positive Benedict's test. These characteristics persist throughout life. As patients grow older their ligaments and cartilages tend to become dark blue in color, and the patients are prone to develop osteoarthritis.

The normal metabolism of tyrosine, which may be formed from phenylalanine as previously discussed, includes the formation of **homogentisic acid,** which is oxidized to fumaric acid and finally to acetoacetic acid. This oxidation is catalyzed by the enzyme **homogentisic acid oxidase.** The metabolic error is a block in the breakdown of the homogentisic acid, caused by a decrease in the concentration of the oxidase in the tissues; normal individuals can readily oxidize increased quantities of this acid without producing pigmented urine specimens. The formation of homogentisic acid and its metabolites is shown in the following scheme. At present there is no specific treatment for this rare genetic disease.

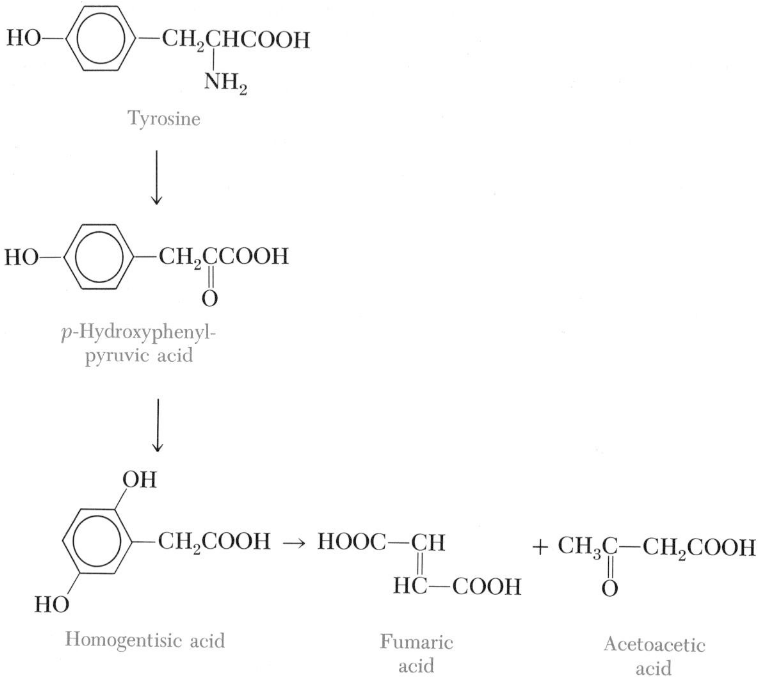

$$HO-\langle\bigcirc\rangle-CH_2CHCOOH$$
$$|$$
$$NH_2$$

Tyrosine

$$\downarrow$$

$$HO-\langle\bigcirc\rangle-CH_2CCOOH$$
$$\|$$
$$O$$

p-Hydroxyphenyl-
pyruvic acid

$$\downarrow$$

$$\begin{array}{c} OH \\ \langle\bigcirc\rangle-CH_2COOH \rightarrow HOOC-CH \quad + CH_3C-CH_2COOH \\ HO \qquad\qquad\qquad HC-COOH \qquad\qquad O \end{array}$$

Homogentisic acid   Fumaric   Acetoacetic
                            acid   acid

TYROSINEMIA

Tyrosinemia is a rare inborn error of tyrosine metabolism seen primarily in premature infants. In the scheme shown above, the metabolic block associated with this defect occurs between *p*-hydroxyphenylpyruvic acid and homogentisic acid. The enzyme required for this step, **p-hydroxyphenylpyruvate oxidase,** has been found to be in low concentration in the liver of premature infants. In infantile tyrosinemia, kidney defects and a nodular cirrhosis of the liver occur, causing rickets, thrombocytopenia, darkening of the skin, and slight mental retardation. The urine contains large amounts of tyrosine and metabolites, including *p*-hydroxyphenylpyruvic, *p*-hydroxyphenylacetic, and *p*-hydroxyphenyllactic acids. The disease may be acute or chronic; the acute cases are characterized by diarrhea, failure to thrive, and death from liver failure in the first seven months of life.

Ascorbic acid is the coenzyme of *p*-hydroxyphenylpyruvate oxidase, and the disease may be alleviated in premature infants by feeding excess ascorbic acid. Longer term treatment for tyrosinemia consists of a diet low in tyrosine and phenylalanine. If the dietary treatment is started early in life, both liver and kidney damage may be prevented.

IMPORTANT TERMS AND CONCEPTS

alkaptonuria	genes	inborn errors of metabolism	replication
chromosomes	genetic engineering	operon	repressor
DNA polymerase	germ cells	phenylketonuria	somatic cells

QUESTIONS

1. What features of single cells such as bacteria are useful in the study of genetics?

2. Describe the characteristics of germ cells and somatic cells.

3. What evidence suggested that DNA is involved in genetic processes?

4. Discuss the close relationship between genes and chromosomes of the cell.

5. Explain with an illustration the process of replication of DNA.

6. Describe the function of DNA polymerase and RNA polymerase in replication and transcription of DNA.

7. Outline a mechanism for the regulation of enzyme synthesis in the cell.

8. What is the central dogma of genetics? Explain.

9. Why is the detection of phenylketonuria so important in newborn infants?

10. Briefly discuss the condition of alkaptonuria.

11. What is genetic engineering and how may it be of value in the future?

SUGGESTED READING

Brown: The Isolation of Genes. Scientific American, Vol. 229, No. 2, p. 20, 1973.

Cooper and Lawton: The Development of the Immune System. Scientific American, Vol. 231, No. 5, p. 58, 1974.

Friedmann: Prenatal Diagnosis of Genetic Disease. Scientific American, Vol. 225, No. 5, p. 34, 1971.

Jerne: The Immune System. Scientific American, Vol. 229, No. 1, p. 52, 1973.

Mazia: The Cell Cycle. Scientific American, Vol. 230, No. 1, p. 54, 1974.

McKusick: The Mapping of Human Chromosomes. Scientific American, Vol. 228, No. 3, p. 34, 1973.

Notkins and Koprowski: How the Immune Response to a Virus Can Cause Disease. Scientific American, Vol. 228, No. 1, p. 22, 1973.

Ptashne and Gilbert: Genetic Repressors. Scientific American, Vol. 222, No. 6, p. 36, 1970.

Ruddle and Kucherlapati: Hybrid Cell and Human Genes. Scientific American, Vol. 231, No. 1, p. 36, 1974.

Stein, Stein, and Kleinsmith: Chromosomal Proteins and Gene Regulation. Scientific American, Vol. 232, No. 2, p. 46, 1975.

Temin: RNA-Directed DNA Synthesis. Scientific American, Vol. 226, No. 1, p. 24, 1972.

BIOCHEMISTRY
OF DRUGS

CHAPTER 33

The *objectives* of this chapter are to enable the student to:

1. Explain the use of aspirin in so many drug preparations.
2. Compare the action of cholinergic and anticholinergic drugs.
3. Recognize the importance of epinephrine and its functions in the body.
4. Explain the increased use of amphetamine-type drugs in today's society.
5. Recognize the similarities and differences between the many drugs used as tranquilizers or sedatives.
6. Recognize the difference in action and potency of marijuana, LSD, and heroin.
7. Describe the mechanism of action of the antifertility drugs.
8. Explain the use of cytotoxic drugs in the treatment of cancer.

If we accept a single definition of disease as *dis ease*, headaches, minor aches and pains, malnutrition, dietary deficiencies, metabolic abnormalities, endocrine disturbances, infections, and malignant cancer all qualify as diseases. Volumes have been written on diseases of the cell, tissues and organs of the body, and the therapeutic agents or drugs used to combat the disease process. It is becoming more and more apparent that all disease has a biochemical basis. The biochemistry of normal and abnormal heredity, deficiency diseases, errors of metabolism, and the process of infection is receiving considerable attention, study, and research.

ANALGESIC DRUGS

To aid in the understanding of the recent emphasis on health-related research, examples of different types of diseases and their treatment will be discussed. So many people are occasionally inconvenienced with headaches and minor aches and pains that their cause and treatment with analgesic drugs should be of interest. In general, these pains are caused by swelling of tissue, resulting in pressure on peripheral nerves, and also minor inflammation of tissues, accompanied by an increase in body temperature which affects nerve endings. A common analgesic drug that serves as the basis for a multitude of headache, cold, and flu remedies is acetylsalicylic acid or **aspirin** (p. 243). When combined with **phenacetin** and **caffeine,** the resultant preparation is the common APC tablets. The compound **N-acetyl-p-amino phenol** is a metabolic product of phenacetin and has replaced this drug in several preparations. Phenacetin is an organic amine derived from acetanilide, a compound originally used as an antipyretic drug (p. 251). Aspirin is an analgesic drug in that it reduces inflammation and swelling of tissues, exerts an antipyretic action in reducing fever, and probably exerts a chemical action on the peripheral nerves. Phenacetin and N-acetyl-p-aminophenol exhibit some of the effects of aspirin and are synergistic with respect to its action. The antipyretic effect of aspirin is thought to be related to the salicylic acid portion of the structure, whereas the antipyresis exhibited by phenacetin and N-acetyl-p-aminophenol depends on the amino-benzene portion of

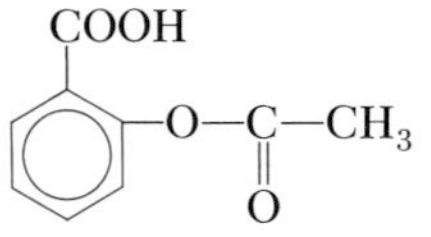

Aspirin 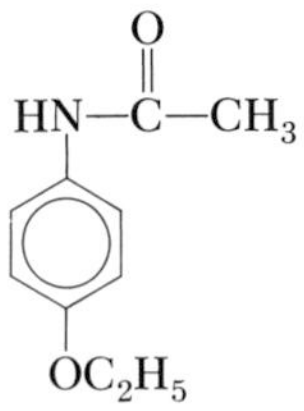 Phenacetin 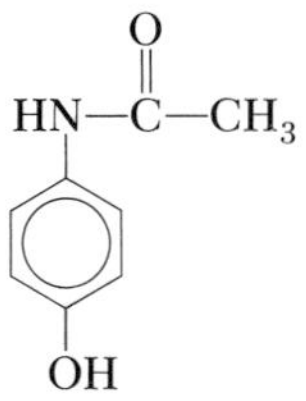 N-acetyl-p aminophenol 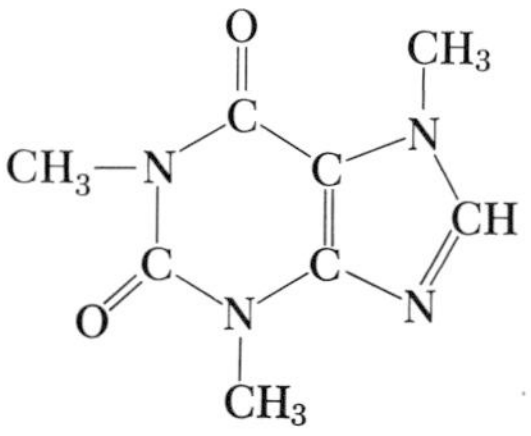 Caffeine

the compound. Caffeine is a diuretic and assists the kidney in excretion of the drug and the circulatory system in the transport of the drug. The addition of buffering agents to aspirin has been found to speed the absorption of the drug into the blood and tissues and has resulted in products such as Bufferin.

TRANQUILIZERS, SEDATIVES, AND HALLUCINOGENIC DRUGS

Diseases of the brain and nervous system ranging from undue concern and nervousness through inability to sleep, anxiety symptoms, neurotic behavior, and pathology of brain tissue are common conditions in these accelerated times. Drugs such as sedatives, tranquilizers, and psychic energizers are too commonly prescribed and used. Mind-expanding and hallucinogenic drugs are receiving considerable attention, as we know. Another drug, L-dopa, provides substantial relief in Parkinson's disease, a condition involving changes in the brain tissue.

Phenothiazine, which has a three-ringed structure in which two benzene rings are linked by a sulfur atom and a nitrogen atom, forms the basis for a group of drugs that are potent adrenergic blockers. The phenothiazines are widely used in the treatment of pyschiatric patients and in the treatment of nausea and vomiting. There is a close relationship between the chemical structure and the activity of the drug. Substitution of a chlorine or methoxy group in position R′ in the basic structure increases the potency of the drug for depressing motor activity and altering psychotic behavior in patients. A CF_3 substitution in this position increases antiemetic and antipsychotic potency. One of the most potent phenothiazines has a CF_3 on position R′ and a piperazine group on position R″. Chlorpromazine is the most frequently prescribed phenothiazine, while fluphenazine is one of the most potent members of this group. Their relationship to the basic structure is shown as follows:

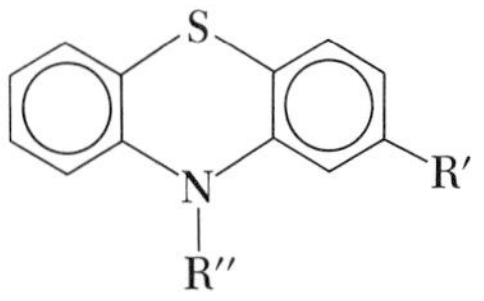

Phenothiazine (Basic structure) 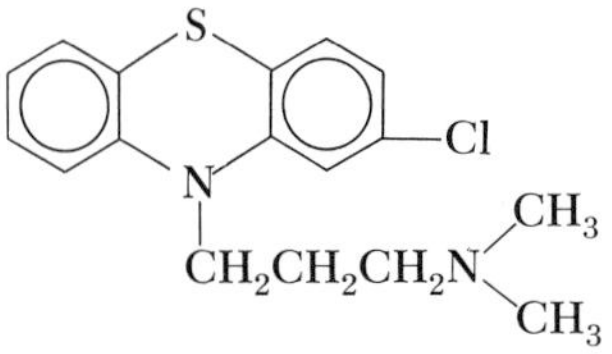 Chlorpromazine 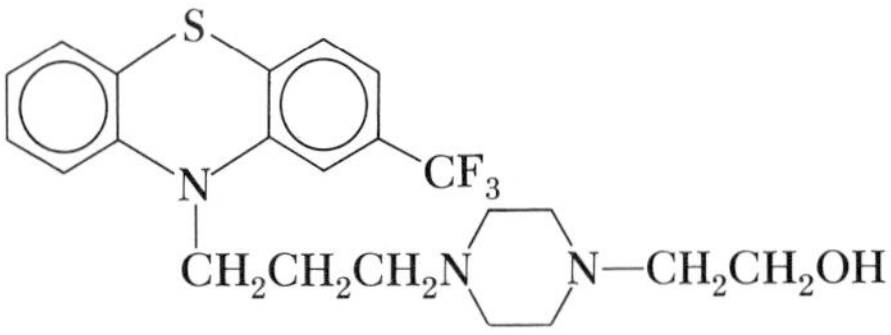 Fluphenazine

Chlorpromazine also has a sedative effect and reduces the blood pressure.

Other compounds used as mild tranquilizers or sedatives, such as barbiturates and meprobamate, are related to the pyrimidines and to urea. These drugs depress the central nervous system. In the case of barbiturates their speed of action and duration of effect depend on an increased lipid solubility and an increase in the length of the side chain, as typified by the structure of Seconal versus that of Barbital. The pharmacological effects of meprobamate are very similar to those of the barbiturates.

Barbital   Seconal   Meprobamate

Continued use of mild sedatives or tranquilizers, like barbiturates, leads to a dependency on the drug. Overdosage or combinations of alcoholic beverages and barbiturates may lead to coma and accidental death. The more potent tranquilizers are subjected to strict control by physicians and are not as widely prescribed. When drugs with strong analgesic and sedative properties are required, narcotic drugs, such as **morphine** and **Demerol,** are employed. Morphine, first isolated from the opium poppy, has a complex chemical structure containing a phenanthrene nucleus and a piperidine nucleus. **Codeine,** which is also commonly used as a weaker narcotic, is a methyl ester of morphine. **Heroin** is the diethyl ester of morphine and is used by drug addicts because it is more lipid-soluble and faster-acting than morphine. Demerol was synthesized as a substitute for morphine but proved to be no less habit-forming.

Morphine   Demerol

Antidepressant or mood-elevating drugs, sometimes called psychic energizers, are being prescribed in increasing quantities. Derivatives of hydrazine directly affect brain and nervous system function by inhibiting the enzyme monoamine oxidase, which apparently results in an antidepressant action. Iproniazid and **isocarboxazid** are examples of potent drugs in this family. Amphetamine, also called benzedrine, and related drugs such as dextroamphetamine, methamphetamine, and ephedrine all exert powerful central nervous system-stimulating actions. These include a decreased sense of fatigue, increased initiative and ability to concentrate, often an elevation of mood, elation and euphoria, and increased motor activity. Such properties have resulted in the use of these drugs as psychic energizers by truck drivers on long trips and by students cramming for exams. Unfortunately, prolonged use of these drugs is almost always followed by mental depression and fatigue. The drugs are also used as appetite depressants in obesity and to counteract depressive syndromes and behavioral syndromes.

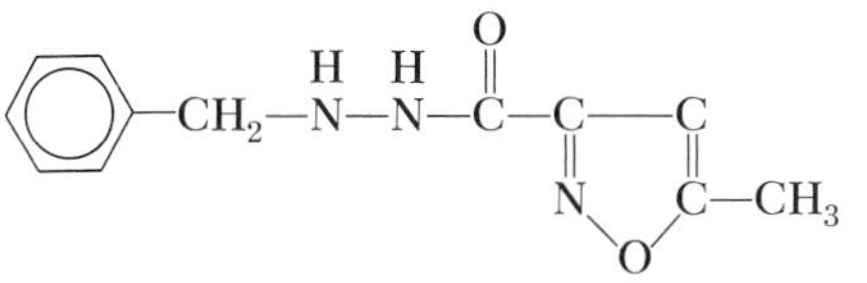

Isocarboxazid

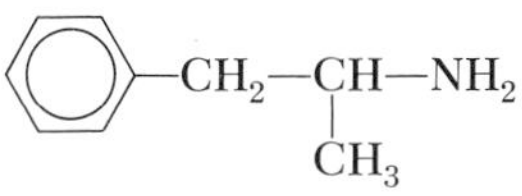

Amphetamine

A long step further in the use of mind-influencing drugs is represented by the **psychedelic** or **hallucinogenic drugs.** It is difficult to characterize the medical effect of these drugs, although many nonmedical experiments are being conducted. In view of the change in personality and mental state of an individual taking this type of drug, the results of ingestion are often unpredictable. **Marijuana** represents a mild type of hallucinogen and is obtained from the *Cannabis sativa* or hemp plant. The most potent marijuana is obtained from the yellow resin produced from the flowers of the ripe plant and is called **hashish.** Chemically the drug is a derivative of an alcohol, **cannabinol,** and the active constituent is believed to be a delta-L form, which has recently been synthesized. **Mescaline** and **lysergic acid diethylamide (LSD)** are examples of more potent hallucinogens.

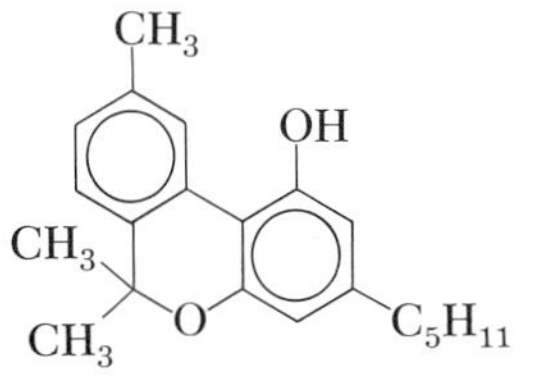

Cannabinol

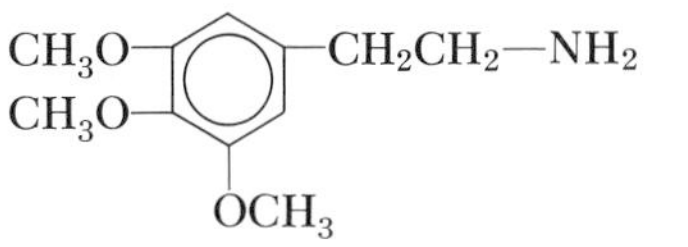

Mescaline

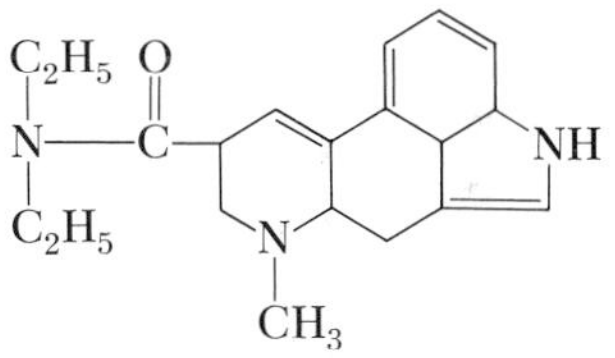

Lysergic acid diethylamide (LSD)

For years, there had been little advance in the treatment of **Parkinson's disease.** This is a disease of the brain affecting the metabolism of dopamine, epinephrine, and norepinephrine known as paralysis agitans, since it involves both a progressive paralytic rigidity and tremors of the extremities. Recently it was found that the dopamine concentration of certain areas of the brain was markedly deficient in chronic patients who had died of the disease. Since dopamine will not penetrate the brain tissue when carried in the blood, the precursor L-dihydroxyphenylalanine, L-dopa, was administered and found to increase the dopamine concentration in the target areas. The use of the drug **L-dopa** provides considerable relief and improvement of symptoms for these patients and has been hailed as a breakthrough in Parkinson's disease therapy.

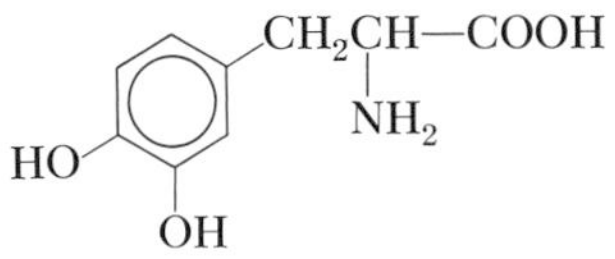

L-Dopa
(L-dihydroxyphenylalanine)

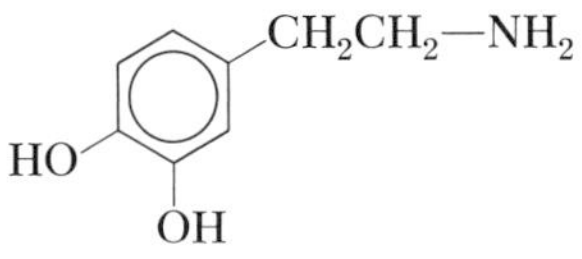

Dopamine

ANTIHISTAMINES

Histamine, which is formed by the decarboxylation of the amino acid histidine (see p. 270), is a powerful pharmacological agent with effects on the vascular system, smooth muscles, and exocrine glands, especially the gastric glands. The administration or release of histamine causes dilatation of capillaries and small blood vessels with a subsequent drop in systemic blood pressure; the dilatation of cerebral vessels results in a histamine headache which may be very severe. Smooth muscles, especially the bronchioles, are stimulated by histamine and may cause respiratory problems in persons suffering from bronchial asthma and other pulmonary diseases. Histamine is a powerful gastric secreto-gogue and produces a copious secretion of gastric juice of high acidity. It also stimulates nerve endings and causes itching when introduced into the superficial layers of the skin.

Antihistamines are drugs that antagonize the pharmacological actions of histamine and also reduce the intensity of allergic reactions. A portion of the chemical structure of various antihistamines is similar to that in histamine, and these drugs act as competitive antagonists to histamine. Apparently they occupy the receptor sites on the effector cells and exclude histamine from these sites. The common core of the chemical structure in both histamine and antihistamines is a substituted amine, i.e., ethylamine. It is believed that it is this portion of the molecule that competes with histamine for the receptors.

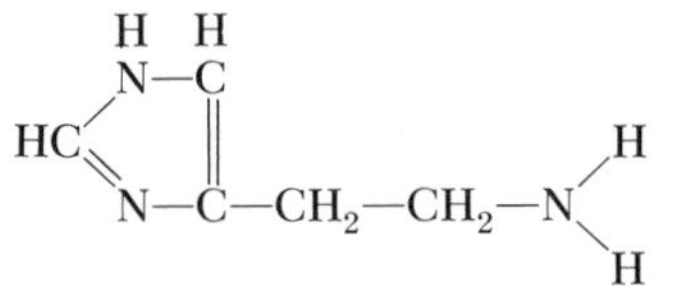

Histamine 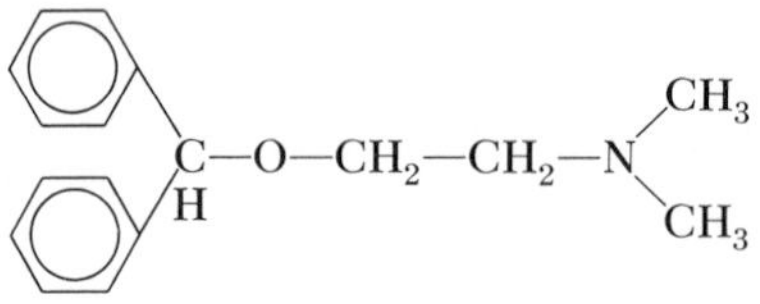   Benadryl (Diphenhydramine)

Therapeutically the antihistamines are most commonly used in the symptomatic treatment of various allergic diseases. Patients with bronchial asthma, hay fever, allergic rhinitis, and chronic rhinitis with superimposed acute colds gain considerable relief by the use of antihistamines. Various types of allergic dermatitis, contact dermatitis, insect bites, and poison ivy are benefited by the topical application of antihistamine-containing lotions. Some of these drugs, especially **Dramamine,** which relies on diphenhydramine as the active agent, are very effective against motion sickness. One common side effect of most antihistamines is their tendency to induce sedation, which restricts their daytime use when it is necessary to operate motor vehicles. **Chlor-trimeton,** shown below, is less prone to produce drowsiness than most other preparations. The prominent hypnotic effect of antihistamines related to **Benadryl** and **Pyribenzamine** has resulted in their use in various proprietary remedies for insomnia, such as Sominex and Nytol.

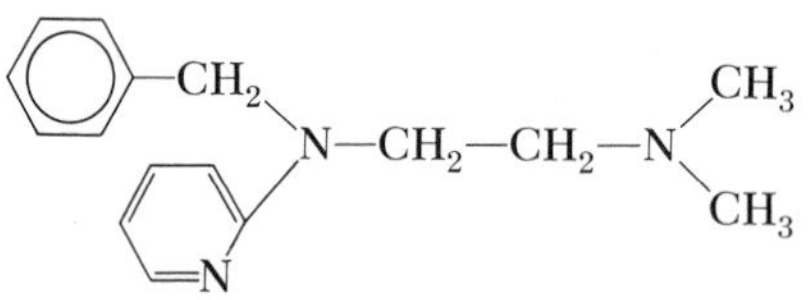

Pyribenzamine (Tripelennamine) 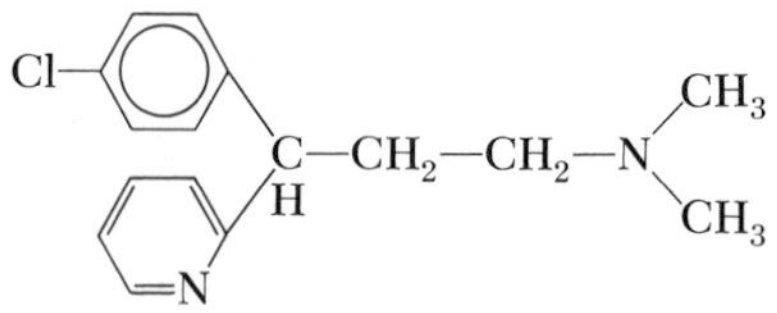   Chlor-trimeton (Chlorpheniramine)

ANTIBACTERIAL AND ANTIBIOTIC DRUGS

The body has to guard against infection by bacteria or viruses from birth to death. The newborn infant is fortified with antibodies against disease upon receiving the gamma globulins in the mother's milk. The layer of skin covering the body, the hydrochloric acid of the gastric juice, the digestive enzymes, and the various phagocytic cells in the circulation, all serve as a first-line defense against infection. Vaccination during childhood stimulates the production of antibodies to certain diseases. Until the mid-1930s, a serious infection or infectious disease was viewed with alarm by physicians and laymen alike. The first major group of chemotherapeutic agents were the sulfonamides or **sulfa drugs,** prepared by the acylation of sulfanilamide. These drugs were antibacterial agents that inhibited the synthesis of a compound like folic acid that was essential for the continued growth of the invading bacteria. Sulfanilamide, for example, acts as a competitive inhibitor of the enzyme that is involved in the utilization of *p*-aminobenzoic acid in the synthesis of tetrahydrofolic acid by the bacteria (see p. 249). **Sulfanilamide** was found to be effective in the treatment of streptococcus infections, pneumonia, puerperal fever,

| Sulfanilamide | Sulfaguanidine | Sulfathiazole | Sulfadiazine |

gonorrhea, and gas gangrene. The drug is only slightly soluble in water and may damage the kidney by accumulation in that organ during excretion. Other toxic reactions, including methemoglobinemia, resulted in the development of other derivatives, such as **sulfaguanidine, sulfathiazole,** and **sulfadiazine.** A thorough study of the therapeutic properties of each sulfa drug resulted in better treatment and control of various infectious diseases. Sulfadiazine, for example, is less toxic than the other sulfa drugs, yet is one of the most effective in the treatment of pneumonia and staphylococcus infections.

A few years after the development of the sulfa drugs, a new type of antimicrobial agent was accidently discovered by Fleming. He observed that a staphylococcus culture on a bacterial plate did not grow around the periphery of a blue-green mold that had contaminated the culture plate. **Penicillin** was isolated from the secretion of the mold and was termed an antibiotic, since it interfered with the growth of the bacteria. A large number of antibiotic agents have been isolated from similar experiments with other molds. It required nine years of intensive research to synthesize penicillin. Other antibiotics, including **streptomycin, tetracycline,** and **prostaphlin,** have been synthesized, and prostaphlin shows considerable promise as an effective control of infections caused by staphylococcal bacteria. As in the case of the sulfa drugs, a family of antibiotics with specific antimicrobial properties has now been developed.

| Penicillin G | Tetracycline |

STEROID DRUGS IN RHEUMATOID ARTHRITIS

There are several types of collagen diseases associated with inflammatory changes in connective tissue which affect mainly the joints, skin, heart, and muscle. **Rheumatoid arthritis** and **lupus erythematosus** are examples of collagen disease. Arthritis is most common and affects men and especially women in their forties and fifties. For many years, aspirin has been used as a mild antiinflammatory agent in arthritis and continues as the maintenance therapy. More recently **cortisone** has been found to reduce inflammation of the joints in arthritis and to reverse the course of this and other collagen diseases. The dosage level required to produce these desirable effects also produced several unwanted side effects. Steroid derivatives of cortisone were developed to decrease the incidence of side effects and increase the therapeutic potency of the drugs. The 9-fluoro-16-methyl derivative of **prednisolone,** a steroid closely related to cortisol, is 100 to 250 times as potent as cortisone in the treatment of rheumatoid arthritis. At present, arthritic patients are maintained with aspirin and given small doses of potent steroid drugs whenever an acute inflammatory process flares up in their joints. The relationship between the structure and the function of cortisone and related steroids has already been discussed in Chapter 25 (see p. 322).

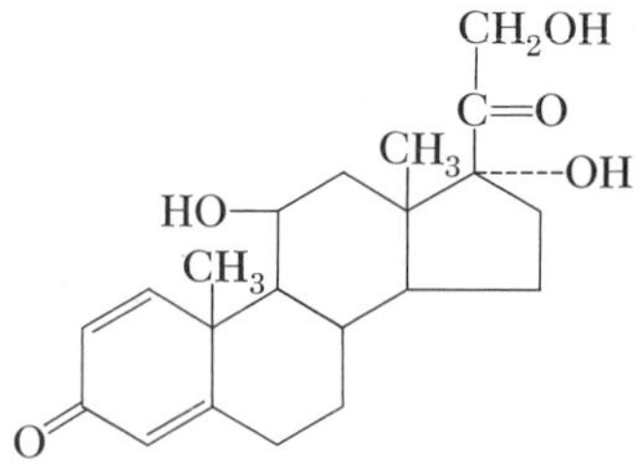

Prednisolone 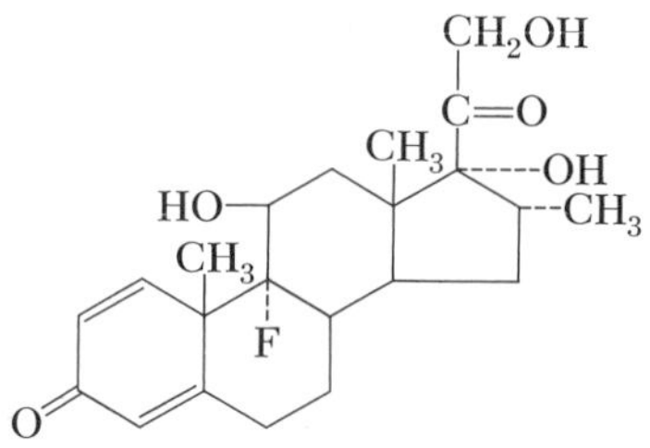   9-Fluoro-16-methyl prednisolone

ANTIFERTILITY DRUGS

Another type of steroid drug related to the sex hormones is the **antifertility drugs.** These drugs are unique in that they are given to inhibit a normal physiological process, whereas the great majority of drugs are used to treat a disease process or to alleviate the symptoms of a disease. As early as 1937 it was shown that the hormone progesterone would inhibit ovulation in rabbits. In the 1950s, several laboratories attempted to develop orally active steroids that possessed the properties of progesterone. **Norethindrone, 17α-ethynyl-19-nortestosterone,** and **norethynodrel,** the progestational component of Enovid, resulted from these studies. These drugs were related to testosterone (see p. 323) and contained a 17α-ethynyl group.

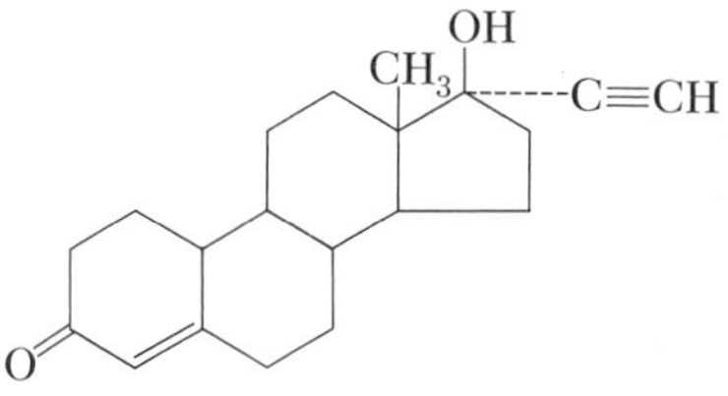

Norethindrone 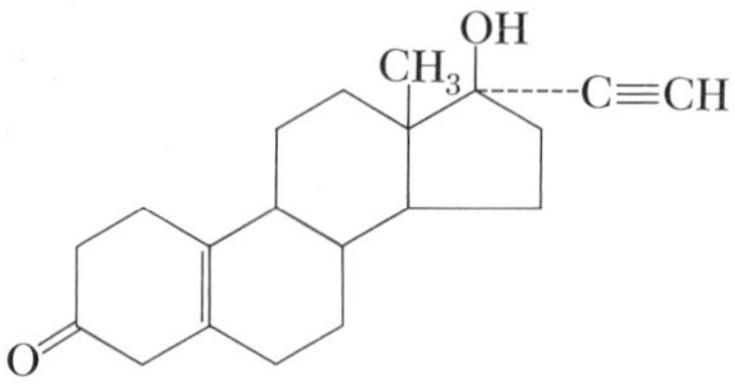   Norethynodrel

Since progesterone is relatively inactive when given orally, derivatives of this compound were studied for oral potency. It was found that 17-acetoxy-progesterone was active orally and that the addition of an α-methyl group to produce **medroxyprogesterone acetate** further enhanced this activity.

In clinical trials in Puerto Rico, Haiti, and the United States, the testosterone and progesterone derivatives described previously were found to effectively suppress ovulation in women. Also, it was discovered that estrogen enhanced the suppressive effect of the progesterone and that the 3-methyl ether of ethynyl estradiol or **mestranol** served as a potent estrogen. A combination of orally effective progesterone and estrogen active drugs is therefore commonly used in "the pill." The original dose was 10 mg of the progesterone

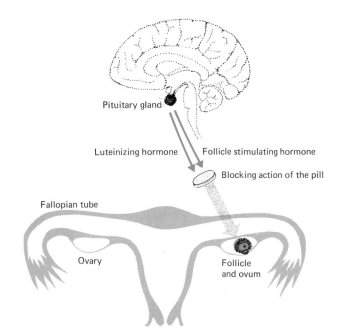

Figure 33–1   The effect of oral contraceptives on the process of ovulation.

Mestranol

Medroxyprogesterone acetate

and 0.15 mg mestranol, but this is being reduced toward 1 mg progesterone and 0.05 mg mestranol to decrease the occurrence of such side effects as nausea, headaches, dizziness, and thrombosis. The drug is usually taken on days five to 25 of the menstrual cycle and then withdrawn to permit normal menstruation.

From a study of the mode of action of oral contraceptives it was concluded that the primary effect is inhibition of follicular maturation (see p. 323), which prevents the occurrence of ovulation.

This mechanism is outlined in Figure 33–1, which illustrates the blocking action of the progesterone derivatives on the hormones of the pituitary gland responsible for development of an ovarian follicle and the subsequent release of an unfertilized ovum by the process of ovulation.

HYPOGLYCEMIC DRUGS IN DIABETES MELLITUS

Diabetes mellitus, because of its frequency, is probably the most important metabolic disease. The fundamental difficulty in the disease is a relative or complete lack of insulin, which is necessary for normal carbohydrate metabolism. Since the metabolic pathways of carbohydrates, fats, and proteins are known to be closely interwoven, any essential fault in carbohydrate metabolism also involves the metabolism of fat and protein, as well as water, electrolyte, and acid-base balance. There is evidence that diabetes is a hereditary disease and that the genetic tendency toward the disease results in the common condition known as prediabetes, which exists in many relatives of diabetics. The chemistry of **insulin** (p. 275) and its function in diabetes (p. 350) have already been discussed. The beta cells of the islet tissue in the pancreas produce insulin, and any condition resulting in hyperglycemia stimulates the pancreas to secrete greater quantities of this hormone. To control diabetes, insulin must be injected into the muscle tissue daily. In view of the discomfort and inconvenience to the patient, many efforts have been made to prolong the action of insulin and to develop drugs with insulin-like activity that can be taken orally. In 1942 it was observed that a sulfamidothiazole compound exhibited a potent hypoglycemic effect. Related compounds were tested, and by 1955 sulfonylurea derivatives such as **tolbutamide** (Orinase) were available as antidiabetic agents. Guanidine derivatives also produced hypoglycemia, and phenethylbiguanide (Phenformin) represents another type of antidiabetic drug. When administered orally these two drugs exhibit different properties.

Orinase (Tolbutamide)   Phenformin (Phenethylbiguanide)

Orinase stimulates the secretion of insulin from the beta cells, while Phenformin stimulates the oxidation of glucose by the peripheral tissues. Both of the drugs are effective in the treatment of diabetic patients over 40 years of age with stable and mild diabetes. Their use is ineffective in young unstable diabetics, especially those prone to ketoacidosis. In patients that do not respond to Orinase alone, the concurrent administration of both drugs is often effective. It has also been found that diabetics with congestive heart failure or severe renal insufficiency should not be treated with either of these drugs.

URICOSURIC DRUGS IN GOUT

Gout is a chronic disease that is characterized by an increased level of uric acid in the blood, hyperuricemia, acute episodes of gouty arthritis, and degenerative changes in the joints. The disease is seen predominantly in men and the initial stages occur in their forties. In the early days, gout appeared to be more prevalent among the aristocracy, and at present, professors and clinicians exhibit the greatest incidence of the disease; however, even vegetarians and lower economic groups show a predilection for the disease.

In the treatment of the disease an attempt is made to decrease the level of uric acid in the blood and in the body stores. **Probenecid** (Benemid), a derivative of benzoic acid, was found to inhibit the enzymes responsible for the reabsorption of uric acid by the kidney. It is a safe drug that lowers serum uric acid by about 50 per cent within two to four days and maintains the reduced level as long as therapy is continued. A more recent therapeutic agent is **allopurinol** that inhibits the enzyme xanthine oxidase and thus reduces the formation of uric acid from its immediate precursors, hypoxanthine and xanthine. Treatment with allopurinol maintains the blood uric acid level at a normal value

Benemid   Allopurinol

and lowers the body pool of uric acid by constant excretion through the kidney. Both of these drugs were first "tailor-made" by pharmacologists for a specific biochemical purpose. Probenecid was developed to decrease the renal tubular secretion of penicillin, and was found to be very effective in decreasing the reabsorption of uric acid by the tubules. Allopurinol was synthesized to serve as a potent inhibitor of xanthine oxidase to prevent the oxidation of 6-mercaptopurine and thus extend its effective action in the treatment of leukemia.

TOPIC OF CURRENT INTEREST

DRUGS IN THE TREATMENT OF CANCER

Cancer is a general term used to describe rapid multiplication of cells and increased growth of certain tissues in the body. Cancers are also called **tumors** and **neoplasms** and are classed as malignant when they spread to other parts of the body and exhibit recurrence of growth after surgery. **Leukemia** is a cancer of blood-forming tissues characterized by abnormal leucocytes or white blood cells; **carcinoma** involves epithelial cells; and **sarcoma** is a tumor of muscle or connective tissue. Considerable research, effort, and money have been expended in extensive studies of cancer in recent years. Tumors in experimental animals have been produced by **carcinogenic agents** such as dimethylbenzanthracene and have been transplanted to aid in the study of the metabolism of cancer cells and the development of therapeutic agents. The presence of possible carcinogenic agents in food additives, public water supplies, and plastics is presently being investigated by environmental pollution specialists and toxicologists. Although surgical removal of tumors and irradiation with x-rays and radioisotopes constitute two major modes of attack on cancer, several classes of chemical compounds, including alkylating agents, antimetabolites, and purine and pyrimidine analogs, show considerable promise in cancer treatment.

Nitrogen-containing compounds related to mustard gas (dichloroethyl sulfide) are alkylating agents that have been used in cancer therapy for several years. To overcome undesirable side effects of these drugs, **cyclophosphamide** was synthesized; it is a cytotoxic agent that has shown good initial results in the treatment of Hodgkin's disease and lymphosarcoma. When used with other drugs, cyclophosphamide has produced complete remissions of acute lymphoblastic leukemia in children for more than five years. An example of an antimetabolite used in therapy is **methotrexate,** a folic acid analog. This drug produces dramatic temporary remissions of leukemia in children and long-lasting remissions in choriocarcinoma (an epithelial tumor that occurs in the uterus at the placental site during pregnancy).

The purine analog **6-mercaptopurine** is believed to suppress the biosynthesis of purines within the cell, thus interfering with the production of RNA and DNA in the tumor cells. This drug, when used in combination with other drugs, is the most

6-Mercaptopurine   Cyclophosphamide

Methotrexate   Fluorouracil valuable purine analog for the treatment of acute leukemia. **Fluorouracil,** a pyrimidine analog, is particularly effective in the treatment of advanced carcinoma, especially of the breast and the gastrointestinal tract.

The structures of some of the cytotoxic drugs used in the treatment of cancer are shown on page 397.

Compounds extracted from natural products, such as **vinblastine** and **vincristine** from the periwinkle plant, are also used in cancer therapy. Vinblastine, a complex alkaloid, has been found to be effective in the treatment of choriocarcinoma in women. An interesting development in the chemotherapy of cancer is the discovery that the enzyme L-asparaginase is an effective agent in the treatment of leukemia.

At present it is difficult to assess the long-range benefits of chemotherapy in cancer. Millions of dollars and uncounted hours are being expended in the search for improved therapy, possible prevention, and ultimately, a cure for cancer.

IMPORTANT TERMS AND CONCEPTS

acetylcholinesterase	antifertility drugs	heroin	sulfa drugs
analgesic	antihistamine	hypoglycemic drugs	tranquilizers
antibiotics	cytotoxic drugs	marijuana	

QUESTIONS

1. Why can it be stated that all disease has a biochemical basis? Explain.

2. Why is aspirin used in so many proprietary drug preparations? Explain.

3. Why are amphetamine-type drugs used in such large quantities today? Explain.

4. Isocarboxazid is an example of a monoamine oxidase inhibitor. Describe the usefulness of the drug.

5. Why are modifications of the phenothiazine structure important therapeutically? Explain.

6. Compare meprobamate and the barbiturates on the basis of their chemical structure, potency, and mode of action.

7. Give two examples of a hallucinogenic drug. Should the use of these drugs be controlled by law? Explain.

8. Discuss the chemical nature and action of L-dopa in Parkinson's disease.

9. Outline the main uses of antihistaminic drugs.

10. Explain the differences and similarities of sulfa drugs and antibiotics such as penicillin.

11. Give an example of a steroid drug used in the treatment of arthritis and explain its action.

12. Describe the mechanism of action of the antifertility drugs.

13. Explain the action of one oral insulin substitute and its advantages.

14. Compare the mechanism of action of Benemid and allopurinol in the treatment of gout.

15. Why is 6-mercaptopurine used in the treatment of leukemia?

16. Compare the chemical structure of allopurinol and 6-mercaptopurine. Should allopurinol be effective in the treatment of leukemia? Explain.

SUGGESTED READING ━━━━━━━━━━━━━━━━━━━━━━━━━━━

Axelrod: Neurotransmitters. Scientific American, Vol. 230, No. 6, p. 58, 1974.

Barron, Jarvick, and Bunnell: The Hallucinogenic Drugs. Scientific American, Vol. 210, No. 4, p. 29, 1964.

Berelson, and Freedman: A Study in Fertility Control. Scientific American, Vol. 210, No. 5, p. 29, 1964.

Bogue: Drugs of the Future. Journal of Chemical Education, Vol. 46, p. 468, 1969.

Braun: The Reversal of Tumor Growth. Scientific American, Vol. 213, No. 5, p. 75, 1965.

Frei and Frereich: Leukemia. Scientific American, Vol. 210, No. 5, p. 88, 1964.

Gates: Analgesic Drugs. Scientific American, Vol. 215, No. 5, p. 131, 1966.

Grinspoon: Marijuana. Scientific American, Vol. 221, No. 6, p. 17, 1969.

Hansch: Drug Research or the Luck of the Draw. Journal of Chemical Education, Vol. 51, No. 6, p. 360, 1974.

Nares and Strickland: Barbiturates. Chemistry, Vol. 47, No. 3, p. 15, 1974.

Nichols: How Opiates Change Behavior. Scientific American, Vol. 212, No. 2, p. 80, 1965.

Weeks: Experimental Narcotic Addiction. Scientific American, Vol. 210, No. 3, p. 46, 1964.

TABLE OF
LOGARITHMS

TABLE OF LOGARITHMS

	0	1	2	3	4	5	6	7	8	9
1.0	.0000	.0043	.0086	.0128	.0170	.0212	.0253	.0294	.0334	.0374
1.1	.0414	.0453	.0492	.0531	.0569	.0607	.0645	.0682	.0719	.0755
1.2	.0792	.0828	.0864	.0899	.0934	.0969	.1004	.1038	.1072	.1106
1.3	.1139	.1173	.1206	.1239	.1271	.1303	.1335	.1367	.1399	.1430
1.4	.1461	.1492	.1523	.1553	.1584	.1614	.1644	.1673	.1703	.1732
1.5	.1761	.1790	.1818	.1847	.1875	.1903	.1931	.1959	.1987	.2014
1.6	.2041	.2068	.2095	.2122	.2148	.2175	.2201	.2227	.2253	.2279
1.7	.2304	.2330	.2355	.2380	.2405	.2430	.2455	.2480	.2504	.2529
1.8	.2553	.2577	.2601	.2625	.2648	.2672	.2695	.2718	.2742	.2765
1.9	.2788	.2810	.2833	.2856	.2878	.2900	.2923	.2945	.2967	.2989
2.0	.3010	.3032	.3054	.3075	.3096	.3118	.3139	.3160	.3181	.3201
2.1	.3222	.3243	.3263	.3284	.3304	.3324	.3345	.3365	.3385	.3404
2.2	.3424	.3444	.3464	.3483	.3502	.3522	.3541	.3560	.3579	.3598
2.3	.3617	.3636	.3655	.3674	.3692	.3711	.3729	.3747	.3766	.3784
2.4	.3802	.3820	.3838	.3856	.3874	.3892	.3909	.3927	.3945	.3962
2.5	.3979	.3997	.4014	.4031	.4048	.4065	.4082	.4099	.4116	.4133
2.6	.4150	.4166	.4183	.4200	.4216	.4232	.4249	.4265	.4281	.4298
2.7	.4314	.4330	.4346	.4362	.4378	.4393	.4409	.4425	.4440	.4456
2.8	.4472	.4487	.4502	.4518	.4533	.4548	.4564	.4579	.4594	.4609
2.9	.4624	.4639	.4654	.4669	.4683	.4698	.4713	.4728	,4742	.4757
3.0	.4771	.4786	.4800	.4814	.4829	.4843	.4857	.4871	.4886	.4900
3.1	.4914	.4928	.4942	.4955	.4969	.4983	.4997	.5011	.5024	.5038
3.2	.5051	.5065	.5079	.5092	.5105	.5119	.5132	.5145	.5159	.5172
3.3	.5185	.5198	.5211	.5224	.5237	.5250	.5263	.5276	.5289	.5302
3.4	.5315	.5328	.5340	.5353	.5366	.5378	.5391	.5403	.5416	.5428
3.5	.5441	.5453	.5465	.5478	.5490	.5502	.5514	.5527	.5539	.5551
3.6	.5563	.5575	.5587	.5599	.5611	.5623	.5635	.5647	.5658	.5670
3.7	.5682	.5694	.5705	.5717	.5729	.5740	.5752	.5763	.5775	.5786
3.8	.5798	.5809	.5821	.5832	.5843	.5855	.5866	.5877	.5888	.5899
3.9	.5911	.5922	.5933	.5944	.5955	.5966	.5977	.5988	.5999	.6010
4.0	.6021	.6031	.6042	.6053	.6064	.6075	.6085	.6096	.6107	.6117
4.1	.6128	.6138	.6149	.6160	.6170	.6180	.6191	.6201	.6212	.6222
4.2	.6232	.6243	.6253	.6263	.6274	.6284	.6294	.6304	.6314	.6325
4.3	.6335	.6345	.6355	.6365	.6375	.6385	.6395	.6405	.6415	.6425
4.4	.6435	.6444	.6454	.6464	.6474	.6484	.6493	.6503	.6513	.6522
4.5	.6532	.6542	.6551	.6561	.6571	.6580	.6590	.6599	.6609	.6618
4.6	.6628	.6637	.6646	.6656	.6665	.6675	.6684	.6693	.6702	.6712
4.7	.6721	.6730	.6739	.6749	.6758	.6767	.6776	.6785	.6794	.6803
4.8	.6812	.6821	.6830	.6839	.6848	.6857	.6866	.6875	.6884	.6893
4.9	.6902	.6911	.6920	.6928	.6937	.6946	.6955	.6964	.6972	.6981
5.0	.6990	.6998	.7007	.7016	.7024	.7033	.7042	.7050	.7059	.7067
5.1	.7076	.7084	.7093	.7101	.7110	.7118	.7126	.7135	.7143	.7152
5.2	.7160	.7168	.7177	.7185	.7193	.7202	.7210	.7218	.7226	.7235
5.3	.7243	.7251	.7259	.7267	.7275	.7284	.7292	.7300	.7308	.7316
5.4	.7324	.7332	.7340	.7348	.7356	.7364	.7372	.7380	.7388	.7396
5.5	.7404	.7412	.7419	.7427	.7435	.7443	.7451	.7459	.7466	.7474
5.6	.7482	.7490	.7497	.7505	.7513	.7520	.7528	.7536	.7543	.7551
5.7	.7559	.7566	.7574	.7582	.7589	.7597	.7604	.7612	.7619	.7627
5.8	.7634	.7642	.7649	.7657	.7664	.7672	.7679	.7686	.7694	.7701
5.9	.7709	.7716	.7723	.7731	.7738	.7745	.7752	.7760	.7767	.7774

TABLE OF LOGARITHMS (Continued)

	0	1	2	3	4	5	6	7	8	9
6.0	.7782	.7789	.7796	.7803	.7810	.7818	.7825	.7832	.7839	.7846
6.1	.7853	.7860	.7868	.7875	.7882	.7889	.7896	.7903	.7910	.7917
6.2	.7924	.7931	.7938	.7945	.7952	.7959	.7966	.7973	.7980	.7987
6.3	.7993	.8000	.8007	.8014	.8021	.8028	.8035	.8041	.8048	.8055
6.4	.8062	.8069	.8075	.8082	.8089	.8096	.8102	.8109	.8116	.8122
6.5	.8129	.8136	.8142	.8149	.8156	.8162	.8169	.8176	.8182	.8189
6.6	.8195	.8202	.8209	.8215	.8222	.8228	.8235	.8241	.8248	.8254
6.7	.8261	.8267	.8274	.8280	.8287	.8293	.8299	.8306	.8312	.8319
6.8	.8325	.8331	.8338	.8344	.8351	.8357	.8363	.8370	.8376	.8382
6.9	.8388	.8395	.8401	.8407	.8414	.8420	.8426	.8432	.8439	.8445
7.0	.8451	.8457	.8463	.8470	.8476	.8482	.8488	.8494	.8500	.8506
7.1	.8513	.8519	.8525	.8531	.8537	.8543	.8549	.8555	.8561	.8567
7.2	.8573	.8579	.8585	.8591	.8597	.8603	.8609	.8615	.8621	.8627
7.3	.8633	.8639	.8645	.8651	.8657	.8663	.8669	.8675	.8681	.8686
7.4	.8692	.8698	.8704	.8710	.8716	.8722	.8727	.8733	.8739	.8745
7.5	.8751	.8756	.8762	.8768	.8774	.8779	.8785	.8791	.8797	.8802
7.6	.8808	.8814	.8820	.8825	.8831	.8837	.8842	.8848	.8854	.8859
7.7	.8865	.8871	.8876	.8882	.8887	.8893	.8899	.8904	.8910	.8915
7.8	.8921	.8927	.8932	.8938	.8943	.8949	.8954	.8960	.8965	.8971
7.9	.8976	.8982	.8987	.8993	.8998	.9004	.9009	.9015	.9020	.9026
8.0	.9031	.9036	.9042	.9047	.9053	.9058	.9063	.9069	.9074	.9079
8.1	.9085	.9090	.9096	.9101	.9106	.9112	.9117	.9122	.9128	.9133
8.2	.9138	.9143	.9149	.9154	.9159	.9165	.9170	.9175	.9180	.9186
8.3	.9191	.9196	.9201	.9206	.9212	.9217	.9222	.9227	.9232	.9238
8.4	.9243	.9248	.9253	.9258	.9263	.9269	.9274	.9279	.9284	.9289
8.5	.9294	.9299	.9304	.9309	.9315	.9320	.9325	.9330	.9335	.9340
8.6	.9345	.9350	.9355	.9360	.9365	.9370	.9375	.9380	.9385	.9390
8.7	.9395	.9400	.9405	.9410	.9415	.9420	.9425	.9430	.9435	.9440
8.8	.9445	.9450	.9455	.9460	.9465	.9469	.9474	.9479	.9484	.9489
8.9	.9494	.9499	.9504	.9509	.9513	.9518	.9523	.9528	.9533	.9538
9.0	.9542	.9547	.9552	.9557	.9562	.9566	.9571	.9576	.9581	.9586
9.1	.9590	.9595	.9600	.9605	.9609	.9614	.9619	.9624	.9628	.9633
9.2	.9638	.9643	.9647	.9652	.9657	.9661	.9666	.9671	.9675	.9680
9.3	.9685	.9689	.9694	.9699	.9703	.9708	.9713	.9717	.9722	.9727
9.4	.9731	.9736	.9741	.9745	.9750	.9754	.9759	.9763	.9768	.9773
9.5	.9777	.9782	.9786	.9791	.9795	.9800	.9805	.9809	.9814	.9818
9.6	.9823	.9827	.9832	.9836	.9841	.9845	.9850	.9854	.9859	.9863
9.7	.9868	.9872	.9877	.9881	.9886	.9890	.9894	.9899	.9903	.9908
9.8	.9912	.9917	.9921	.9926	.9930	.9934	.9939	.9943	.9948	.9952
9.9	.9956	.9961	.9965	.9969	.9974	.9978	.9983	.9987	.9991	.9996

ANSWERS TO QUESTIONS

Chapter 1

1. Thermal, electrical, chemical, sonic, light, and energy of motion. Thermal energy or heat is the easiest to measure.
3. $80,000
5. 104°F
7. Because atoms individually are too small to weigh and thus an expression of their relative weights is used.
9. (a) heterogeneous (b) homogeneous (c) heterogeneous (d) heterogeneous
11. PCl_5, 208.5 amu; $Al(C_2H_5)_3$, 114 amu
13. 4.89 mol Ca
15. 3.740×10^{24} molecules NH_3
17. (a) NO_2 (b) C_5H_4 (c) FeC_4O_4 or $Fe(CO)_4$ (d) $Al_2S_3O_{12}$ or $Al_2(SO_4)_3$

Chapter 2

1. The electron has a charge of 4.8×10^{-10} electrostatic units as does the proton, but the charges are opposite so a -1 charge was assigned to the electron and a $+1$ charge to the proton.
3. Add the number of protons to the number of neutrons in the nucleus to obtain the atomic weight. The number of electrons equals the atomic number.
5. 19 amu; 9 electrons; $:\overset{\cdot\cdot}{\underset{\cdot}{X}}\cdot$
7. $:\overset{\cdot\cdot}{\underset{\cdot}{F}}\cdot + 1e^- \rightarrow [:\overset{\cdot\cdot}{\underset{\cdot\cdot}{F}}:]^{-1};$ Na$\cdot \rightarrow$ Na$^+$ + 1e$^-$
8. As atomic size increases within a family of elements, electron affinity decreases.
11. Atomic number. Electron configuration, which gives rise to specific properties, varies periodically with atomic number.
13. The grouping of 8 electrons in the outer shell of the structure is common to both ions and inert gases.
15. (a) upper right corner   (b) upper right corner
    (c) left side   (d) right side
    (e) in the vertical column at the extreme right
17. When a halogen atom gains an electron to form a negative ion, it attains the electronic configuration of an inert gas giving additional stability. Consequently, this process is highly favored. All the halogens show this tendency, which is illustrated in general for the atomic halogens as follows:

$$:\overset{\cdot\cdot}{\underset{\cdot\cdot}{X}}\cdot + 1e^- \rightarrow :\overset{\cdot\cdot}{\underset{\cdot\cdot}{X}}:^{-1}$$

19. The time necessary for half the weight of a sample of a radioactive element to disintegrate.
21. When uranium, $^{238}_{92}$U, was bombarded with neutrons, elements with higher atomic weights were produced.
23. (a) $^1_0$n (b) $^1_1$H (c) $^{214}_{83}$Bi
25. 1.14×10^8 kcal

Chapter 3

1. The valence electrons are those in the outer shell of the atom, or, more specifically, those in the outer shells.
3. In covalent compounds the electrons are not always shared equally between two atoms with a resultant unequal electric charge distribution in the covalent bond. Even though the molecule is electrically neutral, the charge distribution produces a dipole whose ends are oppositely charged.
5.

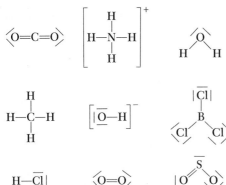

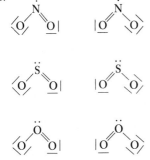

7. (a) triangular planar
   (b) tetrahedral
   (c) square-based pyramid
   (d) trigonal bipyramidal
9. (a) triple carbon-carbon bond   (b) double nitrogen-nitrogen bond
   (c) only single bonds
11. (a) 2   (b) 4   (c) 3   (d) 3   (e) 1   (f) 1   (g) 2   (h) 0
13.

Chapter 4

2. The impact of the molecules on the walls of a container accounts for the pressure exerted by a gas.
3. At higher elevations where the pressure is less than one atmosphere, a balloon has greater volume than at sea level. This is predicted by Boyle's law.
5. 956 ml
7. 1.25 g N_2/l
9. (a) 7.39 atm   (b) 0.10 mol   (c) 15.16 l
12. Because molecules continually enter and leave the liquid but the pressure of the vapors remains constant.
15. (a) large polarizability   (b) large atoms
   (c) low boiling points   (d) free flowing
17. Increase
18. False. This statement is false because the boiling point of any material is determined by

the pressure of the surrounding atmosphere. Water boils at 100°C only at a prevailing pressure of one atmosphere.

19. They play a major role since the enthalpy of fusion or vaporization is largely the energy which must be given to molecules in order that they can overcome intermolecular forces.

Chapter 5

3. Hydrochloric acid   Perchloric acid
     Phosphoric acid   Sulfurous acid
     Nitrous acid   Nitric acid
     Sulfuric acid   Calcium hydroxide
     Ammonium hydroxide   Carbonic acid

5. (a) oxidation-reduction
     (b) metathesis
     (c) oxidation-reduction
     (d) metathesis

6. (a) acid   (b) base   (c) base   (d) acid   (e) acid

8.

	Reducing Agent	Oxidizing Agent
(a)	Zn	H_2SO_4
(b)	I^-	Fe^{+3}
(c)	H_2S	HNO_3
(d)	Ca	H_2O

9. (a) $+4$ (b) $+3$ (c) $+5$ (d) $+7$ (e) $+6$ (f) $+3$

Chapter 6

3. Hydrolysis is the process of splitting a compound into two parts with the help of water. For example, ammonium carbonate would be hydrolyzed as shown:

$$(NH_4)_2CO_3 + 2\,HOH \rightleftarrows 2\,NH_4OH + H_2CO_3$$

4. Crystalline copper sulfate $CuSO_4 \cdot 5H_2O$ has five molecules of water of crystallization loosely attached to the molecule while anhydrous copper sulfate $CuSO_4$ has none.

5. (a) By boiling it for 10 to 15 minutes.
     (b) By the process of distillation.

7. (a) Temporary hardness in water may be removed by boiling the water.
     (b) Permanent hardness requires the addition of chemical compounds such as sodium carbonate that will convert the soluble calcium, magnesium, or iron salts into insoluble compounds which may then be removed by filtration.

8. Colloidal solution; c and e.
     True solution; a, b, and c.
     Suspension; d and f.

10. The colligative properties of electrolytes reflect the presence of ions in their solutions. For example, the freezing point depression for a 1 M solution of NaCl is almost twice that for a solution of glucose ($-3.31°$ versus $-1.86°$).

11. 70 g glucose

12. 297 g K_3PO_4

14. $HA \rightarrow H^+ + A^-$ Strong Acid
     $HA \rightleftarrows H^+ + A^-$ Weak Acid
     The strong acid is effectively 100 per cent dissociated, whereas the weak acid is only partially dissociated.

16. The nature of the solute and solvent, temperature of the solution.

18. 0.0467

20. 0.288 M

22. The size of the particles that make up the solution or "guest" phase.

23. The Tyndall effect is exhibited by colloidal particles when a strong beam of light is passed through a colloidal solution. The light is reflected from the surface of the dispersed particles outlining the path of the beam. The particles in a true solution are completely dissolved and cannot reflect the light beam.

Chapter 7

1. High activation energies.
3. This is because a larger fraction of the collisions involves molecules which have sufficient energy to form the transition state.
5. c, e
6. $18.8 \dfrac{l^2}{mol^2}$

8. (a) $K_i = \dfrac{[K^+][Cl^-]}{[KCl]}$ (b) $K_i = \dfrac{[H^+][CH_3CO_2^-]}{[CH_3CO_2H]}$

   (c) $K_i = \dfrac{[Ca^{+2}][OH^-]^2}{[Ca(OH)_2]}$ (d) $K_i = \dfrac{[Ba^{+2}][SO_4^{-2}]}{[BaSO_4]}$

10. 4×10^{-14} M
11. (a) 6   (b) 2.60   (c) 0.301   (d) 7.38   (e) -1 (f) 1.01
13. The extent of dissociation of an acid or base decreases as the strength decreases. Ka and Kb become smaller as acid and base strength decrease.
14. b, d, f, g, h, i
16. 3.75
17. The pH of a neutral aqueous solution at $25°C$ is 7, whereas that of a basic solution is greater than 7 and that of an acidic solution is less than 7.
20. b
21. 31.7 g Cu
23. Concentrations, temperature, and all other factors which influence equilibrium constants.
24. When charging, the lead storage cell is acting as an electrolytic cell; whereas it acts as a voltaic cell when discharging.

Chapter 8

2. a, c, f, h
3. $4HCl(aq) + MnO_2(s) \rightleftarrows 2H_2O + MnCl_2(aq) + Cl_2(g)$
5. $NaF(s) + H_2SO_4(l) \rightarrow Na_2SO_4(s) + HF(g)$
   $NaCl(s) + H_2SO_4(l) \rightarrow Na_2SO_4(s) + HCl(g)$
   $NaBr(s) + H_3PO_4(l) \rightarrow NaH_2PO_4(s) + HBr(g)$
   $NaI(s) + H_3PO_4(l) \rightarrow NaH_2PO_4(s) + HI(g)$
6. Sulfuric acid is manufactured most commonly by the contact process; the reactions of which are as follows:

$$S(s) + O_2(g) \rightleftarrows SO_2(g)$$

$$SO_2(g) + O_2(g) \xrightarrow[400-500°C]{catalyst} SO_3(g)$$

$$SO_3(g) + H_2SO_4(aq)\ 98\% \rightleftarrows H_2S_2O_7(l)$$

$$H_2S_2O_7(l) + H_2O \rightarrow H_2SO_4(aq)$$

8. Metal amides of the general formula MNH_2 and hydrogen.
9. Nitrous acid HNO_2
   Nitric acid HNO_3
11. The two crystalline allotropes of carbon are graphite and diamond (see Fig. 8–7).
12. Because of its tendency to accept 2 electrons from both nonmetals and metals.
14. Because it can act as a metal and readily lose an electron or as a nonmetal and gain an electron.

Chapter 9

1. The three subdivisions of the metals are representative, transition, and innertransition. The representative metals have either empty or completely filled d-orbitals and partially filled s- and p-orbitals. The transition metals have from one to 10 electrons in d-orbitals and no

valence shell p-orbitals. Innertransition metals have a partially filled or completely filled set of f-orbitals.

3. Metals in general are found in chemical combination in nature but a few occur in the elemental states. These include Cu, Ag, Pt, Au, and Hg.

5. Electrolytic processes, melting and recrystallization, or distillation.

8. a, c, e, h

9. (a) y   (b) w   (c) z   (d) x

10. Acidic $MO_2(S) + 4H^+(aq) + 4H_2O(l) \rightleftharpoons [M(H_2O)_6]^{+4}(aq)$
    Basic $MO_2(S) + 2OH^-(aq) + 2H_2O(l) \rightleftharpoons [M(OH)_6]^{-2}(aq)$

Chapter 10

1. a and d, b and e

3.
$$\text{H}:\overset{\overset{\displaystyle H}{\cdot\cdot}}{\underset{\underset{\displaystyle H}{\cdot\cdot}}{\text{C}}}\cdot\cdot\overset{\overset{\displaystyle H}{\cdot\cdot}}{\underset{\underset{\displaystyle H}{\cdot\cdot}}{\text{C}}}\cdot\cdot\overset{\overset{\displaystyle H}{\cdot\cdot}}{\underset{\underset{\displaystyle H}{\cdot\cdot}}{\text{C}}}\cdot\cdot\overset{\overset{\displaystyle H}{\cdot\cdot}}{\underset{\underset{\displaystyle H}{\cdot\cdot}}{\text{C}}}\cdot\cdot\text{H}$$

5. a; c; d; e

7. (a) $-\overset{\overset{\displaystyle O}{\|}}{C}-$, ketone   (b) no functional group, alkane   (c) $-C-O-C-$, ether
   (d) $-C=C-$, alkene   (e) $-NH_2$, amine   (f) $-OH$, alcohol
   (g) $-COOH$, carboxylic acid   (h) $-\overset{\overset{\displaystyle O}{\|}}{C}-$, ketone   (i) $-Br$, alkyl halide
   (j) $-C\equiv C-$, alkyne

Chapter 11

1. Five isomers.

3. (a)

   (b) $CH_3CH_2CH_2CH[CH(CH_3)_2]CH(CH_3)CH_2CH_2CH_2CH_2CH_3$
   (c) $CH_3(CH_2)_5CH_3$ (d) $(H_3C)_3CBr$
   (e) $CH_3CH_2CH(CH_3)CH(CH_2CH_3)CH[CH(CH_3)_2]CH_2CH_2CH_3$

5. (a) Longest chain contains six carbons; 3-ethylhexane
   (b) Longest chain contains five carbons; 2,3-dimethylpentane
   (c) Substituents do not have the lowest numbers; 2-methyl-3-isopropylhexane
   (d) Substituent numbers must be repeated; 2,2-dimethylbutane
   (e) Substituents do not have the lowest numbers; 1,1,3-trimethylcyclobutane

Chapter 12

1. (a) $\begin{array}{c} H_3C \\ \end{array} \overset{\displaystyle C=C}{\underset{}{}} \begin{array}{c} H \\ CH_2CH_2CH_3 \end{array}$
   (b) $\begin{array}{c} H_3C \\ Cl \end{array} C=C \begin{array}{c} CH_3 \\ Cl \end{array}$

   (c)

   (d)

(e) $H_2C=CHCH_2CH(CH_2CH_3)CH_2CH_2CH_2CH_3$ (f) $CH_3CH_2C\equiv CCH_2CH_3$

(g)

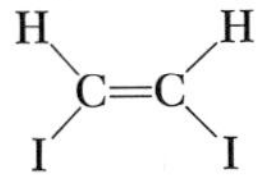

(h) $CH_3C(CH_3)=CHCH_3$

3. Six straight chain isomers and five cyclic isomers: 1-pentene; *cis-* and *trans-*2-pentene; 2-methyl-1-butene; 3-methyl-1-butene; 2-methyl-2-butene; cyclopentane; methylcyclobutane; *cis-* and *trans-*1,2-dimethylcyclopropane; 1,1-dimethylcyclopropane

5. (a) Double bond must be given lowest possible number; 2-methyl-2-pentene
   (b) Substituents must be given lowest possible number; 3-methylcyclohexene
   (c) Longest chain contains four carbons; 2,3-dimethyl-1-butene
   (d) Substituents must be given lowest possible number; 3,5-dimethylcyclohexene
   (e) Double bond must be given lowest possible number; *cis-*3-heptene

8. (a) $CH_3CHICH_3 + KOH + \text{alcohol} \rightarrow CH_3CH=CH_2$

   (b) $CH_3CHICH_3 + KOH + \text{alcohol} \rightarrow CH_3CH=CH_2 \xrightarrow{Br_2}$

   $CH_3CHBrCH_2Br \xrightarrow{KOH} CH_3C\equiv C^-K^+ \xrightarrow{H^+} CH_3C\equiv CH$

   (c) $CH_3CHICH_3 + KOH + \text{alcohol} \rightarrow CH_3CH=CH_2 \xrightarrow{HCL} CH_3CHClCH_3$

   (d) $CH_3C\equiv CH$ (as in b) $\xrightarrow{2HBr} CH_3CBr_2CH_3$

Chapter 13

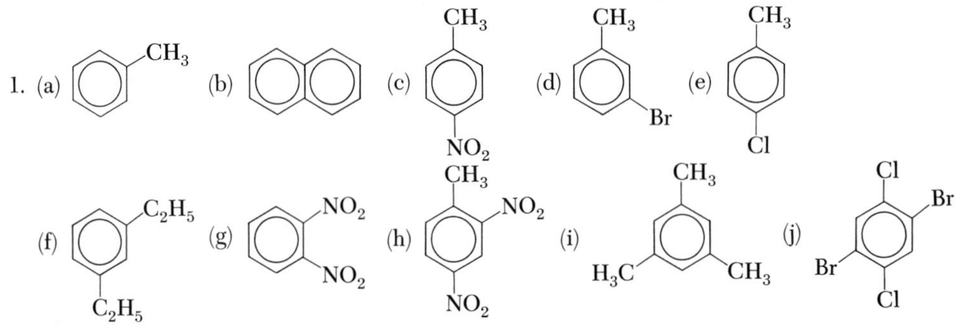

1. (a)

3. (a) Isopropyl benzene (b) *Ortho-* and *para-*bromotoluene (c) *ortho-* and *para-*methylbenzenesulfonic acid (d) *Meta-*dinitrobenzene (e) *Ortho-* and *para-*bromochlorobenzene (f) *Meta-*bromobenzenesulfonic acid (g) *Ortho-* and *para-*nitrotoluene (h) *Ortho-* and *para-*ethylbenzenesulfonic acid

5. Three

Chapter 14

1. Eight isomers: 1-pentanol; 2-pentanol; 3-pentanol; 3-methyl-1-butanol; 3-methyl-2-butanol; 2-methyl-2-butanol; 2-methyl-1-butanol; 2,2-dimethyl-1-propanol

3. (a) 4-methyl-1-pentanol   (b) 3-ethyl-3-pentanol
   (c) 1-methylcyclohexanol   (d) 3-methyl-5-ethyl-4-heptanol
   (e) *meta-*nitrophenol   (f) 1,2,3-trihydroxybutane
   (g) 3-bromo-4-methyl-1-pentanol

5. (a) $(CH_3)_2CHCH_2OH$ (b) $CH_3CHOHCH_3$ (c) $(CH_3)_3COH$

   (d) $CH_3CH_2CHOHCH_2CH_3$ (e)

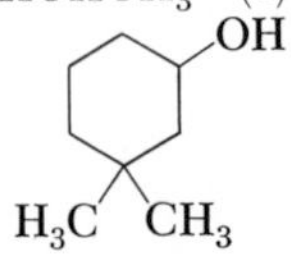

(f) 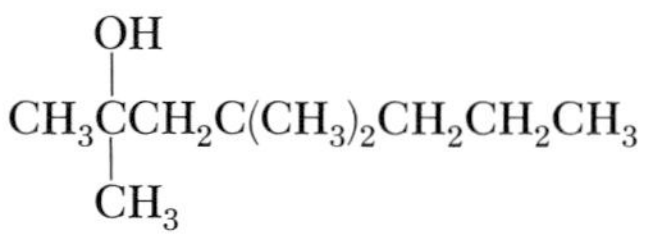 $CH_3\overset{\underset{\displaystyle CH_3}{|}}{\underset{}{C}}CH_2C(CH_3)_2CH_2CH_2CH_3$ (g)

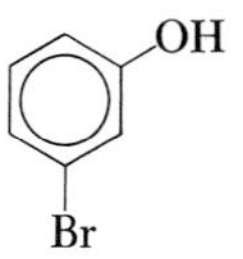

(h) $BrCH_2CH(CH_3)CHOHCH_2CH_2CH_3$ (i)

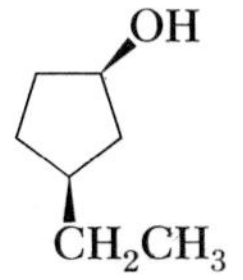

(j) 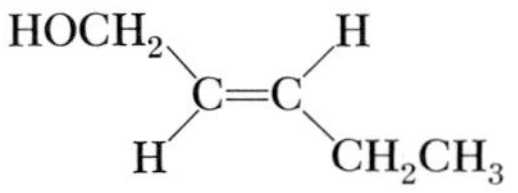 $\underset{H}{\overset{HOCH_2}{\diagdown}}C=C\underset{CH_2CH_3}{\overset{H}{\diagup}}$

6. (a) Is oxidized to a carboxylic acid
   (b) Is oxidized to a ketone
   (c) Is *not* easily oxidized
   (d) Is oxidized to a carboxylic acid

8. (a) $CH_3OCH(CH_3)_2$
   (b) 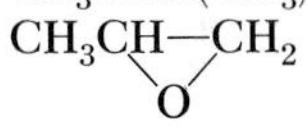 $CH_3CH\!-\!CH_2$ over O

   (c)

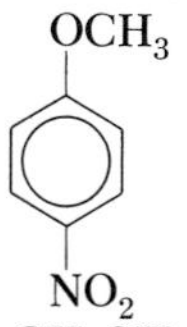

   (d) $CH_3CH_2OCH_2CH_3$
   (e) $CH_3CHBrCH_2CH(OCH_2CH_3)CH_2CH_3$

10. (a) $C_6H_5OCH_3 + HI$ (excess) $\xrightarrow{\text{heat}} C_6H_5OH + CH_3I$
    (b) $C_6H_5OCH_3 + KMnO_4 \rightarrow$ No reaction
    (c) $C_6H_5OCH_3 + Na \rightarrow$ No reaction
    (d) $C_6H_5OCH_3 + HNO_3 + H_2SO_4 \rightarrow$ *ortho-* and *para*-nitrophenols

12. (a)  $CH_2\!-\!CH_2$ over O $\xrightarrow[\text{H+}]{CH_3CH_2CH_2OH} HOCH_2CH_2OCH_2CH_2CH_3$

    (b) $\xrightarrow{CH_3O\overset{-+}{N}a}$ $HOCH_2CH_2OCH_3$

    (c) $\xrightarrow{CH_3NH_2}$ $HOCH_2CH_2NHCH_3$

    (d) $\xrightarrow{HOCH_2CH_2OH}$ $HOCH_2CH_2OCH_2CH_2OH$

    (e) $\xrightarrow{HOCH_2CH_2NH_2}$ $HOCH_2CH_2NHCH_2CH_2OH$

Chapter 15

1. (a) $(CH_3)_3CBr$

   (b)

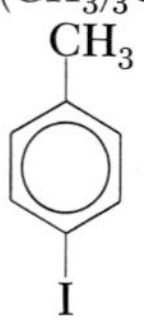

   (c)

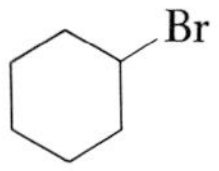

   (d) $CF_2ClCFCl_2$
   (e) CBr_4

3. (a) Substituents must be given the lowest possible number; 2,2,4-trichloropentane
   (b) Longest chain is six carbons; 2-chloro-4-methylhexane
   (c) Substituents must be given the lowest possible number; 1,2,4-tribromobenzene

(d) Double bond must be given the lowest possible number; *trans*-2,3-dichloro-2-pentene
(e) Substituents must be given the lowest possible number; 3-chlorotoluene

Chapter 16

1. (a) $CH_3(CH_2)_5CHO$ (b) $CH_3COCH_2CH_2CH_2CH_3$

(c) 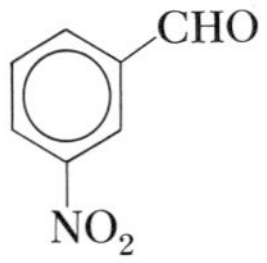   (d) 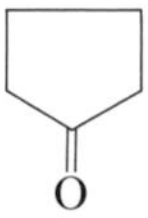   (e) $BrCH_2CH_2CH(CH_3)CH_2CHO$

(f) 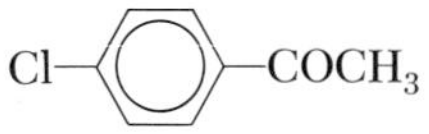   (g)

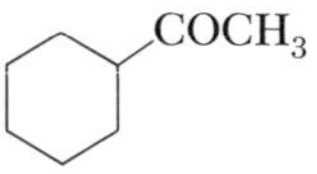

(h)

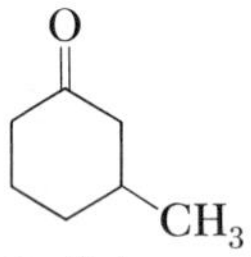

3. (a) Cu^{+2} (tartrate) complex   (b) Cu^{+2} (citrate) complex
   (c) $Ag(NH_3)_2^+$

Chapter 17

1. (a) $CH_3CH_2CO_2^-K^+$ (b) $(CH_3)_2CHCOCl$ (c) $C_6H_5CO_2CH_2CH_3$

(d) 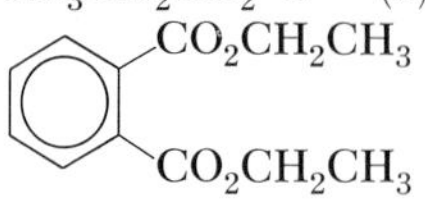   (e) $CH_3CH_2CH_2CONH_2$

(f) NaO_2CCO_2Na (g) 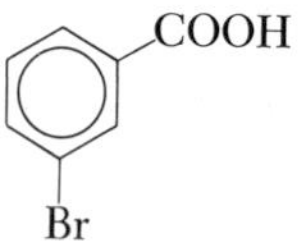   (h) HCO_2CH_3

3. $CH_3CH_2CH_2COOH$:
   (a) $CH_3CH_2OH/H^+ \rightarrow CH_3CH_2CH_2CO_2CH_2CH_3$

   (b) $SOCl_2 \rightarrow CH_3CH_2CH_2COCl$

   (c) $NH_3 \xrightarrow{heat} CH_3CH_2CH_2CONH_2$

   (d) $KOH \rightarrow CH_3CH_2CH_2CO_2^-K^+$

Chapter 18

1. (a) $(CH_3CH_2)_2NH$ (b) $(CH_3CH_2CH_2)_3N$ (c) $C_6H_5NH_2$

(d) 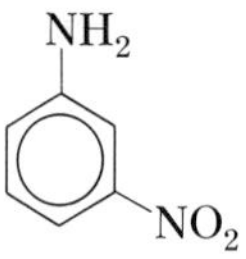   (e) $(CH_3)_2CHCH_2CH_2CH_2NH_2$

(f) 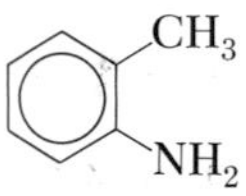   (g) 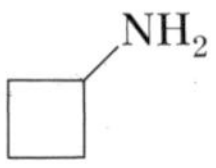   (h)

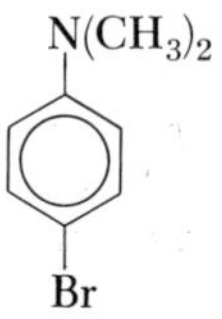

3. The acid converts the amine to the nonvolatile salt, $(CH_3)_2\overset{+}{N}H_2Cl^-$
5. $p\text{-}BrC_6H_4NH_2$:
    (a) $CH_3COCl \rightarrow p\text{-}BrC_6H_4NHCOCH_3 + HCl$
    (b) $NaNO_2 + HCl \rightarrow p\text{-}BrC_6H_4\overset{+}{N}_2Cl^-$
    (c) pdt. of (C) $+ C_6H_5N(CH_3)_2 \rightarrow p\text{-}BrC_6H_4N{=}N{-}C_6H_4N(CH_3)_2\text{-}p$
    (d) $C_6H_5SO_2Cl \rightarrow p\text{-}BrC_6H_4NHSO_2C_6H_5 \xrightarrow{\bar{O}H} [p\text{-}BrC_6H_4NSO_2C_6H_5]^-Na^+$

Chapter 19

1. An addition polymer has the same elements present in the polymer as were present in the monomer, whereas a condensation polymer repeating unit is composed of two different elements joined together via a chemical reaction.
3. $\text{---}[CF_2\text{---}CFCl]_n\text{---}$
5. See contents of Chapter 19.

Chapter 20

1. The development of the electron microscope.
3. Yes, it provides rapid access to the microsomal region for any compounds in the nucleus, such as the RNA molecules involved in protein synthesis in the cell.
5. Rough endoplasmic reticulum is characterized by the ribosomes that are adsorbed to its surface. One of the major functions of the rough form is the synthesis of protein within the cell.
7. The lysosomes contain hydrolytic enzymes that are used to digest cellular debris in the process of phagocytosis. They help clear the tissues of dead cells.

Chapter 21

1. Carbon 53 per cent, hydrogen 7 per cent, oxygen 23 per cent, nitrogen 16 per cent, and sulfur 1 per cent. Since carbohydrates and fats contain mainly carbon, hydrogen, and oxygen, the relatively high (16 per cent) concentration of nitrogen in protein serves as a point of differentiation.
3. $CH_3CHCOOH$ Name: alpha amino propionic acid (alanine)
       |
      NH_2
5. CH_2COO^-
  |
  NH_3^+
   The molecule is electrically neutral, with equal positive and negative charges. It would not migrate in solution in an electrical field and therefore represents the isoelectric point.
7. (a) First the dipeptide could be hydrolyzed with HCl to form the free amino acids. The amino acid mixture could then be subjected to paper chromatography to separate each amino acid according to its R_f value. By running known amino acids on the same strip of paper, the unknown amino acids could be identified as alanine and glycine. (b) Add Sanger's reagent, identify yellow DNP-alanine.
9. The primary structure of a protein is concerned with amino acid sequence and the peptide linkage holding the amino acids in a polypeptide chain. The tertiary structure involves the spatial relationships of the polypeptide chains and the type of bonds between these chains.
11. $R{-}CH{-}COO^- + Ag^+ \rightarrow R{-}CH{-}COOAg$
       | |
      NH_2 NH_2
   $R{-}CH{-}COOH + \text{tannic acid} \rightarrow R{-}CH{-}COOH$
       | |
      NH_3^+ $NH_3{-}\text{tannate}$

Chapter 22

1. Proteins that are characterized by their content of basic amino acids. Nucleotides, nucleosides, purines, pyrimidines, H_3PO_4, and a pentose are formed by complete hydrolysis of nucleic acids.

3. The structure of DNA, for example, consists of chains of nucleotides linked together with phosphoric acid molecules that connect carbon atom number 3 of one sugar molecule to the number 5 carbon of the next sugar.

5. The chains consist of deoxyribose nucleotides joined together by phosphate groups with the bases projecting perpendicularly from the chain into the center of the helix. Because of the base pairing, thymine to adenine with two hydrogen bonds and cytosine to guanine with three hydrogen bonds, the two chains are not identical and do not run in the same direction with respect to the linkages between deoxyribose and the base. The chains are therefore considered antiparallel.

Chapter 23

1. An enzyme is a protein, formed by a living cell, which catalyzes a reaction that is thermodynamically possible by lowering the activation energy so the reaction can take place within the cell.

3. A substrate is the compound on which an enzyme acts. Proteins are acted on by pepsin, or, in modern nomenclature, sucrose is split by sucrase.

5. The enzyme-substrate complex, ES, represents the transition state in enzyme reactions and the formation of the ES complex permits the overall reaction to proceed at a lower energy of activation.

7. Refer to Figure 23–1 on page 294. At low [S], most of the enzyme molecules would be free and not combined in the complex ES. The change in activity or velocity of the reaction would be directly proportional to the substrate concentration. As higher [S] is reached, the enzyme exists mainly in the ES complex and the velocity of the reaction is proportional to [ES] and not to [S], causing a flattening of the curve.

9. In the reaction $E + S \underset{k_{-1}}{\overset{k_1}{\rightleftharpoons}} ES \overset{k_2}{\longrightarrow} E + P$, the Michaelis constant K_m is related to k_1, k_{-1}, and k_2, the three velocity constants.

$$K_m = \frac{k_2 + k_{-1}}{k_1}$$

The K_m is approximately equal to the dissociation constant of the ES complex, and $1/K_m$ is a measure of the affinity of an enzyme for its substrate.

11. Cyanide forms compounds with metals essential for enzyme action; e.g., it can remove copper from an enzyme.

Chapter 24

1. See page 303 for structure. This compound would exhibit optical activity because the middle carbon atom is an asymmetric carbon atom.

3.

D-Fructose	L-Fructose
CH_2OH	CH_2OH
$C{=}O$	$C{=}O$
$HO{-}C^*{-}H$	$H{-}C^*{-}OH$
$H{-}C^*{-}OH$	$HO{-}C^*{-}H$
$H{-}C^*{-}OH$	$HO{-}C^*{-}H$
CH_2OH	CH_2OH

5. An aldose is a sugar that contains an aldehyde group. A hexose is a monosaccharide that contains 6 carbon atoms. A pentose is a monosaccharide that contains 5 carbon atoms. A ketose is a sugar that contains a ketone group. A disaccharide is a combination of two monosaccharides linked together by splitting out a molecule of water.

7. The OH group on carbon 1 in the pyranose and furanose ring forms (carbon 2 in fructo-

furanose), extends downward below the plane of the ring in the α isomers and upward above the plane of the ring in the β isomers.

9. Enediols and sugar acids are products formed in addition to Cu_2O. The formation of Cu_2O is important because it is a visible orange precipitate indicating the presence of a reducing sugar.

11. Refer to the structure on page 308. Since the linkage ties up the reducing groups of both glucose (carbon-1) and fructose (carbon-2), sucrose will not reduce Benedict's solution.

13. The maltose $(1,4\,\alpha)$ type of structure in amylose is responsible for the properties of starch. It is hydrolyzed by enzymes to produce water-soluble compounds that serve as a source of dietary carbohydrates and will reduce Benedict's solution. The cellobiose $(1,4\,\beta)$ type of structure in cellulose is insoluble in water, will not reduce Benedict's solution, and is not attacked by enzymes in the human digestive tract.

Chapter 25

1. $CH_3(CH_2)_{14}COOH \qquad CH_3(CH_2)_{16}COOH \qquad CH_3(CH_2)_7C=C(CH_2)_7COOH$

   Palmitic $\qquad\qquad$ Stearic $\qquad\qquad\qquad$ Oleic

3.

   Glycerol $\qquad$ Butyric acid $\qquad\qquad$ Tributyrin

5. When glycerol is heated with a dehydrating agent such as $KHSO_4$, the aldehyde acrolein is formed. Acrolein can be detected by its pungent penetrating odor. Since all fats contain glycerol they will form acrolein when heated with $KHSO_4$.

7. Refer to equation on page 317. If KOH and tripalmitin were used, the products would be glycerol and potassium palmitate.

9. Insoluble soaps are calcium and magnesium salts of fatty acids that are insoluble in water. Soft soaps are potassium salts of fatty acids and hard soaps are sodium salts of fatty acids. Ordinary cake soaps are sodium soaps, while potassium soaps are used in tincture of green soap.

11. See page 320. Phosphatidyl ethanolamine is a cephalin which is essential in the blood clotting process in the body.

13. The major structural difference between aldosterone and the other steroid hormones is the presence of an aldehyde group on carbon 13, instead of the usual methyl group.

Chapter 26

1. A deficiency of ascorbic acid in the diet results in the disease known as scurvy. Symptoms include loss of weight, anemia, and fatigue, followed by swollen, bleeding gums and hemorrhages under the skin. Eventually the teeth loosen, and the bones become brittle.

3. The water-soluble B vitamins often serve as part of coenzyme molecules, since they are small organic molecules that are readily separated from the protein portion of the enzyme.

5. Refer to the structure on page 330. The component parts include adenine-ribose-phosphate, dihydroxydimethylbutyric acid, β-alanine, and the acetyl group.

7. The Δ^{11} cis-retinal is vitamin A with the end secondary alcohol group oxidized to an aldehyde group, and the configuration of the double bond between carbon 11 and carbon 12 is cis. The relation between these two compounds and the visual process is shown on page 334.

9. The tocopherols are antioxidants that protect unsaturated fatty acids, vitamin A, and mitochondrial systems from oxidation in the body. They are used in food products for the prevention of oxidative rancidity.

Chapter 27

1. The cell produces chemical, heat, and mechanical energy.
3. Phosphoenolpyruvic acid is a high-energy compound because like other enolic phosphates it is readily hydrolyzed to a product that undergoes a spontaneous tautomeric rearrangement with the release of large amounts of energy.
5. Three moles of ATP may be generated by passage of electrons from a mole of substrate through NADH to molecular oxygen, but only two moles may be generated if electrons are transferred directly from FADH because this transfer bypasses the first phosphorylation site.

Chapter 28

1. Starch is split into dextrins and maltose by the action of ptyalin in the saliva in the mouth and in the stomach until the acid contents inactivate the ptyalin. In the small intestine, amylopsin completes the digestion of starch and dextrins to form maltose. The maltose is then split into glucose molecules by the action of maltase. Lactose and sucrose are hydrolyzed into monosaccharides by lactase and sucrase, respectively. The end products of digestion—glucose, fructose, and galactose—are then absorbed directly into the blood stream through the capillary blood vessels of the villi.
3. The enzyme pepsin in gastric juice splits native proteins to proteoses, which are then hydrolyzed to polypeptides, dipeptides, and eventually amino acids in the small intestine by the action of trypsin, chymotrypsin, carboxypeptidase, aminopeptidase, and dipeptidase. The end products of protein digestion, the amino acids, are absorbed through the intestinal mucosa directly into the bloodstream.
5. Refer to Figure 28–1, which presents an overview of intermediary metabolism.

Chapter 29

1. The factors are:
   (1) Storage as glycogen or as fat.
   (2) Oxidation to produce energy.
   (3) Excretion by the kidneys.
   Storage as glycogen replenishes the ready stores of glucose in the liver; storage as fat helps remove excess glucose from the blood. Oxidation is a complex process that is essential to produce energy for body function. Excretion by the kidney only occurs when the hyperglycemia is excessive and prolonged.
3. Glucagon, epinephrine, adrenal cortical hormones, and anterior pituitary hormones are involved in the control of the blood sugar level. Glucagon causes a rise in the blood sugar level by increasing the activity of phosphorylase, which converts liver glycogen to free glucose.
5. Cyclic -3′,5′-AMP is a derivative of adenylic acid. Refer to the formula on page 352. Both glucagon and epinephrine stimulate the formation of this compound which activates phosphorylase, an enzyme that catalyzes the conversion of glycogen into glucose-1-phosphate.
7. See outline on page 354.
9. See outline on page 355.
11. See outline on page 356.
13. Photosynthesis is a process by which plants convert the energy of sunlight to form food material. The light reaction involves the conversion of light energy into chemical energy by the process of photophosphorylation. The iron-containing protein ferredoxin and the cytochrome pigments are associated with chlorophyll in this reaction in which ATP and NADPH are formed. The dark reaction is not dependent on light energy and utilizes the combination of CO_2 with ribulose-1,5-diphosphate followed by the formation of triose phosphates and fructose-di-PO_4 and finally to glucose.

Chapter 30

1. Phospholipids, 200; triglycerides, 150; cholesterol, 160; and total lipid content of 510 mg/100 ml.
3. See scheme on page 362.
5. See scheme on page 363.
7. The liver contains an enzyme that converts acetoacetyl CoA to acetoacetic acid. Also, acetyl CoA can condense with acetoacetyl CoA followed by a cleavage of the product to acetoacetic acid and acetyl CoA.

9. Atherosclerosis is commonly seen in older persons who exhibit increased blood cholesterol levels and deposition of cholesterol plaques in the aorta and other blood vessels. Heart failure and circulatory problems result from atherosclerosis.

Chapter 31

1. Instead of storage depots, there is a temporary pool of amino acids that is available to all tissues and may be used for synthesis of new protein, hormones, enzymes, and nonprotein nitrogenous substances. The excess amino acids in the pool undergo catabolism to form urea and CO_2, H_2O, and energy.
3. An essential amino acid is one that cannot be synthesized by the body and therefore must be supplied by the diet if the synthesis of tissue protein is to occur. A growing child deprived of adequate amounts of essential amino acids would stop growing because tissue protein synthesis could not continue in a normal fashion.
5. The specific programming of the amino acids on the ribosomes to synthesize a specific protein is called translation and is illustrated in Figure 31–1.
7. On the m-RNA there is a specific site consisting of three consecutive bases that binds a particular amino acid; this is the codon. The t-RNA for the same amino acid has a complementary set of bases called an anticodon which binds the t-RNA to the site on the m-RNA.
9. $NH_3 + CO_2 + ATP \rightarrow$ carbamyl phosphate + ornithine $\rightarrow$ citrulline + aspartic acid $\rightarrow$ arginosuccinic acid $\rightarrow$ fumaric acid + arginine $\rightarrow$ urea + ornithine.
11. See outline on page 376.

Chapter 32

1. They are simple systems that contain a smaller amount of genetic information and fewer genes than the cells of man or animals.
3. Haploid cells contain only half the amount of DNA as diploid cells, and the DNA content of diploid cells is fairly constant from one type of cell to another. Also, the DNA from cells of a single species has the same composition of bases which remains constant.
5. The replication of DNA occurs by the separation of its complementary strands, each becoming a template for the synthesis of a complementary daughter strand. Refer to Figure 32–2.
7. A complete outline will be found in Figure 32–3.
9. To initiate early treatment, which consists of restricting the amount of phenylalanine in the diet of the infant.
11. Genetic engineering is the application of the results of genetic research studies to reverse cancer growth, repair inborn errors of metabolism, to improve the quality of livestock, and so on.

Chapter 33

1. Because disease of any type is characterized by adverse changes in biochemical compounds or systems in the cell.
3. Because they stimulate the central nervous system, resulting in a decreased sense of fatigue, increased initiative, elevation of mood, elation and euphoria, and an increase in motor activity.
5. Substitution of groups such as the methoxy group or chlorine modifies the action of the drug and increases the potency for depressing motor activity or altering psychotic behavior, for example.
7. Marijuana and lysergic acid diethylamide are examples of a mild and a potent hallucinogenic drug. Since the results of ingestion of these drugs are unpredictable in different individuals, the legal implications of their use are being carefully evaluated.
9. They are generally used in allergic conditions such as bronchial asthma, hay fever, rhinitis, dermatitis, and for motion sickness. In addition, they are present in remedies for insomnia, insect bites, and poison ivy.
11. Prednisolone is often used in the treatment of arthritis. It reduces inflammation of the joints and reverses the inflammatory changes in connective tissue that occur in arthritis.
13. Tolbutamide (Orinase) acts on the beta cells of the pancreas to stimulate the secretion of insulin. Since this compound may be taken orally, it has an immediate advantage over insulin injections.
15. The drug 6-mercaptopurine suppresses the synthesis of RNA and DNA in the tumor cells and exerts a cytotoxic action on the tumor tissue.

INDEX